PLANT RESISTANCE TO ARTHROPODS

T0217829

Plant Resistance to Arthropods
Molecular and Conventional Approaches

by

C. MICHAEL SMITH

Kansas State University,
Manhattan, KS, U.S.A.

 Springer

A C.I.P. Catalogue record for this book is available from the Library of Congress.

ISBN-13   978-90-481-6934-4 (PB)
ISBN-13   978-1-4020-3702-3 (e-book)

Published by Springer,
P.O. Box 17, 3300 AA Dordrecht, The Netherlands.

www.springeronline.com

Printed on acid-free paper

Dedication

To Rita, Segen, and Sonder -
the melodies of my life.

Contents

Preface

The theory, study, and practice of plant resistance to arthropods has matured greatly since the publication of R. H. Painter's classic *Plant Resistance to Insects* in 1951. In *Plant Resistance to Arthropods - Molecular and Conventional Approaches,* I have attempted to update the literature in this continually expanding area of arthropod pest management and to synthesize new information about transgenic arthropod resistant crop plants, the molecular bases of arthropod resistance in crop plants, and the use of molecular markers to breed arthropod resistant plants. The information is presented in a step-by-step manner that introduces and describes of the study of plant resistance for students, researchers, and educators.

I am thankful to colleagues around the world for their encouragement and support during the development of this book over the past three years. Special thanks are expressed to Elena Boyko, Steve Clement, Xuming Liu and Lieceng Zhu for their critical reviews of the manuscript. I appreciate the efforts of Clayton Forgey and Annie Nordhus, who formatted different revisions of the text. The project benefited greatly from the efforts of two talented artists. Andrea Kohl, a graphic artist at Northern Michigan University designed the book cover. Robert Holcombe, a graphic designer in the Department of Communications at Kansas State University prepared several line drawings for the text. I appreciate the help of Skylar Martin-Brown, who developed the figures of plant allelochemical structures. Thanks are also due to Sharon Starkey for assistance in location of bibliographic materials and in the preparation of photographic images. I acknowledge the support of the administration of the Kansas State University College of Agriculture for granting me sabbatical leave during which I completed a working draft of the book.

I express special thanks to Frank Davis, for his friendship, guidance and inspiration for the past 30 years. Finally, I thank the many students I have had the privilege to educate during the past 25 past years for their eagerness to learn and their constructive criticisms of my concepts of plant resistance to arthropods.

C. Michael Smith
Manhattan
March 7, 2005

CHAPTER 1

INTRODUCTION

1. WHAT IS PLANT RESISTANCE TO ARTHROPODS?

1.1. Uses of Plant Resistance to Arthropods

The cultivation of plants expressing some form of inherited resistance to an arthropod pest has been practiced for several hundred years. Before the domestication of plants for agricultural purposes, those susceptible to arthropods died before they could produce seed or before their damaged seeds could germinate. In effect, resistant plants survived subject to the laws of adaptation and natural selection. It is probable that early indigenous agricultural systems selected and utilized plants resistant to arthropod pests, since these systems developed production practices based on different crop species, and within species, selected different strains and land races of crops.

Crop domestication began about 10,000 years ago with the cultivation of potato, *Solanum* spp., in South America, and the production of maize, *Zea mays* L., about 6,000 years ago in Central America. Humans began to cultivate curcubits and sunflower, *Helianthus* spp., in North America during this time. By 7,000 BC, wheat, *Triticum aestivum* L., barley, *Hordeum vulgare* L., and lentil, *Lens culinaris* Medik., had become major domestic food crops in the Fertile Crescent of present day Iran, Iraq, Israel, Jordan, Syria, and Turkey, and by 5,000 BC agricultural communities had spread through much of what is now China (Garofalo 1999, Smitha 1998). Approximately 3,000 years ago, physiological differences had developed between a number of cultivars and their wild relatives (Anderson 2000). With the advent of these crop plant domestication systems and the use of rudimentary agricultural practices, farmers selected the seeds to use for future crops. Dicke (1972) describes the selection and development of mulberry, *Morus rubra* L., that yielded high populations of the silkworm moth, *Bombyx mori* (L.), and fine quality silk. In this instance, arthropod susceptibility was actually selected for instead of resistance.

During the eighteenth and early nineteenth centuries, insect resistant cultivars of wheat and apples were first developed and cultivated in the United States. As early as 1788, early maturing wheat cultivars were grown in the United States to avoid infestation by the Hessian fly, *Mayetiola destructor* (Say) (Chapman 1826). A few years later, Havens (1792) identified resistance to the Hessian fly in the wheat cultivar 'Underhill' in New York. Lindley (1831) made recommendations for the cultivation of the apple, *Malus* spp., cultivars 'Winter Majetin' and 'Siberian BitterSweet', because of their resistance to the wooly apple aphid, *Eriosoma lanigerum* (Hausmann).

1

In the mid-nineteenth century, plant resistance to an insect played an important role in Franco-American relations. The grape phylloxera, *Daktulosphaira vitifoliae* (Fitch), had been accidentally introduced from North America into the French wine-producing areas about 1860. Within 25 years, *D. vitifoliae* had destroyed nearly one-third (~10 million ha) of the French wine grapes and the French wine industry was devastated. The famous entomologist Charles Valentine Riley recognized that native American grapes, *Vitis labrusca*, were resistant to *D. vitifoliae*. Working with a colleague, J. E. Planchon in France, Riley led efforts to graft French scions of *Vitis vinifera* to resistant *V. labrusca* rootstocks from the Midwestern United States. Planchon's efforts were successful, and the industry recovered. For his efforts, Riley received the French Grand Gold Medal and was named a Chevalier of the Legion of Honor in 1884.

The breeding of arthropod resistant plants became more formalized in the late 19th century, with the rediscovery of Gregor Mendel's basic tenets of heredity and plant hybridization. However, fewer than 100 reports of plant resistance to arthropods were published in the United States during the 19th and early 20th century. In one of the earliest comprehensive reviews of plant resistance to arthropods, Snelling (1941) identified 163 publications dealing with plant resistance in the United States from 1931 until 1940. Since then, numerous reviews have chronicled the progress and accomplishments of scientists conducting research on plant resistance (Beck 1965, Green and Hedin 1986, Harris 1980, Hedin 1978, 1983, Maxwell et al. 1972, Painter 1958, 1968, Smith 1999, Stoner 1998).

The first book on the subject, *Plant Resistance to Insect Pests* was written by R. H. Painter (1951), the founder of this area of research in the United States. In Russia, Chesnokov (1953) published the first review of techniques on the subject, *Methods of Investigating Plant Resistance to Pests*. In the late 1970's, research activities in plant resistance intensified and several additional books on the subject were published, including those of Rosetto (1973) *Resistencia de plantas a insetos,* Russell (1978) *Plant Breeding for Pest and Disease Resistance*, Lara (1979) *Principos de resisencia de plants a insetos,* Panda (1979) *Principles of Host-Plant Resistance to Insects,* Maxwell and Jennings (1980) *Breeding Plants Resistant to Insects*, and the first edition of this text. In one of the few publications of its type, Mattson et al. (1988) developed *Mechanisms of Woody Plant Defenses Against Insects: Pattern for Search.* In 1994, I collaborated with Z. R. Khan and M. D. Pathak to publish an updated techniques book, *Techniques for the Evaluation of Insect Resistance in Crop Plants* (Smith et al. 1994).

In the last decade alone, the treatment of the subject of plant resistance has broadened to include several new perspectives. These include the evolutionary responses of pathogens and pests to plant resistance (Fritz and Simms 1992) and an edited volume completely dedicated to the economic benefits of resistance (Wiseman and Webster 1999). Panda and Khush (1995) developed an excellent updated overview of the literature, while Ananthakrishnan (2001) compiled a contemporary collection of contributions dealing specifically with the allelochemistry of resistant plants. The area of induced plant defense to herbivore and pathogen challenge has expanded greatly with an increasing number of reports of the identification of expressed resistance genes and gene products (see Chapter 9). These have been exceptionally well documented in reviews edited by Agrawal et al. (1999), Baldwin (1994), Chadwick and Goode (1999), Karban and Baldwin (1997) and Kessler and Baldwin (2002).

Research involving the development and use of arthropod resistant crop cultivars has led to significant crop improvements in the major food producing areas of the world in the past 50 years. These improvements include significantly improved food production, contributions toward the alleviation of hunger and improved human nutrition (Khush 1995). One of the most spectacular successes of the use of arthropod resistant crops occurred during the "Green Revolution" in tropical Asia during the 1960s, when high-yielding pest-resistant cultivars of rice, *Oryza sativa* (L.), were introduced into production agriculture. The continued growth of such cultivars has made significant improvements to the economies of several south and Southeast Asian countries, such that many countries that were previously food importers are now food exporters. One cultivar, IR36, developed and produced during the 1970's in Southeast Asia, provided approximately $1 billion of additional annual income to rice producers (Khush and Brar 1991).

Over 500 cultivars, plant material lines, or parent lines of food and fiber crops have been developed and registered in the United States since 1975 (reviewed in Smith 1989 and Stoner 1996). This germplasm has been produced through the cooperative efforts of entomologists and plant breeders employed by state agricultural experiment stations, the United States Department of Agriculture Agricultural Research Service, and private industry. Presently, hundreds of resistant cultivars are grown in the United States other major crop production areas of the World. Over one-half of the cultivars developed are maize, wheat, and *Sorghum bicolor* (L.) Moench, - the major world cereal grain food crops. For example, over one-half of all U. S. commercial maize cultivars have some resistance to the corn leaf aphid, *Rhopalosiphum maidis* (Fitch) (Barry 1969). Over 75% of the maize cultivars have some resistance to the first and second generation of the European corn borer, *Ostrinia nubilalis* Hubner (Barry and Darrah 1991). Most of the U. S. cultivars of soybean, *Glycine max* (L.) Merr., have resistance to the potato leafhopper, *Empoasca fabae* (Harris), and many alfalfa cultivars of alfalfa, *Medicago sativa* L., have resistance to a complex of pest aphids (Wilde 2002).

Many of the cultivars described above were developed in collaborations with by researchers at International Agricultural Research Centers that comprise the Consultative Group for International Agricultural Research (Figure 1.1). Hundreds of cultivars of maize, potato, *Solanum tuberosum* L., rice, sorghum, and wheat, and have been developed at these centers, and many possess resistance to the major arthropod pests of each crop. Often, detailed knowledge about the type and nature of resistance has been determined. Clement and Quisenberry (1999) reviewed an outstanding comprehensive collection of the existing global genetic resources in arthropod-resistant crop plants (see Chapter 5).

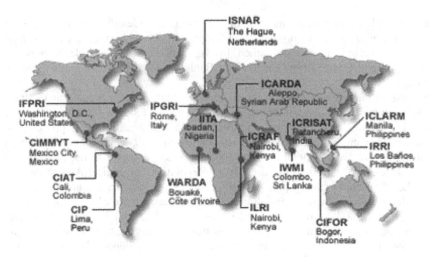

Figure 1.1. Locations of International Agricultural Research Centers comprising the Consultative Group for International Agricultural Research.

2. ADVANTAGES OF PLANT RESISTANCE TO ARTHROPODS

The economic advantage that arthropod resistant cultivars offer producers is genetically incorporated arthropod control for the cost of the seed alone. Even if only moderate levels of resistance are combined with pesticide applications, the costs of insecticidal control and insecticide residue problems are greatly reduced. Schalk and Ratcliffe (1976) estimated that approximately 319,000 tons of insecticides (approximately 37% of the total insecticides applied during the 1960s) were saved annually through the planting of insect resistant cultivars of alfalfa, barley, maize and sorghum in the U. S. This amount is now likely greater, because the greatly increased use of transgenic cultivars of insect-resistant maize and cotton, *Gossypium hirsutum* L., has further reduced pesticide applications in several countries (see Chapter 10). For example, insecticide use for *O. nubilalis* control in the U. S. dropped approximately 30% after the commercialization of *Bt* maize in North America (Rice and Pilcher 1998). Added ecological benefits of such reduced or eliminated pesticide applications are cleaner water supplies and reduced mortality of beneficial arthropod populations.

Arthropod resistance is of practical value even if improved resistant cultivars are not developed. Wightman et al. (1995) studied responses of chickpea, *Cicer arietinum* L., to *Helicoverpa armigera* (Hubner) larval feeding damage in southern India, in the presence or absence of insecticides. Although a *Helicoverpa* resistant landrace does not yield as much as a susceptible landrace or susceptible cultivar, when insecticides are applied the resistant landrace provides profits to producers when they cannot afford to purchase insecticides (Figure 1.2). In some cases, there is no synergistic benefit from insecticides on net crop yield or value and the need for insecticides is eliminated (Buntin et al. 1992, van den Berg et al. 1994).

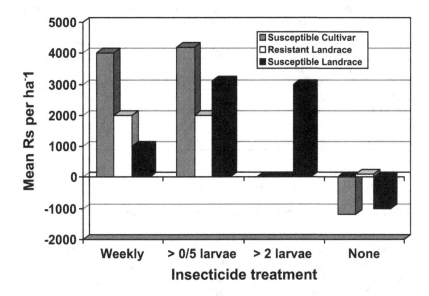

Figure. 1.2. Net Indian farm income for three cultivars of Cicer arietinum grown under four insecticide treatments to control Helicoverpa armigera. Reprinted from Crop Protection, Vol. 14, Wightman, J. A., M. M. Anders, V. R. Rao, and L. M. Reddy. 1995. Management of Helicoverpa armigera (Lepidoptera: Noctuidae) on chickpea in southern India: thresholds and the economics of host plant resistance and insecticide application. Pages 37 - 46, Copyright 1995, Butterworth Heinemann, Inc., with permission from Elseiver.

Regardless of the level of resistance, pest economic threshold levels and economic injury levels require adjustment when resistant cultivars are employed in integrated pest management systems, in order to make insecticide use decisions involving pest resistant cultivars (Teetes 1994, Sharma 1993).

Arthropod resistant cultivars also yield much higher returns per dollar invested than those spent on insecticide development. Luginbill (1969) demonstrated that arthropod resistant varieties of alfalfa, maize and wheat produced in the U. S. during the 1960s yielded a 300% return per research dollar invested. More specifically, *M. destructor*-resistant U. S. wheat cultivars developed during the same period of time yielded a 120-fold higher return of return than did the development of insecticides (Painter 1968). More contemporary research indicates that *M. destructor* resistance in wheat cultivars in Morocco provides a 9:1 return on investment of research funds (Azzam et al. 1997).

Recent economic analyses have provided additional information about the value of plant resistance. The value of research to develop sorghum hybrids with resistance to the greenbug, *Schizaphis graminum* (Rondani), biotype E in the U. S. was estimated to be from $113 million to $389 million per year, depending on whether the provisions of 1989 U. S. farm legislation were considered (Eddleman et al. 1999).

Table 1.1. Estimated annual value of producing Phaseolus lines with tolerance to Empoasca
krameri in Central American agriculture [a]

Phaseolus line	Value ($ per hectare)		
	Net benefit of control	Losses (no control)	Benefit of resistance
EMP187	1,094	317	559
EMP188	1,184	438	526
EMP186	990	302	469
Susceptible	880	662	---

[a] from Heinrichs and Adesina (1999). Reprinted with permission from Thomas Say
Publications in Entomology: Proceedings. Copyright 1999, Entomological Society of
America.
[b] ($ benefit of resistance - $ benefit of susceptibility) + ($ loss on susceptibility - $ loss on
resistance)

In Asia, Africa, and Latin America, the value of arthropod resistant cultivars of
and chickpea, sorghum and pearl millet, *Pennisetum glaucum* (L.) R. Br., is
estimated to be worth more than $580 million per year (Heinrichs and Adensina 1999).
Cardona and Cortes (1991) estimated resistance in *Phaseolus* to the leafhopper
Empoasca krameri Ross and Moore, in Latin America to be approximately $500
per acre per year (Table 1.1). Based on a survey of U. S. alfalfa production, Berberet
et al. (1999) estimated that the increase in annual gross income to producers from the
use of multiple disease and arthropod resistant alfalfa to be approximately $300
million per year. The current economic value of all arthropod resistant cultivars of
wheat is slightly more than $250 million per year (Smith et al. 1999). The value of
transgenic resistant crops is just beginning to be recognized. Eddleman (1995)
estimated that the global economic benefit of commercial cotton cultivars containing
the toxin gene from the bacteria *Bacillus thuringiensis*, (see Chapter 10) to be valued at
between $570 million and $730 million per year, depending on whether insecticide
use continued for secondary species of pest Lepidoptera. The current total estimated
global value of arthropod resistant cultivars is approximately $1.18 billion (Table
1.2).

The effects of deploying resistance genes accumulate over time. In general, the
longer they remain effective, the greater the benefits of their use (Robinson 1996).
These effects were thoroughly documented as arthropod resistant *O. sativa* cultivars
were placed into production in Southeast Asia in the 1970s. In both the Philippines
and Indonesia, yield losses of crops planted with arthropod resistant cultivars were
approximately one-half of the losses in crops planted with non-resistant cultivars
(Panda 1979, Waibel 1987). Wiseman (1999) demonstrated that even low-level resistance
to the fall armyworm, *Spodoptera frugiperda* (J. E. Smith), in the silks of maize
significantly reduced *S. frugiperda* growth and fecundity in only five generations.

Resistant cultivars improve the efficiency of predators, parasites and arthropod
pathogens by decreasing the vigor and physiological state of the pest arthropod. The
effects of many resistant crop cultivars have no detrimental effects and in some
cases, have additive or synergistic effects on the actions of pest arthropod predators
and parasites. (Eigenbrode and Trumble 1994, Quisenberry and Schotzko 1994).

Table 1.2. Net annual global economic value of arthropod resistant crop cultivars

Crop	Pest(s)	Location	$ (x million)
Gossypium hirsutum	Lepidoptera larvae	World	570 [b]
Medicago sativa	Aphids	United States	300 [a]
Oryza sativa	Leafhoppers	Asia	1,000
Sorghum bicolor	Schizaphis graminum	United States	113 [c]
Triticum aestivum	Aceria tosichella	North America	150
	Mayetiola destructor	United States	17
	Diuraphis noxia	United States	13
	Cephus cinctus	United States	12 [d]

[a] Assumes a 60% area of planting multiple pest (disease and insect) resistant cultivars
[b] Assumes insecticide use remains constant for non-target pest species
[c] Assumes no 1989 U. S. farming legislation provisions
[d] Assumes 3.5% annual rate of inflation of 1948 estimate of $3.8 million

The additive effects of resistance genes and arthropod pathogens have been reviewed previously (Smith 1999). For example, results of Wiseman and Hamm (1993) demonstrate how nuclear polyhedrosis viruses increase the mortality of corn earworm, *Helicoverpa zea* Boddie, larvae fed silk tissue of a resistant maize cultivar (Figure 1.3). The use of resistant cultivars to maximize cultural control tactics such as early-planted cultivars, trap crops, and early maturing cultivars is well documented in several crops (Maxwell 1991). The planting of early-maturing, arthropod-resistant cultivars has been shown to reduce populations of several key pests in rice. Trap cropping, a practice used to attract pest arthropod populations and then destroy them, is synergistic when used in combination with arthropod resistant cultivars of cotton, rice and soybean.

In addition to synergizing traditional pest management tactics, there are also several advantages of resistant plants themselves over biological, chemical and cultural controls. Insecticides applied at recommended rates often kill biological control organisms, but resistant cultivars do not, and are compatible with insecticide use. Where biocontrol organisms depend on the sustained density of hosts or prey to remain effective, resistant cultivars function independently of arthropod density and operate at all pest population levels (Panda and Khush 1995). Expanded discussions of the integration of resistant cultivars with biological control organisms, chemical control, and cultural control tactics in integrated pest management systems are presented in Chapter 12.

Resistant cultivars have also been shown to impede the spread of arthropod-vectored plant diseases, by reducing the population growth of disease vectors (see reviews by Kennedy 1976, Gibson and Plumb 1977, Maramorosch 1980). In a 9 year study, Harvey et al. (1994) demonstrated how resistance in *Triticum aestivum* to the wheat curl mite, *Aceria tosichella* Keifer, the vector of wheat streak mosaic virus, reduced virus incidence by as much as 50%. Kishaba et al. (1992) demonstrated similar results (31% - 74% reduction) in the reduction of the transmission of water-melon mosaic virus through the use of breeding lines of muskmelon, *Cucumis melo* L., resistant to the melon aphid, *Aphis gossypii* Glover. Kobayashai et al. (1993) evaluated the resistance of several *Oryza* species to the green rice leaf-hopper, *Nephotettix virescens* (Distant), and the green rice leafhopper, *Nephotettix nigropictus* (Stål), as vectors of rice tungro virus. In several of the species evaluated, reduced infection by the virus was related to the resistance to either of the two vectors.

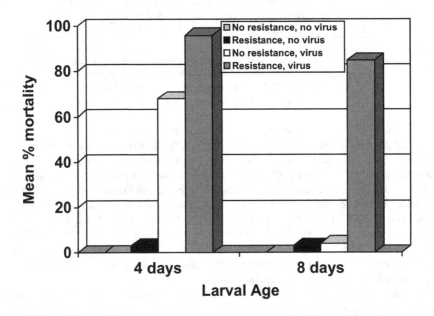

Figure 1.3. Mortality of Helicoverpa zea larvae feeding on artificial diets with (Resistance) and without (No Resistance) silks of the resistant maize cultivar 'Zapalote Chico' and exposed to Elcar (no virus - 0, virus - 1330 polyhedral occlusion bodies). Reprinted from Biological Control, Vol. 3. Wiseman, B. R., and J. J. Hamm. 1993. Nuclear polyhedrosis virus and resistant corn silks enhance mortality of corn earworm (Lepidoptera, Noctuidae) larvae. Pages 337-342, Copyright 1993, Academic Press Inc., with permission from Elseiver.

3. DISADVANTAGES OF PLANT RESISTANCE TO ARTHROPODS

Arthropod resistant cultivars have some disadvantages. In some cases, the level of resistance is incompatible with biological control agents (Bottrell et al. 1998). In arthropod-resistant cultivars of some crops, high densities of plant trichomes and high concentrations of resistance-bearing allelochemicals have been shown to impart detrimental effects on the biology of beneficial arthropod predators and parasites, as well as arthropod pathogens. Negative interactions between insecticides and some resistant cultivars also exist. When arthropods are fed foliage containing high levels of allelochemicals that mediate resistance in some crops, they are better able to detoxify insecticides and miticides (Ghidiu et al. 1990).

Developing a cultivar resistant to a single arthropod species traditionally has required three to five years, and may require ten years or longer for a complex of several arthropods. These intervals have shortened somewhat with the now common use of tropical and subtropical winter nurseries to increase the number of plant generations that can be produced each year. Many crops are grown over broadly diverse geographic ranges, soil types and environmental conditions, necessitating the deployment of different resistant cultivars for different geographic production regions. From the plant breeder's perspective, even with winter nurseries, regional crop resistance to arthropods may be an expensive and time-consuming objective.

Resistance is commonly identified in wild, undomesticated species of plants or landraces that may have only distant taxonomic relations to the crop species under improvement. It is not unusual for these plants to have poor yield, poor plant type, or disease susceptibility. Some of these problems may be eliminated with the adaptation and use of embryo rescue and related plant tissue culture techniques. For the most part, however, the incorporation of resistance from wild species of plants into domestic crop plants is a long-term process.

Arthropod resistant cultivars that rely on the effects of a single, major gene often promote the development of populations of individuals possessing genes virulent to plant resistance genes (see Chapter 11). The use of monogenic resistance often leads to a pattern of sequential gene release, with each new cultivar possessing a different gene or gene arrangement, in order to stay a step ahead of the continuously mutating genetic machinery of the pest arthropod. The development of cultivars with polygenic resistance, to delay biotype development, requires years longer to accomplish. Cultivars with moderate levels of multigene resistance to stem boring Lepidoptera have been used in Asian rice production and North American maize production for many years without the development of resistance-breaking borer biotypes (Heinrichs 1986). Different aspects of the gene-for-gene interaction between arthropod biotypes and plant resistance genes are fully discussed in Chapter 11.

4. DEFINITIONS

By definition, **plant resistance to arthropods** is the sum of the constitutive, genetically inherited qualities that result in a plant of one cultivar or species being less damaged than a susceptible plant lacking these qualities. Plant resistance to arthropods must always be measured on a relative scale, with the degree of resistance based on comparison to susceptible control plants that are more severely damaged or killed under similar experimental conditions, as well as resistant control plants with a known, predetermined level of resistance. Relative measurements are necessary, since resistance is influenced by environmental fluctuations occurring over both time and space. In the terms of the plant resistance researcher, susceptibility is the inability of a plant to inherit qualities that express resistance to arthropods.

Induced resistance to arthropods is expressed in plants damaged by pest feeding or oviposition. This damage activates **defense response (DR)** genes and the redirection of normal cell maintenance genes to plant defense. Damaged plants produce **elicitors** that activate plant gene expression and the synthesis of volatile and non-volatile allelochemicals such as proteinase inhibitors, phenolics, and enzymes involved in the different types of plant defense (Agrawal et al. 1999). Several plant signaling pathways, including those driven by jasmonic acid, salicylic acid, ethylene and abcissic acid orchestrate the induction of plant defenses to arthropod attack (Walling 2000). Induced arthropod resistance, first demonstrated in apple by Bramstedt (1938), has been demonstrated in over 100 species of the major plant taxa (Karban and Kuc 1999). A complete discussion of induced resistance and the plant genes expressed as a result of the induction process is provided in Chapter 9.

Pseudo- or false resistance may occur in normally susceptible plants for several different reasons. Plants may avoid arthropod attack due to earlier than normal phenological development and resultant unsuitability for arthropod development. False resistance may occur as a result as a result of temporary variations in temperature, daylength, soil chemistry, plant or soil water content, or internal plant metabolism. Finally, normally susceptible plants may appear resistant as a result of simply escaping damage due to incomplete arthropod infestations.

Associational resistance occurs through the practice of intercropping, when normally susceptible plants grow in association with a resistant plant, and derive protection from arthropod predation. The diversionary or delaying actions of mixtures of plant species can help slow the development of pest arthropod populations in general, and may also help prevent the development of arthropod biotypes (Chapter 11) that develop virulence to plant resistance genes. A specialized type of associational resistance has been shown to exist in Graminaceous crops infected with fungal endophytes that produce alkaloids that kill or delay the development of pest arthropods (Breen et al. 1994, Clement et al. 1994). An in-depth discussion of endophyte-arthropod resistance interactions is provided in Chapter 7.

Overall, intercropping has positive implications for arthropod resistance. Thrips populations are much lower in polycultures of sorghum and cowpea, mungbean, than in either crop grown in monoculture (Ampong-Nyarko et al. 1994). Similar reductions have been reported for pest aphid populations in sorghum-soybean mixtures and *Phaseolus* spp.-maize mixtures (Bottenberg and Irwin 1991, 1992), and for populations of the flea beetle, *Phyllotreta cruciferae* (Goeze), on mixtures of *Vicia* and broccoli, *Brassica oleracea* L. (Garcia and Altieri 1992). Khan et al. (1997) developed a very specialized intercropping system consisting of molasses grass, *Melinis minutiflora,* and maize for management of the maize stem borers *Busseola fusca* Fuller and *Chilo suppressalis* (Walker). The molasses grass crop repels borer larvae and adults, and attracts significantly more parasites to borers infesting maize, resulting in significant (~10 fold) reductions in borer damage.

Because plant-arthropod-environment interactions vary widely, no single management tactic, including plant resistance, is universally effective. In at least one instance, intercropping has been shown to have a negative effect on a resistant cultivar, where intercropping of cowpea and maize diminishes resistance in cowpea to the cowpea borer, *Maruca testulalis* (Geyer) (Gethi et al. 1993). From a practical standpoint, associational resistance may be imitated by the development of plant cultivars based on several different sources of resistance, or mixtures of resistant and susceptible cultivars.   An expanded discussion of crop and gene mixtures is presented in Chapter 8.

5. RESISTANCE CATEGORIES

Three types of plant resistance to arthropods are commonly referred to in plant resistance literature. These resistance types were originally defined by Painter (1951) as mechanisms (Figure 1.4), and were more accurately termed functional categories by Horber (1980). Although I originally termed these categories functional modalities of resistance (Smith 1989) there are several reasons for them to be referred to as categories. By definition, a category is a general class or group, and a modality is a classification or form.   Conversely, a mechanism is a fundamental physical or chemical process involved in or responsible for an action, reaction or other natural phenomenon. The term basis refers to the foundation or principal component of anything. Thus, the terms category and modality refer to the way a group of items are classified, while the terms basis and mechanism denote the principal process governing a natural phenomenon.

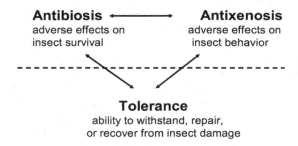

Figure 1.4. The antibiosis, antixenosis and tolerance categories (originally described as mechanisms) of plant resistance to insects. (from Painter 1951)

Many examples demonstrate how resistant plants categorized or classified as exhibiting antibiosis or antixenosis, while the plants themselves demonstrate tolerance as a third type of resistance, independent of arthropod effects. In contrast to Painter's original use, the term mechanism should be used to describe the underlying chemical or morphological plant processes that, where known, are responsible for the (negative) reaction of arthropods to resistant plants. To describe the outcome of arthropod-plant interactions, the term category should be used to refer to antibiosis, antixenosis and other as of yet undefined types of plant-arthropod interactions, observed as responses of arthropods to a plant resistance mechanism. The effects of resistant plants on arthropods can be manifested as **antibiosis**, in which case the biology of the pest arthropod is adversely affected; or as **antixenosis**, in which the plant acts as a poor host and the pest then selects an alternate host plant. The inherent genetic qualities of the plant itself may provide it the ability to withstand or recover from insect damage, in which case it is said to express **tolerance** to the pest. The inter-relationship of the three categories is illustrated in Figure 1.4. As indicated previously, these terms have been accepted because of conceptual convenience, but they are not always biologically discrete entities.

Often the antibiosis and antixenosis categories of resistance overlap, because of the difficulty involved in designing experiments to delineate between the two. For example, if an arthropod confined to a resistant plant fails to gain weight at the rate it normally does on a susceptible plant, it might be assumed that the lack of weight gain is due to the presence of antibiotic properties in the plant. However, the lack of weight gain may also be due to the presence of a physical or chemical feeding deterrent with strong antixenotic properties. This deterrent may initiate aberrant behavior in the test arthropod that results in a weakened physiological condition that could be assumed to be the result of an antibiotic effect. In Painter's words "There is increasing evidence that many examples formerly thought to be antibiosis actually are extremely high levels of nonpreference. It has been impossible to determine whether young, tiny insects have starved to death or been poisoned." (Painter 1968).

Combinations of antibiosis and antixenosis are reported often, as a result of researchers conducting very detailed experiments that have delineated the contributions of each category of resistance. A cursory survey of the literature indicates that antibiosis and antixenosis occur together across many plant taxa, including major cereal crops, food legumes, forages, fruits, ornamental plants and vegetables. In a few instances, all three categories have been shown to operate simultaneously. Aphid resistance in barley, sorghum, wheat and sugarcane, a complex hybrid of *Saccharum* species, involves antibiosis, antixenosis and tolerance (Castro et al. 1996, Hawley et al. 2003, White 1990). Resistance in maize to the pink stem borer, *Sesamia nonagrioides* Lef., also involves each of the three categories (Butrón et al. 1998). Bodnaryk and Lamb (1991) noted that all three categories of resistance operate in the resistance of yellow mustard, *Sinapis alba* L., to the flea beetle, *Phyllotreta cruciferae* (Goeze). Detailed descriptions and discussions of each of the three categories and the methodologies involved in investigating them are presented in Chapters 2, 3, 4, and 6.

The practice of identifying and cultivating plants with arthropod-resistant qualities is an ancient one that we continue to use and improve for use in modern crop pest management systems. The use of resistant crop cultivars has been and continues to be necessitated by the continual development of arthropod populations with genetic resistance to chemical pesticides and plant resistance genes, and by a continual need to produce crops with fewer pesticides and at lower production costs.

A major advantage in the cultivation of arthropod-resistant crops is that their production costs are lower, due to the fact that some or all of the arthropod control costs are incorporated into the seeds or clones themselves. In the following chapters we will investigate how this control is identified, how it is inherited, the techniques used to manipulate it, and how it can ultimately be used to manage arthropod populations in crop pest management systems.

REFERENCES CITED

Agrawal, A. A., S. Tuzun and E. Bent (Eds.). 1999. Induced Plant Defenses against Pathogens and Herbivores: Biochemistry, Ecology, and Agriculture. APS Press, St. Paul. 390 pp.

Ampong-Nyarko, K., K. V. S. Reddy, R. A. Nyangor, and K. N. Saxena. 1994. Reduction of insect pest attack on sorghum and cowpea by intercropping. Entomol. Exp. Appl. 70:179-184.

Ananthakrishnan, T. N. (Ed.). 2001. Insect and Plant Defense Dynamics. Science Publishers, Enfield. 253 pp.

Anderson, D. 2000. A Brief History of Agriculture. http://www.aces.uiuc.edu/~sare/.

Azzam, A., S. Azzam, S. Lhaloui, A. Amri, M. El Bouhssini, and M. Moussaoui. 1997. Economic returns to research in Hessian fly (Diptera: Cecidomyiidae) resistant bread-wheat varieties in Morocco. J. Econ. Entomol. 90:1-5.

Baldwin, I. T. 1994. Chemical Changes Rapidly Induced by Folivory. In: Insect - Plant Interactions V, E. Bernays (Ed.). CRC Press, Boca Raton. pp. 1 –23.

Barry, D. 1969. European corn borer and corn leaf aphid resistance in corn hybrids. Ohio Agric. Res. Dev. Cent. Res. Circ. 174.

Barry, D., and L. L. Darrah. 1991. Effect of research on commercial hybrid maize resistance to European corn borer (Lepidoptera: Pyralidae). J. Econ. Entomol. 84:1053-1059.

Beck, S. D. 1965. Resistance of plants to insects. Ann. Rev. Entomol. 10:107-232.

Berberet, R. C., M. E. Payton, and C. E. Ward. 1999. Value of host plant resistance in improving profitability of alfalfa forage production. In: Wiseman, B. R. and J. A. Webster (Eds.), Economic, Environmental, and Social Benefits of Resistance in Field Crops. Thomas Say Publications in Entomology: Proceedings. Entomological Society of America, Lanham, MD. pp. 115-130.

Bodnaryk, R. P., and R. J. Lamb. 1991. Mechanisms of resistance to the flea beetle, *Phyllotreta cruciferae* (Goeze) in mustard seedlings, *Sinapis alba* L.. Can. J. Plant Sci. 71:13-20.

Bottenberg, H., and M. E. Irwin. 1991. Influence of wind speed on residence time of *Uroleucon ambrosiae alatae* (Homoptera: Aphididae) on bean plants in bean monocultures and bean-maize mixtures. Environ. Entomol. 20:1375-1380.

Bottenberg, H., and M. E. Irwin. 1992. Canopy structure in soybean monocultures and soybean-sorghum mixtures: Impact on aphid (Homoptera: Aphididae) landing rates. Environ. Entomol. 21:542-548.

Bottrell, D. G., P. Barbosa and F. Gould. 1998. Manipulating natural enemies by plant variety selection and modification: A realistic strategy? Ann. Rev. Entomol. 43:347-367.

Bramstedt, F. 1938. Der nachweis der blutlausunanfalligkeit der apfelsorten auf histologischer grundlage. Z. Pflanzenkrank Pflanzenschutz. 48:480-488.

Breen, J. P. 1994. *Acremonium* endophyte interactions with enhanced plant resistance to insects. Ann. Rev. Entomol. 39:401–423.

Buntin, G. D., S. L. Ott, and J. W. Johnson. 1992. Integration of plant resistance, insecticides, and planting date for management of the Hessian fly (Diptera: Cecidomyiidae) in winter wheat. J. Econ. Entomol. 85:530–538.

Butrón, A., R. A. Malvar, P. Velasco, P. Revilla, and A. Ordás. 1998. Defense mechanisms of maize against pink stem borer. Crop Sci. 38:1159–1163.

Cardona, C., and M. L. Cortes. 1991. Evaluacion economica de la tolerancia de variedades de frijol allorito verde, *Empoasca kraemeri* Ross & Moore (Homoptera: Cicadellidae). Rev. Colomb. Entomol. 17:19–23.

Castro, A. M., A. Martin, and L. M. Martin. 1996. Location of genes controlling resistance to greenbug (*Schizaphis graminum* Rond.) in *Hordeum chilense*. Plant Breed. 115:335–338.

Chadwick, D. J., and J. A. Goode (Eds.). 1999. Insect Plant Interactions and Induced Plant Defense. John Wiley, Chichester. 281 pp.

Chapman, I. 1826. Some observations on the Hessian fly; written in the year 1797. Mem. Phil. Soc. Prom. Agr. 5:143–153.

Chesnokov, P. G. 1953. *Methods of Investigating Plant Resistance to Pests*. National Science Foundation, Washington, DC. 107 pp. (Israel Program for Scientific Translation).

Clement, S. L., W. J. Kaiser, and H. Eischenseer. 1994. Acremonium endophytes in germplasms of major grasses and their utilization for insect resistance. In: Bacon, C. W. and J. F. White, (Eds.), Biotechnology of Endophytic Fungi of Grasses, CRC Press, Boca Raton. pp. 185 – 199.

Clement, S. L., and Quisenberry, S. S. (Eds.) 1999. Global Plant Genetic Resources for Insect Resistant Crops. CRC Press, Boca Raton. 295 pp.

Dicke, F. F. 1972. Philosophy on the biological control of insect pests. J. Environ. Qual. 1:249–254.

Eddleman, B. R., C. C. Chang, and B. A. McCarl. 1999. Economic benefits from grain sorghum variety improvement in the United States. In: B. R. Wiseman and J. A. Webster (Eds.), Economic, Environmental, and Social Benefits of Resistance in Field Crops. Thomas Say Publications in Entomology: Proceedings. Entomological Society of America, Lanham, MD. pp. 17–44.

Eddleman, B. R., D. Dearmont, Q. He, and B. A. McCarl. 1995. Potential economic benefits to society from Bt cotton. In: D. J. Wiseman and D. R. Archer (Eds.), Proc. Beltwide Cotton Conference, San Antonio, TX, 4–7 Jan, 1995. National Cotton Council of America, Memphis, TN.

Eigenbrode, S. D., and J. T. Trumble. 1994. Host plant resistance to insects in integrated pest management in vegetable crops. J. Agric. Entomol. 11:201–224.

Fritz, R. S., and E. L. Simms (Eds.). 1992. Plant Resistance to Herbivores and Pathogens: Ecology, Evolution and Genetics. University of Chicago Press, Chicago. 590 pp.

Garcia, M. A., and M. A. Altieri. 1992. Explaining differences in flea beetle *Phyllotreta cruciferae* Goeze densities in simple and mixed broccoli cropping systems as a function of individual behavior. Entomol. Exp. Appl. 62: 201–209.

Garofalo, M. P. 1999. The History of Gardening: A Timeline From Ancient Times to the 20th Century. Version 2.2.7. http://www.gardendigest.com/timegl.htm.

Gethi, M., E. O. Omolo, and J. M. Mueke. 1993. The effect of intercropping on relative resistance and susceptibility of cowpea cultivars to *Maruca testulalis* Geyer when in mono and when intercropped with maize. Insect Sci. Applic. 14:305–313.

Ghidiu, G. M., C. Carter, and C. A. Silcox. 1990. The effect of host plant on Colorado potato beetle (Coleoptera: Chrysomelidae) susceptibility to pyrethroid insecticides. Pestic. Sci. 28: 259–270.

Gibson, R. W., and R. T. Plumb. 1977. Aphid resistance and the spread of viruses, In: Harris, K. F. and K. Maramorosch (Eds.), Aphids as Virus Vectors. Academic Press, New York, pp. 488–491.

Green, M. B., and P. A. Hedin. 1986. Natural Resistance of Plants to Pests. ACS Symposium Series 296. American Chemical Society. Washington, D.C. 243 pp.

Harris, M. O. 1980. Biology and Breeding for Resistance to Arthropods and Pathogens in Agricultural Plants. Texas Agric. Exp. Sta. Publ. MP–1451.

Harvey, T. L., T. J. Martin, and D. L. Seifers. 1994. Importance of plant resistance to insect and mite vectors in controlling virus diseases of plants: resistance to the wheat curl mite (Acari: Eriophyidae). J. Agric. Entomol. 11:271–277.

Havens, J. N. 1792. Observations on the Hessian fly. Trans. New York Soc. Agron. Pt. 1:89–107.

Hawley, C. J., F. B. Peairs, and T. L. Randolph. 2003. Categories of resistance at different growth stages in Halt, a winter wheat resistant to the Russian wheat aphid (Homoptera: Aphididae). J. Econ. Entomol. 96: 214–219.

Hedin, P. A. (Ed.) 1978. Plant Resistance to Insects. ACS Symposium Series 62. American Chemical Society. Washington, D. C. 286 pp.

Hedin, P. A. (Ed.) 1983. Plant Resistance to Insects. ACS Symposium Series 208. American Chemical Society. Washington, D.C. 374 pp.

Heinrichs, E. A. 1986. Perspectives and directions for the continued development of insect-resistant rice varieties. Agric., Ecosyst. Environ. 18:9–36.

Heinrichs, E. A., and A. A. Adensina. 1999. Contribution of multiple-pest resistance to tropical crop production. In: Wiseman, B. R. and J. A. Webster. (Eds.), Economic, Environmental, and Social Benefits of Resistance in Field Crops. Thomas Say Publications in Entomology: Proceedings. Entomological Society of America, Lanham, MD. pp. 149–189.

Horber, E. 1980. Types and classification of resistance. In: F. G. Maxwell and P. R. Jennings (Eds.). Breeding Plants Resistant to Insects. John Wiley, New York. pp. 15–21.

Karban, R., and I. T. Baldwin. 1997. Induced Responses to Herbivory. University of Chicago Press, Chicago.

Karban, R., and J. Kuc. 1999. Induced resistance against pathogens and herbivores: An overview. In: Agrawal, A. A., Tuzun, S., and Bent, E. (Eds.), Induced Plant Defenses against Pathogens and Herbivores Biochemistry, Ecology and Agriculture. pp. 1–16.

Kennedy, G. G. 1976. Host plant resistance and the spread of plant diseases. Environ. Entomol. 5:827–832.

Kessler, A., and I. T. Baldwin. 2002. Plant responses to insect herbivory: The emerging molecular analysis. Ann. Rev. Plant Biol. 53: 299–328.

Khan, Z. R., K. Ampong-Nyarko, P. Chiliswa, A. Hassanali, S. Kimani, W. Lwande, W. A. Overholt, J. A. Pickett, L. E. Smart, L. J. Washams, and C. M. Woodcock. 1997. Intercropping increases parasitism of pests. Nature. 388:631–632.

Khush, G. S. 1995. Modern cultivars - their real contribution to food supply and equity. Geojourn. 35:275–284.

Khush, G. S., and D. S. Brar. 1991. Genetics of resistance to insects in crop plants. Adv. Agron. 45:223–274.

Kishaba, A. N., S. J. Castle, D. L. Coudriet, J. D. McCreight, and G. W. Bohn. 1992. Virus transmission by *Aphis gossypii* Glover to aphid–resistant and susceptible muskmelons. J. Amer. Soc. Hort. Sci. 117:248–254.

Kobayashi, N., R. Ikeda, I. T. Domingo, and D. A. Vaughn. 1993. Resistance to infection of rice tungro viruses and vector resistance in wild species of rice (*Oryza* sp.). Japan J. Breed. 43:377–387.

Lara, F. M. 1979. Principios de Resistancia de Plantas a Insectos. (Portugese). Piracicaba, Livroceres, Brazil, 207 pp.

Lindley, G. 1831. A guide to the orchard and kitchen garden. 597 pp.

Luginbill, P., Jr., 1969. Developing resistant plants - the ideal method of controlling insects. U. S. Dep. Agric. ARS Prod. Res. Rep. 111:1–14.

Maramorosch, K. 1980. Insects and plant pathogens. In: F. G. Maxwell and P. R. Jennings (Eds.), Breeding Plants Resistant to Insects. John Wiley, New York. pp. 138–155.

Mattson, W. J., J. Levieux, and C. Bernard-Dagan. (Eds.). 1988. Mechanisms of Woody Plant Defenses against Insects Search for Patterns. Springer, New York. 416 pp.

Maxwell, F. G. 1991. Use of insect resistant plants in integrated pest management programmes. FAO Plant Prot. Bull. 39:139–146.

Maxwell, F. G., J. N. Jenkins, and W. L. Parrott. 1972. Resistance of plants to insects. Adv. Agron. 24:187–265.

Maxwell, F. G., and P. R. Jennings (eds.). 1980. Breeding Plants Resistant to Insects. John Wiley, New York. 683 pp.

Painter, R. H. 1951. Insect Resistance in Crop Plants. University of Kansas Press. Lawrence. 520 pp.

Painter, R. H. 1958. Resistance of plants to insects. Ann. Rev. Entomol. 3:267–290.

Painter, R. H. 1968. Crops that resist insects provide a way to increase world food swupply. Kansas Agric. Exp. Sta. Bulletin 520, 22 pp.

Panda, N. 1979. Principles of Host-Plant Resistance to Insect Pests. Allanheld, Osmun & Co. and Universe Books, New York. 386 pp.

Panda, N., and G. S. Kush. 1995. Host Plant Resistance to Insects. CAB / International Rice Research Institute, Wallingford, Oxon, UK. 431 pp.

Quisenberry, S. S. and D. J. Schotzko. 1994. Integration of plant resistance with pest management methods in crop production systems. J. Agric. Entomol. 11:279–290.

Rice, M. E., and C. D. Pilcher. 1998. Potential benefits and limitations of transgenic Bt corn for management of the European corn borer (Lepidoptera: Crambidae). Am. Entomol. 44:75–78.

Robinson, R. A. 1996. Return To Resistance: Breeding Crops To Reduce Pesticide Dependency. Fertile Ground Books, Davis, CA, 480 pp.

Rosetto, C. J. 1973. Resistencia de plantas a insetos. Piracicaba. ESALQ, Piracicaba, Brasil (mimeografado) (Portugese). 110 pp.

Russell, G. E. 1978. Plant Breeding for Pest and Disease Resistance. Butterworth Publishers, Inc. Boston. 485 pp.

Schalk, J. M., and R. H. Ratcliffe. 1976. Evaluation of ARS program on alternative methods of insect control: Host plant resistance to insects. Bull. Entomol. Soc. Am. 22:7–10.

Sharma, H. C. 1993. Host-plant resistance to insects in sorghum and its role in integrated pest management. Crop Protection. 12:11–34.

Smith, C. M. 1989. Plant Resistance to Insects – A Fundamental Approach. John Wiley, New York. 286 pp.

Smith, C. M. 1999. Plant Resistance to Insects. In: Rechcigl, J. and Rechcigl, N. (Eds.) Biological and Biotechnological Control of Insects. Lewis Publishers, Boca Raton, FL. pp. 171–205.

Smith, C. M., S. S. Quisenberry, and F. du Toit. 1999. The Value of Conserved Wheat Germplasm Possessing Arthropod Resistance. In: S. L. Clement and S. S. Quisenberry. (Eds.) Global Plant Genetic Resources for Insect Resistant Crops. CRC Press, Boca Raton, Florida. pp. 25 – 49.

Smith, C. M., Z. R. Khan, and M. D. Pathak. 1994. Techniques for Evaluating Insect Resistance in Crop Plants. Lewis Publ. Co. 320 pp.

Smitha, F. E. 1998. Antiquity Online: Civilizations, philosophies and changing religions. http://www.fsmitha.com/h1/index.html.

Snelling, R. O. 1941. Resistance of plants to insect attack. Bot. Rev. 7:543–586.

Stoner, K. A. 1996. Plant resistance to insects: A resource available for sustainable agriculture. Biol. Agric. Hort. 13:7–38.

Teetes, G. L. 1994. Adjusting crop management recommendations for insect-resistant crop varieties. J. Agric. Entomol. 11:191–200.

van den Berg, J., G. D. J. van Rensburg, and M. C. van der Westhuizen. 1994. Host-plant resistance and chemical control of Chilo partellus (Swinhoe) and Busseola fusca (Fuller) in an integrated pest management system on grain sorghum. Crop Protection. 13: 308–310.

Waibel, H. 1987. The Economics of Integrated Pest Control in Irrigated Rice. A Case Study from the Philippines. Springer. Berlin. 196 pp.

Walling, L. L. 2000. The myriad plant responses to herbivores. J. Plant Growth Regul. 19:195–216.

White, W. H. 1990. Yellow sugarcane aphid (Homoptera: Aphididae) resistance mechanisms in selected sugarcane cultivars. J. Econ. Entomol. 83:2111–2114.

Wightman, J. A., M. M. Anders, V. R. Rao, and L. M. Reddy. 1995. Management of *Helicoverpa armigera* (Lepidoptera: Noctuidae) on chickpea in southern India: thresholds and the economics of host plant resistance and insecticide application. Crop Protection. 14: 37 – 46.

Wilde, G. 2002. Arthropod Host Plant Resistance. Encyclopedia of Pest Management. Marcel Dekker, New York. pp. 33 – 35.

Wiseman, B. R. 1999. Cumulative effects of antibiosis on five biological parameters of the fall armyworm. Fla. Entomol. 82:277–283.

Wiseman, B. R., and J. J. Hamm. 1993. Nuclear polyhedrosis virus and resistant corn silks enhance mortality of corn earworm (Lepidoptera: Noctuidae) larvae. Biol. Control 3:337–342.

Wiseman, B. R., and J. A. Webster. (Eds.). 1999. Economic, Environmental, and Social Benefits of Resistance in Field Crops. Thomas Say Publications in Entomology: Proceedings. Entomological Society of America, Lanham, MD. 189 pp.

CHAPTER 2

ANTIXENOSIS - ADVERSE EFFECTS OF RESISTANCE ON ARTHROPOD BEHAVIOR

1. DEFINITIONS AND CAUSES

Antixenosis is a term derived from the Greek word xeno (guest) that describes the inability of a plant to serve as a host to an arthropod. If this situation exists in a plant-arthropod interaction, a potential pest chooses to select an alternate host plant. The term antixenosis resistance was developed by Marcos Kogan and Eldon Ortman (1978) to more accurately describe the nonpreference reaction of arthropods to a resistant plant, and to complement the terminology for antibiosis resistance to arthropods (see Chapter 3). Nonpreference was originally defined by Painter (1951) (see Figure 1.5) as the group of plant characters and arthropod responses that lead to a plant being less damaged than another plant lacking these characters and the arthropod responses to them. Both antixenosis and nonpreference denote the presence of morphological or chemical plant factors that adversely alter arthropod behavior, resulting in the selection of an alternate host plant. Physical barriers such as thickened plant epidermal layers, waxy deposits on leaves, stems, or fruits, or a change in the density of trichomes (plant hairs) on normally susceptible plants may force arthropods to abandon their efforts to consume, ingest or oviposit on an otherwise palatable plant. Resistant plants may also be devoid of or lack sufficient levels of phytochemicals to stimulate arthropod feeding or oviposition, and allow them to escape consumption. Arthropod resistant plants may also possess unique phytochemicals that repel or deter herbivores from feeding or ovipositing. Finally, resistant plants may also contain chemicals that are toxic to arthropods after digestion of plant parts.

2. ARTHROPOD SENSORY SYSTEMS INVOLVED IN HOST SELECTION

Understanding how antixenosis functions in the resistance of a plant to an arthropod requires developing a perception of the arthropod's sensory environment. By taking this approach, we can gain some appreciation of the basic factors governing arthropod perception and integration of external stimuli detected by an arthropod's olfactory, visual, tactile, and gustatory receptors. The following sections describe each type of stimuli. More detailed discussions should be consulted and studied in reviews by Bernays (1992).

2.1. Olfaction

In order to perceive the odors emitted by potential host plants, arthropods rely on an olfactory guidance system controlled by cuticular sense organs known as sensilla basiconica, located on the antennae. Basiconic sensilla are porous, thin-walled structures ranging in length from 10 to 20 um (Figure 2.1). Great diversity exists in the number and arrangement of these sensilla on the antennae of various arthropods.

The olfactory sensitivity of different arthropod species is instinctively tuned to and controlled by a given qualitative and quantitative blend of odors. Most plant species are unique in their composition of volatile phytochemicals produced by fruiting structures, leaves, roots and stems. Specific groups of odor components in foliage of vegetables such as carrot, leek, onion and potato play important roles in directing arthropod movement to their host plants (Guerin et al. 1983, Leconte and Thibout 1981, Matsumoto 1970, Pierce et al. 1978, Visser et al. 1979). Mustaparta (1975) was one of the researchers to suggest that specific olfactory sensilla respond to specific odor components of a plant's odor "bouquet". Results of more recent research indeed has revealed how arthropods employ olfactory discrimination to determine the differences between unacceptable resistant pants and acceptable susceptible plants (Lapis and Borden 1993, Seifelnasr 1991)

Vincent Dethier developed original definitions for the effects of phytochemicals based on the responses they elicit in arthropods (Dethier et al. 1960), and the plant resistance research community has incorporated these terms into their working vocabulary. Odors emitted by plants that stimulate arthropod olfactory receptors and cause long-range arthropod movement toward the odors are **attractants**. In the opposite situation, plants exhibiting antixenosis may produce olfactory **repellents** that cause arthropods to move away from the plants producing the odor. Susceptible plants also emit **arrestant** odors that cause arthropods to stop movement when in close proximity to the odor source. The interplay between the odors emitted by host and non-host plant sources, the regulation of these odors by environmental factors, the perception of the odors by arthropods, and the resultant arthropod behaviors were discussed by Visser (1986) and are summarized in Figure 2.2. Additional experiments (Dickens et al. 1993, Thiery and Visser 1987) have demonstrated many more specifics about the olfactory perception of green leaf volatiles by arthropods.

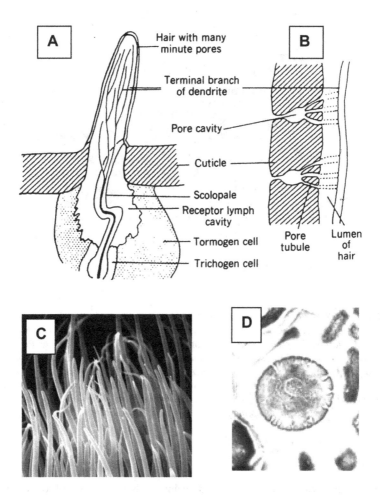

Figure 2.1. Insect sensilla basiconica (A) diagram of sensillum, (B) exploded diagram of hair wall showing pores through which stimulating molecules reach the nerve (dendrite), (C) sensilla basiconica on the antennal club of Lissorhoptrus oryzophilus (4,088X), (D) cross section of sensillum basiconica on the antennal club of L. oryzophilus showing pores in sensillum wall (35,200X). Figures 2.1.a & 2.1.b reprinted from Chapman and Blaney. How animals perceive secondary compounds. In G. A. Rosenthal and D. H. Janzen (Eds.) Herbivores: Their Interaction with Secondary Metabolites, Pages 161-198, Copyright 1979, Academic Press Inc., with permission from Elseiver.

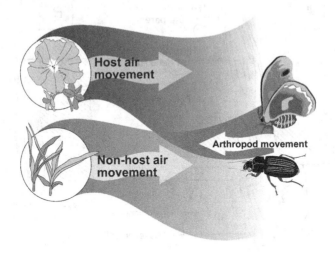

Figure 2.2. Schematic representation of the release of plant volatiles, their dispersion and perception by arthropods, and resultant arthropod behaviors. Host and non-host plants release volatile odor plumes that differ both quantitatively and qualitatively. Perception of the plume by an arthropod is dependent on the olfactory tuning of peripheral antennal sensory receptors. Arthropod orientation and movement to an odor source is governed by a positive anemotaxis (reaction to wind), as well as a positive chemotaxis (directed movement to or away from a chemical stimulus). Positive contact chemoreception of the plant results in the arthropod accepting the plant, followed by feeding and oviposition.

2.2. Vision

Prokopy and Owens (1983) described vision in herbivorous arthropods as being governed by their perception of the spectra quality of light stimuli, (i. e. brightness, hue, and saturation of various wavelengths) as well as the dimensions of the objects viewed, and the pattern or shape of the object. During orientation to potential host plants, arthropods simultaneously perceive visual and chemical stimuli (Green et al. 1994). During long-range orientation, an arthropod may use vision for recognition of the shape of an object and utilize olfaction to perceive plant attractants. After approaching the immediate location of the plant, movement to the plant surface is most likely guided by perception of the plant outline. Final contact with the plant surface by many arthropods is due to a positive response to yellow or yellow-green pigments in plant foliage that occurs in the spectral range of from 500 to 580 nm. Roessingh and Stadler (1990) demonstrated a combination of plant shape and color in studies of the oviposition behavior of the cabbage root fly, *Delia radicum* (L.).

In a direct application of this behavior, Broadbent et al. (1990) studied the relationship between flower color in *Chrysanthemum* spp. cultivars and resistance to the western flower thrips, *Frankiniella occidentalis* (Pegande). Thrips displayed a pronounced preference for yellow-flowered cultivars over white- flowered cultivars. Moharramipour et al. (1997) found similar results, noting that barley, *Hordeum vulgare* L., hybrids with green waxless foliage were less preferred by the corn leaf aphid, *Rhopalosiphum maidis* (L.), than hybrids with yellow waxless foliage. Antixenosis resistance has been achieved by genetically altering the color of plant foliage to reflect different wavelengths of light. Some curcubit cultivars with silver leaves reflect more blue and ultraviolet wavelengths of light than green-leaved cultivars, and are resistant to aphids and aphid-vectored diseases (Shifriss 1981).

The red leaf color in cotton, *Gossypium hirsutum* L., is a heritable character that causes antixenotic reactions in adult boll weevils, *Anthonomous grandis* Boheman (Iseley 1928, Jones et al. 1981). Red foliage in some cultivars of cabbage, *Brassica oleracea* var. *capitata* L., imparts antixenosis to alates of the cabbage aphid, *Brevicoryne brassicae* (L.), but this resistance is ephemeral and declines over the life of the plant (Singh and Ellis 1993). The same trend was noted in evaluations of cultivars of crabapple, *Malus ioensis*, (Wood) Britt., for resistance to Japanese beetle, *Popillia japonica* Newman, by Spicer et al. (1995). Cultivars with young red leaves that turned green with maturity were much more susceptible than cultivars with completely green leaves. Reinert et al. (1983) noted a similar preference in oviposition of the larger canna leafroller, *Calpodes ethlius* (Stoll), for cultivars of *Canna* spp. with red foliage over cultivars with green foliage. Fiori and Craig (1987) used the color intensity of birch leaf supernatants to determine degrees of resistance in birch, *Betula lutea* F. Michx., to oviposition by the birch leafminer, *Fenusa pusila* (Lepeletier). Birch species with high levels of oviposition have lower leaf supernatant spectrophotometric absorption rates than species that are resistant.

2.3. Thigmoreception

After an arthropod contacts the plant surface, trichoid sensilla on the body, tarsi, head, and antennae (Figure 2.3) perceive tactile stimuli and supply information about host plant morphology to the arthropod nervous system. Stimuli are received from the leaf or stem surface, or from trichomes, epidermal ridges, or leaf margin notches that trigger genetically controlled sequences of arthropod feeding or oviposition behavior. Plant morphological features may promote positive mechanical stimuli and act as feeding or oviposition stimulants. Changes in the shape, size and number of such plant morphological features may also prevent or disrupt the normal mechanoreceptive process, resulting in deterrency of feeding or oviposition.

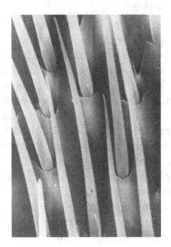

Figure 2.3. Sensilla trichoidea on the antennal club of Hypera meles (2,000X). Reprinted from J. Insect Morphol. & Embryol., Vol. 5. Smith, C. M., J. L. Frazier, L. B. Coons and W. E. Knight. 1976. Antennal morphology of the clover weevil, Hypera meles (F.) Int. J. Insect Morph. and Embryol. Pages 349-355, Copyright 1976, Pergamon Press, Inc., with permission from Elseiver.

A classic example of changing plant morphology to establish arthropod resistance is the breeding of the 'frego' or twisted bract character (Figure 2.4a) from wild genotypes of *Gossypium* into cotton cultivars. Normal cotton buds are tightly enclosed in bracts (Figure 2.4) that create a favorable environment for oviposition and feeding of the boll weevil. The open, twisted condition of the frego bract imparts weevil resistance by removing the positive stimuli that promote the use of normal bracts (Mitchell et al. 1973).

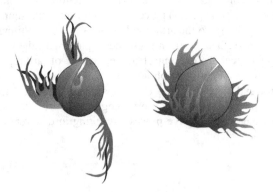

Figure 2.4. Frego (twisted) bract character of a cotton cultivar with resistance to Anthonomous grandis *(left) and enclosed bracts of a susceptible cultivar (right).*

2.4. Gustation

Arthropods taste plant tissues with contact chemosensory sensilla styloconica, maxillary palpi, and labral gustatory receptors that transmit sensory information to the arthropod central nervous system. At the tip of these sensilla, a single pore, open to the environment, receives chemosensory stimuli from molecules of plant allelochemicals in the liquid or vapor phase (Figure 2.5). Stimuli provided by these molecules are then electrochemically transduced and transmitted to the arthropod central nervous system (Hanson 1983). Several different types of gustatory receptors determine the qualitative and quantitative differences in the chemical content of the plant tissues tasted. The response spectrum of the arthropod gustatory receptors depends on the distribution of the different types of phytochemicals in the plants within an arthropod's host range. The receptors of generalist (oligophagous) arthropods normally have a wider response spectrum than the receptors of specialist (monophagous) arthropods (Visser 1983). In reviewing the sensory response of the mouthpart sensilla of grasshoppers, Chapman (1988) concluded that some sensilla need not be stimulated for distinction of acceptable versus unacceptable food and that monophagous grasshoppers require fewer sensilla to select food than polyphagous species.

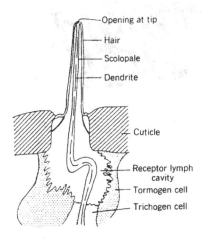

Figure 2.5. Longitudinal section through the tip of an insect contact chemoreceptor. Reprinted from Chapman and Blaney. How animals perceive secondary compounds. In G. A. Rosenthal and D. H. Janzen (Eds.) Herbivores: Their Interaction with Secondary Metabolites, Pages 161-198, Copyright 1979, Academic Press Inc., with permission from Elseiver.

Excellent examples of the types of monophagous responses described by Chapman exist in research with *Delia* spp. flies infesting various cruciferous crops. Neural responses of contact sensilla on the tarsi and labellum of the turnip root fly,

Delia floralis (Fallen), are positively correlated to fly oviposition behavior (Simmonds et al. 1994). These sensilla are highly responsive to specific glucosinolate chemical compounds on the leaf surface, and only minute changes in the composition of the glucosinolate mixture can affect the neural responses of the fly. Certain types of sensilla are also finely tuned to glucosinolate perception. Type D tarsal sensilla are more responsive than type A tarsal sensilla or labellar sensilla. A similar relationship is demonstrated by research of Roessingh et al. (1992) in studies of the oviposition behavior of *D. radicum* ovipositing on cauliflower, *Brassica oleracea* var. *botrytis* L. *D. radicum* type D sensilla on tarsal segments 3 and 4 are highly receptive to specific glucosinolates compared to other tarsal segments. As with *D. floralis*, changes in the glucosinolate molecule, in this case, the side chain length, mediate changes in *D. radicum* oviposition. To further illustrate the nuances of monophagy and arthropod-host plant specificity, both *Delia* species respond to some common glucosinolates, but each species responds maximally to different blends of specific glucosinolates.

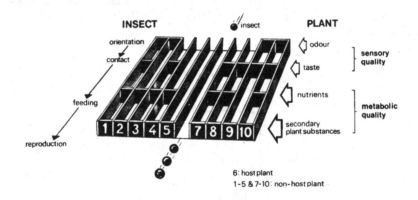

Figure 2.6. "Marble box" perception of plant and environmental stimuli by arthropods and resulting host (row 6) or non-host (rows 1-5, 7-10) selection. (From Visser 1983. Reprinted with permission from Plant Resistance to Insects. Copyright 1983 American Chemical Society)

Phytochemical stimuli that elicit continued arthropod feeding after being tasted are known as **phago (feeding) stimulants**, while those stimuli that prevent feeding are referred to as **phagodeterrents**. As illustrated by the examples of *Delia spp.* oviposition above, the same terms apply to phytochemicals affecting oviposition, with oviposition stimulants promoting oviposition and oviposition deterrents preventing oviposition. Dethier et al. (1960) also coined the terms **feeding incitants** and **suppressants** to denote plant stimuli that initiate or prevent feeding, respectively. Visser (1983) developed a schematic description of arthropod perception of plant stimuli and the resulting arthropod responses (Figure 2.6). This model demonstrates that the summed sensory input from chemical, visual, and tactile plant stimuli determine the ultimate degree of acceptance and utilization of a plant by an arthropod.

3. THEORETICAL CONCEPTS OF ARTHROPOD HOST SELECTION

Humans have observed the behavior and life history of arthropods for centuries. Not surprisingly, several theories have been advanced since the beginning of the twentieth century about how an arthropod locates a plant. The first to offer a postulate on arthropod host selection was C. T. Brues (1920), who proposed a Botanical Instinct Theory, which suggested that arthropods select host plants that meet specific nutritional and ecological requirements of an arthropod not offered by other plant species. Brues viewed the acceptance of a particular plant by an arthropod to be related to the nutritional composition and ecological niche of the plant in an arthropod's environment. The Token Stimuli Theory, proposed by Geoffery Fraenkel (1959), reasoned that specific "secondary plant substances" or phytochemicals, such as glycosides, phenols, tannins, alkaloids, terpenoids, and saponins determine arthropod host plant selection. Fraenkel suggested that these chemicals are non-nutritional and that their sole purpose is to defend plants against phytophagous arthropods and diseases. He theorized that arthropods evolve from polyphagy to monophagy to overcome the adverse effects of secondary plant substances and when they do, they use these compounds as beneficial cues to locate acceptable host plants. Studies by Seigler and Price (1976) and Heftmann (1975) later demonstrated that these "secondary" plant substances have high rates of metabolic turnover in plants, that they are closely associated with primary metabolic functions, and that they function as regulators of important biochemical plant processes. Thus, the functions of phytochemicals affecting arthropod resistance are more primary than secondary (Tuomi 1992).

Prior to discoveries dealing with the metabolism of secondary plant substances, J. S. Kennedy (1965) and A. J. Thorsteinson (1960) proposed that host selection be based on arthropod responses to both non-nutrient and nutrient phytochemicals. Their theories stemmed from the fact that many arthropods can be stimulated to feed by nutrient chemicals such as amino acids, carbohydrates, and vitamins (House 1969, Hsiao 1969). The Dual Discrimination Theory, proposed by Kennedy, continues to guide contemporary thinking about arthropod host plant selection. In 1970, Roger Whittaker synthesized the concepts of nutrient and non-nutritive plant allelochemicals, as well as the different ideas about their primary or secondary function, by introducing the term **allelochemical** to replace secondary plant substances (Whittaker 1970). He defined an allelochemical as a "non-nutritional chemical produced by an individual of one species that affects the growth, health, behavior, or population biology of another species". Narratives in following sections of this chapter and in Chapter 3 will describe the many different types of allelochemicals that affect arthropod feeding behavior, oviposition, and survival in either a positive or negative manner.

Allelochemicals can function as **allomones**, which benefit the producing organism, in this case a host plant, and they also act as **kairomones**, where they benefit the arthropod recipient. In terms of plant resistance to arthropod, allomones represent plant-produced deterrents, repellents, or inhibitors of feeding and oviposition. At the other extreme of host acceptability, attractants, arrestants, and feeding or oviposition stimulants in susceptible plants, such as those mentioned for the Brassiceae spp.-*Delia* spp. interactions mentioned in Section 2.2.4, function as kairomones.

4. EFFECTS OF PLANT DEFENSES ON ARTHROPOD BEHAVIOR

4.1. Morphological Defenses

4.1.1. Plant trichomes

Where present, plant trichomes are the first structure with which arthropods come in contact after making the decision to alight or walk on the plant surface. The role of trichomes in plant defense have been reviewed extensively (Jeffree 1986, Levin 1973, Peter et al. 1995, Peter and Shanower 2001), and it is clear that trichome-based antixenosis is a very broad-based defense, apparent in crop plant representatives from major taxa such as legumes, crucifers, grasses, and solanaceous plants (Table 2.1). Trichomes may be simple, erect hairs, hook-shaped hairs, or complex multi-celluar glandular structures. Simple trichomes limit the ability of arthropods to attach themselves to the plant surface, in order to initiate and maintain feeding, while hooked and glandular trichomes either entrap or impale the arthropod body, causing it to desiccate and die. As mentioned previously, the effects manifested by trichomes can be measured as either antixenosis or antibiosis. The effects of the toxic secretions from glandular trichomes are described and discussed as examples of antibiosis resistance in Chapter 3.

In several leguminous crop plants, "more is better", and dense growths of simple erect trichomes on leaves of some cultivars of alfalfa, *Medicago sativa* L., and soybean, *Glycine max* (L.) Merr., deter feeding by *E. fabae* (Taylor 1956, Lee 1983). Dense leaf pubescence also contributes to the feeding antixenosis of some cultivars of soybean (Figure 2.7) to the cabbage looper, *Trichoplusia ni* (Hubner), (Khan et al. 1986) the Mexican bean beetle, *Epilachna varivestis* (Mulsant), (Gannon and Bach 1996), the two-spotted spider mite, *Tetranychus urticae* (Elden 1997) and agromyzid beanflies (Chiang and Norris 1983). Yet in some of the same germplasm, Lambert et al. (1992) noted a lack of antixenosis to oviposition by three additional species of lepidoptera on densely pubescent soybean lines. Talekar and Hu (1993) bioassayed agromyzid beanflies for feeding and oviposition antixenosis with mungbean, *Vigna radiata* (L.), and found a similar result in a glaborous (trichome free) wild relative, *Vigna glabrescens* (Marechal, Mascherpa & Stainier) that exhibited antixenosis to agromyzids.

Glandular haired alfalfa species also exhibit antixenosis to the potato leafhopper (Shade et al. 1979), the alfalfa weevil, *Hypera postica* Gyllenhal (Johnson et al. 1980a,b, Danielson et al. 1987), the alfalfa seed chalcid, *Bruchophagus roddi* (Gussakovsky) (Brewer et al. 1983), the spotted alfalfa aphid, *Therioaphis maculata* (Buckton) (Ferguson et al. 1982), the alfalfa blotch leafminer, *Agromyza frontella* (Rondani) (MacLean and Byers 1983) and the pea aphid, *Acyrthosiphon pisum* (Harris) (Shade and Kitch 1983).

Table 2.1. Arthropods affected by resistant plants possessing leaf and stem trichomes

Plant or genus	Trichome type	Arthropod affected	References
Alnus	Simple	Chrysomelidae	Baur et al. 1991
Brassica napus	Simple	*Phyllotreta cruciferae*	Palaniswamy & Bodnaryk 1994
Euphorbia pulcherrima	Simple	*Bemisia* spp.	Bilderback & Mattson 1977
Fragaria	Simple	*Otiorhynchus sulcatus*	Doss et al. 1987
Glycine max	Simple	Agromyzidae	Chiang & Norris 1983
		Trichoplusia ni	Khan 1986
		Epilachna varivestis	Gannon & Bach 1996
		Empoasca fabae	Lee 1983
		Tetranychus spp.	Elden 1997
Gossypium	Simple	Leafhopper	Reed 1974
		Anthonomus grandis	Wessling 1958
		Spodoptera littoralis	Kamel 1965
		Lygus lineolaris	Meredith & Schuster 1979
		Lygus Hesperus	Benedict et al. 1983
Helianthis	Glandular	*Homoeosoma electellum*	Rogers et al. 1987
Lycopersicon	Glandular	*Liriomyza trifolii*	Hawthorne et al. 1992
	Glandular	*Phytoseiulus ersimilis*	Nihoul 1994
	Simple	*Bemisia tabaci*	Heinz & Zalom 1995
Medicago sativa	Simple	*Empoasca fabae*	Shade et al. 1979
	Glandular	*Hypera postica*	Danielson et al. 1987
		Bruchophagus roddi	Brewer et al. 1983
		Therioaphis maculata	Ferguson et al. 1982
		Acyrthosiphon pisum	Shade & Kitch 1983
		Agromyza frontella	MacLean & Byers 1983
Phaseolus	Simple	*Liriomyza trifolii*	Quiring et al. 1992
Saccharum officinarum	Simple	Pyralidae	Verma & Mathur 1950

Table 2.1. continued

Solanum	Simple	*Empoasca fabae*	Taylor 1956
	Glandular	*Myzus persicae*	LaPointe & Tingey 1986
Sorghum biocolor	Simple	*Atherigona soccata*	Maiti & Gibson 1983
Triticum aestivum	Simple	*Rhopalosiphum padi*	Roberts & Foster 1983
		Oulema melanopus	Hoxie et al. 1975
		Mayetiola destructor	Roberts et al. 1979
		Sipha flava	Webster et al. 1994
Vigna	Simple	*Maruca testulalis*	Oghiakhe et al. 1992

Glandular trichomes (Figure 2.8) on the foliage of the wild potato, *Solanum neocardenasii*, adversely affect the feeding behavior of the green peach aphid, *Myzus persicae* (Sulzer), and the leafminer *Liriomyza trifolii* (Burgess), by delaying the amount of time required to begin feeding (Hawthorne et al. 1992, LaPointe and Tingey 1986). Tingey and Laubengayer (1986) demonstrated that removing the pubescence from foliage resulted in an increase in feeding by *E. fabae*.

High densities of trichomes on the buds of some cotton cultivars also deter feeding and oviposition by the boll weevil, *Anthomomus grandis grandis* Boheman (Wessling 1958). Kamel (1965) determined that cotton cultivars with increased trichome density on lower leaf surfaces were more resistant to the cotton leafworm,

Figure 2.7. Simple trichomes of soybean leaves cause antixenosis to Trichoplusia ni larvae (6X). Reprinted from Khan, Z. R., J. T. Ward, and D. M. Norris. Role of trichomes in soybean resistance to cabbage looper, Trichoplusia ni. Entomol. Exp. Appl.42:109-117. Copyright 1986, D. W. Junk Publ. Co., with kind permission of Springer Science and Business Media.

Spodoptera littoralis (Boisd.). Pubescent cotton cultivars also exhibit antixenosis to the tarnished plant bug, *Lygus lineolaris* Palisot de Beauvois, (Meredith and Schuster

1979, Wilson and George 1986), the western plant bug, *Lygus hesperus* Knight, (Benedict et al. 1983), and *T. ni* (George et al. 1977). The combined effect of both increased trichome density and length reduce populations of both *Bemisia* and *Empoasca* on some cotton cultivars (Butler et al. 1992). However, pubescent cotton promotes population growth of the cotton leafhopper, *Pseudatomoscelis seriatus* (Reuter) (Lukefahr et al. 1970) and tobacco budworm, *Heliothis virescens* (F.) (Lukefahr et al. 1971). Cotton cultivars with smooth-leafed foliage with little or no pubescence are resistant to these arthropods (Robinson et al. 1980).

Similarly, a glaborous isoline of pearl millet, *Pennisetum glaucum* (L.) R. Br., has a negative effect on oviposition and feeding by fall armyworm, *Spodoptera frugiperda* (J. E. Smith), adults and larvae (Burton et al. 1977). Maize selections of the 'Antigua' ancestry, with reduced trichome density are less preferred for oviposition by *H. zea* moths and possess resistance to larval feeding (Wiseman et al. 1976, Widstrom et al. 1979). The oviposition behavior of the silverleaf whitefly, *Bemisia argentifolii* Bellows & Perring, also favors a pubescent substrate and glaborous cultivars of both soybean and tomato, *Lycopersicon esculentum* Mill., and are much less preferred for *B. tabaci* oviposition (Heinz and Zalom 1995, McAuslane 1996).

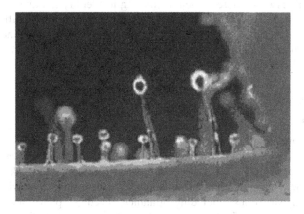

Figure 2.8. Glandular trichomes of potato foliage.
http://www.cgiar.org/who/wwa_potatoes.html

Some cultivars of wheat, *Triticum aestivum* L., and *Avena* species with dense growth of long, erect trichomes deter oviposition by the cereal leaf beetle, *Oulema melanopus* (L.), much more than cultivars with sparse growth of short trichomes (Hoxie et al. 1975, Papp et al. 1992, Wallace et al. 1974). The pubescent wheat cultivar 'Vel' also exhibits antixenosis to adults and larvae of the Hessian fly, *Mayetiola destructor* (Say) (Roberts et al. 1979). Adult flies lay fewer eggs, egg hatch is reduced, and larval mobility is impaired. Some pubescent wheat cultivars also have antixenotic effects on *R. padi*, an important vector of barley yellow dwarf virus (Roberts and Foster 1983), as well as the yellow sugarcane aphid, *Sipha flava* (Forbes) (Webster et al. 1994). As with cotton, however, pubescent wheats are susceptible to non-target pests such as the wheat curl mite, *Aceria tosichella* Keifer, and the wheat streak mosaic virus vectored by the mite (Harvey and Martin 1980).

Foliar pubescence is also responsible for antixenosis in cultivars of sorghum, *Sorghum bicolor* (L.) Moench, that are resistant to the sorghum shootfly, *Atherigona soccata* (Rondani) (Maiti and Gibson 1983), and in cultivars of poinsettia, *Euphorbia pulcherrima* Willd. ex Klotzsch, that are resistant to the whitefly, *Trialeurodes vaporariorum* (Westwood) (Bilderback and Mattson 1977). Doss et al. (1987) determined that the resistance in clones of strawberry, *Fragaria* x *ananassa* Duchesne, to feeding and oviposition by the black vine weevil, *Otiorhynchus sulcatus* (F.), is related to the density of simple trichomes on the underside of leaves. Johnson and Lewis (1993) made similar determinations in studies of antixenosis to grape phylloxera, *Daktuloshaira vitifoliae* (Fitch), in grape (*Vitus*) cultivars, and demonstrated that increased pubescence density on the underside of leaves decreased the numbers of *D. vitifoliae* galls.

The species-to-species variation in trichome-based antixenosis to different arthropod pests in the same crop remind us that one pest's poison (a dense trichome field) is another's passion. At the same time, these differences point out one of the major difficulties encountered when attempting to identify multiple pest resistance germplasm. If the behavioral idiosyncrasies of each pest are significant enough, multiple pest resistance may need to be based on more that one behavior - modifying characteristic and may need to involve additional physical or allelochemical characters as well.

4.1.2. Surface waxes

Once an arthropod successfully deals with leaf or stem trichomes, it encounters the leaf or stem epidermal layer of the plant surface. Plant leaves are protected against desiccation, arthropod predation, and disease by a layer of surface waxes produced over the leaf epicuticle. Plant waxes are esters formed by the linkage of a long chain fatty acid and an aliphatic alcohol. Waxes occur in a wide variety of morphological variations, including granules, needles, plates and ribbons, as described in the review of Jeffree (1986), a long-standing source of information about the structural aspects of plant surface waxes. Eigenbrode and Espelie (1995) produced an excellent overview of the growing number of diverse effects of surface waxes on arthropods in which they classified the differences in surface wax content, appearance and arthropod susceptibility of plants. Glossy, bloomless and glazed or waxy genotypes have all been shown to have differing effects on arthropod behavior (antixenosis) and development (see Chapter 3).

Foliar wax coatings play an important role in the resistance of some crop cultivars to arthropod attack (Table 2.2) when sense organs on the arthropod tarsi and mouthparts perceive negative chemical and tactile stimuli from the leaf surface. One of the earliest examples of the involvement of waxes in plant resistance to arthropods was produced by Anstey and Moore (1954), who demonstrated that wax blooms on the leaves of some cruciferous crops deter feeding of the cabbage flea beetle, *Phyllotreta albionica* (LeConte). Stork (1980) demonstrated similar results with the effects of waxy leaves of Brussels sprouts, *Brassica oleracea* var. *gemmifera* L., to the mustard beetle, *Phaedon cochleariae* (F.). The dense wax coating consists of vertical rods and dendritic plates that interfere with adhesion of the tarsal setae of the beetle to the leaf surface.

In the late 1980s, *Brassica oleracea* lines of broccoli, Brussels sprouts, cabbage, cauliflower, collards, and kale were identified that produce leaves devoid of the wax bloom with a glossy, reflective green appearance (Fig 2.9). Compared to waxy genotypes, glossy lines have reduced lipid microstructure and quantity, as well as altered chemical composition (Eigenbrode et al. 1991). These effects were noted early on by Thompson (1963) and Way and Murdie (1965). Glossy-leafed kale and Brussels sprouts sustain less feeding by the cabbage aphid, *Brevicoryne brassicae* (L.), and the cabbage whitefly, *Aleurodes brassicae* (Walker) than waxy-leafed cultivars. The glossy genotypes cause antixenosis reactions in larvae of the diamondback moth, *Plutella xylostella* (L.) (Eigenbrode and Shelton 1990, Stoner 1990).

Evaluations of several plant introductions of glossy-leafed oilseed rape, *Brassica napus* L., by Ramachandran et al. (1998) failed to demonstrate resistance to larval feeding and oviposition to *P. xylostella*, but did identify a novel source of resistance that is apparently independent of surface wax composition.

Table 2.2. *Effects of plant surface waxes on antixenosis resistance to arthropods*

Plant	Arthropod	Wax effect(s)	Reference
Brassica oleracea var. *acephala*	*Brevicoryne brassicae*	Stimulant	Thompson 1963
Brassica oleracea var. *gemmifera*	*Phaedon cochleariae*	Adhesion	Stork 1980
	Phyllotreta albionica	Deterrent	Anstey & Moore 1957, Bodnaryk 1992, Stoner 1990
	Therioaphis maculata	Deterrent	Bergman et al. 1991
Medicago sativa			
Rubus phoenicolasius	*Amphorophora rubi*	Barrier	Lupton 1967
Sorghum biocolor	*Locusta migratoiodes*	Deterrent	Atkin & Hamilton 1982
	Schizaphis graminum	Stimulant	Weibel & Starks 1986
Zea mays	*Spodoptrea frugiperda*	Deterrent	Yang et al. 1993

In Graminaceous crops, glossy-leafed plants also impart antixenotic resistance in sorghum to feeding by *Schizaphis graminum* (Rhodani) (Peiretti et al. 1980, Weibel and

Starks 1986), and the stem borer, *Chilo partellus* (Swinhoe) (Chapman et al. 1983). Resistance in sorghum to the shoot fly, *Atherigona soccata* Rondani, is also a function of leaf glossiness. Resistant cultivars have a smooth, amorphous wax layer with few wax crystals, while susceptible waxy cultivars have significantly more wax in the epicuticle. The leaves of susceptible plants produce a dense meshwork of crystalline wax that allows water droplets, the *A. soccata* oviposition site, to spread to the leaf edges (Nwanze et al. 1992, Sree et al. 1994).

Just as plant trichomes may not affect the expression of resistance, the presence of glossy leaves of some plants may have no effect on arthropod resistance. In some cases, waxes actually stimulate arthropod feeding and oviposition. For example, Lamb et al. (1993) found that no significant effects were derived from glossy-leafed plants of rape, kale or collards, compared to waxy-leafed mutants of each, in resistance to the aphid *Lipaphis eryisimi* (Kaltenbach). In a survey of *Brassicaceae*

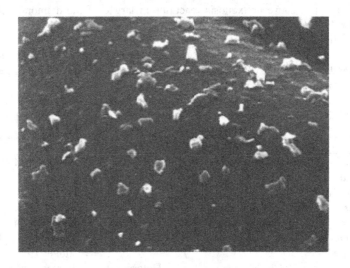

Figure 2.9. Glossy leaf surface of Brassica *devoid of surface wax bloom. (From Eigenbrode et al. 1991. Reproduced, with permission, from the Journal of Economic Entomology, Vol. 84, Copyright 1991 by the Entomological Society of America)*

crop plants, Bodnaryk (1992) determined that flea beetles, *Phyllotreae cruciferae* (Goeze), actually feed significantly more over the entire leaf surface of glossy-leafed *Brassicas,* compared to waxy-leafed mutants, where beetles feed only on leaf edges.

Removing the glossy surface by mechanically rubbing leaves of Brussels sprouts, cauliflower and oil seed rape plants render them susceptible. Beetle feeding is altered by a change of as little as 33% in the quantity of epicuticular wax in oil seed rape. Glossy-leafed barley cultivars harbor much higher populations of aphids than populations on waxy cultivars (Tsumuki et al. 1989).

Lupton (1967), demonstrated how the foliage of raspberry, *Rubus phoenicolasius* Maxim., also produces heavy wax secretions that resist feeding by the raspberry beetle, *Byturus tomentosus* Barber, and the rubus aphid, *Amphorophora rubi* (Kaltenbach). Some 30 years later, the resistance of red raspberry, *Rubus idaeus* L., to the large raspberry aphid, *Amphorophora idaei* (Borner) is linked to markedly higher levels of cycloartenyl and α-amyryl esters, sterols and branched alkanes in wax from resistant cultivars (Shepherd et al. 1999a, 1999b). Wax production reaches maximum levels in mature leaves.

In sorghum, wax production is a key to resistance to the migratory locust, *Locusta migratoiodes* (R&F.) (Atkin and Hamilton 1982), but the opposite trend in plant maturity exists. Foliar wax from young plants is more deterrent to locust feeding than that from older leaves (Figure 2.10). Foliar surface waxes of plants may also

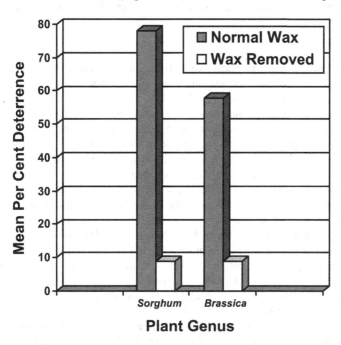

Figure 2.10. Deterrence (precent insects stopping at palpation) of Locusta migratoria *nymphs by surface waxes on plant foliage. (Adapted from Woodhead & Chapman 1986. Reprinted with permission of Edward Arnold Publishing Co.)*

contain allelochemicals that negatively affect the behavior of arthropods by repelling them prior to feeding, or deterring their feeding and oviposition (see Section 2.4.2.2).

Wax extracts from glossy-leafed cabbage cultivars deter feeding of *P. xylostella* larvae, causing them to wander over the leaf surface more and feed less than larvae exposed to extracts of susceptible plants with normal wax components (Eigenbrode and Pillai 1998).

Increased levels of wax-related alcohols in several plants are also responsible for arthropod resistance. In selected alfalfa plants with resistance to *T. maculata* the leaf surface of resistant plants produces significantly higher levels of leaf triacontanol than aphid susceptible plants (Bergman et al. 1991). Cultivars of tobacco, *Nicotiana* spp., resistant to *H. virescens* contain increased levels of the alcohol docosanol (Johnson and Severson 1984), while some sorghum cultivars with greatly increased levels of triterpenols are resistant to aphids (Heupel 1985). In azalea, *Rhododendron* spp., resistance to the azalea lace bug, *Stephantis pyrioides* (Scott), is linked to increased levels of the triterpenoid compounds α-amyrin and β-amyrin (Baldson et al. (1995).

4.1.3. Tissue thickness

Modifications to the plant physical structure are also responsible for arthropod resistance, and the thickness of various plant tissues determines the degree of resistance in some crop cultivars (Table 2.3). According to Chesnokov (1962) "There is a large group of phenomena of immunity, conditioned by the anatomic-morphological peculiarities of species and varieties, that creates conditions that exclude penetration of the damaging arthropod to the site of nutrition." In one of the earliest references to a physical or chemical factor mediating arthropod resistance, Sakharov (1925) demonstrated how some sunflower cultivars rely on a thickened phytomelanin layer in the pericarp of the sunflower seed coat to resist feeding by larvae of the sunflower moth, *Homoeosoma nebulella* Hubner (Figure 2.11). The resistance of several cultivars of sunflower, *Helianthus annuus* L., to penetration by larvae of the new world sunflower moth, *Homoeosoma electellum* (Hulst), is related to the same phenomenon (Rogers and Kreitner 1983). In the case of *H. electellum*, early lignification of the pericarp sclerenchymal cell wall also acts synergistically with the phytomelanin layer to increase sunflower pericarp hardness.

The foliar toughness of several cruciferous crops adversely affects the feeding behavior of mustard beetles (Tanton 1962). Cultivars of sugarcane, a complex hybrid of *Saccharum* species, with thick layers of leaf epidermis and parenchyma cells deter feeding by the top shoot borer, *Scirpophaga nivella* (F.) (Chang and Shih 1959). In evaluating the resistance of *Lycopersicon* species to *F. occidentalis*, Krishna Kumar et al. (1995) found that accessions with strong antixenosis exhibited shriveled and distorted epidermal cells apparently sacrificed during feeding, but the underlying mesophyll cells were unaffected. The mechanical strength of cotyledons of certain subterranean clover, *Trifolium subterraneum* (L.), cultivars is directly responsible for antixenosis of those cultivars to feeding by the redlegged earth mite, *Halotydeus destructor* Tucker (Jiang and Ridsdill-Smith 1996). The upper epidermal cells of susceptible cultivars collapse during mite feeding, while those of the resistant cultivar buckle locally, but remain intact.

Stems thickened by increased layers of epidermal cells deter or limit entrance of stem damaging arthropods of some cultivars of rice, sugarcane, and wheat (Fiori and Dolan 1981, Patanakamjorn and Pathak 1967, Martin et al. 1975, Wallace et al. 1974). Thick cortex layers in the stems of the wild tomato, *Lycopersicon hirsutum* Dunal, deter feeding by the potato aphid, *Macrosiphum euphorbiae* (Thomas) (Quiras et al. 1977). Bergvinson et al. (1994a,b, 1995) has shown that resistance in maize to the European corn borer, *Ostrinia nubilalis* Hubner, is dependent on a number of physiochemical factors that result in increased leaf and stem toughness in maize. In the interaction of *Boronia megastigma* (Nees) with the psyllid, *Ctenarytaina thysanura,* boronia cultivars with oviposition antixenosis to *C. thysanura* have harder terminal shoots (Mensah and Madden 1991).

The arrangement of stem tissues is critical to the resistance of some crop plants. Tightly packed vascular bundles play a role in the resistance of lettuce, *Lactuca sativa* L., to *B. argentifolii* (Cohen et al. 1996), and physically-based stem resistance in alfalfa to *E. fabae* (Brewer et al. 1986) is dependent on numerous, large vascular bundles, as illustrated in Figure 2.12. Resistance in sorghum to *A. soccata* is also related to thickened cells surrounding the leaf vascular bundles (Blum 1968).

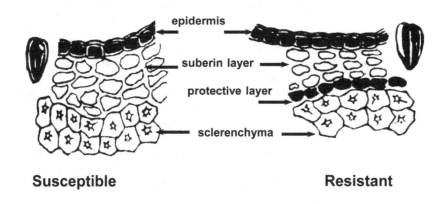

Susceptible   **Resistant**

Figure 2.11. Structure of the seed coats of Helianthus *cultivars susceptible and resistant to larval feeding damage by* Homoeosoma nebulella. *(Chesnokov 1962)*

Highly thickened, lignified pod walls in some cultivars of southern pea, *Pisum sativum* var. *macrocarpon* L., deter oviposition and larval feeding of the cowpea curculio, *Chalcodermus aeneus* Boheman (Fery and Cuthbert 1979). Bruchid resistance in mungbean, *Vigna radiata* (L.) is also related to a thickened seed coat, as well as small seed size (Lambrides and Imrie 2000). Thickened seed walls are also related to antixenosis in *Medicago* species with resistance to *B. roddi* (Springer et al. 1990) and in soybean cultivars resistant to the soybean pod borer, *Grapholitha glicinvorella* (Matsumura) (Nishijima 1960).

Table 2.3. Plant tissue thickness as an antixenosis factor

Plant or genus	Arthropod(s)	Tissue	References
Betula pendula	Geometrid moth	Stem	Mutikainen et al. 1996
Boronia megastigma	*Ctenarytaina thysanura*	Shoots	Mensah & Madden 1991
Crucifers	*Phaedon cochleariae*	Leaf	Tanton 1962
Helianthus	*Homoeosoma electellum*	Pericarp	Rogers & Kreitner 1983
Lactuca sativa	*Bemisia argentifolii*	Stem	Cohen et al. 1996
Lycopersicon hirsutum	*Macrosiphum euphorbiae*	Stem	Quiras et al. 1977
Medicago sativa	*Hypera postica*	Stem	Fiori & Dolan 1981
	Empoasca fabae	Stem	Brewer et al. 1996
	Bruchophagus roddi	Pod wall	Springer et al. 1990
	Grapholitha glicinvorella	Pod wall	Nishijima 1960
Oryza sativa	*Chilo suppressalis*	Stem	Patanakamjorn & Pathak 1967
Saccharum officinarum	*Scirpophaga novella*	Leaf	Chang & Shih 1959
	Diatraea saccharalis	Stem	Martin et al. 1975
Sorghum biocolor	*Atherigona varia soccata*	Leaf	Blum 1968
Triticum aestivum	*Cephus cinctus*	Stem	Wallace et al. 1974
Vigna unguiculata	*Chalcodermus aeneus*	Pod wall	Fery & Cuthbert 1979
Zea mays	*Ostrinia nubilalis*	Leaf	Beeghly et al. 1997, Bergvinson 1994b
	Spodoptera frugiperda	Leaf	Davis et al. 1995

Modified fruiting structures. Antixenosis is also apparent in cultivars of some crops with modifications in fruiting structures, such as the glume or panicle, and the resulting grain hull or husk. Sharma et al. (1994) found that sorghum cultivars with antixenosis to the head bug, *Eurystylus immaculatus* Odh., have an increased panicle glume length and increased grain hardness, compared to several susceptible cultivars. In eggplant, *Solanum melongena* L., antixenosis resistance to the fruit and shoot borer, *Leuinodes orbonalis* Guen., is highly correlated to tight seed arrangement in the fruit mesocarp (Lal 1991).

Tight-hulled rice cultivars are less susceptible to the Angoumois grain moth, *Sitotroga cerealella* (Olivier) and the rice weevil, *Sitophilus oryzae* (L.), than those with loose or gaping palea and lemma (Russell and Cogburn 1977, Rosetto et al. 1973). In maize plants, tight-husked ears are resistant to penetration by *H. zea, S. oryzae* and the maize weevil, *Sitophilus zeamais* Motchulsky (Singh et al. 1972, Wiseman et al 1974, Wiseman et al. 1977).

4.2. Chemical defenses

As indicated previously, allelochemicals may act as repellents during the olfactory orientation of an arthropod to a resistant plant or as feeding deterrents or feeding inhibitors when an arthropod tastes a resistant plant. Chapman (1974) suggested the use of the general term **feeding inhibitor**, due to the great diversity of the types of bioassays conducted and the variety of terms applied to the results of the bioassay of plant allelochemicals.

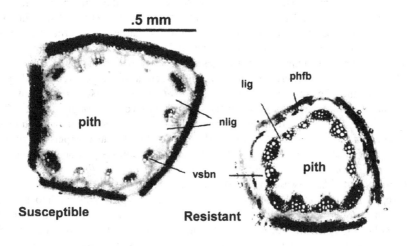

Figure 2.12. Physically-based resistance to Empoasca fabae *in stems of alfalfa. (From Brewer et al. 1986. Reprinted with permission from J. Econ. Entomol., Vol. 79: 1249-1253. Copyright 1986., Entomological Society of America.) lig = lignified area; nlig = nonlignified area; phfb = phloem fibers; vsbn = vascular bundles*

Many of the underlying concepts pertaining to such compounds and their perception by arthropods were reviewed by Schoonhoven (1982), who described both the deterrent and inhibitory plant allelochemicals functioning in antixenosis as antifeedants, and by Frazier et al. (1986), who referred to them as feeding inhibitors. The review of Schoonhoven et al. (1992) explored the principles of arthropod neural coding in relation to feeding deterrents and the correlation between chemoreceptor activity and arthropod feeding behavior. Interest in the chemical bases of plant resistance, primarily in antifeedants and feeding deterrents, has steadily increased since the mid 1960's. Antixenosis resistance based on plant allelochemical content exists in numerous crops.

4.2.1. Repellents

Volatile hydrocarbons emitted by the foliage of resistant plants comprise a great variety of arthropod repellents (Table 2.4). Bordasch and Berryman (1977) determined that several monoterpenes from the resin vapors in foliage of grand fir, *Abies grandis* (Dougl. ex D.Don) Lindl., repel the fir engraver beetle, *Scolytus ventralis* LeConte, and Perttunen (1957) detected a similar reaction by the bark beetles *Hylurgops palliatus* Gyll., and *Hylastes ater* Payk. Repellents in arthropod resistant rice cultivars mediate antixenosis to feeding by the green rice leafhopper, *Nephottetix virescens* Distant (Khan and Saxena 1985), the brown planthopper, *Nilaparvata lugens* Stal (Saxena and Okech 1985), and the whitebacked planthopper, *Sogatella furcifera* (Horv.) (Kahn and Saxena 1986). Volatile hydrocarbons from the surface of raspberry leaves have also been linked to antixenosis resistance to *A. idaei* by linear discriminate analysis of gas chromatographic data (Robertson et al. 1991).

Volatiles from strawberry species with high essential oil content repel feeding by *T. urticae* and the strawberry spider mite, *Tetranychus turkestani* Ugarov and Nikolsik (Dabrowski and Rodriguez 1971). A unique volatile organic acid, 2,3-dihydrofarnesoic acid, produced by the glandular trichomes of wild tomato, *Lycopersicon hirsutum* f. *glabratum* (Humb. & Bonpl.), also repels *T. urticae* feeding (Guo et al. 1993, Snyder et al. 1993). Volatile compounds produced on the leaf surface of another wild tomato species, *Lycopersicon penellii* (Corr.), repel the leafminer, *L. trifolii* (Hawthorne et al. 1992) and *M. euphorbiae* (Goffreda et al. 1989).

Repellency may also be due to the lack of perception of volatile allelochemical attractants. Guerin and Ryan (1984) determined that a decrease in the production of root volatiles by some cultivars of carrot, *Daucus carota* L., plays a role in resistance to the carrot rust fly, *Psila rosae* (F.). Decreased production of diterpenes in some tobacco plant introductions also mediates resistance to *H. virescens* (Jackson et al. 1986) and to *M. persicae* (Johnson and Severson 1982).

4.2.2. Deterrents

Allelochemicals in many plants deter arthropod feeding and oviposition. Those that most frequently cause deterrency include alkaloids, flavonoids, terpene lactones, and phenols, produced and stored in leaf cell walls, vacuoles or specialized structures such as the trichomes and waxes discussed previously (Table 2.5). As will be discussed in greater detail in Chapter 9, these compounds exist constitutively, or are expressed by *de novo* synthesis following tissue damage due to arthropod feeding or oviposition.

Table 2.4. The role of repellent allelochemicals in antixenosis

Plant or genus	Arthropod	Effect	Reference
Abies amabilis	*Scolytus ventralis*	Repellent	Bordasch & Berryman 1977
Daucus carota	*Psila rosae*	Escape	Guerin & Ryan 1984
Fragaria ananassa	*Tetranychus urticae*	Repellent	Dabrowski & Rodriguez 1971
Lycopersicum esculentum	*T. urticae*	Repellent	Snyder et al. 1993
Nicotiana attenuate	*Myzus persicae*	Escape	Johnson & Severson 1982
	Heliothis virescens	Escape	Jackson et al. 1986
Oryza sativa	*Nephottetix virescens*	Repellent	Khan & Saxena 1985
	Nilaparvata lugens	Repellent	Saxena & Okech 1985
	Sogatella furcifera	Repellent	Saxena & Okech 1985

Many experiments have been conducted to determine the precise roles of phenolics in plant defense, either as antifeedants or growth inhibitors (Appel 1993). As of yet however, a direct cause and effect relationship has not been established for phenolics, due in part to problems in separating then from other phytochemicals. Several studies have shown a partial role of various phenolic compounds in antixenosis; and in some cases, they play a direct role in arthropod resistance.

In interactions between phenolics and different species of beetles, phenolics serve as a more direct factor in resistance. The blue willow beetle, *Phyllodecta vulgatissima* (L.), and the brown willow beetle, *Galerucella lineola* (Fab.), are strongly deterred from feeding on willow, *Salix alba* L., cultivars that have high concentrations of phenolic glucosides (Kelly and Curry 1991, Kendall et al. 1996). *Epicauta* spp. blister beetles and the vegetable weevil, *Lisstrodes costirostris obliqus* (Klug), are deterred from feeding on the leaves of yellow sweetclover, *Melilotus officinalis* Pursh, by the phenolic compound coumarin (Figure 2.13) (Matsumoto 1962, Gorz et al 1972). Though coumarin is the deterring allelochemical, it occurs after the hydrolysis of trans-o-hydroxy-cinnamic acid to coumarin.

In grass crop plants, phenolic compounds, such as rutin, chlorogenic acid and flavonoids seem to act as partial resistance factors. *Schizaphis graminum* is deterred from feeding by the phenolics procyanidin, *p*-hydroxybenzaldehyde, and dhurrin (Figure 2.14) in resistant sorghum cultivars (Dreyer et al. 1981, Reese 1981). However, resistance is also related to reduced levels of the pectic feeding stimulant arabinogalactan, increased amounts of the feeding inhibitor pectic fructan, and increased amounts of pectin methoxy (Campbell and Dreyer 1985, Dreyer and Campbell 1983).

Phenolic acids, metabolites of *p*-hydroxybenzaldehyde, have been implicated as factors in the resistance of sorghum foliage to *S. gregaria* (Woodhead and Cooper-Driver 1979) and *p*-hydroxybenzaldehyde alone in high amounts in the surface waxes of sorghum seedlings imparts some resistance to locusts (Woodhead 1982).

Table 2.5 Arthropod feeding feeding detererents involved in plant antixenosis

Plant or genus	Arthropods	Chemical	References
Gossypium hirsutum	Heliothis virescens, Helicoverpa zea, Tetranychus urticae	Glucoside Phenols	Hedin et al. 1983 Elliger et al. 1978 Lane & Schuster 1981
Helianthus maxmiliani	Homoeosoma electellum	Sesquiterpene lactones	Gershenzon et al. 1985
Lotus pedunculatus	Costelytra zealandica	Isoflavan	Russell et al. 1978
Lupinus spp.	Sitona lineatus	Alkaloid	Cantot & Papineau 1983
L. hirsutum f. glabratum	Scrobipalpuloides absoluta	Methyl ketone	Maluf et al. 1997
Medicago sativa	C. zealandica	Saponin	Sutherland 1975
Melilotus officinalis	Epicauta spp., Listrodes costirostris obliqus	Coumarin	Gorz et al. 1972 Matsumoto 1962
Populus tremuloides	Spodoptera eridania	Tannins	Manuwoto et al. 1985
Oryza sativa	Chilo suppressalis	Phenolics	Das 1976
Rhododendron	Sciopithes obscurus	Sesquiterpene lactones	Doss et al. 1980
Salix spp.	Phyllodecta vulgatissima	Chlorogenic acid	Matsuda & Matsuo 1985
	Galerucella lineola	Glucosides	Kendall et al. 1996
Solanum tuberosum	Empoasca fabae Agriotes obscurus	Alkaloids Alkaloids	Sinden et al. 1986 Jonasson & Olsson 1994
Sorghum biocolor	Schizaphis graminum	Phenolics, Pectic fructan	Campbell & Dreyer 1985 Reese 1981
	Locusta migratoria migratorioides	Phenolics, Benzaldehyde	Woodhead 1982
	Atherigona soccata	Tannins	Sharma et al. 1991 Kumar & Singh 1998
Trifolium subterraneum	Halotydeus destructor, Rhopalosiphum padi	Isoflavones, Phenolics	Wang et al. 1998 Leszczynski et al. 1985

Table 2.5 continued

Triticum	Sitbion avenae		Dreyer & Jones 1981
aestivum	S. graminum		
Zea mays	Ostrinia nubilalis,	DIMBOA	Robinson et al. 1982,
	Diabrotica virgifera	DIBOA	Xie et al. 1990
	virgifera		

In resistant wheat cultivars, dihydroxyphenols are associated with feeding deterrency in wheat to *R. padi, S. graminum* and the English grain aphid, *Sitobion avenae* (F.) (Dreyer and Jones 1981, Leszczynski et al. 1985) and in rice resistant to the striped stemborer, *Chilo suppressalis* (Walker) (Das 1976). However, Johnson et al. (2002) evaluated *S. frugiperda* feeding among several species of grasses varying in armyworm antixenosis.

Although major qualitative and quantitative phenolic differences exist among grass species, there is no correlation between antixenosis and the levels of chlorogenic acid, flavonoids or total phenolics for the grass species evaluated. Finally, Bi et al. (1997) used transgenic tobacco lines differing in phenylalanine lyase activity to test the effects of phenolics such as chlorogenic acid, rutin and total flavonoids on a specialist lepidoptera, the tobacco hornworm, *Manduca sexta* (L.), and a generalist lepidoptera, *H. virescens*. There are no correlations of resistance between any of the phenolics involved with resistance to either arthropod.

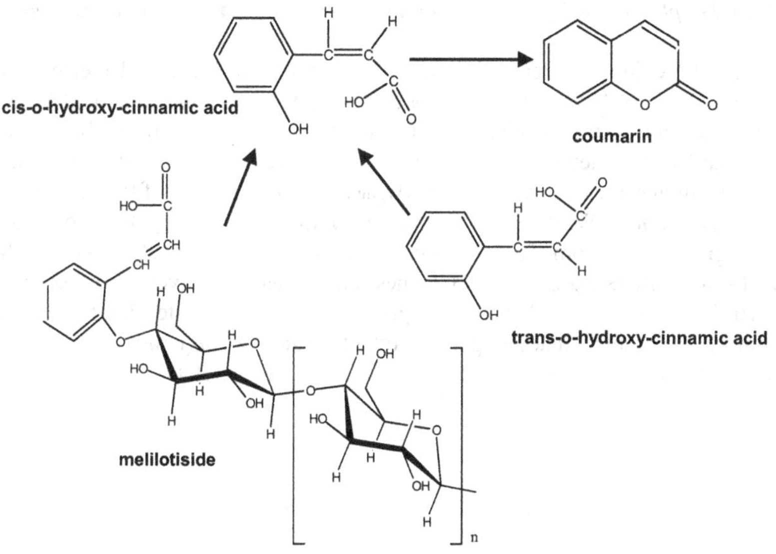

Figure 2.13. Coumarin, an allelochemical produced by Melilotus officinalis *that deters feeding of* Listroderes costirostris obliqus, Epicauta fabricii *and* Epicauta vittata. *(Gorz et al. 1982, Matsumoto 1962)*

In many instances, the presence of an allelochemical imparts some degree of feeding or oviposition deterrence. However, antixenosis resistance may also be due to the lack of an attractant or stimulant that normally occurs in a susceptible plant. As mentioned previously, glucosinolates from the leaves of cruciferous crops play a role in *Delia spp.* root fly oviposition behavior. For *D. radicum*, glucosinolates are major oviposition stimulants, while non-glucosinolates are the major oviposition stimuli for *D. floralis* (Baur et al. 1996, Hopkins et al. 1997). Reduced amounts of both types of compounds account for fly oviposition antixenosis among various cultivars of kale. A similar relationship exists for the resistance of some cultivars of sweet potato, *Ipomoea batatas* [(L.) Lam.], the sweet potato weevil, *Cylas formicarius elegantulus* (Summers). Levels of the weevil oviposition stimulant boehmeryl acetate are approximately 10-fold lower in the resistant cultivar than in the susceptible cultivar (Son et al. (1991).

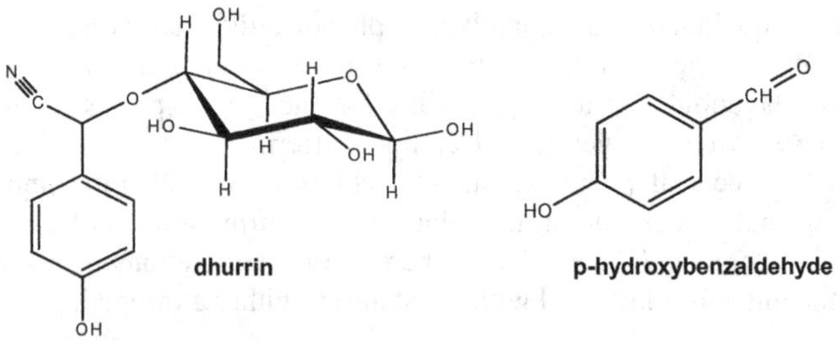

dhurrin   **p-hydroxybenzaldehyde**

Figure 2.14. The phenolic arthropod feeding deterrents dhurrin and p-hydroxybenzaldehyde.

Arthropod feeding deterrents also occur in several forage crops. Deterrency in subterranean clover to *H. destructor* is closely linked to isoflavone content. The free isoflavones formononetin, genisten and biochanin A, along with their 7-*O*-glucosides and biosynthetic precursors all occur in deterrent concentrations in a mite-resistant subterranean clover cultivar (Wang et al. 1998). Larvae of the grass grub, *Costelytra zealandica* (White), are also deterred from feeding by the isoflavone vestitol (Figure 2.15) from *Lotus pedunculatus* Cav. (Russell et al. 1978) and by saponins from alfalfa (Sutherland 1975). Coumestrol, a related isoflavone from soybean foliage will be discussed in Chapter 3. Saponins, triterpenes or steroids linked to a sugar moiety, deter the Guadeloupean leaf cutter ant, *Acromyrmex octospinosis* (Reich) in some species of yam, *Dioscorea villosa* L. (Febvay et al. 1983).

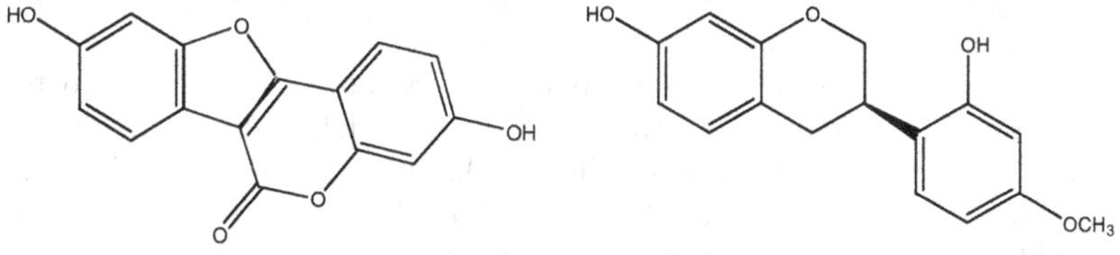

coumestrol   **vestistol**

Figure 2.15. Isoflavone allelochemicals from leguminous crop plants. Vestitol from Medicago sativa *and* Lotus pedunculatus, *a feeding deterrent to* Costelytra zealandica; *coumestrol from* Glycine max *foliage, an antibiotic compound to* Pseudoplusia includens.

The relationship between antixenosis and plant glycoalkaloid content is much more distinct. Foliar glycoalkaloids in wild *Solanum* species deter feeding of *E. fabae* (Sinden et al. 1986). A similar relationship exists between the wireworm, *Agriotes obscurus* (L.), feeding on *Solanum* tubers, where higher amounts of the alkaloids chalcone and solanine are avoided during *A. obscurus* feeding (Jonasson and Olsson 1994). In addition, adult pea leaf weevils, *Sitona lineatus* (L.), are known to avoid feeding on several *Lupinus* species with high alkaloid content (Cantot and Papineau 1983). The methyl ketone 2-tridecanone, from the tips of glandular trichomes in *Lycopersicon hirsutum* f. *glabratum* C. H. Mull., is a known toxin to several pests of cultivated tomato and will be discussed in Chapter 3. Maluf et al. (1997) determined that oviposition and feeding of the tomato pinworm, *Scrobipalpuloides absoluta* (Meyrick, 1917) are significantly deterred on plants with increased 2-tridecanone content.

An aglucone organic acid in the foliage of maize, 2, 4-dihydroxy-7-methoxy-2H-1, 4-benzoxazin-3 (4H)-one, (DIMBOA) is one of the most widely studied plant allelochemicals affecting arthropod resistance (Bergvinson 1997). When normal, healthy maize foliage is mechanically damaged, the glucoside 2-0-glucosyl-4-hydroxy-1, 4-benzoxazin-3-one is enzymatically converted to DIMBOA (Figure 2.16) (Wahlroos and Virtanen 1959). DIMBOA and its 2,-α-glucoside (DIBOA) deter feeding by *O. nubilalis* (Robinson et al. 1982), the western corn rootworm, *Diabrotica virgifera virgifera* (LeConte) (Xie et al. 1990) and the Asian corn borer, *Ostrinia furnacalis* (Gunenée) (Yan et al. 1999). MBOA, a stable end product of DIMBOA degradation is also biologically active to several maize arthropod pests (see Chapter 3), yet serves as an attractant to *D. virgifera* (Bjostad and Hibbard 1992).

Terpene lactones deter the feeding of several different arthropods. The sesquiterpene lactone 8, ß-sarracinoyloxycumambranolide, from the arthropod resistant sunflower, *Helianthus maximiliani* Schrad., deters feeding by the southern armyworm, *Spodoptera eridania* (Cramer); the migratory grasshopper, *Melanoplus sanguinipes* (F.); and *H. electellum* (Gershenzon et al 1985). Arthropod-resistant *Rhododendron* species also contain high levels of the sesquiterpene lactone germacrone, which deters feeding by the obscure root weevil, *Sciopithes obscurus*

Horn (Doss et al. 1980). Norditerpene dilactones in the foliage of weeping podocarpus, *Podocarpus gracilior* Pilg., deter feeding by *H. zea*, *S. frugiperda*, and the pink bollworm, *Pectinophora gossypiella* (Saunders) (Kubo et al. 1984).

Plant tannins have been suggested to inhibit arthropod growth, due to their presumed action in binding with proteins to form insoluble, digestion-inhibiting complexes. Martin et al. (1987) and Martin and Martin (1982) however, indicate that there is little evidence to suggest that tannins inhibit arthropod digestion, and present evidence that tannins act as feeding deterrents.

The phenols quercitin, rutin, and procyanidin (condensed tannin) in arthropod - resistant cotton cultivars deter the feeding of *H. virescens*, *H. zea*, *P. gossypiella* and *T. urticae*, resulting in restricted growth of these arthropods (Chan et al. 1978, Elliger et al. 1978, Lane and Schuster 1981, Lukefahr and Martin 1966). Here again is a case of antixenosis and antibiosis being closely intertwined. Sharma et al. (1991) and Kumar and Singh (1998) have detected a similar relationship in high tannin lines of sorghum resistant to *C. sorghicola*.

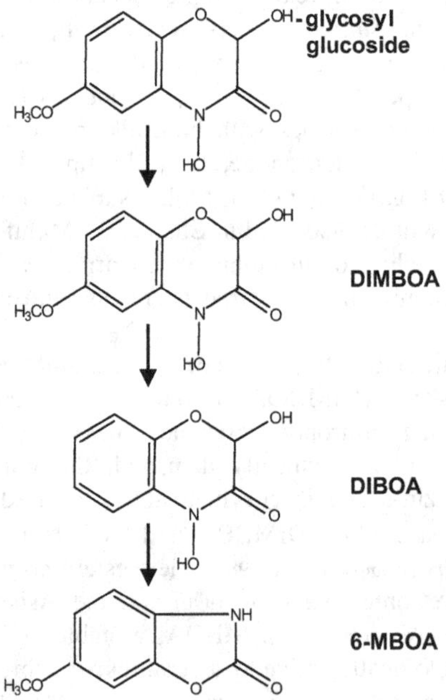

Figure 2.16. Production of DIMBOA (2,4-dihydroxy-8-methoxy-2H-1, 4-benzoxazin-3(4H)-one), DIBOA (2,4-dihydroxy-1,4-benoxazin-3-one), and 6-MBOA (6-methoxybenzoxazolinone) by enzymatic hydrolysis of mechanically damaged maize foliage.

The flavonoid chrysanthemin (cyanidin-3-,β-glucoside) from cotton cultivars with red floral pigmentation also deters feeding of *H. virescens* larvae (Hedin et al. 1983). Condensed tannin in the leaves of quaking aspen, *Populus tremuloides* Michx., also deters feeding by the southern armyworm, *Spodoptera eridania* (Cramer), and causes suppression of larval growth (Manuwoto et al. 1985).

5. COEVOLUTION OF PLANT DEFENSES AND ARTHROPOD BEHAVIOR

Changes in arthropod host plant selection depend on the dynamic equilibrium between an arthropod and it's potential host plants at various points in co-evolutionary time. The outcome of this equilibrium depends on the respective genetic potentials of the plant and the arthropod and their rates of change relative to one another, along with the tempering influences of the external environment. At any given point in evolutionary time, either the plant or the arthropod will have the genetic "upper hand". The position of the plant will be determined by the degree of success of physical and chemical defenses developed in response to arthropod herbivory, while the success of an arthropod will depend on the physiological and morphological "countermeasures" developed to overcome plant resistance factors (Figure 2.17).

Modern arthropod life paralleled the development of flowering plants in the Jurassic period of the Mesozoic Era. Recent fossil evidence involving the specialized feeding of rolled-leaf hispine Chrysomelid beetles on gingers indicates that a close association between plants and their herbivores occurred in the Late Cretaceous, at least 20 million years earlier than suggested by other arthropod body fossils (Wilf et al. 2000). Nevertheless, many arthropods living prior to the Jurassic period were polyphagous, and the arthropod segment of plant-arthropod coevolution during this period is viewed by many as a shift from general, polyphagous feeding to specialized oligophagous feeding.

The occurrence of many plant defensive allelochemicals such as alkaloids and terpenoids about 200 million years ago coincides with the development of modern arthropod life, and suggests one possible reason for their development.

Jermy (1991) has rejected the "coevolutionary feedback" theory as an explanation of plant defense adaptation and coevolution, and proposed an alternative hypothesis. Jermy (1993) proposed that heritable mutations in the arthropod genome cause changes in the arthropod central nervous system that ultimately alter plant recognition and host plant selection. As supporting evidence, Jermy cites Chapman's recent host plant specialization hypothesis (Jermy 1993) that a non-evolved, wide-ranging array of allelochemical deterrent cells exist in phytophagous arthropods, and that arthropod specificity evolves as arthropods loose sensitivity to deterrents and become receptive to attractive phagostimulants. Arthropod feeding deterrents have evolved with increasing chemical complexity. In all of the major biosynthetic pathways, more complex organic molecules are present as arthropod feeding deterrents, than those present in the past (Figure 2.18).

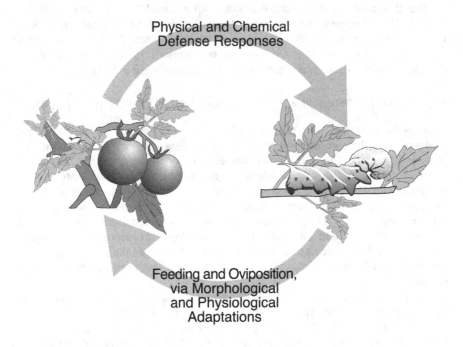

Physical and Chemical
Defense Responses

Feeding and Oviposition,
via Morphological
and Physiological
Adaptations

Figure 2.17. Coevolution of plants and arthropods. Plant-Lycopersicon lycopersicum,
Arthropod-Manduca sexta larvae. (from USDA-NRCS PLANTS Database / Britton, N.L., and
A. Brown. 1913. Illustrated flora of the northern states and Canada. Vol. 3:168)

Regardless of the outcome of plant-arthropod coevolution, it is obvious that
arthropods are at best only one of the selectional forces affecting the physical and
chemical changes that plants have undergone throughout evolutionary time. Plant
changes in response to herbivory, disease and abiotic stresses, as well as the
behavioral, metabolic and genetic changes that arthropods have undergone in order
to adapt to new host plants underscore the genetic plasticities of plants and
arthropods. These flexibilities also suggest that from a practical standpoint, the
development of plant resistance to arthropods based heavily on either antixenosis or
antibiosis may be, and in some cases has been short-lived, because of the ability of
arthropods to counteract or overcome such changes in plants. No one category of
resistance is totally self-sufficient, as will be further discussed in Chapters 3 and 4.

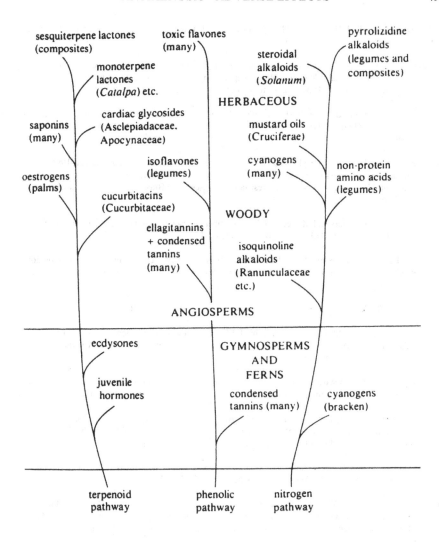

Figure 2.18. Evolution of insect feeding deterrents. Reprinted from Harborne, J. B. Introduction in Ecological Biochemistry. Copyright 1982, Academic Press Inc., with permission from Elseiver.

REFERENCES CITED

Anstey, T. H. and J. F. Moore. 1954. Inheritance of glossy foliage and cream petals in green sprouting broccoli. J. Hered. 45:39–41.

Appel, H. M. 1993. Phenolics in ecological interactions: the importance of oxidation. J. Chem. Ecol. 19:1521–1552.

Atkin, D. S. J., and R. J. Hamilton. 1982. The effects of plant waxes on insects. J. Nat. Prod. 45: 694–696.

Baldson, J. A., K. E. Espelie, and S. K. Braman. 1995. Epicuticular lipids from Azalea (*Rhododendron* spp.) and their potential role in host plant acceptance by azalea lace bug, Stephanitis pyrioides (Heteroptera: Tingidae). Biochem. System. Ecol. 23:477–485.

Baur, R., S. Binder, and G. Benz. 1991. Nonglandular leaf trichomes as short-term inducible defense of the grey alder, *Alnus incana* (L), against the chrysomelid beetle, *Agelastica alni* L. Oecologia. 87:219–226.

Baur, R., A. N. E. Birch, R. J. Hopkins, D. W. Griffiths, M. S. J. Simmonds, and E. Stadler. 1996. Oviposition and chemosensory stimulation of the root flies *Delia radicum* and *D.floralis* in response to plants and leaf surface extracts from resistant and susceptible Brassica genotypes. Entomol. Exp. Appl. 78:61–75.

Beeghly, H. H., J. G. Coors and M. Lee. 1997. Plant fiber composition and resistance to European corn borer in four maize populations. Maydica. 42:297–303.

Benedict, J. H., T. F. Leigh, and A. H. Hyer. 1983. *Lygus hesperus* (Heteroptera: Miridae) oviposition behavior, growth, and survival in relation to cotton trichome density. Environ. Entomol. 12:331–335.

Bernays, E. 1992. Insect - Plant Interactions, Volume IV. CRC Press, Boca Raton, FL.

Bergman, D. K., J. W. Dillwith, A. A. Zarrabi, J. L. Caddel, and R. C. Berberet. 1991. Epicuticular lipids of alfalfa relative to its susceptibility to spotted alfalfa aphids (Homoptera, Aphididae). Environ. Entomol. 20:781–785.

Bergvinson, D. J., J. T. Arnason, and L. N. Pietrzak. 1994a. Localization and quantification of cell wall phenolics in European corn borer resistant and susceptible maize inbreds. Can. J. Bot. 72:243–1249.

Bergvinson, D. J., J. T. Arnason, R. I. Hamilton, J. A. Mihm, and D. C. Jewell. 1994b. Determining leaf toughness and its role in maize resistance to the European corn borer (Lepidoptera:Pyralidae). J. Econ, Entomol, 87:1743–1748.

Bergvinson, D. J., R. I. Hamilton, and J. T. Arnason. 1995. Leaf profile of maize resistance factors to European corn borer, *Ostrinia nubilalis*. J. Chem. Ecol. 21:343–354.

Bergvinson, D. J. 1997. Windows of maize resistance. In: J. A. Mihm (Ed.), Insect Resistant Maize: Recent Advances and Utilization, Proceedings of an International Symposium, International Maize and Wheat Improvement Center (CIMMYT), Mexico, D. F., 1994, CIMMYT, El Batan, Mexico. 117–126.

Bi, J. L., J. B. Murphy, and G. W. Felton. 1997. Antinutritive and oxidative components as mechanisms of induced resistance in cotton to *Helicoverpa zea*. J. Chem. Ecol. 23:97–117

Bilderback, T. E., and R. H. Mattson. 1977. Whitefly host preference associated with selected biochemical and phenotypic characteristics of poinsettias. J. Amer. Soc. Hort. Sci. 102:327–331.

Bjostad, L. B., and B. E. Hibbard. 1992. 6-methoxy-2-benzoxazolinone: a semichemical for host location by western corn rootworm larvae. J. Chem. Ecol. 18:931–944.

Blum, A. 1968. Anatomical phenomena in seedlings of sorghum varieties resistant to the sorghum shoot fly (*Atherigona varia soccata*). Crop Sci. 8:388–390.

Bodnaryk, R. P. 1992. Leaf epicuticular wax, an antixenotic factor in Brassicaceae that affects the rate and pattern of feeding of flea beetles, *Phyllotreta cruciferae* (Goeze). Can. J. Plant Sci. 72:1295–1303.

Bordasch, R. P., and A. A. Berryman. 1977. Host resistance to the fir engraver beetle, *Scolytus ventralis* (Coleoptera: Scolytidae). 2. Repellency of *Abies grandis* resins and some monoterpenes. Can. Entomol. 109:95–100.

Brewer, G. J., E. L. Sorensen, and E. K. Horber. 1983. Trichomes and field resistance of *Medicago* species to the alfalfa seed chalcid (Hymenoptera: Eurytomidae). Environ. Entomol. 12:247–252.

Brewer, G. J., E. L. Sorensen, and E. K. Horber, and G. L. Kreitner. 1986. Alfalfa stem anatomy and potato leafhopper (Homoptera: Cicadellidae) resistance. J. Econ. Entomol. 79:1249–1253.

Britton, N.L., and A. Brown. 1913. Illustrated flora of the northern states and Canada. 3:168.

Broadbent, A. B., J. A. Matteoni, and W. R. Allen. 1990. Feeding preferences of the western flower thrips, *Frankliniella occidentalis* (Pergande) (Thysandoptera: Thripidae), and incidence of tomato spotted wilt virus amoung cultivars of florist's chrysanthemum. Can. Entomol. 122:1111–1117.

Brues, C. F. 1920. The selection of food plants by insects with specific reference to lepidopterous larvae. Am. Natur. 54:313–22.

Burton, G. W., W. W. Hanna, J. C. Johnson, Jr., D. B. Leuck, W. G. Monson, J. B. Powell, H. D. Wells, and N. W. Widstrom. 1977. Pleiotropic effects of the tr. trichomeless gene in pearl millet transpiration, forage quality and pest resistance. Crop Sci. 17:613–616.

Butler, G. D., Jr., F. D. Wilson and G. Fisher. 1992. Cotton leaf trichomes on popybica and *Bemisia tabaci*. Crop Protection. 10:461–464.

Campbell, B. C., and D. L. Dreyer. 1985. Host-plant resistance of sorghum: Differential hydrolysis of sorghum pectic substances by polysaccharases of greenbug biotypes (*Schizaphis graminum*, Homoptera: Aphididae). Arch. Insect Biochem. Physiol. 2:203–215.

Cantot, P., and J. Papineau. 1983. Discrimination of lupine with low alkaloid content by adult *Sitona lineatus* L. Agronomie. 3:937–940.

Chan, B. G., A. C. Waiss and M. J. Lukefahr. 1978. Condensed tannin, an antibiotic chemical from *Gossypium hirsutum* L. J. Insect Physiol. 24:113–118.

Chang, H., and C. Y. Shih. 1959. A study on the leaf mid-rib structure of sugarcane as related with resistance to the top borer (*Scirpophaga nivella* F.). Taiwan Sugar Exp. Sta. Rep. 19:53–56.

Chapman, R. F. 1974. The chemical inhibition of feeding by phytophagous insects: a review. Bull. Entomol. Res. 64: 339–363.

Chapman, R. F. 1988. Sensory aspects of host-plant recognition by Acridoidea: Questions associated with the multiplicity of receptors and variability of response. J. Insect Physiol. 34:167–174.

Chapman, R. F., and W. M. Blaney. 1979. How animals perceive secondary compounds. In: G. A. Rosenthal and D. H. Janzen (Eds.), Herbivores: Their Interaction with Secondary Metabolites. Academic Press, London. pp. 161–198.

Chapman, R. F., S. Woodhead, and E. A. Bernays. 1983. Survival and dispersal of young larvae of *Chilo partellus* (Swinhoe) (Lepidoptera: Pyralidae) in two cultivars of sorghum. Bull. Entomol. Res. 73:65–74.

Chesnokov, P. G. 1962. Methods of Investigating Plant Resistance to Pests. National Science Foundation, Washington, DC. (Israel Program for Scientific Translation). 107 pp.

Chiang, H. S., and D. M. Norris. 1983. Morphological and physiological parameters of soybean resistance to agromyzid beanflies. Environ. Entomol. 12:260–265.

Cohen, A. C., T. J. Henneberry, and C. C. Chu. 1996. Geometric relationships between whitefly feeding behavior and vascular bundle arrangements. Entomol. Exp.Appl.78:135–142.

Dabrowski, Z. T., and J. G. Rodriguez. 1971. Studies on resistance of strawberries to mites. 3. Preference and nonpreference responses of *Tetranychus urticae* and *T. turkestani* to essential oils of foliage. J. Econ. Entomol. 64:387–391.

Danielson, S. D., G. R. Manglitz, and E. L. Sorensen. 1987. Resistance of perennial glandular-haired *Medicago* species to oviposition by alfalfa weevils (Coleoptera: Curculionidae). Environ. Entomol. 16:195–197.

Das, Y. T. 1976. Some factors of resistance to *Chilo suppressalis* in rice varieties. Entomol. Exp. Appl. 20:131–134.

Davis, F. M., G. T. Baker, and W. P. Williams. 1995. Anatomical characteristics of maize resistant to leaf feeding by southwestern corn borer (Lepidoptera: Pyralidae) and fall armyworm (Lepidoptera: Noctuidae). J. Agric. Entomol. 12: 55–65.

Dethier, V. G., L. Barton–Browne, and C. N. Smith. 1960. The designation of chemicals in terms of the responses they elicit from insects. J. Econ. Entomol. 53:134–6.

Dickens, J. C., G. D. Prestwich, C. S. Ng, and J. H. Visser. 1993. Selectively fluorinated analogs reveal differential olfactory reception and inactivation of green leaf volatiles in insects. J. Chem. Ecol. 19:1981–1991.

Doss, R. P., R. Luthi, and B. F. Hrutfiord. 1980. Germacrone, a sesquiterpene repellent to obscure root weevil from *Rhododendron edgeworthii*. Phytochem. 19:2379–2380.

Doss, R. P., C. H. Shanks, Jr., J. D. Chamberlain, and J. K. L. Garth. 1987. Role of leaf hairs in resistance of a clone of strawberry, *Fragaria chiloensis*, to feeding by adult black vine weevil, *Otiorhynchus sulcatus* (Coleoptera: Curculionidae). Environ. Entomol. 16:764–766.

Dreyer, D. L., and B. C. Campbell. 1983. Association of the degree of methylation of intercellular pectin with plant resistance to aphids and with induction of aphid biotypes. Experimentia. 40:224–226.

Dreyer, D. L., and K. C. Jones. 1981. Feeding deterrency of flavonoids and related phenolics towards *Schizaphis graminum* and *Myzus persicae*: Aphid feeding deterrents in wheat. Phytochem. 20:2489–2493.

Dreyer, D. L., J. C. Reese, and K. C. Jones. 1981. Aphid feeding deterrents in sorghum. Bioassay, isolation and characterization. J. Chem. Ecol. 7:273–83.

Eigenbrode, S. D., and K. E. Espelie. 1995. Effects of epicuticular lipids on insect herbivores. Ann. Rev. Entomol. 40:171–194.

Eigenbrode, S. D., and S. K. Pillai. 1998. Neonate *Plutella xylostella* responses to surface wax components of a resistant cabbage. J. Chem. Ecol. 24:1611–1627.

Eigenbrode, S. D., and A. M. Shelton. 1990. Behavior of neonate diamondback moth larvae (Lepidopytera: Plutellidae) on glossy-leafed resistant *Brassica oleracea* L. Environ. Entomol. 19:1566–1571.

Eigenbrode, S. D., K. A. Stoner, A. M. Shelton, and W. C. Kain. 1991. Glossy leaf waxes associated with resistance to diamondback moth (Lepidoptera, Plutellidae) in *Brassica oleracea*. J. Econ. Entomol. 84:1609–1618.

Elden, T. C. 1997. Influence of soybean lines isogenic for pubescence type on two spotted spider mite (Acarina: Tetranychidae) development and feeding damage. J. Entomol. Sci. 32:296–302.

Elliger, C. A., B. G. Chan, and A. C. Waiss, Jr. 1978. Relative toxicity of minor cotton terpenoids compared to gossypol. J. Econ. Entomol. 71:161–164.

Febvay, G., P. Bourgeois, and A. Kermarrec. 1985. Antifeedants for an attine ant, *Acromyrmex octospinosus* (Reich) (Hymenoptera-Formicidae), in yam (Dioscoreaceae) leaves. Agronomie. 5:439-444.

Ferguson, S., E. L. Sorensen and E. K. Horber. 1982. Resistance to the spotted alfalfa aphid (Homoptera: Aphididae) in glandular-haired *Medicago* species. Environ. Entomol. 11:1229–1232.

Fery, R. L., and F. P. Cuthbert, Jr. 1979. Measurement of podwall resistance to the cowpea curculio in the southern pea, *Vigna unguiculata* (L.) Walp. Hort. Sci. 14:29–30.

Fiori, B. J., and D. D. Dolan. 1981. Field tests for *Medicago* resistance against the potato leafhopper (Homoptera: Cicadellidae). Can. Entomol. 113:1049–1053.

Fiori, B. J., and D. W. Craig. 1987. Relationship between color intensity of leaf supernatatants from resistant and susceptible birch trees and rate of oviposition by the birch leafminer (Hymenoptera: Tenthredinidae) J. Econ. Entomol. 80:1331–1333.

Fraenkel, G. 1959. Evaluation of our thoughts on secondary plant substances. Science. 129:1466–70.

Frazier J. L. 1986. The perception of allelochemicals that inhibit feeding. In: L. B. Brattsten and S. Ahmad (Eds.), Molecular Aspects of Insect-Plant Associations. Plenum, New York pp. 1–18.

Gannon, A. J., and C. E. Bach. 1996. Effects of soybean trichome density on Mexican bean beetle (Coleoptera: Coccinellidae) development and feeding preference. Environ. Entomol. 25:1077–1082.

George, B. W., R. S. Seay, and D. L. Coudriet. 1977. Nectariless and pubescent characters in cotton: effect on the cabbage looper. J. Econ. Entomol. 77:267–269.

Gershenzon, J., M. Rossiter, T. J. Mabry, C. E. Rogers, M. H. Blust, and T. L. Hopkins. 1985. Insect antifeedant terpenoids in wild sunflower. A possible source of resistance to the sunflower moth. In: P. A. Hedin (Ed.), Bioregulators for Pest Control. ACS Symposium Series 276, American Chemical Society, Washington, DC. pp. 433–446.

Goffreda, J. C., M. A. Mutschler, D. A. Ave, W. M. Tingey, and J. C. Steffens. 1989. Aphid deterrence by glucose esters in glandular trichome exudate of the wild tomato, *Lycopersicon pennellii*. J. Chem. Ecol. 15:2135–2147.

Gorz, H. J., F. A. Haskins, and G. R. Manglitz. 1972. Effect of coumarin and related compounds on blister beetle feeding in sweetclover. J. Econ. Entomol. 65:1632–5.

Green, T. A., R. J. Prokopy, and D. W. Hosmer. 1994. Distance of response to host tree models by female apple maggot flies, *Rhagoletis pomonella* (Walsh) (Diptera: Trephritidae): Interaction of visual and olfactory stimuli. J. Chem. Ecol. 20:2393–2413.

Guerin, P. M., E. Stadler, and H. R. Buser. 1983. Identification of host plant attractants for the carrot fly, *Psila rosae*. J. Chem. Ecol. 9:843–861.

Guerin, P. M., and M. F. Ryan. 1984. Relationship between root volatiles of some carrot cultivars and their resistance to the carrot fly, *Psila rosae*. Entomol. Exp. Appl. 38:317–224.

Guo, Z., P. A. Weston, and J. C. Snyder. 1993. Repellency to two-spotted spider mite, *Tetranychus urticae* Koch, as related to leaf surface chemistry of *Lycopersicon hirsutum* accessions. J. Chem. Ecol. 19:2965–2579.

Hanson, F. E. 1983. The behavioral and neurophysiological basis of food plant selection by lepidopterous larvae. In: S. Ahmad (Ed.), Herbivorous Insects Host Seeking Behavior and Mechanisms. Academic Press, New York, pp. 3–23.

Harborne, J. B. 1982. Introduction in Ecological Biochemistry. Academic Press, New York, 278 pp.

Harvey, T. L., and T. J. Martin. 1980. Effects of wheat pubescence on infestations of wheat curl mite and incidence of wheat streak mosaic. J. Econ. Entomol. 73:225–227.

Hawthorne, D. J., J. A. Shapiro, W. M. Tingey, and M. A. Mutschler. 1992. Trichome-borne and artificially applied acylsugars of wild tomato deter feeding and oviposition of the leafminer *Liriomyza trifolii*. Entomol. Exp. Appl. 65:65–73.

Hedin, P. A., J. N. Jenkins, D. H. Collum, W. H. White, and W. L. Parrott. 1983. Multiple factors in cotton contributing to resistance to the tobacco budworm, *Heliothis virescens* F. In: P. A. Hedin (Ed.), Plant Resistance to Insects. ACS Symposium Series. 208. American Chemical Society, Washington, D. C. pp. 347–365.

Heftmann, E. 1975. Functions of steroids in plants. Phytochem. 14:891–901.

Heinz, K. M. and F. G. Zalom. 1995. Variation in trichome-based resistance to *Bemisia argentifolii* (Homoptera: Aleyrodidae) oviposition on tomato. J. Econ. Entomol. 88:1494–1502.

Heupel, R. C. 1985. Varietal similarities and difference in the polycyclic isopentenoid composition of sorghum. Phytochem. 24:2929–2937.

Hopkins, R. J., A. N. E. Birch, D. W. Griffiths, R. Baur, E. Stadler, and R. G. Mckinlay. 1997. Leaf surface compounds and oviposition preference of turnip root fly *Delia floralis*: The role of glucosinolate and nonglucosinolate compounds. J. Chem. Ecol. 23:629–643.

House, H. L. 1969. Effects of different proportions of nutrients on insects. Entomol. Exp. Appl. 12:651–69.

Hoxie, R. P., S. G. Wellso, and J. A. Webster. 1975. Cereal leaf beetle response to wheat trichome length and density. Environ. Entomol. 4:365–370.

Hsiao, T. H. 1969. Chemical basis of host selection and plant resistance in oligophagous insects. Entomol. Exp. Appl. 12:777–788.

Iseley, D. 1928. The relation of leaf color and leaf size to boll weevil infestation. J. Econ. Entomol. 21:553–559.

Jackson, D. M., R. F. Severson, A. W. Johnson, and G. A. Herzog. 1986. Effects of cuticular duvane diterpenes from green tobacco leaves on tobacco budworm (Lepidoptera: Noctuidae) oviposition. J. Chem. Ecol. 12: 1349–1359.

Jeffree, C. E. 1986. The cuticle, epicuticular waxes and trichomes of plants, with reference to their structure, functions and evolution. In: B. E. Juniper and T. R. E. Southwood Eds.), Insects and the Plant Surface. Edward Arnold. London. pp. 23–64.

Jermy, T. 1991. Evolutionary interpretations of insect-plant relationships - a closer look. Symp.Biol Hung. 39:301–311.

Jermy, T. 1993. Evolution of insect-plant relationships - A devil's advocate approach. Entomol. Exp. Appl. 66:3–12.

Jiang, Y., and T. J. Ridsdill-Smith. 1996. Examination of the involvement of mechanical strength in antixenotic resistance of subterranean clover cotyledons to the redlegged earth mite (*Halotydeus destructor*) (Acarina: Penthaleidae). Bull. Entomol. Res. 86:263–270.

Johnson, D. T., and B. A. Lewis. 1993. Grape phylloxera (Homoptera: Phylloxeridae): comparison of leaf damage to grape cultivars grownin Arkansas. J. Entomol. Sci. 28:447–452.

Johnson, A. W., and R. F. Severson. 1982. Physical and chemical leaf surface characteristics of aphid resistant and susceptible tobacco. Tob. Sci. 26:98–102.

Johnson, A. W., and R. F. Severson. 1984. Leaf surface chemistry of tobacco budworm resistant tobacco. J. Agric. Entomol. 1:23–32.

Johnson, A. W., M. E. Snook, and B. R. Wiseman. 2002. Green leaf chemistry of various turfgrasses: Differentiation and resistance to fall armyworm. Crop Sci. 42: 2004–2010.

Johnson, K. J. R., E. L. Sorensen, and E. K. Horber. 1980a. Resistance in glandular haired annual *Medicago* species to feeding by adult alfalfa weevils. Environ. Entomol. 9:133–136.

Johnson, K. J. R., E. L. Sorensen, and E. K. Horber. 1980b. Resistance of glandular haired *Medicago* species to oviposition by alfalfa weevils (*Hypera postica*). Environ. Entomol. 9:241–245.

Jonasson, T., and K. Olsson. 1994. The influence of glycoalkoloids, chlorogenic acid and sugars on the susceptibility of potato tubers to wireworm. Pot Res. 37:205–216.

Jones, J. E., and J. W. Brand. 1981. Influence of red plant color genes on boll weevil nonpreference and agronomic performance. Proc. Beltwide Cotton Prod. Res. Conf. p. 95.

Kamel, S. A. 1965. Relationship between leaf hairiness and resistance to cotton leafworm. Emp. Cotton Grow. Rev. 42:41–48.

Kelly, M. T., and J. P. Curry. 1991. The influence of phenolic compounds on the suitability of three *Salix* species as hosts for the willow beetle *Phratora vulgatissima*. Entomol. Exp. Appl. 61:25–32.

Kendall, D. A, T. Hunter, G. M. Arnold, J. Liggitt, T. Morris, and C. W. Wiltshire. 1996. Susceptibility of willow clones (*Salix* spp.) to herbivory by *Phyllodecta vulgatissima* (L.) and *Galerucella lineola* (Fab.) (Coleoptera, Chrysomelidae). Ann. Appl. Biol. 129:379–390.

Kennedy, J. S. 1965. Mechanisms of host plant selection. Ann. Appl. Biol. 56:317–22.

Khan, Z. R., and R. C. Saxena. 1985. Effect of steam distillate extract of a resistant rice variety on feeding behavior of *Nephotettix virescens* (Homoptera : Cicadellidae). J. Econ. Entomol. 78: 562–566.

Khan, Z. R., and R. C. Saxena. 1986. Effect of steam distillate extracts of resistant and susceptible rice cultivars on behavior of *Sogatella furcifera* (Homoptera : Delphacidae). J. Econ. Entomol. 79: 928–935.

Khan, Z. R., J. T. Ward, and D. M. Norris. 1986. Role of trichomes in soybean resistance to cabbage looper, *Trichoplusia ni*. Entomol. Exp. Appl. 42: 109–117.

Kogan, M., and E. E. Ortman. 1978. Antixenosis-a new term proposed to replace Painter's "Non-preference" modality of resistance. Bull. Entomol. Soc. Am. 24:175–176.

Krishna Kumar, K. N., D. E. Ullman, and J. J. Cho. 1995. Resistance among Lycopersicon species *Frankliniella occidentalis* (Thysanoptera: Thripidae). J. Econ. Entomol. 88:1057–1065.

Kubo, I., T. Matsumoto, and J. A. Klocke. 1984. Multichemical resistance of the conifer *Podocarpus gracilion* (Podocarpaceae) to insect attack. Chem. Ecol. 10:547–560.

Kumar, S., and R. Singh. 1998. Inheritance of tannin in relation to shootfly resistance in sorghum. Cereal Res. Commun. 26:271–273.

Lal, O. P. 1991. Varietal resistance in the eggplant, *Solanum melongena*, against the shioot and fruit borer, *Leucinodes orbonalis* Guen. (Lepidoptera: Pyralidae). J. Plant Dis. Prot. 98:405–410.

Lamb, R. J. M. A. H. Smith, and R. P. Bodnaryk. 1993. Leaf waxiness and the performance of *Lipaphis erysimi* (Kaltenbach) (Homoptera: Aphididae) on three Brassica crops. Can. Entomol. 125:1023–1031.

Lambert, L., R. M. Beach, T. C. Kilen, and J. W. Todd. 1992. Soybean pubescence and its influence on larval development and oviposition preference of lepidopterous insects. Crop Sci. 32:463–466.

Lambrides, C. J., and B. C. Imrie. 2000. Susceptibility of mungbean varieties to the bruchid species *Callosobruchus maculatus* (F.), *C. phaseoli* (Gyll.), *C. chinensis* (L.), and *Acanthoscelides obtectus* (Say.) (Coleoptera : Chrysomelidae). Australian J. Agric. Res. 51:85–89.

Lane, H. C., and M. F. Schuster. 1981. Condensed tannins of cotton leaves. Phytochem. 20:425–427.

Lapis, E. B., and J. H. Borden. 1993. Olfactory discrimination by *Heteropsylla cubana* (Homoptera, Psyllidae) between susceptible and resistant species of *Leucaena* (Leguminosae). J. Chem. Ecol. 19:83–90.

Lapointe, S. L., and W. M. Tingey. 1986. Glandular trichomes of *Solanum neocardenasii* confer resistance to green peach aphid (Homoptera: Aphididae). J. Econ. Entomol. 79:1264–1268.

Leconte, C., and E. Thibout. 1981. Attraction of the leek moth, *Acroleprosis assectella*, in an olfactometer, by volatile allelochemical compounds found in the leek, *Allium porrum*. Entomol. Exp. Appl. 30:293–300.

Lee, Y. I. 1983. The potato leafhopper, *Empoasca fabae*, soybean pubescence, and hopperburn resistance. Ph.D. Thesis, University of Illinois.

Leszczynski, B., J. Warchol, and S. Niraz. 1985. The influence of phenolic compounds on the preference of winter wheat cultivars by cereal aphids. Insect Sci. Applic. 6: 157–158.

Levin, D. A. 1973. The role of trichomes in plant defense. Quart. Rev. Biol. 48:3–15.

Lukefahr, M. J., and D. F. Martin. 1966. Cotton-plant pigments as a source of resistance to the bollworm and tobacco budworm. J. Econ. Entomol. 59:176–179.

Lukefahr, M. J., C. B. Cowan, and J. E. Houghtaling. 1970. Field evaluations of improved cotton strains resistant to the cotton fleahopper. J. Econ. Entomol. 63:1101–1103.

Lukefahr, M. J., J. E. Houghtaling, and H. M. Graham. 1971. Suppression of *Heliothis* populations with glabrous cotton strains. J. Econ. Entomol. 64:486–488.

Lupton, F. G. H. 1967. The use of resistant varieties in crop protection. World Rev. Pest-Control. 6:47–58.

MacLean, P. S., and R. A. Byers. 1983. Ovipositional preferences of the alfalfa blotch leafminer (Diptera: Agromyzidae) among some simple and glandular-haired *Medicago* species. Environ. Entomol. 12:1083–1086.

Maiti, R. K., and P. T. Gibson. 1983. Trichomes in segregating generations of sorghum matings. II. Association with shootfly resistance. Crop Sci. 23:76–79.

Maluf, W. R., L. V. Barbosa, and L. V. C. Santacecilia. 1997. 2-tridecanone-mediated mechanisms of resistance to the South American tomato pinworm *Scrobipalpuloides absoluta* (Meyrick, 1917) (Lepidoptera-Gelechiidae) in *Lycopersicon* spp. Euphytica. 93:189–194.

Manuwoto, S., J. M. Scriber, M. T. Hsia, and P. Sunarjo. 1985. Antibiosis/antixenosis in tulip tree and quaking aspen leaves against the polyphagous southern armyworm, *Spodoptera eridania*. Oecologia. 67:1–7.

Martin, J. S., and M. M. Martin. 1982. Tannin assays in ecological studies: Lack of correlation between phenolics, proanthocyanidins and protein-precipitating constituents in mature foliage of six oak species. Oecologia. 54:205–211.

Martin, J. S., M. M. Martin, and E. A. Bernays. 1987. Failure of tannic acid to inhibit digestion or reduce digestibility of plant protein in gut fluids of insect herbivores: Implications for theories of plant defense. J. Chem. Ecol. 13:605–621.

Martin, G. A., C. A. Richard, and S. D. Hensley. 1975. Host resistance to *Diatraea saccharalis* (F.): Relationship of sugarcane node hardness to larval damage. Environ. Entomol. 4:687–688.

Matsuda, K., and H. Matsuo. 1985. A flavonoid, luteolin-7-glucoside, as well as salicin and populin, stimulating the feeding of beetles attacking salicaceous plants. Appl. Entomol. Zool. 20:305–313.

Matsumoto, Y. A. 1970. Volatile organic sulfur compounds as insect attractants with special reference to host selection. In: D. L. Wood, R. M. Silverstein, and M. Nakajima (Eds.), Control of Insect Behavior by Natural Products. Academic Press, New York. pp. 133–160.

Matsumoto, Y. A. 1962. A dual effect of coumarin, olfactory attraction and feeding inhibition on the vegetable weevil adult, in relation to the uneatability of sweet clover leaves. Japan J. Appl. Entomol. Zool. 6:141–9.

McAuslane, H. J. 1996. Influence of leaf pubescence on ovipositional preference of *Bemisia argentifolii* (Homoptera: Aleyrodidae) on soybean. Environ. Entomol. 25:834–841.

Mensah, R. K., and J. L. Madden. 1991. Resistance and susceptibility of *Boronia megastigma* cultivars to infestations by the psyllid *Ctenarytaina thysanura*. Entomol. Exp. Appl. 61:189–198.

Meredith, W. R., Jr., and M. F. Schuster. 1979. Tolerance of glabrous and pubescent cottons to tarnished plant bug. Crop Sci. 19:484–488.

Mitchell, H. C., W. H. Cross, W. L. McGovern, and E. M. Dawson. 1973. Behavior of the boll weevil on frego bract cotton. J. Econ. Entomol. 66:677–680.

Moharramipour, S., H. Tsumuki, K. Sato, S. Murata, and K. Kanehisa. 1997. Effects of leaf color, epicuticular wax amount and gramine content in barley hybrids on cereal aphid populations. Appl. Entomol. Zool. 32:1–8.

Mustaparta, H. 1975. Behavioral responses of the pine weevil, *Hylobius abietis* L. (Col.: Curculionidae) to odors activating different groups of receptor cells. J. Comp. Physiol. 102:57–63.

Mutikainen, P., M. Walls, and J. Ovaska. 1996. Herbivore-induced resistance in *Betula pendula*: The vole of plant vascular architecture. Oecologia. 108:723–727.

Nihoul, P. 1994. Phenology of glandular trichomes related to entrapment of *Phytoseiulus ersimilis* A. - H. in the glasshouse tomato. J. Hort. Sci. 69:783–789.

Nishijima, Y. 1960. Host plant preference of the soybean pod borer, *Grapholitha glicinivorella* (Matsumura) (Lep., Encosmidae). Entomol. Exp. Appl. 3:38–47.

Nwanze, K. F., R. J. Pring, P. S. Sree, D. R. Butler, Y. V. R. Reddy, and P. Soman. 1992. Resistance in sorghum to the shoot fly, *Atherigona soccata* - Epicuticular wax and wetness of the central whorl leaf of young seedlings. Ann. Appl. Biol. 120:373–382.

Oghiakhe, S., L. E. N. Jackai, W. A. Makanjuola, and C. J. Hodgson. 1992. Morphology, distribution, and the role of trichomes in cowpea (*Vigna unguiculata*) resistance to the legume pod borer, *Maruca testulalis* (Lepidoptera, Pyralidae). Bull. Entomol. Res. 82:499–505.

Painter, R. H. 1951. Insect Resistance in Crop Plants. University of Kansas Press, Lawrence. 521 pp.

Palaniswamy, P., and R. P. Bodnaryk. 1994. A wild Brassica from Sicily provides trichome-based resistance against flea beetles, *Phyllotreta cruciferae* (Goeze) (Coleoptera: Chrysomelidae). Can. Entomol. 126:1119–1130.

Papp, M., J. Kolarov, and A. Mesterhazy. 1992. Relation between pubescence of seedling and flag leaves of winter wheat and its significance in breeding resistance to the cereal leaf beetle (Coleoptera: Chrysomelidae). Environ. Entomol. 21:700–705.

Patanakamjorn, S., and M. D. Pathak. 1967. Varietal resistance of rice to Asiatic rice borer, *Chilo suppressalis* (Lepidoptera : Crambidae), and its association with various plant characters. Ann. Entomol. Soc. Am. 60:287–292.

Peiretti, R. A., I. Amini, D. W. Weibel, K. J. Starks, and R. W. McNew. 1980. Relationship of "Bloomless" (bmbm) sorghum to greenbug resistance. Crop Sci. 20:173–176.

Perttunen, V. 1957. Reactions of two bark beetle species, *Hylurgops palliatus* Gyll. and *H. asteaster* Payk. (Col., Scoytidae) to the terpene alpha-pinene. Ann. Entomol. Fenn. 23:101–100.

Peter, A. J., T. G. Shanower, and J. Romeis. 1995. The role of plant troichomes in insect resistance: a selective review. Phytophaga. 7:41–64.

Peter, A. J., and T. G. Shanower. 2001. Role of plant surface in resistance to insect herbivores. In: T. N. Ananthakrishnan (Ed.), Insects and Plant Defense Dynamics. Science Publishers, Enfield, NH USA pp. 107–132.

Pierce, H. D., Jr., R. S. Vernon, J. H. Borden, and A. C. Oehlschlager. 1978. Host selection by *Hylemya antiqua* (Meigen): Identification of three new attractants and oviposition stimulants. J. Chem. Ecol. 4:65–72.

Prokopy, R. J. and E. D. Owens. 1983. Visual detection of plants by herbivorous insects. Ann. Rev. Entomol. 28:337–364.

Quiras, C. F., M. A. Stevens, C. M. Rick, and M. K. Kok-Yokomi. 1977. Resistance in tomato the pink form of the potato aphid, *Macrosiphum euphorbiae* (Thomas): The role of anatomy, epidermal hairs and foliage composition. J. Am. Soc. Hortic. Sci. 102:166–171.

Quiring, D. T., P. R. Timmins, and S. J. Park. 1992. Effect of variations in hooked trichome densities of *Phaseolus vulgaris* on longevity of *Liriomyza trifolii* (Diptera, Agromyzidae) adults. Environ. Entomol. 21:1357–1361.

Ramachandran, S., G. D. Buntin, J. N. All, and P. L. Raymer. 1998. Diamondback moth (Lepidoptera: Plutellidae) resistance of *Brassica napus* and *B. oleracea* lines with differing leaf characteristics. J. Econ. Entomol. 91:987–992.

Reed, W. 1974. Selection of cotton varieties for resistance to insect pests in Uganda. Cotton Grow. Rev. 51:106–123.

Reese, J. C. 1981. Insect dietetics: Complexities of plant-insect interactions. In: G. Bhaskaran, S. Friedman, and J. G. Rodriguez (Eds.), Current Topics in Insect Endocrinology and Nutrition. Plenum, New York. pp. 317–335.

Reinert, J. A., T. K. Broschat, and H. M. Donselman. 1983. Resistance of *Canna* sp. to the skipper butterfly, (*Calpodes ethlius*) (Lepidoptera: Hesperidae). Environ. Entomol. 12:1829–1832.

Roberts, J. J. and J. E. Foster. 1983. Effect of leaf pubescence in wheat on the bird cherry oat aphid (Homoptera: Aphidae). J. Econ. Entomol. 76:1320–1322.

Roberts, J. J., R. L. Gallun, F. L. Patterson, and J. E. Foster. 1979. Effects of wheat leaf pubescence on the Hessian fly. J. Econ. Entomol. 72:211–214.

Robertson, G. W., D. W. Griffiths, A. N. E. Birch, A. T. Jones, J. W. Nichol, and J. E. Hall. 1991. Further evidence that resistance in raspberry to the virus vector aphid *Amphorphora idaei*, is related to the chemical composition of the leaf surface. Ann. Appl. Biol. 119:443–449.

Robinson, J. F., J. A. Klun, W. D. Guthrie, and T. A. Brindley. 1982. European corn borer (Lepidoptera: Pyralidae) leaf feeding resistance: DIMBOA bioassays. J. Kansas Entomol. Soc. 55: 357–364.

Robinson, S. H., D. A. Wolfenbarger and R. H. Dilday. 1980. Antixenosis of smooth leaf cotton to the ovipositional response of tobacco budworm. Crop Sci. 20:646–649.

Roessingh, P. and E. Stadler. 1990. Foliar form, colour and surface characteristics influence oviposition behaviour in the cabbage root fly *Delia radicum*. Entomol. Exp. Appl. 57:93–100.

Roessingh, P., E. Stadler, G. R. Fenwick, L. A. Lewis, J. K. Nielsen, J. Hurter, and T. Ramp. 1992. Oviposition and tarsal chemoreceptors of the cabbage root fly are stimulated by glucosinolates and host plant extracts. Entomol. Exp. Appl. 65:267–282.

Rogers, C. E., and G. L. Kreitner. 1983. Phytomelanin of sunflower achemes: a mechanism for pericarp resistance to abrasion by larvae of the sunflower moth (Lepidoptera: Pyralidae). Environ. Entomol. 12:277–285.

Rogers, C. E., J. Gershenzon, N. Ohno, T. J. Mabry, R. D. Stipanovic, and G. L. Kreitner. 1987. Terpenes of wild sunflowers (*Helianthus*): An effective mechanism against seed predation by larvae of the sunflower moth, *Homoeosoma electellum* (Lepidoptera: Pyralidae). Environ. Entomol. 16:586–592.

Rosetto, C. J., R. H. Painter, and D. Wilbur. 1973. Resistancia de variodades arroz en ensa a *Sitophilus zeamais* Motchulsky. Histochem. Latineam. 9:10–18.

Russell, M. P. and R. R. Cogburn. 1977. World collection rice varieties: resistance to seed penetration by *Sitotroga cerealella* (Olivier) (Lepidoptera: Gelchiidae). J. Stored Prod. Res. 13:103–106.

Russell. G. B., O. R. W. Sutherland, R. F. N. Hutchins, and P. E. Christmas. 1978. Vestitol: A phytoalexin with insect feeding-deterrent activity. J. Chem. Ecol. 4: 571–579.

Sakharov, N. L. 1925. *Podsolnechnaya metelitsa* (The sunflower moth). Moskva, Izdatel'stvo "Novaya derevnya".

Saxena, R. C., and S. H. Okech. 1985. Role of plant volatiles in resistance of selected rice varieties to brown planthopper, *Nilaparvata lugens* (Stal) (Homptera: Delphacidae). J. Chem. Ecol. 11:1601–1616.

Schoonhoven, L. M. 1982. Biological aspects of antifeedants. Entomol. Exp. Appl. 31:57–69.

Schoonhoven, L. M., W. M. Blaney, and M. S. J. Simmonds. 1992. Sensory coding of feeding deterrents in phytophagous Insects. In: E. Bernays (Ed.), Insect - Plant Interactions IV, CRC Press, Boca Raton. pp. 59–79.

Seifelnasr, Y. E. 1991. Influence of olfactory stimulants on resistance and susceptibility of pearl millet, *Pennisetum americanum* to the rice weevil, *Sitophilus oryzae*. Entomol. Exp. Appl. 59:163–168.

Seigler, D., and P. W. Price. 1976. Secondary compounds in plants: Primary functions. Am. Natur. 110: 101–105.

Shade, R. E., and L. W. Kitch. 1983. Pea aphid (Homoptera: Aphididae) biology on glandular-haired *Medicago* species. Environ. Entomol. 12:237–240.

Shade, R. E., M. J. Doskocil, and N. P. Maxon. 1979. Potato leafhopper resistance in glandular-haired alfalfa species. Crop Sci. 19:287–289.

Sharma, H. C., Y. O. Doumbia, M. Haidara, J. F. Scheuring, K. V. Ramaiah, and N. F. Beninati. 1994. Sources and mechanisms of resistance to sorghum head bug, *Eurystylus immaculatus* Odh. in west Africa. Insect Sci. Applic. 15:39–48.

Sharma, H. C., P. Vidyasagar, and K. Leuschner. 1991. Componental analysis of the factors influencing resistance to sorghum midge, *Contarinia sorghicola* Coq. Insect Sci. Applic. 11:889–898.

Shepherd, T., G. W. Robertson, D.W. Griffiths, and A. N. E. Birch. 1999a. Epicuticular wax ester and triacylglycerol composition in relation to aphid infestation and resistance in red raspberry (*Rubus idaeus* L.) Phytochem. 52:1255–1267.

Shepherd, T., G. W. Robertson, D.W. Griffiths, and A. N. E. Birch. 1999b. Epicuticular wax composition in relation to aphid infestation and resistance in red raspberry (*Rubus idaeus* L.) Phytochem. 52:1239–1254.

Shifriss, O. 1981. Do Curcurbita plants with silvery leaves escape virus infection? Curcurbit Gen. Coop. Rep. 4:42–45.

Simmonds, M. S. J., W. M. Blaney, R. Mithen, A. N. E. Birch, and J. Lewis. 1994. Behavioural and chemosensory responses of the turnip root fly (*Delia floris*) to glucosinolates. Entomol. Exp. Appl. 71:41–57.

Sinden, S. L., L. L. Sanford, W. W. Cantelo, and K. L. Deahl. 1986. Leptine glycoalkaloids and resistance to the Colorado potato beetle (Coleoptera : Chrysomelidae) in *Solanum chacoense*. Environ. Entomol. 15:1057–1062.

Singh, K., N. S. Agarwal, and G. K. Girish. 1972. The oviposition and development of *Sitophilus oryzae* (L.) (Coleoptera: Curculionidae) in different maize hyrids and corn parts. Indian J. Entomol. 34:148–154.

Singh, R., and P. R. Ellis. 1993. Sources, mechanisms and bases of resistance in Cruciferae to the cabbage aphid, *Brevicoryne brassicae*. IOBC/WPRS Bull. 16:21–35.

Smith, C. M., J. L. Frazier, L. B. Coons and W. E. Knight. 1976. Antennal morphology of the clover weevil, *Hypera meles* (F.) Int. J. Insect Morph. and Embryol. 5:349–355.

Snyder, J. C., Z. H. Guo, R. Thacker, J. P. Goodman, and J. Stpyrek. 1993. 2,3 - dihydrofarnesoic acid, a unique terpene from trichomes of *Lycopersicon hirsutum*, repels spider mites. J. Chem. Ecol. 19:2981–2997.

Son, K.- C., R. F. Severson, and S. J. Kays. 1991. Pre- and postharvest changes in sweetpotato root surface chemicals modulating insect resistance. HortSci. 26:1514–1516.

Spicer, P. G., D. A. Potter, and R. E. Metcalf. 1995. Resistance of flowering crabapple cultivars to defoliation by the Japanese beetle. J. Econ. Entomol. 88:979–985.

Springer, T. L., S. D. Kindler, and E. L. Sorensen. 1990. Comparison of pod-wall characteristics with seed damage and resistance to the alflalfa seed chalcid (Hymenoptera: Eurytomidae) in *Medicago* species. Environ. Entomol. 19:1614–1617.

Sree, P. S., K. F. Nwanze, D. R. Butler, D. D. R. Reddy, and Y. V. R. Reddy. 1994. Morpholoical factors of the central whorl leaf associated with leaf surface wetness and resistance in sorghum to shoot fly, *Atherigona soccata*. Ann. Appl. Biol. 125:467–476.

Stoner, K. A. 1990. Glossy leaf wax and host-plant resistance to insects in *Brassica oleracea* L. under natural infestations. Environ. Entomol. 19:730–739.

Stork, N. E. 1980. Role of waxblooms in preventing attachment to brassicas by the mustard beetle, *Phaedon cochleariae*. Entomol. Exp. Appl. 28:99–106.

Sutherland, O. R. W., N. D. Hood, and J. R. Hiller. 1975. Lucerne root saponins: a feeding deterrent for the grass grub *Costelytra zealandica* (Coleoptera : Scarabaeidae). N. Z. J. Zool. 2:93–100.

Talekar, N. S. and W. J. Hu. 1993. Morphological characters in *Vigna glabrescens* resistant to agromyzids (Diptera: Agromyzidae). J. Econ. Entomol. 86:1287–1290.

Tanton, M. T. 1962. The effect of leaf "toughness" on the feeding of larvae of the mustard beetle, *Phaedon cochleariae* Fab. Entomol. Exp. Appl. 5:74–78.

Taylor, N. L. 1956. Pubescence inheritance and leafhopper resistance relationships in alfalfa. Agron. J. 48:78–81.

Thiery, D. and J. H. Visser. 1987. Misleading the Colorado potato beetle with an odor blend. J. Chem. Ecol. 13:1139–1146.

Tingey, W. M., and J. E. Laubengayer. 1986. Glandular trichomes of a resistant hybrid potato alter feeding behavior of the potato leafhopper (Homoptera: Cicadellidae). J. Econ. Entomol. 79: 1230–1234.

Thompson, K. F. 1963. Resistance to the cabbage aphid, (*Brevicoryna brassicae*) in Brassica plants. Nature. 198:209.

Thorsteinson, A. J. 1960. Host selection in phytophagous insects. Ann. Rev. Entomol. 5:193–218.

Tsumuki, H., K. Kanehisa, and K. Kawada. 1989. Leaf surface wax as a possible resistance factor of barley to cereal leaf aphids. Appl. Entomol. Zool. 24:295–301.

Tuomi, J. 1992. Toward integration of plant defense theories. Trend. Ecol. Evol. 7:365–367.

Visser, J. H. 1983. Differential sensory preceptions of plant compounds by insects. In: P. A. Hedin (Ed.), Plant Resistance to Insects. ACS Symposium Series 208, American Chemical Society. Washington, DC pp. 216–243.

Visser, J. H. 1986. Host odor perception in phytophagous insects. Ann. Rev. Entomol. 31:121–44.

Visser, J. H., S. Van Straten, and T. Maarse. 1979. Isolation and identification of volatiles in the foliage of potato, *Solanum tuberosum*, a host plant of the Colorado beetle, *Leptinotarsa decimlineata*. J. Chem. Ecol. 5:11–23.

Wahlroos, V., and A. I. Virtanen. 1959. The precursors of 6 MBOA in maize and wheat plants, their isolations and some of their properties. Acta Chem. Scan. 13:1906–1908.

Wallace, L. E., F. H. McNeal, and M. A. Berg. 1974. Resistance to both *Oulema melanopus* and *Cephus cinctus* in pubescent-leaved and solid stemmed wheat selections. J. Econ. Entomol. 67:105–110.

Wang, S. F., T. J. Ridsdill-Smith, and E. L. Ghisalberti. 1998. Role of isoflavonoids in resistance of subterranean clover trifoliates to redlegged earth mite, *Halotydeus destructor*. J. Chem. Ecol. 24:2089–2100.

Way, M. J., and G. Murdie. 1965. An example of varietal variations in resistance of Brussels sprouts. Ann. Appl. Biol. 56:326–328.

Webster, J. A., C. Inayatullah, M. Hamissou, and K. A. Mirkes. 1994. Leaf pubescence effects in wheat on yellow sugarcane aphids and greenbugs (Homoptera: Aphididae). J. Econ. Entomol. 87:231–240.

Weibel, D. E., and K. J. Starks. 1986. Greenbug nonpreference for bloomless sorghum. Crop Sci. 26: 1151–1153.

Wessling, W. H. 1958. Genotypic reactions to boll weevil attack in upland cottons. J. Econ. Entomol. 51:508–512.

Whittaker, R. H. 1970. The biochemical ecology of higher plants. In: E. Sondheimer and J. B. Simeone (Eds.), Chemical Ecology. Academic Press, New York, pp. 43–70.

Widstrom, N. W., W. W.McMillian, and B. R. Wiseman. 1979. Ovipositional preference of the corn earworm and the development of trichomes on two exotic corn selections. Environ. Entomol. 8:833–839.

Wilf, P., C. C. Labandeira, W. J. Kress, C. L. Staines, D. M. Windsor, A. L. Allen, and K. R. Johnson. 2000. Timing the radiations of leaf beetles: Hispines on gingers from latest Cretaceous to recent. Science. 289: 291–294.

Wilson, F. D., and B. W. George. 1986. Smoothleaf and hirsute cottons: Response to insect pests and yield in Arizona. J. Econ. Entomol. 79:229–232.

Wiseman, B. R., W. W. McMillian, and N. W. Widstrom. 1974. Techniques, accomplishments, and future potential of breeding for resistance in corn to the corn earworm, fall armyworm and maize weevil, and in sorghum to the sorghum midge. In: F. G. Maxwell and F. M. Harris (Eds.), Proc. Summer Inst. Biol. Control Plant Insects Dis. Univ. Mississippi Press, Jackson, MS. pp. 381–393.

Wiseman, B. R., W. W. McMillian, and N. W. Widstrom. 1976. Feeding of corn earworm in the laboratory on excised silks of selected corn entries with notes on Orius insidiosus. Fla. Entomol. 59:305–308.

Wiseman, B. R., W. W. McMillian, and N. W. Widstrom. 1977. Ear characteristics and mechanisms of resistance among selected corns to corn earworm. Fla. Entomol. 60:97–103.

Woodhead, S. 1982. p-hydroxybenzaldehyde in the surface wax of sorghum: its importance in seedling resistance to acridids. Entomol. Exp. Appl. 31:296–302.

Woodhead, S., and R. F. Chapman. 1986. Insect behavior and the chemistry of plant surface waxes. In: B. Juniper and R. Southwood (Eds.), Insects and the Plant Surface. Edward Arnold, London. pp.123–135.

Woodhead, S., and G. Cooper-Driver. 1979. Phenolic acids and resistance to insect attack in Sorghum bicolor. Biochem. Syst. Ecol. 7:309–310.

Xie, Y. S., J. T. Arnason, B. J. R. Philogene, and J. D. H. Lambert. 1990. Role of 2, 4- dihydroxy-7-methoxy-1,4-benzoazin-3-one (DIMBOA) in the resistance of maize to western corn rootworm, Diabrotica virgifera vigifera (Leconte) (Coleoptera:Chrysomelidae). Can. Entomol. 122:1177–1186.

Yan, F., X. Liang, and X. Zhu. 1999. The role of DIMBOA on the feeding of Asian corn borer, Ostrinia furnacalis (Guenee) (Lep., Pyralidae). J. Appl. Entomol. 123:49–53.

Yang, G., B. R. Wiseman, D. J. Isenhour, and K. E. Espelie. 1993. Chemical and ultrastructural analysis of corn cuticular lipids and their effect on feeding by fall armyworm larvae. J. Chem. Ecol. 19:2055–2074.

CHAPTER 3

ANTIBIOSIS - ADVERSE EFFECTS OF RESISTANCE ON ARTHROPOD BIOLOGY

1. GENERAL

The antibiosis category of plant resistance occurs when the negative effects of a resistant plant affect the biology of an arthropod attempting to use that plant as a host (Figure 1.5). The antibiotic effects of a resistant plant range from mild to lethal, and may result from both chemical and morphological plant defensive factors. Lethal effects may be acute, in which case they often affect young larvae and eggs. The chronic effects of antibiosis often lead to mortality in older larvae and prepupae that fail to pupate, and in pupae and adults which fail to eclose. Individuals surviving the direct effects of antibiosis may also suffer the debilitating effects of reduced body size and weight, prolonged periods of development in the immature stages, and reduced fecundity. A discussion about the types of techniques used to identify and measure antibiosis under experimental conditions is presented in section 6.2.1.

The expression of resistance as antibiosis is widespread across many genera of crop plants. In cereal crops, antibiosis is a major component in maize to several pest Lepidoptera, including the African armyworm, *Spodoptera exempta* (Walker) (Okello-Ekochu and Wilkins 1996), the fall armyworm, *Spodoptera frugiperda* (J. E. Smith) (Williams and Davis 1997), the stems borers *Busseola fusca* (Fuller), *Chilo partellus* (Swinhoe), *Sesamia nonagrioides* Lefevbre, the southwestern corn borer, *Diatraea grandiosella* Dyar (Kumar 1993, Williams and Davis 1997, Ordas et al. 2002), the European corn borer, *Ostrinia nubilalis* (Hubner), and the carmine spider mite, *Tetranychus cinnabarinus* (Boisduval) (Tadmor et al. 1999). In wheat, *Triticum aestivum* L., antibiosis functions in resistance to the greenbug, *Schizaphis graminum* (Rondani), (Webster and Porter 2000, Lage et al. 2003, Smith and Starkey 2003), the Russian wheat aphid, *Diuraphis noxia* Mordvilko, (Smith et al. 1991, Hein 1992, Hawley et al. 2003), and the grain aphid, *Sitobion avenae* (F.) (Lowe 1984). Antibiosis is also expressed in the resistance of sorghum, *Sorghum bicolor* (L.) Moench, to *S. graminum* (Tonet and DaSilva 1994), to *S. frugiperda* (Meckenstock et al. 1991) and the sorghum midge, *Contarinia sorghicola* Coq. (Sharma et al. 1993).

In forage crops, antibiosis is well documented as a factor in the resistance of groundnut, *Apios americana* Medik., to the tobacco armyworm, *Spodoptera litura* (Fab.) (Stevenson et al 1993a); mungbean, *Vigna radiata* (L.) R.Wilczek, resistance to the adzuki bean weevil, *Callosobruchus chinensis* (L.) (Talekar and Lin 1992);

65

resistance in chickpea, *Cicer arietinum* L., to *Heliothis armigera* (Hubner) (Srivastava and Srivastava 1990); resistance in cowpea, *Vigna unguiculata* L. Caupí., to the pod-sucking bug, *Clavigralla tomentosicella* Stal., (Olatunde and Odebiyi 1991); resistance in pea, *Pisum sativum* L., to the pea weevil, *Bruchus pisorum* (L.) (Clement et al. 2002); and resistance in soybean to the corn earworm, *Helicoverpa zea* (Boddie) (Mebrahtu et al. 2002).

Antibiosis operates in the resistance of both deciduous and coniferous trees (Larsson and Strong 1992, Sahota et al. 1998). van Helden et al. (1993) detected antibiosis in cultivars of lettuce, *Lactuca sativa* L., resistant to the aphid *Nasonovia ribisnigri* (Mosley) and Mansour et al. (1994) identified lines of melon, *Cucumis melo* L., expressing antibiosis to *T. cinnabarinus*. Antibiosis also exists in different citrus rootstocks to several pests, including the stinkbug, *Biprorulus bibax* (Breddin) (James 1992), the Diaprepes root weevil, *Diaprepes abbreviatus* (L.) (LaPointe and Bowman 2002), and the citrus leafminer, *Phyllocnistis citrella* (Stainton) (Jacas et al. 1997).

2. DEFENSES OF ARTHROPOD RESISTANT CROP PLANTS IMPARTING ANTIBIOSIS

2.1. Allelochemicals

2.1.1. Toxins

Allelochemicals such as alkaloids, ketones, and organic acids are toxic to arthropods. The toxic nature of alkaloids produced by plants is well known, and a review by Wink (1993) provides detailed discussions of their roles in plant-arthropod interactions. Several alkaloids mediate resistance to arthropods in agricultural plants. The glycoalkaloid content of *Solanum* species resistant to the potato leafhopper, *Empoasca fabae* (Harris), is directly correlated with hopper survival (Raman et al. 1978). Leptine glycoalkaloids in *S. chalcoense*, as well as the alkaloid solanocardenine in *S. neocardenasii* are toxic to the Colorado potato beetle, *Leptinotarsa decemlineata* (Say) (Sinden et al. 1986, 1991, Lorenzen et al. 2001) (Table 3.1).

Castanera et al. (1996) identified clones of potato, *Solanum tuberosum* L., with very high levels of activity of polyphenol oxidase, an enzyme whose quinone products reduce food digestion. The same clones greatly reduce the survival of young *L. decemlineata* larvae. The alkaloid α-tomatine plays a role in the resistance of wild species of potato and tomato. Elliger et al. (1981) found that the α-tomatine content of the wild tomato, *Lycopersicon hirsutum* f. *glabratum* C. H. Mull., is 3–4 times greater than that of tomato cultivars susceptible to *H. zea*. α-tomatine is also partially responsible for resistance to *L. decemlineata* in tomato (Sinden et al. 1978) and potato (Dimock et al. 1986, Sinden et al. 1991). However, there is no strong correlation between the level of arthropod resistance and α-tomatine content in tomato, due presumably to the interaction of α-tomatine with free foliar sterols (Campbell and Duffey 1981). The alkaloids lupanin and 1–hydroxylupanine in narrow-leafed lupin,

Table 3.1. *Toxic allelochemicals involved in antibiosis plant resistance to arthropods*

Plant	Toxin	Arthropod(s) Affected	Reference(s)
Citrus	Linalool	*Anastrepha suspensa*	Greany et al. 1983
Daucus	Chlorogenic acid	*Psilia rosae*	Cole 1985
Geranium	Organic acids	*Tetranychus urticae*	Gerhold et al. 1984
Helianthus	Sesquiterpene lactones, diterpenes	*Homoeosoma electellum*	Rogers et al. 1987
Lactuca	Isochlorogenic acid	*Pemphigus bursarius*	Cole 1984
Lupinus	Lupanin, lupanine	*Myzus persicae*	Berlandier 1996
Lycopersicon hirsutum f. *typicum*	α-tomatine	*L. decemlineata*	Sinden et al. 1978
		Helicoverpa zea	Elliger et al. 1981
	2–tridecanone	*L. decemlineata*	Kennedy & Sorensen 1985
		H. zea	Dimmock & Kennedy 1983
		Manduca sexta	Kennedy & Yamamoto 1979 Williams et al. 1980
		Spodoptera exigua	Lin et al. 1987
	2–undecanone	*H. zea*	Farrar & Kennedy 1988
		S. exigua	Lin et al. 1987
		Keiferia lycopersicalla	
Phaseolus	α-amylase	*Callosobruchus maculatus*	Gatehouse et al. 1979
Vigna		*Callosobruchus chinensis*	Ishimoto & Kitamura 1989
Phaseolus	Arcelin protein	*Zabrotes subfasciatus*	Minney et al. 1990
	Phaseolin protein	*C. maculatus*	Moraes et al. 2000
Solanum	Glycoalkaloids	*Leptinotarsa decemlineata*	Dimmock et al.1986
	Solanocardenine		Lorenzen et al. 2001
	α-tomatine		Sinden et al. 1991
Vigna	Vicilin protein	*C. maculatus*	Xavier-Filho et al. 1996
Zea mays	DIMBOA, 6–MBOA	*Rhopalosiphum maidis*	Long et al. 1977
		D. virgifera virgifera	Assabgui et al.1995, Xie et al. 1992
	DIMBOA	*Ostrinia nubilalis*	Klun et al. 1970
		Sesamia nonagriodes	Gutierrez & Castanera 1986
	DIBOA	*Schizaphis graminum*	Argandona et al. 1983

Lupinus angustiflolius, have been implicated in lupin resistance to the green peach aphid, *Myzus persicae* (Sulzer) (Berlandier 1996).

The leaf content of the indole alkaloid gramine (N, N—dimethyldimethyl indole) has been correlated with the resistance of barley, *Hordeum vulgare* L., to *S. graminum* and the bird cherry oat aphid, *Rhopalosiphum padi* (L.), in the field, greenhouse and laboratory (Zuniga et al. 1985, Zuniga and Corcuera 1986, Kanehisa et al. 1990). Gramine deters aphid feeding when incorporated into artificial diet at concentrations similar to those in plants (Zuniga et al. 1988). However, gramine has not been recovered from the honeydew of aphids feeding on barley (Dreyer and Campbell 1987). Independent experiments in Japan and Sweden attempted to confirm gramine as a breeding marker in barley populations segregating for *R. padi* resistance, but in neither case is there a correlation of aphid resistance to gramine content. Moharramipour et al. (1997) found no correlation between *R. padi* population levels and gramine content in a barley population from a cross between high gramine content parents of wild barley, *Hordeum vulgare* subsp. *spontaneum* and the cultivated barely *Hordeum vulgare* subsp. *vulgare.* Ahman et al. (2000), using the same parents, developed six different segregating populations and obtained similar results, further indicating that gramine does not confer barley resistance to *R. padi.*

The methyl ketones 2–tridecanone and 2–undecanone (Figure 3.1) play a key role in the defense of *L. hirsutum* f. *glabratum* foliage against arthropod defoliation. Both are produced in vacuoles on the tip of foliar glandular trichomes (Figure 2.8). 2–tridecanone, produced in much higher quantities in the foliage of resistant cultivars than susceptible cultivars, is toxic to *H. zea* and to the tobacco hornworm, *Manduca sexta* (L.) (Kennedy and Yamamoto 1979, Williams et al. 1980, Dimock and Kennedy 1983), and at least partially responsible for resistance to the *L. decemlineata* (Kennedy and Sorenson 1985). 2–undecanone causes mortality of *H. zea* larvae by inhibition of pupation, but has no effect on *M. sexta* larvae (Farrar and Kennedy 1987, 1988). Both 2–tridecanone and 2–undecanone are toxic to the tomato fruitworm, *Keiferia lycopersicalla* (Walsingham), and the beet armyworm, *Spodoptera exigua* (Hubner) (Lin et al. 1987). As indicated previously in Chapter 2, methyl ketones also have significant antixenotic effects.

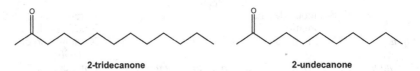

2-tridecanone   **2-undecanone**

Figure 3.1. The methyl ketone toxins –tridecanone and –undecanone produced in the glandular trichomes of Solanum hirsutum *f.* glabratum.

Organic acids in arthropod resistant plants have antibiotic effects. The cyclic hydroxamic acid DIMBOA and its decomposition product 6–MBOA (Figure 2.14) in maize foliage and roots have antibiotic effects to *O. nubilalis* (Klun et al. 1967, 1970), the stalk corn borer, *Sesamia nonagriodes* Lef., (Gutierrez and Castanera 1986) and the corn rootworm, *Diabrotica virgifera virgifera* Leconte (Xie et al. 1992, Assabgui et al.

1995). DIMBOA is also an active component in the resistance of maize to the corn leaf aphid, *Rhopalosiphum maidis* (Fitch) (Long et al. 1977, Beck et al. 1983). The concentrations of DIMBOA in maize cultivars are highly correlated with antibiosis resistance (Klun and Robinson 1969, Barry et al. 1994). DIMBOA acts as a digestive toxin in the gut of feeding *O. nubilalis* larvae, while MBOA acts to reduce the efficiency of conversion of food digested by *O. nubilalis* (Houseman et al. 1992). In larvae of the stalk corn borer, *Sesamia nonagrioides* (Lefebvre), DIMBOA acts as it does in *O. nubilalis*, reducing growth and the efficiency of conversion of ingested food (Ortego et al. 1998).

DIMBOA and 6–MBOA have antibiotic effects to the aphid *Metopolophium dirhodum* (Walker) (Argandona et al. 1980). Otherwise, DIMBOA concentrations in rye and wheat cultivars have only been correlated to *R. padi* and *S. avenae* resistance levels (Argandona et al. 1981, Thackray et al. 1990). Levels of DIBOA (a 7–methyoxylated analogue of DIMBOA) are correlated to *R. padi* population development in wild barley species (Barria et al. 1992). Toxic effects are evident in *S. graminum* fed artificial diets containing DIMBOA or DIBOA (Argandona et al. 1983, Zuniga et al. 1983) and in *R. padi* fed artificial diets containing DIBOA (Barria et al. 1992). Attempts to correlate DIMBOA to *D. noxia* resistance have been inconclusive, as there are no correlations between DIMBOA levels and the total time of phloem feeding or *D. noxia* growth rate (Givovich and Niemeyer 1996, Mayoral et al. 1996). This result may be due to the fact that DIMBOA levels decline very rapidly in wheat seedlings (Argondona et al. 1981). Slesak et al. (2001) reported similar results for the effects of DIMBOA and the duration of *R. padi* feeding. Field experiments remain necessary to validate the actual role of hydroxamic acids in cereal cultivars resistant to aphids.

Organic acids toxic to arthropods occur in resistant cultivars of carrot, *Daucus carota* L., and geranium, *Pelargonium x hortorum* (L.). The exudate of trichomes of geranium cultivars resistant to the twospotted spider mite, *Tetranychus urticae* Koch, consists mainly of anacardic acids, a class of phenolic acids. The anacardic acids romanicardic acid and geranicardic acid (Figure 3.2) are moderately toxic to mites (Gerhold et al. 1984, Grazzini et al. 1995), and the variation in their composition in different geranium cultivars can be monitored by high-pressure liquid chromatography (see Chapter 6). The concentration of chlorogenic acid in carrot cultivars resistant to the carrot rust fly, *Psilia rosae* (F.), is closely correlated to the level of fly population development (Cole 1985). Similarly, concentrations of isochlorogenic acid in lettuce cultivars are correlated to the level of resistance to the lettuce root aphid, *Pemphigus bursarius* (L.) (Cole 1984) (See Section 6.3.1.3).

Terpene metabolites also mediate antibiosis in crop plants. Greany et al. (1983) determined that the resistance of citrus fruit to damage by the Caribbean fruit fly, *Anastrepha suspensa* (Loew), is related to the terpenoid content of the fruit rind. High mortality occurs among young larvae when they attempt to penetrate the oily layer of the citrus peel. Orange and lemon fruits that are more resistant than grapefruit, have a much higher concentration of the terpene alcohol linalool than grapefruit, and a higher total volume of fruit peel oil glands. Several diterpenes and sesquiterpene lactones from the foliage of resistant *Helianthus* species cause antibiotic symptoms (mortality, delayed development, retarded growth) in larvae of

the sunflower moth, *Homoeosoma electellum* (Hulst) (Rogers et al. 1987).   The
sesquiterpene lactone 8, β-sarracinoyloxycumam-branolide, produced by glandular
leaf trichomes, is a prominent example.   Saponins, which are produced in the plant
terpenoid metabolic pathway (see Section 2.4.2.2), are toxic to larvae of the grass
grub, *Costelytra zealandica* (White), feeding on resistant alfalfa and trefoil cultivars
grown in the pastures of New Zealand (Sutherland et al. 1982).

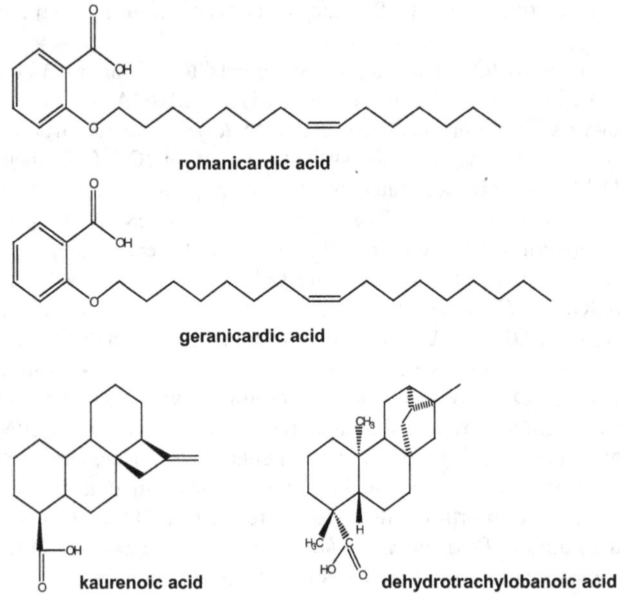

*Figure 3.2. Organic acids from arthropod resistant crop cultivars. Romanicardic acid and
geranicardic acid from a geranium cultivar resistant to* Tetranychus urticae *(Gerhold 1984),
kaurenoic acid and trachylobanoic acid from Helianthus species resistant to*
Homeosoma electellum *(Elliger et al. 1976)*

Plant proteins also exert toxic antibiotic effects on arthropods. Gatehouse et al.
(1979) first attributed the resistance of cowpea cultivars resistant to the cowpea
weevil, *Callosobruchus maculatus* (F.) to increased levels of trypsin or α-amylase
inhibitors, compared to susceptible cultivars.   Ishimoto and Kitamura (1989)
identified an α-amylase inhibitor (α AI-1) in seed of common bean, *Phaseolus
vulgaris* L., that causes antibiosis effects in larvae of *C. maculatus* and *C. chinensis*
but has no effect on larvae of the Mexican bean weevil, *Zabrotes subfasciatus*
(Boh.), or the bean weevil, *Acanthoscelides obtectus* (Say) (Ishimoto and Kitamura 1992).
Arcelin, an additional seed protein identified from a *Z. subfasciatus*-resistant
common bean line (Osborn et al. 1986) was shown to cause antibiosis to weevil larvae
(Minney et al. 1990). Later research by Xavier-Filho et al. (1996) linked *C. maculatus*

resistance to an increased content of vicilin storage proteins (7S globulins) in weevil-resistant cultivars. *C. maculatus* larvae fed resistant seeds high in vicilin content have reduced rates of development and increased mortality, due to reduced digestive enzyme activity (Macedo et al. 1993) and to vicilins tightly binding to larval gut chitin (Sales et al. 1992). Moraes et al. (2000) identified peptides of phaseolin, a vicilin-like 7S storage globulin, in the seed of common beans and showed that phaseolin deterred *C. maculatus* larval development.

2.1.2. Growth inhibition due to the presence of inhibitors

Chronic growth inhibition, due to either the presence of growth inhibitors or the absence of or reduction in the level of plant nutrients, is exhibited in several arthropod resistant cultivars (Table 3.2).

Table 3.2. Antibiotic allelochemicals that inhibit the growth of arthropods

Plant genus	Inhibitor	Arthropod(s) affected	Reference
Arachis	Quercetin diglycosides	*Spodoptera litura*	Stevenson et al. 1993b
Cicer	Oxalic acid	*Helicoverpa armigera*	Yoshida et al. 1995
Glycine	Coumestrol	*Pseudoplusia includens*	Rose et al. 1988 Caballero et al. 1987
Gossypium	Gossypol (sesquiterpene aldehyde)	*Heliothis virescens* *Spodoptera littoralis*	Lukefahr & Martin 1966
		Helicoverpa zea *Earias vittella*	Meisner et al. 1977
		Pectinophora gossypiella	Mohan et al. 1994
	Caryophyllene oxide	*H. virescens*	Stipanovic et al.1986
	Hemigossypol, Heliocides 1 & 2, (terpene quinones)	*H. virescens*	Bell et al. 1975 Hedin et al. 1992 Stipanovic et al. 1977
Helianthus	Kaurenoic acid, Trachylobanoic acid	*Lepidoptera*	Elliger et al. 1976
Manihot	Rutin, kaemferol glycoside	*Phenacoccus manihoti*	Calatayud et al. 1994
Picea	Resin acids	*Pissodes strobi*	Tomlin et al. 1996
Vigna	Isorhamnetin, quercetin	*Aphis fabae*	Lattanzio et al. 2000
Zea	Apimaysin Chlorogenic acid 3'-methoxymaysin Maysin	*H. zea*	Guo et al. 1999 Waiss et al. 1979 Wiseman et al. 1992 Wiseman & Snook 1995

The flavone glycosides maysin, apimaysin, 3'-methoxymaysin and chlorogenic acid (Figure 3.3) are allelochemicals contained in the silks of maize cultivars resistant to *H. zea* (Waiss et al. 1979). Weight gain in *H. zea* larvae is negatively correlated with the concentration of these compounds (Guo et al. 1999, Wiseman et al. 1992, Wiseman and Snook 1995). Maysin, chlorogenic acid and related compounds have also been shown to express resistance in centipedegrass, *Eremochloa ophiuroides* (Munro), to defoliation by *S. frugiperda* larvae (Wiseman et al. 1990).

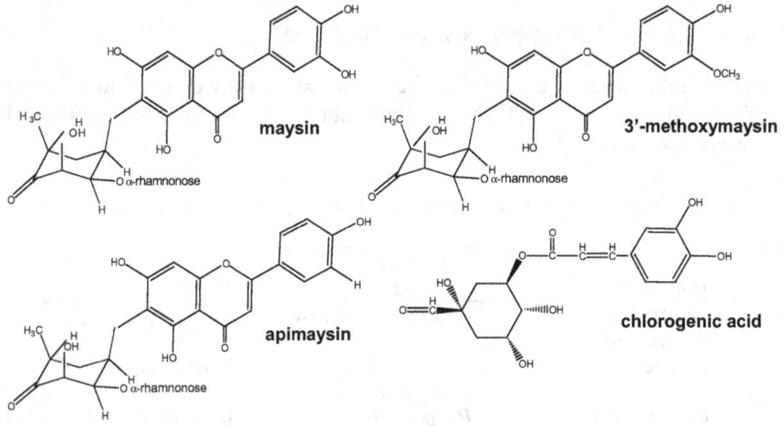

Figure 3.3. Maysin, apimaysin and 3'-methoxymaysin flavone glycosides, and chlorogenic acid from foliage of insect resistant maize cultivars inhibit growth of Helicoverpa zea *larvae. (Guo et al. 1999, Wiseman et al. 1992, Wiseman and Snook 1995)*

Infestation of plants of cassava, *Manihot esculenta* Crantz, resistant to the cassava mealybug, *Phenacoccus manihoti* Matt. Ferr., with mealybugs causes a significant increase in the level of the flavonoid glycosides rutin and kaemferol glycoside (Calatayud et al. 1994). The flavonoid aglycones isorhamnetin and quercetin have a direct involvement in the resistance of cowpea to the cowpea aphid, *Aphis fabae* (Scopoli) (Lattanzio et al. 2000). Ethyl acetate extracts from the foliage of other cowpea cultivars severely limit the growth of the aphid *Aphis craccivora* (Walp.) (Annan et al. 1996). Several quercetin diglycosides and two unique phenolic acid esters from foliage of groundnut species resistant to *S. litura* greatly reduce larval development (Stevenson 1993, Stevenson et al. 1993b).

Coumestrol (Figure 2.15), an isoflavone found in several legumes, displays pronounced estrogenic effects (Harborne 1982) in vertebrates. Coumestrol occurs in high concentration in the foliage of the wild arthropod resistant soybean species *Glycine soja* (Caballero et al. 1987). Larvae of the soybean looper, *Pseudoplusia includens* (Walker), fed diets containing coumestrol at concentrations similar to that occurring in resistant plants suffer significant weight reductions (Rose et al. 1988).

The organic acids kaurenoic acid and trachylobanoic acid (Figure 3.2) are produced in the florets of arthropod resistant sunflower cultivars, and retard the development of larvae of several species of Lepidoptera (Elliger et al. 1976). Further evidence of their growth inhibitory properties is demonstrated by a reversal of growth inhibition when larvae are fed large quantities of cholesterol, a steroid essential to arthropod development. Cortical resin acids occur in significantly larger quantities in Sitka spruce, *Picea sitchensis* (Bong.) Carr., resistant to the white pine weevil, *Pissodes strobi* (Peck), than in susceptible trees (Tomlin et al. 1996). Oxalic acid is exuded from trichomes of chickpea plants (see section 2.2.2.1) and accumulates on leaves and pods of cultivars resistant to *H. armigera*. Oxalic acid ingested by *H. armigera* larvae has both corrosive and anti-digestive effects, resulting in decreased *H. armigera* larval weight and an increased duration of larval development (Yoshida et al. 1995).

Several terpenoids produced in the foliar pigment glands of arthropod resistant cotton cultivars have antigrowth effects on several arthropod pests. The terpenoid aldehyde gossypol (Figure 3.4) was first shown to inhibit growth in larvae of the tobacco budworm, *Heliothis virescens* (F.) (Lukefahr and Martin 1966).

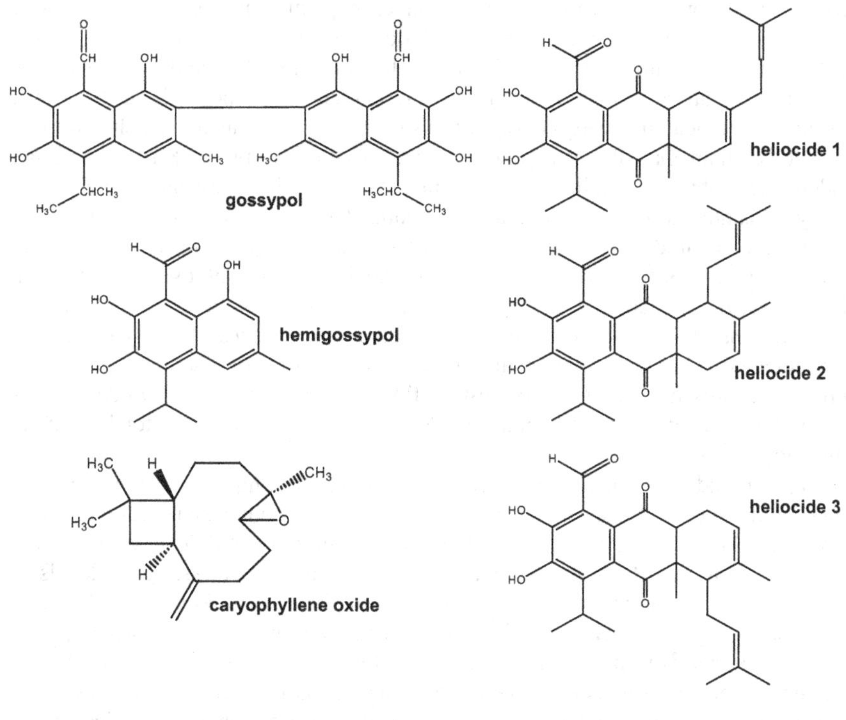

Figure 3.4. Terpenoids produced in insect-resistant cotton cultivars that inhibit the growth of foliar feeding Lepidoptera: gossypol; hemigossypol; heliocides 1, 2 and 3; and caryophyllene oxide. (Stipanovic et al. 1986)

Additional studies have since demonstrated that gossypol adversely affects the Egyptian cotton leafworm, *Spodoptera littoralis* (Boisd.) (Meisner et al. 1977), *H. armigera*, the spotted bollworm, *Earias vittella* (Fab.), and the pink bollworm, *Pectinophora gossypiella* Saunders (Mohan et al. 1994). Additional terpenoid compounds, termed "x" factors for several years because of their structural complexity, are present in certain cotton cultivars in which the gossypol content does not totally explain resistance (Lukefahr et al. 1974). The concentrations of the "x" factors hemigossypolone and "heliocides" 1, 2 and 3 (Figure 3.4) (Bell et al. 1975, Stipanovic et al. 1977), are greatly increased in *H. virescens* resistant lines with normal gossypol concentrations and are thought to be more definitive resistance factors (Hedin et al. 1992). The volatile monoterpene caryophyllene oxide (Figure 3.4) also inhibits *H. virescens* larval growth at high concentrations and synergizes the effects of gossypol (Stipanovic et al. 1986).

2.1.3. Growth inhibition due to reduced levels of nutrients

Growth inhibition in arthropods feeding on crop plant may also be related to a reduction in the nutrient level of ingested food. Penny et al. (1967) determined that maize plants resistant to *O. nubilalis* larvae have an ascorbic acid content that is inadequate for larval growth. The amino acid content of the pea cultivar 'Laurier', resistant to the pea aphid, *Acyrthosiphon pisum* (Harris), has also been shown to be much lower than that of susceptible cultivars (Auclair et al. 1957). Kazemi and van Emden (1992) obtained similar results in investigating wheat resistance to *R. padi*. Highly resistant cultivars contain greatly reduced levels of alanine, histidine and theronine, compared to susceptible cultivars. Reduced glutamic acid content in barley cultivars resistant to *R. padi* is similarly linked to antibiosis (Weibull 1994). Ciepiela (1989) demonstrated wheat antibiosis to *S. avenae* is directly related to the concentrations of free and essential amino acids in plant tissues. Mollema and Cole (1996) showed a similar relationship between accessions of cucumber, *Cucumis sativa* L., expressing antibiosis to western flower thrips, *Frankliniella occidentalis* (Pergande), and total aromatic amino acids. Accessions with lower total aromatic amino acid content supported less *F. occidentalis* growth.

Ciepiela and Sempruch (1999) evaluated the relationship between nonprotein amino acids and antibiosis in wheat to *S. avenae* and found significant negative correlations between the levels of L-DOPA and ornithine to the level of *S. avenae* resistance. Holt and Birch (1984) obtained similar results in contrasting the levels of several nonprotein amino acids in the foliage of faba bean, *Vicia faba* L., with resistance to *A. fabae*. Telang et al. (1999) demonstrated how resistant wheat plants prevent *D. noxia* from nutritionally enhancing the amino acid composition of their ingested phloem, as they do on susceptible plants. 'Mudgo', rice cultivar resistant to the brown planthopper, *Nilaparvata lugens* Stal, contains lower quantities of the amino acids asparagine and glutamic acid than susceptible rice cultivars (Sogawa and Pathak 1970). Antibiosis of the rice cultivars 'Rathu Heenati' to the whitebacked planthopper, *Sogatella furcifera* (Horvath) is linked to reduced plant sucrose and fructose content, relative to a susceptible cultivar, and to corresponding reductions in *S. furcifera* glucosidae activity and weight gain on resistant plants (Guangjie et al. 1995).

2.2. Physical and Morphological Barriers

2.2.1. Hypersensitive plant tissue responses

Rapidly growing plant tissues are often associated with the tolerance of crop plants to arthropod damage, since tissue growth may be related to the vigor of growth of a particular cultivar (Chapter 8). However, rapid growth may be so dramatic that arthropods exhibit antibiotic effects (Painter 1951). In the early 1900's research by Hinds (1906) showed that rapidly growing cotton boll tissues killed larvae of the boll weevil, *Anthonomus grandis grandis* Boheman (Table 3.3). Adkisson (1962) demonstrated similar effects of cotton tissues on *P. gossypiella* larvae. Unfortunately, both traits are linked to undesirable agronomic characters and have been abandoned as breeding characters.

Table 3.3 Crop plants exhibiting hypersensitive responses in tissues that kill eggs and/or larvae of infesting arthropods

Plant genus	Arthropod affected	Reference
Brassica	*Artogeia rapae*	Shapiro & DeVay 1987
Gossypium	*Anthonomus grandis grandis*	Hinds 1906
	Pectinophora gossypiella	Adkisson 1962
Oryza	*Orseolia oryzae*	Bentur & Kalode 1996
Phaseolus	*Apion godmani*	Garza et al. 2001
Picea	*Pissodes strobi*	Alfaro et al. 1995
		Tomlin & Borden 1994
Pinus	*Rhyacionia buoliana*	Harris 1960
	Dendroctonus frontalis	Hodges et al. 1979
Pisum	*Bruchus pisorum*	Doss et al. 2000
Solanum	*Leptinotarsa decemlineata*	Balbyshev & Lorenzen 1997
	Liriomyza pictella	Oatman 1959

Hypersensitive responses in plants have since been observed to be quite widespread. Similar effects were next shown to be expressed by cultivars of eggplant, *Solanum melongena* L., resistant to the melon leaf miner, *Liriomyza pictella* Blanchard, (Oatman 1959) and in *Pinus* species resistant to the pine shoot moth, *Rhyacionia buoliana* (Dennis and Schiffermuller) (Harris 1960). Hypersensitive reactions also occur in foliage from of some mustard, *Brassica nigra* (L.), cultivars to eggs of the imported cabbageworm, *Artogeia rapae* L. (Shapiro and DeVay 1987) (Figure 3.5). Plants produce a necrotic zone around the base of *A. rapae* eggs, causing them to desiccate.

Antibiosis in pea to *B. pisorum* is linked to a similar mechanism. Weevil oviposition stimulates tumor-like growths of undifferentiated cells on pea pod walls under each weevil egg that eventually impede larval development (Doss et al. 2000). A similar antibiosis occurs to bean pod weevils, *Apion godmani* Wagner. Weevils oviposit on pods of resistant lines of *Phaseolus vulgaris* that form pod wall callus to inhibit *A. godmani* development (Garza et al. 2001). After callus formation, necrotic tissues develop concentrically around the oviposition site and the associated eggs and larvae die.

Balbyshev and Lorenzen (1997) discovered a *Solanum* spp. hybrid that exhibits a hypersensitive response to the presence of *L. decemlineata* eggs. As with the mechanisms described for mustard and bean, a necrotic zone containing *L. decemlineata* eggs disintegrates and detaches from the infested leaf. A hypersensitive response also occurs in leaves of rice cultivars resistant to the rice gall midge, *Orseolia oryzae* (Wood-Mason). Symptoms include leaf browning and necrosis of the infested meristem tissue, accompanied by death of the associated *O. oryzae* larvae (Bentur and Kalode 1996). Hoch et al. (2000) noted a hypersensitive response in leaves of an Asian birch tree, *Betula davurica* (Pall.), to oviposition by the birch leafminer, *Fenusa pusilla* (Lepeltier). Eggs of *F. pusilla* placed on *B. davurica* leaves become surrounded by an area of necrotic tissue and die.

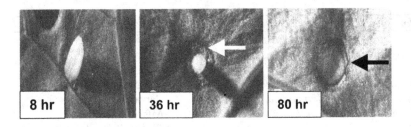

Figure 3.5. Hypersensitive response of Brassica nigra to Artogeia rapae oviposition. Arrows indicate necrotic tissue ring formed around insect egg. Reprinted from Shapiro, A. M. and J. E. DeVay. 1987. Hypersensitivity reaction of Brassica nigra L. (Cruciferae) kills eggs of Pieris butterflies (Lepidoptera: Pieridae). Oecologia. 71:631-632.
Copyright 1987, Springer Verlag, with kind permission of Springer Science and Business Media.

Antibiosis to the cabbage aphid, *Brevicoryne brassicae* (L.) in several *Brassica* species was investigated by Cole (1994) using the electrical penetration graph measurement of aphid feeding (see Chapter 6). In the absence of allelochemical differences between resistant and susceptible plants, empirical evidence suggests that aphid penetration of the plant phloem sieve element releases elicitors that induce a hypersensitive physical response that blocks the phloem sieve pores and the aphid stylet canal.

The oleoresin flow of coniferous trees following arthropod attack is also a hypersensitive response. Hodges et al. (1979) determined that pine trees have higher levels of resistance to the southern pine beetle, *Dendroctonus frontalis* Zimm., due to higher oleoresin flow rate, greater resin viscosity, and more rapid resin crystallization. Resin-based resistance has also been detected to *P. strobi* in white spruce trees (Alfaro et al. 1995) that have more resin canals per unit area of bark than susceptible trees. Sitka spruce tree resistance to *P. strobi* is similarly linked to larger numbers of resin ducts (Tomlin and Borden 1994). All of these physical characters aid coniferous trees in resisting arthropod attack.

2.2.2. Plant structures

Trichomes. Arthropod egg, larval or adult mortality may also occur after contact with plant trichomes on the surface of leaves and stems (Table 3.4). Perhaps the greatest amount of trichome research has occurred with germplasm of the genus *Solanum* in attempts to identify sources of resistance to *L. decemlineata*. The foliage of a wild Bolivian potato, *Solanum berthaultii* Hawkes, is protected by glandular trichomes from damage by many different arthropods (Gibson 1971, Tingey and Gibson 1978, Tingey and Laubengayer 1981, Casagrande 1982, Tingey and Sinden 1982, LaPointe and Tingey 1986, Gregory et al. 1986a,b, Neal et al. 1989, Tingey 1991). The field of trichomes (Figure 2.8) is composed of tall (type B) trichomes with distal glands that exude an adhesive coating unto the tarsi of arthropods attempting to move about the leaf surface. As trapped arthropods struggle to free themselves of the type B trichome adhesive, the heads of shorter type A trichomes are ruptured (Tingey and Laubengayer 1981), releasing the two components of a natural epoxy: a resin, chlorogenic acid, and a catalyst, polyphenol oxidase (Ryan et al. 1982,1983, Kowalski et al. 1992). Individuals trapped in the hardening resin die of starvation. Susceptible *Solanum tuberosum* plants possess trichomes that lack the polyphenol oxidase activity (Kowalski et al. 1992). Type B trichome exudate from *S. berthaultii* is composed of a complex of acyl sugars containing short-chain carboxylic acids (Neal et al. 1990).

When trichomes are removed mechanically or chemically, normally resistant plants are rendered susceptible to *L. decemlineata, M. persicae,* or the potato tuber moth, *Epitrix cucumeris* (Harris) (Malakar and Tingey 2000, Neal et al. 1990, Yencho and Tingey 1994). Trichome exudates from foliage of *Solanum neocardenasii* function in a similar manner in resistance to *M. persicae* (LaPointe and Tingey 1986), and glandular trichomes of *Solanum sisymbriifolium* express resistance to the tortoise beetle, *Conchyloctenia tigrina* Oliver (Hill et al. 1997). Yencho et al. (1994), using a *S. berthaultii* accession with only type A trichomes, demonstrated that *S. berthaultii* foliage is also deterrent to *L. decemlineata* feeding. Sikinyi et al. (1997) identified effective *L. decemlineata* resistance in accessions of several different *Solanum* species unrelated to glandular trichomes or foliar glycoalkaloids. The review of Tingey (1991) provides an excellent account of various antibiotic effects of *Solanum* glandular trichomes.

Some accessions of wild tomato, *Lycopersicon hirsutum* f. *typicum* Humb. & Bonpl., possess foliar trichomes that produce an unknown exudate toxic to *S. exigua* larvae (Eigenbrode et al. 1996). Acylsugars from trichomes of other *L. hirsutum* f. *typicum* accessions also adversely affect oviposition by the leafminer *Liriomyza trifolii* (Burgess) (Hawthorne et al. 1992).

Table 3.4. Plant trichomes mediating antibiosis to arthropods

Trichome Type & Plant	Arthropod(s) affected	References
Glandular		
Cicer arietinum	*Heliothis armigera*	Srivastava & Srivastava 1990, Yoshida et al. 1997
Fragaria chiloensis	*Tetranychus urticae*	Luczynski et al. 1990
Helianthus spp.	*Homoeosoma electellum*	Rogers et al. 1987
Lycopersicon hirsutum	*Spodoptera exigua*	Eigenbrode et al. 1996
f. *typicum*	*Liriomyza trifolii*	Hawthorne et al. 1992
Medicago spp.	*Hypera postica*	Shade et al. 1975, 1979 Ranger & Hower 2001
Solanum berthaultii	*Leptinotarsa decemlineata*	Casagrande 1982
S. neocardensaii	*Myzus persicae*	Gibson 1971
	Empoasca fabae	LaPointe &Tingey1986
	Epitrix cucumeris	Malakar & Tingey 2000 Neal et al. 1990 Tingey & Gibson 1978 Tingey & Laubengayer 1981 Tingey & Sinden 1982 Yencho et al. 1994
S. sisymbriifolium	*Conchyloctenia tigrina*	Hill et al. 1997
Hooked		
Phaseolus	Aphids	Gepp 1977
	Empoasca fabae	Sengonca & Gerlach 1984
	Thrips, *Bemisia* spp.	Pillemer & Tingey 1976, 1978
Simple		
Brassica	*Phyllotreta cruciferae*	Palaniswamy & Bodnaryk 1994
Glycine max	*Epilachna varivestis*	Gannon & Bach 1996
	Trialeurodes abutilonea	Lambert et al. 1995
	Atrachya menetriesi	Kanno 1996
	Etiella zinckenella	Trietsche 1994
Gossypium hirsutum	*Heliothis virescens*	Ramalho et al. 1984
Salix borealis	*Melasoma lapponica*	Zvereva et al. 1998
Tilia tomentosa	*Eotetranychus tiliarium*	Czajkowska & Kielkiewicz 2002
Triticum aestivum	*Oulema melanopus*	Wellso 1979 Papp et al. 1992
Vigna unguiculata	*Maruca testulalis*	Oghiakhe et al. 1992

Decreased trichome density is an effective resistance trait in commercial tomato cultivars to the silverleaf whitefly, *Bemisia argentifolii* Bellows & Perring (Heinz and Zalom 1995). Here trichome density, as opposed to glandular exudate, mediates resistance.

Glandular trichomes on leaves, stems, and reproductive structures of several *Medicago* species exude a sticky secretion when ruptured, composed of various aldehydes, alkanes, and esters (Triebe et al. 1981). The exudate entraps and kills larvae of the alfalfa weevil, *Hypera postica* Gyllenhal, (Kreitner and Sorensen 1979, Shade et al. 1975) and *E. fabae* nymphs (Shade et al. 1979). Elden and McCaslin (1997) examined the association between glandular trichome density and *E. fabae* resistance in alfalfa clones varying widely in glandular trichome density. With this plant material, neither trichome exudate nor physical entrapment of *E. fabae* nymphs was observed. However, Ranger and Hower (2001) described the morphology of both erect and procumbent alfalfa glandular trichomes in a clone of perennial alfalfa highly resistant to *E. fabae* and photographed exudate from procumbent trichomes trapping first instar nymphs (Figure 3.6).

Medicago gland exudate does not entrap larger arthropods but does decrease adult alfalfa weevil feeding and oviposition (Johnson et al. 1980a,b) and reduces populations of the alfalfa seed chalcid, *Bruchophagus roddi* (Gussakovsky) (Brewer et al. 1983). Dense masses of simple trichomes on wild *Medicago* species also limit population development of the spotted alfalfa aphid, *Therioaphis maculata* (Buckton) (Ferguson et al. 1982).

As mentioned in Section 2.1.2, oxalic acid has shown to be the principal exudate from trichomes on leaves and pods of chickpea cultivars resistant to *H. armigera* (Yoshida et al. 1997), and appears to be responsible for the bulk of the antibiosis effect of *H. armigera* resistant cultivars (Srivastava and Srivastava 1990). Glandular trichome exudates of clones from crosses between beach strawberry, *Fragaria chiloensis* (L.) Duchesne, and cultivated strawberry, *Fragaria* x *ananassa* Duchesne, are at least partially responsible for resistance to *T. urticae* (Luczynski et al. 1990). Exudates mechanically trap mites by forming a coating around mite tarsi, which limits fecundity and decreases survival.

Nonglandular trichomes also impart antibiotic effects to arthropods. Hooked trichomes on the foliage of the common bean and lima bean, *Phaseolus lunatus* L., impale *E. fabae* nymphs during movement on bean plant leaves (Figure 3.7) (Pillemer and Tingey 1976). Cultivars with a high density of hooked trichomes are much more resistant than those with low trichome density (Pillemer and Tingey 1978). Hooked bean trichomes also impale aphids, whiteflies (Sengonca and Gerlach 1984), leafminers (Quiring et al. 1992) and thrips (Gepp 1977).

Simple, non-glandular erect trichomes on the leaves and pods of cowpea and soybean plants also express antibiosis. The length and density of non-glandular trichomes on cowpea cultivars resistant to the legume pod borer, *Maruca testulalis* (Geyer) are negatively correlated to borer larval survival (Oghiakhe 1995, Oghiakhe et al. 1992). In the United States, densely pubescent leaves of soybean cause significantly greater larval mortality of the Mexican bean beetle, *Epilachna varivestis* Mulsant, than glaborous or normal-leafed plants (Gannon and Bach 1996). However, soybean cultivars with simple trichomes flattened against the leaf surface are more resistant to *Bemisia argentifolii* and the bandedwing whitefly, *Trialeurodes abutilonea* (Haldeman), than those with normal erect trichomes (Lambert et al. 1995). In Asian soybean production, cultivars with resistance to the false melon beetle, *Atrachya menetriesi* Falderman, have highly pubescent leaves covered with simple trichomes (Kanno 1996). Conversely, reduced trichome density contributes to oviposition resistance in soybean to the lima bean pod borer, *Etiella zinckenella* (Treitschke) (Talekar and Lin 1994).

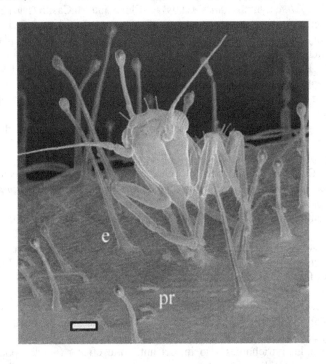

Figure 3.6. Erect (e) and procumbent (pr) alfalfa glandular trichomes surrounding a first instar Empoasca fabae, nymph trapped in trichome exudate on an Empoasca fabae-resistant clone. (Bar = 60 μm, x100. Reprinted from Ranger, C. M. and A. A. Hower. 2001. Glandular morphology from a perennial alfalfa clone resistant to the potato leafhopper. Crop Sci. 41: 1427-1434, Copyright 2001, the Crop Science Society of America, Inc., with permission of the Crop Science Society of America.

Increased densities of simple leaf trichomes are also responsible for resistance in boreal willow, *Salix borealis* (Fries.), to the leaf beetle, *Melasoma lapponica* L. (Zvereva et al. 1998); resistance in wild *Brassica* species to the flea beetle, *Phyllotreta cruciferae* (Goeze) (Palaniswamy and Bodnaryk 1994) and the resistance in the linden species *Tilia tomentosa* to the linden spider mite, *Eotetranychus tiliarium* (Herm.) (Czajkowska and Kielkiewicz 2002).

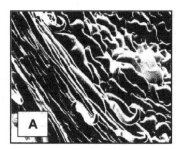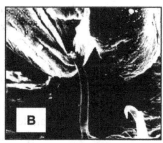

Figure 3.7. Hooked Phaseolus trichomes imparting antibiosis resistance to arthropods. (A) Procumbent hooked trichomes of P. lunatus, (B) Erect hooked trichomes of P. vulgaris impale nymphs of Empoasca fabae Harris. Reprinted from Pillemer, E. A. and W. M. Tingey. 1978. Hooked trichomes and resistance of Phaseolus vulgaris to Empoasca fabae (Harris). Entomol. Exp. Appl. 24:83-94, Copyright 1987 D. W. Junk Publ. Co., with kind permission of Springer Science and Business Media.

Simple trichomes on the foliage of wheat cultivars resistant to the cereal leaf beetle, *Oulema melanopus* (L.), also have antibiotic effects. Eggs deposited on the trichome field rising above the leaf surface suffer mortality due to desiccation and puncture by trichomes (Wellso 1979). *O. melanopus* larvae also die from punctures of the alimentary canal sustained after ingestion of trichome fragments (Wellso 1973). Papp et al. (1992) and Papp (1994) observed a strong negative correlation between increased trichome length and reduced *O. melanopus* feeding damage. High densities of simple leaf trichomes on pubescent cotton cultivars also increase the mortality of *H. virescens* larvae by impairing their movement over the leaf surface and increasing their susceptibility to predation (Ramalho et al. 1984).

Epicuticular lipids. Epicuticular lipids have antibiotic as well as antixenotic effects. Yang et al. (1992) demonstrated a link between maize antibiosis to foliage feeding Lepidoptera and the epicuticular lipids of resistant maize germplasm. *H. zea* larvae fed diet containing silk tissue of resistant plants from which the lipids has been extracted weigh more than larvae fed similar diets with unextracted silk tissue. Similar results were obtained when *S. frugiperda* larvae were fed diets containing extracted and unextracted maize leaf tissues (Yang et al. 1993). Hedin et al. (1993) identified a unique component of the surface lipids of maize lines resistant to *D. grandiosella* larvae as 2–hydroxy-4, 7–dimethoxy-1, 4–benzoxazin-3–one (N-O-Me-DIMBOA). N-O-Me-DIMBOA occurs at a higher concentration than either DIMBOA or 6–MBOA, leaf allelochemicals previously discussed in Section 2.1.1.

In addition to the antixenotic effects described in Chapter 2, Section 4.1.2, epicuticular lipids from certain azalea, *Azalea indica* L., genotypes also exhibit antibiotic effects in the azalea lace bug, *Stephanitis pyrioides* Scott. Increased concentrations of heptadecanoic acid, *n*-hentriacontane, oleanic acid and ursolic acid are highly correlated to reduced development of *S. pyrioides* eggs, nymphs and reduced nymph survival (Wang et al. 1999). Oleanic acid is the major epicuticular lipid of olive fruits and plays a major role in the resistance of olive to oviposition by the olive fruit fly, *Bactrocera oleae* (Gmelin) (Kombargi et al. (1998). The lipids α-amyrin, β-amyrin and tricontanol in susceptible azalea genotypes are also highly correlated to enhanced *S. pyrioides* survival.

Physical tissue strength. Additional physical barriers in the plant epidermis exert antibiotic effects on arthropods. Wheat stem sawflies of the genus *Cephus* damage wheat in northern regions of North America, Africa and in western Asia. Larvae feed in the hollow stems of susceptible plants, reducing nutrient flow to developing grain heads. Resistance was first reported in solid-stemmed wheat cultivars to *C. pygmaeus* (L.) and *C. tabidu* (F.) in Russia (Painter 1951). Platt and Farstad (1941) described how the mechanical barrier of the solid stem restricted feeding and growth of *C. cinctus* (Norton) larvae attempting to move between nodes of solid stemmed plants. Solid stemmed wheat cultivars with *Cephus* resistance are presently produced in North America and North Africa (Carlson et al. 1997, Miller et al. 1993, Morrill et al. 1992).

The leaf sheaths and stems of most graminaceous crop plants contain silica. The greatly increased content of silica-containing cells in plants of some rice cultivars is closely tied to their resistance to some species of stalk boring Lepidoptera. Djamin and Pathak (1967) determined that increased stem silica content significantly increased resistance of rice to the striped stem borer, *Chilo suppressalis* (Walker). Larvae feeding on cultivars with high stem silica content experience extreme mandibular abrasion during feeding, resulting in a loss of mandibular teeth (Figure 3.8).

Rice cultivars resistant to the African striped borer, *Chilo zacconius* Blesz., and the yellow rice borer, *Tryporyza incertulas* (Walker), also contain high levels of stem silica (Panda et al. 1975, Ukwungwu and Odebiyi 1985) that presumably affects larvae of these species in a similar manner. Rice cultivars resistant to the rice leaffolder, *Cnaphalocrocis medinalis* (Guenée), have closely arranged leaf silica cells that also cause excessive wear on the mandibles of leaffolder larvae (Ramachandran and Khan 1991). Moore (1984) demonstrated a positive correlation between cultivars of Italian ryegrass, *Lolium multiflorium* Lam., resistant to the frit fly, *Oscinella frit* (L.), and the density of costal silica bodies in ryegrass leaf sheaths. Increased silica content in maize, sorghum, and wheat cultivars also contributes to arthropod resistance in (Rojanaridpiched et al. 1984, Miller et al. 1960, Blum 1968).

Tissue toughness is also a mechanical component of antibiosis in maize to stalk boring Lepidoptera. Experiments conducted by Buendgen et al. (1990) and Beeghly et al. (1997) with maize inbred line populations have demonstrated significant correlations between *O. nubilalis* tunneling and the maize stalk fiber components neutral detergent fiber, acid detergent fiber and lignin. Beeghly et al. (1997) predicted that relatively small changes in stalk fiber content required for *O. nubilalis* resistance would have little impact on the value of maize silage for ruminant consumption.

Bergvinson et al. (1994a) detected a strong correlation between leaf toughness and leaf feeding damage by *O. nubilalis* larvae in several multiple borer resistant maize cultivars in Canada. In additional experiments with an *O. nubilalis* resistant maize synthetic line, Bergvinson et al. (1994b, 1995) found that leaf toughness, fiber content and epidermal cell wall absorbance of bound hydroxycinnamic acids such as (*E*)-ferulic acid or (*E*)-p-coumaric acid are strong indicators of *O. nubilalis* resistance. Increased (*E*)-ferulic acid content is linked to wheat resistance to the orange wheat blossom midge, *Sitodiplosis mosellana* (Gehin) (Abdel et al. 2001) and barley resistance to *R. padi* (Cabrera et al. 1995).

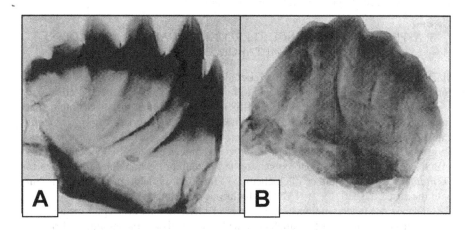

Figure 3.8. Abrasion of the incisor mandible region of Chilo suppressalis (Walker), larvae fed rice plants of differing silica silica content (A) Mandible of larva fed susceptible plant of normal silica content, (B) Mandible of larva fed resistant plant with high silica content. [from Djamin and Pathak (1967), reproduced with the permission of the Entomological Society of America]

Dowd and Norton (1995) investigated antibiosis in callus tissue of a maize line resistant to *H. zea* larvae. Older, browner callus is more resistant to larval feeding than younger callus and contains higher accumulations of ferulic and *p*-coumaric acid than younger callus. Total hydroxycinnamic acid levels in clones of Scots pine, *Pinus sylvestris* L., have also been closely linked to pine antibiosis to the sawfly, *Diprion pini* L. (Auger et al. 1991).

Maize resistance to *S. frugiperda* and *D. grandiosella* larvae also has a mechanical component. Davis et al. (1997) first noted that maize inbred lines with antibiosis to both arthropods have much thicker leaf cuticle and epidermal cell walls than susceptible inbred lines. These observations support data that demonstrate that this resistant germplasm has much greater fiber content than susceptible lines, as increased hemicellulose and hemicellulose crosslinking have been detected in maize tissues with *S. frugiperda* resistance (Hedin et al. 1996). Increased cellulose levels have been correlated to both *S. frugiperda* and *D. grandiosella* larval resistance (Williams et al. 1998).

Changes in plant fruiting structures. The development of pest arthropods may also be disrupted significantly by physically rearranging the plant fruiting structures through plant breeding. Sorghum cultivars resistant to *C. sorghicola* or the sorghum shootfly, *Atherigona soccata* Rondani, have relatively shorter florets than susceptible cultivars that disrupt the reproductive biology of both pests (Omari et al. 1983, Sharma et al. 1990). Sorghum antibiosis to *H. armigera* and the head bug, *Calocoris angustatus* Leth., occurs in cultivars with panicles that form more loosely than those of susceptible cultivars (Sharma et al. 1994).

3. AFFECTS OF RESISTANCE ALLELOCHEMICALS ON ARTHROPOD METABOLISM

Maize germplasm with resistance to Lepidoptera based on hydroxybenzoic acid content has been used as a tool to begin to understand some of the specific effects of resistance allelochemicals on arthropod metabolism. Yan et al. (1995) found that consumption of DIMBOA by larvae of the Asian corn borer, *Ostrinia furnacalis* Guenée, inhibits activity of acetycholinesterase and esterase detoxification enzymes. Feng et al. (1992a,b) evaluated both the *in vitro* and *in vivo* effects of DIMBOA and 6–MBOA on polysubstrate monooxygenase and glutathione s-transferase activities in *O. nubilalis* larvae. Neither allelochemical increased cytochrome P450 levels but both increased the activity of NADH-cytochrome c reductase.

Our meager understanding of the effects of resistance on arthropod metabolism is complicated by the fact that allelochemicals from resistant plants induce different types of arthropod enzymes in different ways. Dowd et al. (1983) observed contrasting levels of hydrolytic esterase activity in larvae of the cabbage looper, *Trichoplusia ni* (Hübner), and *P. includens* fed diets containing leaf extracts of *P. includens* resistant-and susceptible soybean cultivars. Esterase activity is reduced in midgut tissue of *P. includens* larvae fed diet containing leaf extract from resistant cultivars compared to those fed diets containing extracts of leaves from a susceptible cultivar. Esterase activity is greater than normal in *Trichoplusia ni* larvae, due presumably to a more diverse host range of this arthropod. *P. includens* larvae fed diets containing the isoflavone coumestrol, an allelochemical involved in looper resistance (Caballero et al. 1987) have lower rates of hydrolysis of the pyrethroid insecticide fenvalerate than those fed control diet (Dowd et al 1986), and suffer enhanced fenvalerate toxicity (Rose et al. 1988).

Predictions of the effects of allelochemicals from arthropod resistant plants with insecticides are difficult, since different detoxification enzymes may act on them and occur at different concentrations in resistant plants. Kennedy and Farrar (1987) compared the effect of 2–tridecanone (see Section 2.1.1), on fenvalerate-resistant and fenvalerate-susceptible *L. decemlineata* populations and found that both populations were equally affected. Ingestion of 2–tridecanone by *H. zea* larvae however, induces enhanced tolerance to carbaryl (Kennedy 1984).

REFERENCES CITED

Abdel, E.-S. M., P. Hucl, F. W. Sosulski, R. Graf, C. Gillot, and L. Pietrzak. 2001. Screening spring wheat for midge resistance in relation to ferulic acid content. J. Agric. Food Chem. 49:3559–3566.

Adkisson, P. L. 1962. Cotton stocks screened for resistance to the pink bollworm, 1960–1961. Tex. Agric. Exp. Stn. Misc. Publ. 606.

Alfaro, R. I. 1995. An induced defense reaction in white spruce to attack by the white pine weevil, *Pissodes strobi.* Can. J. Forest Res. 25:1725–1730.

Annan, I. B., W. M. Tingey, G. A. Schaefers, and K. N. Saxena. 1996. Reproductive performance and population dynamics of cowpea aphid (Homoptera: Aphididae) on leaf extracts of resistant and susceptible cowpeas. J. Chem. Ecol. 22:1345–1354.

Ahman, I., S. Tuvesson, and M. Johansson. 2000. Does indole alkaloid gramine confer resistance in barley to aphid *Rhopalosiphum padi*? J. Chem. Ecol. 26:233–255.

Argandona, V. H., J. G. Luza, H. M. Niemeyer, and L. J. Corcuera. 1980. Role of hydroxamic acids in the resistance of cereals to aphids. Phytochem. 19:1665–68.

Argandona, V. H., H. M. Niemeyer, and L. J. Corcuera. 1981. Effect of content and distribution of hydroxamic acids in wheat on infestation by the aphid *Schizaphis graminum*. Phytochem. 20:637–76.

Argandona, V. H., L. J. Corcuera, H. M. Niemeyer, and B. C. Campbell. 1983. Toxicity and feeding deterrency of hydroxamic acids from gramineae in synthetic diets against the green bug, *Schizaphis graminum.* Entomol. Exp. Appl. 34:134–138.

Assabgui, R. A., J. T. Arnason, and R. I. Hamilton. 1995. Field evaluations of hydroxamic acids and antibiosis factors in elite maize inbreds to the western corn rootworm (Coleoptera: Chrysomelidae). J. Econ. Entomol. 88:1482–1493.

Auclair, J. L., J. B. Maltais, and J. J. Cartier. 1957. Factors in resistance of peas to the pea aphid, *Acyrthosiphon pisum* (Harris) (Homoptera: Aphididae). II. Amino acids. Can. Entomol. 10:457–464.

Auger, Par M. A., C. Jay-Allemand, C. Bastien, and C. Geri. 1991. Foliage edibility of Scots pine clones for *Diprion pini* L. (Hym., Diprionidae). II. Relationships between the content of phenols within the needles and the mortality of *D. pini* larvae. J. Appl. Entomol. 111:78–85.

Balbyshev, N. F., and J. H. Lorenzen. 1997. Hypersensitivity and egg drop: A novel mechanism of host plant resistance to Colorado potato beetle (Coleoptera: Chrysomelidae). J. Econ. Entomol. 90:652–657.

Barria, B. N., S. V. Copaja, and H. M. Niemeyer. 1992. Occurrence of DIBOA in wild *Hordeum* species and its relation to aphid resistance. Phytochem. 31:89–91.

Barry, D., D. Alfaro, and L. L. Darrah. 1994. Relation of European corn borer (Lepidoptera: Pyralidae) leaf-feeding resistance and DIMBOA content in maize. Environ. Entomol. 23:177–182.

Beck, D. L., G. M. Dunn, D. G. Routley, and J. S. Bowman. 1983. Biochemical basis of resistance in corn to the corn leaf aphid. Crop Sci. 23:995–998.

Beeghly, H. H., J. G. Coors, and M. Lee. 1997. Plant fiber composition and resistance to European corn borer in four maize populations. Maydica. 42:297–303.

Bell, A. A., R. D. Stipanovic, C. R. Howell, and P. R. Fryxell. 1975. Antimicrobial terpenoids of Gossypium: Hemigossypol, 6-methoxygossypol and 6-deoxyhemigossypol. Phytochem. 14:225–234.

Bentur, J. S., and M. B. Kalode. 1996. Hypersensitive reaction and induced resistance in rice against the Asian rice gall midge *Orseolia oryzae*. Entomol. Exp. Appl. 78:77–81.

Bergvinson, D. J., J. T. Arnason, R. L. Hamilton, J. A. Mihm, and D. C. Jewell. 1994a. Determining leaf toughness and its role in maize resistance to the European corn borer (Lepidoptera: Pyralidae). J. Econ. Entomol. 87:743–1748.

Bergvinson, D. J., J. T. Arnason, and L. N. Pietrzak. 1994b. Localization and quantification of cell wall phenolics in European corn borer resistant and susceptible maize inbreds. Can. J. Bot. 72:243–1249.

Bergvinson, D. J., R. L. Hamilton, and J. T. Arnason. 1995. Leaf profile of maize resistance factors to European corn borer, *Ostrinia nubilalis*. J. Chem. Ecol. 21: 343–354.

Berlandier, F. A. 1996. Alkaloid level in narrow-leafed lupin, *Lupinus angustifolius*, influences green peach aphid reproductive performance. Entomol Exp. Appl. 79:19–24.

Blum, A. 1968. Anatomical phenomena in seedlings of sorghum varieties resistant to the sorghum shoot fly (*Atherigona varia soccata*). Crop Sci. 8:388–391.

Brewer, G. J., E. L. Sorensen, and E. K. Horber. 1983. Trichomes and field resistance of *Medicago* species to the alfalfa seed chalcid (Hymenoptera: Eurytomidae). Environ. Entomol. 12:247–252.

Buendgen, M. R., J. G. Coors, A. W. Grombacher, and W. A. Russell. 1990. Relationship between European corn borer resistance and plant cell wall composition in the BS9, WFISIHI, and WFISILO maize populations. Crop Sci. 30:505–510.

Caballero, P., C. M. Smith, F. R. Fronczek, and N. H. Fischer. 1987. Isoflavonoids from soybean with potential insecticidal activity. J. Nat. Prod. 49:1126–1129.

Cabrera, H. M., O. Munoz, G. E. Zuniga, L. J. Corcuera, and V. H. Argondona. 1995. Changes in ferulic acid and lipid content in aphid-infested barely. Phytochem. 39:1023–1029.

Calatayud, P. A., Y. Rahbé, B. Delobel, F. Khuong-Huu, M. Tertuliano, and B. Le Rü. 1994. Influence of secondary compounds in the phloem sap of cassava on expression of antibiosis towards the mealybug *Phenacoccus manihoti*. Entomol. Exp. Appl. 72:47–57.

Campbell, B. C., S. S. Duffey. 1981. Alleviation of α-tomatine-induced toxicity to the parasitoid, *Hyposoter exiguae*, by phytosterols in the diet of the host, *Heliothis zea*. J. Chem. Ecol. 7:927–946.

Carlson, G. R., P. L. Bruckner, J. E. Berg, G. D. Kushnak, D. M. Wichman, J. L. Eckhoff, K. A. Tilley, G. F. Stallknecht, R. N. Stougaard, H. H. Bowman, W. C. Morrill, G. A. Taylor, and E. A. Hockett, 1997. Registration of Vanguard wheat. Crop Sci. 37:291.

Casagrande, R. A. 1982. Colorado potato beetle resistance in a wild potato, *Solanum berthaultii*. J. Econ. Entomol. 75:368–372.

Castanera, P., J. C. Steffens, and W. M. Tingey. 1996. Biological performance of Colorado potato beetle larvae on potato genotypes with differing levels of polyphenol oxidase. J. Chem. Ecol. 22:91–101.

Ciepiela, A. 1989. Biochemical basis of winter wheat resistance to the grain aphid, *Sitobion avenae*. Entomol. Exp. Appl. 51:269–275.

Ciepiela, A. P., and C. Sempruch. 1999. Effect of L-3, 4-dihydroxyphenylalanine, ornithine and gamma-aminobutyric acid on winter wheat resistance to grain aphid. J. Appl. Entomol. 123:285–288.

Clement, S. L., D. C. Hardie, and L. R. Elberson. 2002. Variation among accessions of *Pisum fulvum* for resistance to pea weevil. Crop Sci. 42: 2167–2173.

Cole, R. A. 1984. Phenolic acids associated with the resistance of lettuce cultivars to the lettuce root aphid. Ann. Appl. Biol. 105:129–145.

Cole, R. A. 1985. Relationship between the concentration of chlorogenic acid in carrot roots and the incidence of carrot fly larval damage. Ann. Appl. Biol. 106: 211–217.

Cole, R. A. 1994. Locating a resistance mechanism to the cabbage aphid in two wild Brassicas. Entomol. Exp. Appl. 71:23–31.

Czajkowska, B., and M. Kielkiewicz. 2002. Linden-leaf morphology and the host-plant susceptibility to *Eotetranychus tiliarium* (Hermann) (Acaridae: Tetranychidae)., In F. Bernini, R. Nannelli, G. Nuzzaci, E. de Lillo (eds.) *Acarid Phylogeny and Evolution. Adaptations in mites and ticks.* Kluwer Academic Publ., Netherlands, pp. 435–440.

Davis, F. M., G. T. Baker, and W. P. Williams. 1997. Anatomical characteristics of maize resistance to leaf feeding by southwestern corn borer and fall armyworm. J. Agric. Entomol. 12:55–65.

Dimock, M. A., and G. G. Kennedy. 1983. The role of glandular trichomes in the resistance of *Lycopersicon hirsutum* f. *glabratum* to *Heliothis zea*. Entomol. Exp. Appl. 33:263–268.

Dimock, M. B., S. L. LaPointe, and W. M. Tingey. 1986. *Solanum neocardensaii*: A new source of potato resistance to the Colorado potato beetle (Coleoptera: Chrysomelidae). J. Econ. Entomol. 79:1269–1275.

Djamin, A., and M. D. Pathak. 1967. Role of silica in resistance to Asiatic borer (rice) *Chilo suppressalis* in rice varieties. J. Econ. Entomol. 60:347–351.

Doss, R. P., J. Oliver, W. M. Proebsting, S. W. Potter, S. Kuy, S. L. Clement, R. T. Williamson, J. R. Carney, and E. D. DeVilbiss. 2000. Bruchins: Insect-derived plant regulators that stimulate neoplasm formation. Proc. Natl. Acad. Sci. USA. 97:6218–6223.

Dowd, P. F., and R. A. Norton. 1995. Browning-associated mechanisms of resistance to insects in corn callus tissue. J. Chem. Ecol. 21:583–600.

Dowd, P. F., C. M. Smith, and T. C. Sparks. 1983. Influence of soybean leaf extracts on ester cleavage in cabbage and soybean loopers (Lepidoptera: Noctuidae). J. Econ. Entomol. 76:700–703.

Dowd, P. F., R. R. Rose, C. M. Smith, and T. C. Sparks. 1986. Influence of extracts from soybean (*Glycine max* (L.) Merr.) leaves on hydrolytic and glutathione s-transferase activity in the soybean looper (*Pseudoplusia includens* (Walker)). J. Agric. Food Chem. 34:444–447.

Dreyer, D. L., and B. C. Campbell. 1987. Chemical basis of host-plant resistance to aphids. Plant Cell Environ. 10:353–361.

Eigenbrode, S. D., J. T. Trumble, and K. K. White. 1996. Trichome exudates and resistance to beet armyworm (Lepidoptera: Noctuidae) in *Lycopersicon hirsutum* f. *typicum* accessions. Environ. Entomol. 25:90–95.

Elden, T. C., and M. McCaslin. 1997. Potato leafhopper (Homoptera: Cicadellidae) resistance in perennial glandular-haired alfalfa clones. J. Econ. Entomol. 90:842–847.

Elliger, C. A., D. F. Zinkel, B. G. Chan, and A. C. Waiss, Jr. 1976. Diterpene acids as larval growth inhibitors. Experientia 32:1364–1366.

Elliger, C. A., Y. Wong, B. G. Chan and A. C. Waiss, Jr. 1981. Growth inhibitors in tomato (*Lycopersicon*) to tomato fruitworm (*Heliothis zea*). J. Chem. Ecol. 7:753–758.

Farrar, R. R. Jr., and G. G. Kennedy. 1987. 2-Undecanone, a constituent of the glandular trichomes of *Lycopersicon hirsutum* f. *glabratum*: Effects on *Heliothis zea* and *Manduca sexta* growth and survival. Entomol. Exp. Appl. 43:17–23.

Farrar, Jr., R. R. and G. G. Kennedy. 1988. 2-Undecanone a pupal mortality factor in *Heliothis zea*: sensitive larval stage and *in planta* activity in *Lycopersicon hirsutum* f. *glabratum*. Entomol. Exp. Appl. 47:205–210.

Feng, R., J. G. Houseman, and A. E. R. Downe. 1992a. Effect of ingested meridic diet and corn leaves on midgut detoxification processes in the European corn borer, *Ostrinia nubilalis*. Pest Biochem. Physiol. 42:203–210.

Feng, R., J. G. Houseman, A. E. R. Downe, J. Atkinson, and J. T. Arnason. 1992b. Effects of 2, 4-Dihydroxy-7- methoxy-1, 4-benzoxazin-3-one (DIMBOA) and 6-methoxybenzoxazolinone (MBOA) on the detoxification processes in the larval midgut of the European corn borer. Pest Biochem. Physiol. 44:147–154.

Ferguson, S., E. L. Sorensen, and E. K. Horber. 1982. Resistance to the spotted alfalfa aphid (Homoptera: Aphididae) in glandular-haired *Medicago* species. Environ. Entomol. 11:1229–1232.

Gannon, A. J., and C. E. Bach. 1996. Effects of soybean trichome density on Mexican bean beetle (Coleoptera: Coccinellidae) development and feeding preference. Environ. Entomol. 25:1077–1082.

Garza, R., C. Cardona, and S. P. Singh. 2001. Hypersensitive response of beans to *Apion godmani* (Coleoptera: Curculionidae). J. Econ. Entomol. 94:958–962.

Gatehouse, A. M. R., J. A. Gatehouse, P. Dobie, A. M. Kilminster, and D. Boulter. 1979. Biochemical basis of insect resistance in *Vigna unguiculata*. J. Sci. Food Agric. 30:948–958.

Gerhold, D. L., R. Craig, and R. O. Mumma. 1984. Analysis of trichome exudate from mite-resistant geraniums. J. Chem. Ecol. 10:713–722.

Gepp, V. J. 1977. Hindrance of arthropods by trichomes of bean plants (*Phaseolus vulgaris* L.). Anz. Schaedlingskd. Pflanzenschutz Umweltschutz. 50:8–12.

Gibson, R. W. 1971. Glandular hairs providing resistance to aphids in certain wild potato species. Ann. Appl. Biol. 68:113–119.

Givovich, A., and H. M. Niemeyer. 1996. Role of hydroxamic acids in the resistance of wheat to the Russian wheat aphid, *Diuraphis noxia* (Mordvilko) (Homoptera, Aphididae). J. Appl. Entomol. 120:537–539.

Grazzini, R. A., D. Hesk, E. Yerger, D. Cox-Foster, J. Medford, R. Craig, and R. O. Mumma. 1995. Distribution of anacardic acids associated with small pest resistance among cultivars of *Pelargonium x hortorum*. J. Am. Hort. Soc. 120:343–346.

Greany, P. D., S. C. Styer, P. L. Davis, P. E. Shaw, and D. L. Chambers. 1983. Biochemical resistance of citrus to fruit flies. Demonstration and elucidation of resistance to the Caribbean fruit fly, *Anastrepha suspensa*. Entomol. Expl. Appl. 34:40–50.

Gregory, P., W. M. Tingey, D. A. Ave, and P. Y. Bouthyette. 1986a. Potato glandular trichomes: A physiochemical defense mechanism against insects. In: M. B. Green and P. A. Hedin (eds.). *Natural Resistance of Plants to Pests*. Role of Allelochemicals. ACS Symposium Series 296. American Chemical Society, Washington, DC. pp. 160–167.

Gregory, P., D. A. Ave, P. Y. Bouthyette, and W. M. Tingey. 1986b. Insect-defensive chemistry of potato glandular trichomes. In: B. E. Juniper and T. R. E. Southwood (eds.). *Insects and the Plant Surface*. Edward Arnold Ltd. London. pp. 181–191.

Guangjie, L., R. M. Wilkins, and R. C. Saxena. 1995. Utilization of sugars from susceptible and resistant rice varieties by the white-blacked planthopper, *Sogatella furcifera* (Horvath) (Homoptera: Delphacidae). Acta Entomol. Sinica. 38:421–427.

Guo, B. Z., N. W. Widstrom, B. R. Wiseman, M. E. Snook, R. E. Lynch, and D. Plaisted. 1999. Comparison of silk maysin, antibiosis to corn earworm larvae (Lepidoptera: Noctuidae), and silk browning in crosses of dent x sweet corn. J. Econ. Entomol. 92:746–753.

Gutiérrez, C., and P. Castañera. 1986. Efecto de los tejidos de maíz con alto y bajo contenido en DIMBOA sobre la biologia del taladro *Sesamia nonagrioides* Lef. (Lepidoptera: Noctuidae). Investigaciones Agrarias: Producción y Protección vegetal. 1:109–119.

Harris, P. 1960. Production of pine resin and its effects on survival of *Rhyacionia buoliana* (Schiff). Can. J. Zool. 38:121–130.

Harborne, J. B. 1982. Introduction to Ecological Biochemistry. Academic Press. New York. 278 pp.

Hawley, C. J., F. B. Peairs, and T. L. Randolph. 2003. Categories of resistance at different growth stages in Halt, a winter wheat resistant to the Russian wheat aphid (Homoptera: Aphididae). J. Econ. Entomol. 96:214–219.

Hawthorne, D. J., J. A. Shapiro, W. M. Tingey, and M. A. Mutschler. 1992. Trichome-borne and artificially applied acylsugars of wild tomato deter feeding and oviposition of the leafminer *Liriomyza trifolii*. Entomol. Exp. Appl. 65:65-73.

Hedin, P. A., W. L. Parrott, and J. N. Jenkins. 1992. Relationships of glands, cotton square terpenoid aldehydes, and other allelochemicals to larval growth of *Heliothis virescens* (Lepidoptera: Noctuidae). J. Econ. Entomol. 85:359–364.

Hedin, P. A., F. M. Davis, and W. P. Williams. 1993. 2-hydroxy-4,7- dimethoxy-1, 4-benzoxazin-3-one (N-)-ME-DIMBOA), a possible toxic factor in corn to the southwestern corn borer. J. Chem. Ecol. 19:531–542.

Hedin, P. A., F. M. Davis, W. P. Williams, R. P. Hicks, and T. H. Fisher. 1996. Hemicellulose is an important leaf-feeding resistance factor in corn to the fall armyworm. J. Chem. Ecol. 22:1655–1668.

Hein, G.L. 1992. Influence of plant growth on Russian wheat aphid, *Diuraphis noxia* (Homoptera: Aphididae). Reproduction and damage symptom expression. J. Kansas Entomol. Soc. 65:369–376.

Heinz, K. M., and F. G. Zalom. 1995. Variation in trichome-based resistance to *Bemisia argentifolii* (Homoptera: Aleyrodidae) oviposition on tomato. J. Econ. Entomol. 88:1494-1502.

Hill, M. P., P. E. Hulley, J. Allsopp, and G. Vanharmelen. 1997. Glandular trichomes on the exotic *Solanum sisymbriifolium* Lamarck (Solanaceae): Effective deterrents against an indigenous South African herbivore. Afr. Entomol. 5:41–50.

Hinds, W. E. 1906. Proliferation as a factor in the natural control of the Mexican cotton boll weevil. U. S. Dept. Agr. Bur. Ent. Bull. 59. 45p.

Hoch, W. A., E. L. Zeldin, and B. H. McCown. 2000. Resistance to the birch leafminer, *Fenusa pusilla* (Hymenoptera: Tenthredinidae) within the genus *Betula*. J. Econ. Entomol. 93:1810–1813.

Hodges, J. D., W. W. Elam, W. F. Watson, and T. E. Nebeker. 1979. Oleoresin characteristics and susceptibility of four southern pines to southern pine beetle (Coleoptera, Scolytidae) attacks. Can. Entomol. 111:889–896.

Holt, J., and N. Birch. 1984. Taxonomy, evolution and domestication of *Vicia* in relation to aphid resistance. Ann. Appl. Biol. 105:547–565.

Houseman, J., F. Campos, N. M. P. Thie, B. J. R. Philogène, J. Atkinson, P. Morand, and J. T. Arnason. 1992. Effect of the maize-derived compounds DIMBOA and MBOA on growth and digestive processes of European corn borer (Lepidoptera: Pyralidae). J. Econ. Entomol. 85:669–674.

Ishimoto, M., and K. Kitamura. 1989. Growth inhibitory effects of an α-amylase inhibitor from kidney bean, *Phaseolus vulgaris* (L.) on three species of bruchids (Coleoptera : Bruchidae). Appl. Entomol. Zool. 24:281–286.

Ishimoto, M., and K. Kitamura. 1992. Tolerance to the seed α-amylase inhibitor by the two insect pests of the common bean, *Zabrotes subfasciatus* and *Acanthoscelides obtectus* (Coleoptera: Bruchidae). Appl. Entomol Zool. 27:243–251.

Jacas, J. A., A. Garrido, C. Margaix, J. Forner, A. Alcaide, and J. A. Pina. 1997. Screening of different citrus rootstocks and citrus-related species for resistance to *Phyllocnistis citrella* (Lepidoptera: Gracillariidae). Crop Prot. 16:701–705.

James, D. G. 1992. Effect of citrus variety on oviposition, fecundity and longevity in *Biprorulus bibax* (Breddin) (Heteroptera). Acta Entomol. Bohemoslov. 89:65–67.

Johnson, K. J. R., E. L. Sorensen, and E. K. Horber. 1980a. Resistance in glandular haired annual *Medicago* species to feeding by adult alfalfa weevils. Environ. Entomol. 9:133–136.

Johnson, K. J. R., E. L. Sorensen, and E. K. Horber. 1980b. Resistance of glandular haired *Medicago* species to oviposition by alfalfa weevils (*Hypera postica*). Environ. Entomol. 9:241–245.

Kanehisa, K., H. Tsumuki, K. Kawada, and M. A. Rustamani. 1990. Relations of gramine contents and aphid populations on barley lines. Appl. Entomol. Zool. 25: 251–259.

Kanno, H. 1996. Role of leaf pubescence in soybean resistance to the false melon beetle, *Atrachya menetriesi* Falderman (Coleoptera: Chrysomelidae). Appl. Entomol. Zool. 31:597–603.

Kazemi, M. H. and H. F. van Emden. 1992. Partial antibiosis to *Rhopalosiphum padi* in wheat and some phytochemical correlations. Ann. Appl. Biol. 121:1–9.

Kennedy, G. G. 1984. 2-tridecanone, tomatoes, and *Heliothis zea*: Potential incompatibility of plant antibiosis with insecticidal control. Entomol. Exp. Appl. 35:305–311.

Kennedy, G. G., and R. R. Farrar, Jr. 1987. Response of insecticide- resistant and susceptible Colorado potato beetles, *Leptinotarsa decemlineata* to 2-tridecanone and resistant tomato foliage: the absence of cross-resistance. Entomol. Exp. Appl. 45:187–192.

Kennedy, G. G., and C. F. Sorenson. 1985. Role of glandular trichomes in the resistance of *Lycopersicon hirsutum* f. *glabratum* to Colorado potato beetle (Coleoptera: Chrysomelidae). J. Econ. Entomol. 78:547–551.

Kennedy, G. G., and R. T. Yamamoto. 1979. A toxic factor causing resistance in a wild tomato to the tobacco hornworm and some other insects. Entomol. Exp. Appl. 26:121–126.

Klun, J. A., C. L. Tipton, and T. A. Brindley. 1967. 2,4-Dihydroxy-7-methoxy-1, 4-benzoxazin-3-one (DIMBOA), an active agent in the resistance of maize to the European corn borer. J. Econ. Entomol. 60:1529–1533.

Klun, J. A., and J. F. Robinson. 1969. Concentration of two 1,4 benzoxazinones in dent corn at various stages of development and its relation to resistance of the host plant to the European corn borer. J. Econ. Entomol. 62:214–220.

Klun, J. A., W. D. Guthrie, A. R. Hallauer, and W. A. Russell. 1970. Genetic nature of the concentration of 2,4-dihydroxy -7-methoxy 2H-1, 4 benzoxazin-3 (4H)-one and resistance to the European corn borer in a diallel set of eleven maize inbreds. Crop Sci. 10:87–90.

Kombargi, W. S., S. E. Michelakis, and C. A. Petrakis. 1998. Effect of olive surface waxes on oviposition by *Bactocera oleae* (Diptera: Tephritidae). J. Econ. Entomol. 91:993–998.

Kowalski, S. P., N. T. Eannetta, A. T. Hirzel, and J. C. Steffens. 1992. Purification and characterization of polyphenol oxidase from glandular trichomes of *Solanum berthaultii*. Plant Physiol. 100:677–684.

Kreitner, G. L., and E. L. Sorensen. 1979. Glandular trichomes on *Medicago* species. Crop Sci. 19:380–384.

Kumar, H. 1993. Responses of *Chilo partellus* (Lepidoptera: Pyralidae) and *Busseola fusca* (Lepidoptera: Noctuidae) to hybrids of a resistant and susceptible maize. J. Econ. Entomol. 86:962–968.

Lage, J., B. Skovmand, and S. B. Andersen. 2003. Expression and suppression of resistance to greenbug (Homoptera: Aphididae) in synthetic hexaploid wheats derived from *Triticum dicoccum* x *Aegilops tauschii* crosses. J. Econ. Entomol. 96: 202–206.

Lambert, A. L., R. M. McPherson, and K. E. Espelie. 1995. Soybean host plant resistance mechanisms that alter abundance of whiteflies (Homoptera: Aleyrodidae). Environ. Entomol. 24:1381–1386.

LaPointe, S. L., and K. D. Bowman. 2002. Is there meaningful plant resistance to *Diaprepes abbreviatus* (Coleoptera: Curculionidae) in citrus rootstock germplasm? J. Econ. Entomol. 95:1059–1–65.

Lapointe, S. L., and W. M. Tingey. 1986. Glandular trichomes of *Solanum neocardenasii* confer resistance to green peach aphid (Homoptera: Aphididae). J. Econ. Entomol. 79:1264–1268.

Larsson, S., and D. R. Strong. 1992. Ovipositional choice and larval survival of *Dasineura marginementorquens* (Diptera: Cecidomyiidae) on resistant and susceptible *Salix viminalis*. Ecol. Entomol. 17:227–232.

Lattanzio, V., S. Arpaia, A. Cardinali, D. Di Venere, and V. Linsalata. 2000. Role of endogenous flavonoids in resistance mechanism of *Vigna* to aphids. J. Agric. Food Chem. 48:5316–5320.

Lin, S. Y. H., J. T. Trumble, and J. Kumanoto. 1987. Activity of volatile compounds in glandular trichomes of *Lycopersicon* species against two insect herbivores. J. Chem. Ecol. 13:837–850.

Long, B. J., G. M. Dunn, J. S. Bowman, and D. G. Routley. 1977. Relationship of hydroxamic acid content in corn and resistance to the corn leaf aphid. Crop Sci. 17:55–58.

Lorenzen, J. H., N. F. Balbyshev, A. M. Lafta, H. Casper, X. Tian, and B. Sagredo. 2001. Resistant potato selections contain leptine and inhibit development of the Colorado potato beetle (Coleoptera: Chrysomelidae). J. Econ. Entomol. 94:1260–1267.

Lowe, H. B. J. 1984. Characteristics of resistance to grain aphid *Sitobion avenae* in winter wheat. Ann. Appl. Biol. 105:529–538.

Luczynski, A., M. B. Isman, D. A. Raworth, and C. K. Chan. 1990. Chemical and morphological factors of resistance against the twospotted spider might in beach strawberry. J. Econ. Entomol. 83:564–569.

Lukefahr, M. J., and D. F. Martin. 1966. Cotton-plant pigments as a source of resistance to the bollworm and tobacco budworm. J. Econ. Entomol. 59:176–179.

Lukefahr, M. J., T. N. Shaver, D. E. Cruhm, and J. E. Houghtaling. 1974. Location, transference, and recovery of growth inhibition factor present in three *Gossypium hirsutum* race stocks. Proc. Beltwide Cotton Prod. Res. Conf. p. 93–95.

Macedo, M. L.R., K. V. S. Fernandes, M. P. Sales, and J. Xavier-Filho. 1995. Purification and some properties of storage proteins (vicilins) from cowpea (*Vigna unguiculata*) seeds which are susceptible and resistant to the bruchid beetle *Callosobruchus maculatus*. *Brazil. J. Med. Biol. Res.* 28:183–190.

Malakar, R., and W. M. Tingey. 2000. Glandular trichomes of *Solanum berthaultii* and its hybrid with potato deter oviposition and impair growth of potato tuber moth worm. Entomol. Exp. Appl. 94–249–257.

Mansour, F., Z. Shain, Z. Karchi, and U. Gerson. 1994. Resistance of selected melon lines to the carmine spider mite *Tetranychus cinnabarinus* (Acari: Tetranychidae) – field and laboratory experiments. Bull. Ent. Res. 80:345–347.

Mayoral, A. M., W. F. Tjallingii and P. Castanera. 1996. Probing behaviour of *Diuraphis noxia* on five cereal species with different hydroxamic acid levels. Entomol. Exp. Appl. 78:341–348.

Mebrahtu, T., M. Kraemer, and T. Andebrhan. 2002. Evaluation of soybean breeding lines for corn earworm antibiosis. Crop Sci. 42:1465–1470.

Meckenstock, D. H., M. T. Castro, H. N. Pitre, and F. Gomez. 1991. Antibiosis to fall armyworm in Honduran landrace sorghum. Environ. Entomol. 20: 1259–1266.

Meisner, J., M. Zur, E. Kabonci, and K. R. S. Ascher. 1977. Influence of gossypol content of leaves of different cotton strains on the development of *Spodoptera littoralis* larvae. J. Econ. Entomol. 70:714–716.

Miller, B. S., R. J. Robinson, J. A. Johnson, E. T. Jones, and B. W. X. Ponnaiya. 1960. Studies on the relation between silica in wheat plants and resistance to Hessian fly attack. J. Econ. Entomol. 53:945–949.

Miller, R. H., S. El Masri, and K. Al Jundi. 1993. Plant density and wheat stem sawfly (Hymenoptera: Cephidae) resistance in Syrian wheats. Bull. Ent. Res. 83:95–102.

Minney, B. H. P., A. M. R. Gatehouse, P. Dobie, J. Dendy, C. Cardona, and J. A. Gatehouse. 1990. Biochemical bases of seed resistant to *Zabrotes subfasciatus* (bean weevil) in *Phaseolus vulgaris* (common bean), a mechanism for arcelin toxicity. J. Insect Physiol. 36:757–767.

Mohan, P., P. Singh, S. S. Narayanan, and R. Ratan. 1994. Relation of gossypol- gland density with bollworm incidence and yield in tree cotton (Gossypium arboretum). Indian J. Agr. Sci. 64:691–696.

Moharramipour, S. H. Tsumuki, K. Sato, S. Murata and K. Kanehisa. 1997. Effects of leaf color, epicuticular wax amount and gramine content in barley hybrids on cereal aphid populations. Appl. Entomol. Zool. 32:1–8.

Mollema, C., and R. A. Cole. 1996. Low aromatic amino acid concentrations in leaf proteins determine resistance to *Frankliniella occidentalis* in four vegetable crops. Entomol. Exp. Appl. 78:325–333.

Moore, D. 1984. The role of silica in protecting Italian ryegrass (*Lolium multiflorum*) from attack by stem-boring larvae (e. g. *Oscinella frit*) and related species. Ann. Appl. Biol. 104:161–166.

Moraes, R. A., M. P. Sales, M. S. P. Pinto, L. B. Silva, A. E. A. Oliveira, O. L. T. Machado, K. V. S. Fernandes and J. Xavier-Filho. 2000. Lima bean (*Phaseolus lunatus*) seed coat phaseolin is detrimental to the cowpea weevil (*Callosobruchus maculatus*). Brazil. J. Med. Biol. Res. 33:191–198.

Morrill, W. L., J. W. Gabor, E. A. Hockett, and G. D. Kushnak. 1992. Wheat stem sawfly (Hymenoptera: Cephidae): damage and detection. J. Econ. Entomol. 85:2413–2417.

Neal, J. J., J. C. Steffens and W. M. Tingey. 1989. Glandular trichomes of *Solanum berthaultii* and resistance of the Colorado potato beetle. Entomol. Exp. Appl. 51:133–140.

Neal, J. J., W. M. Tingey, and J. C. Steffens. 1990. Sucrose esters of carboxcyclic acids glandular trichomes of *Solanum berthaultii* deter settling and probing by the green peach aphid. J. Chem. Ecol. 16:487–497.

Oatman, E. R. 1959. Host range studies of the melon leaf miner, *Liriomyza pictella* (Thompson). Ann. Entomol. Soc. Am. 52:739–741.

Oghiakhe, S. 1995. Effect of pubescence in cowpea resistance to the legume pod *borer Maruca testulalis* (Lepidoptera: Pyralidae). Crop Prot. 14:379–387.

Oghiakhe, S., L. E. N. Jackai, W. A. Makanjuola, and C. J. Hodgson. 1992. Morphology, distribution, and the role of trichomes in cowpea (*Vigna unguiculata*) resistance to the legume pod borer, *Maruca testulalis* (Lepidoptera, Pyralidae). Bull. Entomol. Res. 82:499–505.

Okello-Ekochu, E. J. and R. M. Wilkins. 1996. Influence of some maize cultivars/lines (*Zea mays* L.) on biological characters of the African armyworm, *Spodoptera exempta* (Walker) (Lepidoptera:Noctuidae). Internat. J. Pest Management 42:81–87.

Olatunde, G. O., and J. A. Odebiyi. 1991. Some aspects of antibiosis in cowpeas resistant to *Clavigralla tomentosicollis* Stal. (Hemiptera: Coreidae) in Nigeria. Tropical Pest Man. 37:273–276.

Omari, T., B. L. Agrawal, and L. R. House. 1983. Componental analyses of the factors influencing shoot fly resistance in sorghum (*Sorghum bicolor* (L.) Moench.). J. Appl. Res. Quarter. 17:215–218.

Ordas, A. Butron, P. Soengas, A. Ordas, and R. A. Malvar. 2002. Antibiosis of the pith of maize to *Sesamia nonagrioides* (Lepidoptera: Noctuidae). J. Econ. Entomol. 95:1044–1048.

Ortego, F., M. Ruiz, and P. Castanera. 1998. Effect of DIMBOA on growth and digestive physiology of *Sesamia nonagrioides* (Lepidoptera: Noctuidae) larvae. J. Insect Physiol. 44:95–101.

Osborn, T. C., T. Blake, P. Gepts and F. A. Bliss. 1986. Bean arcelin 2. Genetic variation, inheritance and linkage relationships of a novel seed protein of *Phaseolus vulgaris* L. Theor. Appl. Genet. 71:847–855.

Painter, R. H. 1951. *Insect Resistance in Crop Plants*. University of Kansas Press. Lawrence, KS. 521pp.

Palaniswamy, P., and R. P. Bodnaryk. 1994. A wild *Brassica* from Sicily provides trichome-based resistance against flea beetles, *Phyllotreta cruciferae* (Goeze) (Coleoptera: Chrysomelidae). Can. Entomol. 126:1119–1130.

Panda, N., B. Pradhan, A. P. Samalo, and P. S. P. Rao. 1975. Note on the relationship of some biochemical factors with the resistance in rice varieties to yellow rice borer. Indian J. Agric. Sci. 45:499–501.

Papp, M. 1994. Resistance to cereal leaf beetle and bird cherry-oat aphid in winter wheat. Genet. Pol. 35B:105–117.

Papp, M., J. Kolarov and A. Mesterhazy. 1992. Relation between pubescence of seedling and flag leaves of winter wheat and its significance in breeding resistance to the cereal leaf beetle (Coleoptera: Chrysomelidae). Environ. Entomol. 21:700–705.

Penny, L. H., G. E. Scott, and W. D. Guthrie. 1967. Recurrent selection for European corn borer resistance in maize. Crop Sci. 7:407–409.

Pillemer, E. A., and W. M. Tingey. 1976. Hooked trichomes: a physical barrier to a major agricultural pest. Science. 193:482–484.

Pillemer, E. A., and W. M. Tingey. 1978. Hooked trichomes and resistance of *Phaseolus vulgaris* to *Empoasca fabae* (Harris). Entomol. Exp. Appl. 24:83–94.

Platt, A.W., and C. W. Farstad. 1941. The resistance of crop plants to insect attack. Dominion Expt. Sta. Swift Current Saskat. 1–37.

Quiring, D. T., P. R. Timmins, and S. J. Park. 1992. Effect of variations in hooked trichome densities of *Phaseolus vulgaris* on longevity of *Liriomyza trifolii* (Diptera: Agromyzidae) adults. Environ. Entomol. 21:1357–1361.

Ramachandran, R., and Z. R. Khan. 1991. Mechanisms of resistance in wild rice *Oryza brachyantha* to rice leaffolder *Cnaphalacrocis medinalis* (Guenee) (Lepidoptera: Pyralidae). J. Chem. Ecol. 17:41–65.

Ramalho, F. S., W. L. Parrott, J. N. Jenkins, and J. C. McCarty, Jr. 1984. Effects of cotton leaf trichomes on the mobility of newly hatched tobacco budworms (Lepidoptera: Noctuidae). J. Econ. Entomol. 77:619–621.

Raman, K. V., W. M. Tingey, and P. Gregory. 1978. Potato glycoalkaloids: Effect on survival and feeding behavior of the potato leafhopper. J. Econ. Entomol. 72:337–341.

Ranger, C. M., and A. A. Hower. 2001. Glandular morphology from a perennial alfalfa clone resistant to the potato leafhopper. Crop Sci. 41: 1427–1434.

Rojanaridpiched, C. V. E. Gracen, H. L. Everett, J. C. Coors, B. F. Pugh, and P. Bouthyette. 1984. Multiple factor resistance in maize to European corn borer. Maydica 29:305–315.

Rogers, C. E., J. Gershenzon, N. Ohno, T. J. Mabry, R. D. Stipanovic, and G. L. Kreitner. 1987. Terpenes of wild sunflowers *(Heliothis)*: An effective mechanism against seed predation by larvae of the sunflower moth, *Homoeosoma electellum* (Lepidoptera: Pyralidae). Environ. Entomol. 16:586–592.

Rose, R. L., T. C. Sparks, and C. M. Smith. 1988. Insecticide toxicity to larvae of the soybean looper and the velvetbean caterpillar (Lepidoptera: Noctuidae) as influenced by feeding on resistant soybean (PI227687) leaves and coumestrol. J. Econ. Entomol. 81:1288–1294.

Ryan, J. D., P. Gregory, and W. M. Tingey. 1982. Phenolic oxidase activities in glandular trichomes of *Solanum berthaultii*. Phytochem. 21:1885–1887.

Ryan, J. D., P. Gregory, and W. M. Tingey. 1983. Glandular trichomes: Enzymatic browning assays for improved selection of resistance to the green peach aphid. Am. Potato J. 61:861–868.

Sahota, T. S., J. F. Manville, F. G. Peet, E. E. White, A. I. Ibaraki, and J. R. Nault. 1998. Resistance against white pine weevil: Effects on weevil, reproduction and host finding. Can. Entomol. 130:337–347.

Sales, M. P., M. L. R. Macedo, and J. Xavier-Filho. 1992. Digestibility of cowpea (*Vigna unguiculata*) vicilins by pepsin, papain and bruchid midgut proteinases. Comp. Biochem. Physiol. 103B: 945–950.

Sengonca, C. and S. Gerlach. 1984. The influence of leaf surface features on the effectiveness of the predatory thrips, *Scholothrips longicornis* (Thysanoptera: Thripidae). Entomophaga. 29:55–61.

Shade, R. E., T. E. Thompson, and W. R. Campbell. 1975. An alfalfa weevil resistance mechanism detected in *Medicago*. J. Econ. Entomol. 68:399–404.

Shade, R. E., M. J. Doskocil, and N. P. Maxon. 1979. Potato leafhopper resistance in glandular-haired alfalfa species. Crop Sci. 19:287–289.

Shapiro, A. M., and J. E. DeVay. 1987. Hypersensitivity reaction of *Brassica nigra* L. (Cruciferae) kills eggs of *Pieris* butterflies (Lepidoptera: Pieridae). Oecologia. 71:631–632.

Sharma, H. C., P. Vidyasagar, and K. Leuschner. 1990. Componential analysis of the factors influencing resistance to sorghum midge, *Contarinia sorghicola* Coq. Insect Sci. Applic. 11:889–898.

Sharma, H. C., P. Vidyasagar, and V. Subramanian. 1993. Antibiosis component of resistance in sorghum to sorghum midge, *Contarinia sorghicola*. Ann. Appl. Biol. 123:469–483.

Sharma, H. C., V. F. Lopez, and P. Vidyasagar. 1994. Influence of panicle compactness and host plant resistance in sequential plantings on population increase of panicle-feeding insects in *Sorghum bicolor* (L.) Moench. Internat. J. Pest Management. 40:216–221.

Sikinyi, E., D. J. Hannapel, P. M. Imerman, and H. M. Stahr. 1997. Novel mechanism for resistance to Colorado potato beetle (Coleoptera: Chrysomelidae) in wild *Solanum* species. J. Econ. Entomol. 90:689–696.

Sinden, S. L., L. L. Sanford, W. W. Cantelo, and K. L. Deahl. 1986. Leptine glycoalkaloids and resistance to the Colorado potato beetle (Coleoptera: Chrysomelidae) in *Solanum chalcoense*. Environ. Entomol. 15:1057–1062.

Sinden, S. L., L. L. Sanford, W. W. Cantelo, and K. L. Deahl. 1991. Allelochemically mediated host resistance to the Colorado potato beetle, *Leptinotarsa decemlineata* (Say)(Coleoptera: Chrysomelidae). Mem. Ent. Soc. Can. 157:19–28.

Sinden, S. L., J. M. Schalk, and A. K. Stoner. 1978. Effects of daylength and maturity of tomato plants on tomatine content and resistance to the Colorado potato beetle. J. Amer. Soc. Hort. Sci. 103:596–600.

Slesak, E., M. Slesak, and B. Gabrys. 2001. Effect of methyl jasmonate on hydroxamic acid, protease activity, and bird cherry-oat aphid *Rhopalosiphum padi* L. probing behavior. J. Chem. Ecol. 12:2529–2543.

Smith, C. M., D. J. Schotzko, R. S. Zemetra, E. J. Souza, and S. Schroeder-Teeter. 1991. Identification of Russian wheat aphid (Homoptera: Aphididae) resistance in wheat. J. Econ. Entomol. 84:328–332.

Smith, C. M., and S. Starkey. 2003. Resistance to greenbug (Heteroptera: Aphididae) biotype I in *Aegilops tauschii* synthetic wheats. J. Econ. Entomol. 96:1571–1576.

Sogawa, K., and M. D. Pathak. 1970. Mechanisms of brown planthopper resistance in Mudgo variety of rice (Hemiptera: Delphacidae). Appl. Entomol. Zool. 5:145–158.

Srivastava, C. P., and R. P. Srivastava. 1990. Antibiosis in chickpea (*Cicer arietinum* L.) to gram pod borer, *Heliothis armigera* (Hübner) (Noctuidae: Lepidoptera) in India. Entomon. 15:89–93.

Stevenson, P.C. 1993. Biochemical resistance in wild species of groundnut (*Arachis*) to *Spodoptera litura* (Lepidoptera: Noctuidae). Bull. OILB/SROP 16:155–162.

Stevenson, P.C., W. M. Blaney, M. J. S. Simmonds, and J. Wightman. 1993a. The identification and characterization of resistance in wild species of *Arachis* to *Spodoptera litura* (Lepidoptera: Noctuidae). Bull. Entomol. Res. 83:421–429.

Stevenson, P. C., J. C. Anderson, W. M. Blaney, and M. J. S. Simmonds. 1993b. Developmental inhibition of *Spodoptera litura* (Fab.) larvae by a novel caffeoylquinic acid from the wild groundnut, *Arachis paraguariensis* (Chod et Hassl.). J. Chem. Ecol. 19:2917–2933.

Stipanovic, R. D., A. A. Bell, D. H. O'Brien, and M. J. Lukefahr. 1977. Heliocide H2: An insecticidal sesquiterpenoid from cotton (*Gossypium*). Tetrahedron Lett. 6:567–570.

Stipanovic, R. D., H. J. Williams, and L. A. Smith. 1986. Cotton terpenoid inhibition of *Heliothis virescens* development. In: M. B. Green and P. A. Hedin (eds.). Natural Resistance of Plants to Pests. ACS Symposium Series 296. American Chemical Society. Washington, DC. pp. 79–94.

Sutherland, O. R. W., R. F. N. Hutchins, and W. F. Greenfield. 1982. Effects of lucerne saponins and *Lotus* condensed tannins on survival of grass grubs, *Costelytra zealandica*. N. Z. J. Zool. 9:511–514.

Tadmor, Y., E. Lewinsohn, F. AboMoch, A. BarZur, and F. Mansour. 1999. Antibiosis of maize inbred lines to the carmine spider mite, *Tetranychus cinnabarinus*. Phytoparasitica. 27:35–41.

Talekar, N. S., and C. P. Lin. 1992. Characterization of *Callosobruchus chinensis* (Coleoptera: Bruchidae) resistance in mungbean. J. Econ. Entomol. 85:1150–1153.

Talekar, N. S., and C. P. Lin. 1994. Characterization of resistance to limabean pod borer (Lepidoptera: Pyralidae) in soybean. J. Econ. Entomol. 87:821–825.

Telang, A., J. Sandstrom, E. Dyreson, and N. A. Moran. 1999. Feeding damage by *Diuraphis noxia* results in a nutritionally enhanced phloem diet. Entomol. Exp. Appl. 91:403–412.

Thackray, D.J., S. D. Wratten, P. J. Edwards, and H. M. Niemeyer. 1990. Resistance to the aphids *Sitobion avenae* and *Rhopalosiphum padi* in gramineae in relation to hydroxamic acid levels. Ann. Appl. Biol. 116: 573–582.

Tingey, W. M. 1991. Potato glandular trichomes defensive activity against insect attack. In: P. A. Hedin (Ed.), Naturally Occurring Pest Bioregulators, ACS Symposium Series 449, American Chemical Society, Washington, DC pp. 126–135.

Tingey, W. M., and R. W. Gibson. 1978. Feeding and mobility of the potato leafhopper impaired by glandular trichomes of *Solanum berthaultii* and *S. polyadenium*. J. Econ. Entomol. 71:856–858

Tingey, W. M., and J. E. Laubengayer. 1981. Defense against the green peach aphid and potato leafhopper by glandular trichomes of *Solanum berthaultii*. J. Econ. Entomol. 74:721–725.

Tingey, W. M., and S. L. Sinden. 1982. Glandular pubescence, glycoalkaloid composition and resistance to the green peach aphid, potato leafhopper, and potato flea beetle in *Solanum berthaultii*. Am. Potato J. 59:95–106.

Tomlin, E. S. ,and J. H. Borden. 1994. Relationship between leader morphology and resistance or susceptibility of Sitka spruce to the white pine weevil. Can. J. For. Res. 24:810–816.

Tomlin, E. S., J. H. Borden, and H. D. Pierce. 1996. Relationship between cortical resin acids and resistance of Sitka spruce to the white pine weevil. Can. J. Bot. 74:599–606.

Tonet, G. L., and R. F. P. DaSilva. 1994. Antibiosis of wheat genotypes to C-biotypes of *Schizaphis graminum* (Rondani, 1852) (Homoptera, Aphididae). Pesquisa Agr. Brasil. 29:1181–1186.

Triebe, D. C., C. E. Meloan, and E. L. Sorensen. 1981. The chemical identification of the glandular hair exudate for *Medicago scutellata*. 27th Alfalfa Improvement Conference. ARM-NC-19. p. 52.

Ukwungwu, M. N. and J. A. Odebiyi. 1985. Incidence of *Chilo zacconius* Bleszynski on some rice varieties in relation to plant characters. Insect Sci. Applic. 6: 653–656.

van Helden, M., W. F. Tjallingii , and F. L. Dieleman. 1993. The resistance of lettuce (*Lactuca sativa* L.) to *Nasonovia ribisnigri:* bionomics of *N. ribisnigri* on near isogenic lettuce lines. Entomol. Exp. Appl. 66:53–58.

Waiss, A. C, Jr., B. G. Chan, C. A. Elliger, B. R. Wiseman, W. W. McMillian, N. W. Widstrom, M. S. Zuber, and A. J. Keaster. 1979. Maysin, a flavone glycoside from corn silks with antibiotic activity toward corn earworm. J. Econ. Entomol. 72:256–258.

Wang ,Y., K. Braman, C. D. Robacker, and J. G. Latimer. 1999. Composition and variability of epicuticular lipids of azaleas and their relationship to azalea lace bug resistance. J. Amer. Soc. Hort. Sci. 124:239–244.

Webster, J. A., and D. R. Porter. 2000. Plant resistance components of two greenbug (Homoptera: Aphididae) resistant wheats. J. Econ. Entomol. 93:1000–1004.

Weibull, J. 1994. Glutamic acid content of phloem sap is not a good predictor of plant resistance to *Rhopalosiphum padi*. Phytochem. 35:601–602.

Wellso, S. G. 1973. Cereal leaf beetle: larval feeding, orientation, development, and survival on four small-grain cultivars in the laboratory. Ann. Entomol. Soc. Am. 66: 1201–1208.

Wellso, S. G. 1979. Cereal leaf beetle: Interaction with and ovipositional adaptation to a resistant wheat. Environ. Entomol. 8:454–457.

Williams, W. G., G. G. Kennedy, R. T. Yamamoto, J. D. Thacker, and J. Bordner. 1980. 2-tridecanone: A naturally occurring insecticide from the wild tomato *Lycopersicon hirsutum* f. *glabratum*. Science. 207:888–889.

Williams, W. P., F. M. Davis, P. M. Buckley, P. A. Hedin, G. T. Baker, and D. S. Luthe. 1998. Factors associated with resistance to fall armyworm (Lepidoptera: Noctuidae) and Southwestern corn borer (Lepidoptera: Crambidae) in corn at different vegetative stages. J. Econ. Entomol. 91:1471–1480.

Williams, W. P., and F. M. Davis. 1997. Mechanisms and bases of resistance in maize to southwestern corn borer and fall armyworm. In: J. A. Mihm (Ed.), Insect Resistance in Maize: Recent Advances and Utilization. Proceeds Intl. Symp., Intl. Maize & Wheat Impr. Ctr., CIMMYT, Mexico 27. Nov-Dec 1994. CIMMYT, Mex. D. F.

Wink, M. 1993. The role of quinolizidine alkaloids in plant-insect interactions. In: E. A. Bernays (Ed.), Insect-Plant Interactions, Volume IV. CRC Press, Boca Raton. pp. 131–166.

Wiseman, B. R., and M. E. Snook. 1995. Effect of corn silk age on flavone content and development of corn earworm (Lepidoptera: Noctuidae) larvae. J. Econ. Entomol. 88:1795–1800.

Wiseman, B. R., R. C. Gueldner, R. E. Lynch, and R. F. Severson. 1990. Biochemical activity of centipedegrass against fall armyworm larvae. J. Chem. Ecol. 16:2677–2690.

Wiseman, B. R., M. E. Snook, R. L. Wilson, and D. J. Isenhour. 1992. Allelochemical content of selected popcorn silks - Effects on growth of corn earworm larvae (Lepidoptera, Noctuidae). J. Econ. Entomol. 85:2500–2504.

Xavier-Filho, J., M. P. Sales, K. V. S. Fernandes, and V. M. Gomes. 1996. The resistance of cowpea (*Vigna unguiculata*) seeds to the cowpea weevil (*Callosobruchus maculatus*) is due to the association of variant vicilins (7S storage proteins) to chitinous structures in the insect's midgut. Arq. Biol. Technol. 39:693–699.

Xie, Y., J. T. Arnason, B. J. R. Philogene, H. T. Olechowski, and R. I. Hamilton. 1992. Variation of hydroxamic acid content in maize roots in relation to geographic origin of maize germplasm and resistance to western corn rootworm (Coleoptera: Chrysomelidae). J. Econ. Entomol. 85:2478–2485.

Yan, F. M., C. G. Xu, S. G. Li, C. S. Lin, and J. H. Li. 1995. Effects of DIMBOA on several enzymatic systems in Asian corn borer, *Ostrinia furnacalis* (Guenee). J. Chem. Ecol. 21:2047–2056.

Yang, G., B. R. Wiseman, and K. E. Espelie. 1992. Cuticular lipids from silks of seven corn genotypes and their effect on development of corn earworm larvae [(*Helicoverpa zea* (Boddie)]. J. Agri. Food Chem. 40:1058–1061.

Yang, G., B. R. Wiseman, D. J. Isenhour, and K. E. Espelie. 1993. Chemical and ultrastructural analysis of corn cuticular lipids and their effect on feeding by fall armyworm larvae. J. Chem. Ecol. 19:2055–2074.

Yencho, G. C., and W. M. Tingey. 1994. Glandular trichomes of *Solanum berthaultii* alter host preference of the Colorado potato beetle, *Leptinotarsa decemlineata*. Entomol. Exp. Appl. 70:217–225.

Yencho, G. C., J. A. A. Renwick, J. C. Steffens, and W. M. Tingey. 1994. Leaf surface extracts of *Solanum berthaultii* Hawkes deter Colorado potato beetle feeding. J. Chem. Ecol. 20:991–1007.

Yoshida, M., S. E. Cowgill, and J. A. Wightman. 1995. Mechanism of resistance to *Helicoverpa armigera* (Lepidoptera: Noctuidae) in chickpea: The role of oxalic acid in leaf exudate as an antibiotic factor. J. Econ. Entomol. 88:1783–1786.

Yoshida, M., S. E. Cowgill, and J. A. Wightman. 1997. Roles of oxalic and malic acids in chickpea trichome exudates in host-plant resistance to *Helicoverpa armigera*. J. Chem. Ecol. 23:1195–1210.

Zuniga, G. E., and L. J. Corcuera. 1986. Effect of gramine in the resistance of barley seedlings to the aphid *Rhopalosiphum padi*. Entomol. Exp. Appl. 40:259–262.

Zuniga, G. E., V. H. Argandona, H. M. Niemeyer, and L. J. Corcuera. 1983. Hydroxamic acid content in wild and cultivated Gramineae. Phytochem. 22:2665–2668.

Zuniga, G. E., M. S. Salgado, and L. J. Corcuera. 1985. Role of an indole alkaloid in the resistance of barley seedlings to aphids. Phytochem. 24:945–947.

Zuniga, G. E., E. M. Varanda, and L. J. Corcuera. 1988. Effect of gramine on the feeding behavior of the aphids *Schizaphis graminum* and *Rhopalosiphum padi*. Entomol. Exp. Appl. 47:161–165.

Zvereva, E. L., M. V. Kozlov, and P. Niemela. 1998. Effects of leaf pubescence in *Salix borealis* on host-plant choice and feeding behavior of the leaf beetle, *Melasoma lapponica*. Entomol. Exp. Appl. 89:297–303.

CHAPTER 4

TOLERANCE - THE EFFECT OF PLANT GROWTH CHARACTERS ON ARTHROPOD RESISTANCE

1. TOLERANCE

Plants may also be resistant to arthropods via the tolerance category of resistance, as defined in Chapter 1, by possessing the ability to withstand or recover from damage caused by arthropod populations equal to those on susceptible cultivars. The expression of tolerance is determined by the inherent genetic qualities of a plant that enable it to outgrow an arthropod infestation or to recover and add new growth after the destruction or removal of damaged plant fluids or tissues. From an agronomic perspective, the plants of a tolerant cultivar produce a greater amount of biomass than plants of non-tolerant, susceptible cultivars. Strauss and Agrawal (1999) describe five primary factors involved in increased plant tolerance. These include increased net photosynthetic rate, high relative growth rate, increased branching or tillering after apical dominance release, pre-existing high levels of carbon stored in roots, and the ability to shunt stored carbon from roots to shoots. From a plant breeding perspective, this means the selection of genotypes with increased growth and vigor, in order to survive arthropod infestation. Some cultivars of maize, *Zea mays* L., tolerant of damage by the corn earworm, *Helicoverpa zea* (Boddie), and the European corn borer, *Ostrinia nubilalis* Hübner, actually harbor larger larval populations than susceptible cultivars, due presumably to their increased biomass (Wiseman et al. 1972, Hudon et al. 1979), but this does not decrease their effectiveness in providing greater yields than susceptible cultivars.

Unlike antixenosis and antibiosis, tolerance involves only plant characteristics and is not part of an arthropod/plant interaction. However, tolerance often occurs in combination with antibiosis and antixenosis, as indicated by examples later in this chapter. From the perspective of the total effect of the resistant plant on the arthropod population, cultivars with tolerance require less antixenosis or antibiosis than cultivars without tolerance. Tolerance resistance in crop cultivars offers several advantages. Arthropod populations are not reduced from exposure to tolerant plants as they are on plants exhibiting antibiosis and antixenosis. Pest arthropod populations are more likely to remain avirulent to plant resistance genes, because the selection pressure placed on them by high levels of antibiosis is reduced or absent (see Chapter 11).

Tolerance also enhances the effects of beneficial arthropods agents in crop

101

protection systems. Tolerant cultivars do not expose beneficial insects to the adverse effects of plant morphological or allelochemical factors in cultivars that exhibit antibiosis or antixenosis. Chapter 12 includes in depth discussions of the benefits of combining tolerance with other resistance categories.

Because of its unique nature in plant resistance to arthropods, the quantitative assessment of tolerance is accomplished by using different experimental procedures than those used to study antixenosis or antibiosis. The differences in the types of techniques used to evaluate plant material for the three different categories of resistance are discussed at length in Section 6.2.1.

2. OCCURRENCE OF TOLERANCE IN CROP PLANTS

Tolerance exists in cultivars across a wide taxonomic range of plant families (Table 4.1). The following discussion is intended to provide an overview of where and how tolerance occurs. Comprehensive reviews by Snelling (1941), Painter (1951), Velusamy and Heinrichs (1986), and Reese et al. (1994) provide additional literature pertaining to crop tolerance to arthropods.

The most extensive research in the area of plant tolerance has been conducted with cereal grain aphid crops such as barley, *Hordeum vulgare* L., maize, *Zea mays* L., rice, *Oryza sativa* (L.), rye, *Secale cereale* L., sorghum, *Sorghum bicolor* (L.) Moench, and wheat, *Triticum aestivum* L. Snelling and Dahms (1937) first noted tolerance in sorghum to the chinch bug, *Blissus leucopterus leucopterus* (Say), in the 1930's. Dahms (1948) later evaluated the response of barley cultivars for tolerance to damage by *Schizaphis graminum* (Rondani). A standardized mass seedling screening technique to evaluate small grains for resistance to *S. graminum* damage, was developed by Wood (1961) and identified several wheat genotypes tolerance to *S. graminum* damage. Tolerance has been detected in sorghum identified as resistant to *S. graminum* in numerous studies (Schuster and Starks 1973, Girma et al. 1999, Wilde and Tuinstra 2000).

Tolerance has been shown to also play a major role in wheat resistance to different aphid pests. Havlickova (1997) and Papp and Mesterhazy (1993) each identified wheat germplasm exhibiting tolerance to the bird cherry oat aphid, *Rhopalosiphum padi* L. Tolerance is at least a partial component of all current cultivars of wheat with resistance to the Russian wheat aphid, *Diuraphis noxia* (Mordvilko), (duToit 1989, Smith et al. 1992, Miller et al. 2003), as well as wheat germplasm with *S. graminum* resistance (Webster and Porter 2000, Flinn et al. 2001, Lage et al. 2003).

Sources of sorghum tolerance have been detected to the spotted stalk borer, *Chilo partellus* Swinhoe, and the stalk borer, *Busseola fusca* Fuller (Dabrowski and Kidiarai 1983, van den Berg et al. 1994). Macharia and Mueke (1986) also identified tolerance to the barley fly, *Delia flavibasis* Stein, in West African sorghum cultivars. Sharma and Lopez (1993) identified tolerance as a major component of the resistance in sorghum in India to the mirid *Calocoris angustatus* Leithiery, and Mize and Wilde (1986) detected resistance in sorghum to *B. leucopterus*.

Table 4.1. Incidence of tolerance crop plant resistance to arthropods.

Plant	Arthropods	Plant factor	References
Abies	*Choristoneura occidentalis*	Yield	Clancy et al. 1993
Betula	*Agrilus anxius*	yield	Miller et al. 1991
Brassica campestris	*Phyllotreta cruciferae*	growth	Brandt & Lamb 1994
Cucumis melo	*Aphis gossypii*	survival	Bohn et al. 1973
Cucurbita pepo	*Bemisia argentifolii*	chlorophyll	Cardoza et al. 1999
Fragaria	*Tetranychus urticae*	yield	Gimenez-Ferrer et al. 1994, Schuster et al. 1980
Gossypium	*Earias vitella* *Lygus lineolaris* *Thrips spp.*	growth, yield	Sharma & Agarwal 1984, Bowman & McCarty 1997, Meredith & Laster 1975
Grasses	*Zulia entreriana*	yield	Nilakhe 1987, Ferrufino & LaPointe 1989
	Labops hesperus	yield	Hewitt 1980
Hibiscus esculentus	*Amrasca biguttula*	growth	Teli & Dalaya 1981
Hordeum	*Schizaphis graminum*	survival	Dahms 1948
Lycopersicon esculentum	*Tetranychus urticae*	yield	Gilbert et al. 1966
Manihot utilisima	*Mononychellus* sp. mite	yield	Byrne et al. 1982
	Phenacoccus manihoti	yield	Leru & Tertuliano 1993
Medicago	*Acyrthrosiphon kondoi*	size	Stern et al. 1980,
	Acyrthosiphon pisum	yield	Bishop et al. 1982
	Hypera postica	yield	Showalter et al. 1975
	Spissistilus festinus	yield	Moellenbeck et al. 1993
	Therioaphis maculata	yield	Kindler et al. 1971
Oryza sativa	*Chilo suppressalis*	yield	Das 1976
	Lissorhoptorus oryzophilus	yield	Oliver et al. 1972, Grigarick 1984, N'Guessan & Quisenberry 1994

Table 4.1 continued

	Nilaparvata lugens	yield	Ho et al. 1982, Nair et al. 1978
	Orseolia oryzivora	yield	Williams et al. 1999
	Spodoptera frugiperda	yield	Lye & Smith 1988
Phaseolus	*Empoasca fabaei Thrips spp.*	chlorophyll yield	Shaafsma et al. 1998, Cardona et al. 2002
Picea	*Pissodes strobi*	yield	King et al. 1997
Solanum melongena	*Leucinodes orbonalis*	yield	Mukhopadhyay & Mandal 1994
Sorghum	*B. leucopterus leucopterus*	survival	Mize & Wilde 1986
	Calocoris angustatus	yield	Sharma & Lopez 1993
	Chilo partellus	yield	Dabrowski & Kidiarai 1983, van den Berg et al. 1994
	Delia flavibasis	yield	Macharia & Mueke 1986
	Schizaphis graminum	survival	Girma et al. 1999, Schuster & Starks 1973, Wilde & Tuinstra 2000
Turfgrasses	*Prosapia bicincta*	yield	Shortman et al. 2002
	Popillia japonica	yield	Crutchfield & Potter 1995
Triticum aestivum	*Rhopalosiphum padi*	yield	Havlickova 1997 Papp & Mesterhazy 1993
	Schizaphis graminum	yield	Wood 1961
	Diuraphis noxia	chlorophyll	Deol et al. 2001, du Toit et al. 1989, Smith et al. 1992, Webster & Porter 2000, Flinn et al. 2001, Lage et al. 2003, Miller et al. 2003

Table 4.1 continued

Zea mays	*Blissus leucopterus*	survival	Painter et al. 1935
	Chilo partellus	yield	Dabrowski & Nyangiri 1983
	Diatraea grandiosella	yield	Kumar & Mihm 1995
	Diatraea saccharalis		
	Spodoptera frugiperda		
	Diabrotica virgifera	yield	Zuber et al. 1971
	Helicoverpa zea	yield	Wiseman et al. 1972, Wiseman & Widstrom 1992
	Ostrinia nubilalis	yield	Jarvis et al. 1991
	Schizaphis graminum	yield	Wood 1961

Tolerance to *N. lugens* exists in various rice cultivars in south and Southeast Asia (Ho et al. 1982, Nair et al. 1978, Panda and Heinrichs 1983) and in several wild rices (Jung-Tsung et al. 1986). Tolerance to the African gall midge, *Orseolia oryzivora* Harris and Gagne, was detected in an improved indica rice cultivars from Indonesia (Williams et al. 1999). Tolerance also exists in rice cultivars resistant to the rice water weevil, *Lissorhoptrus oryzophilus* Kuschel (Oliver et al. 1972, Grigarick 1984, N'Guessan and Quisenberry 1994) and the striped stem borer, *Chilo suppressalis* (Walker) (Das 1976).

Tolerance in maize cultivars resistant the western corn rootworm, *Diabrotica virgifera virgifera* LeConte, results from greatly increased root volume compared to that of susceptible cultivars (Painter 1968, Zuber et al. 1971, Rogers et al. 1976) (Figure 4.1). Correlations exist between the root volume ratios of *D. virgifera* infested-and uninfested plots, and the physical resistance to pulling the root systems of plants in the two plots out of the ground (Ortman et al. 1968) resulted in the development of several sources of *D. virgifera* resistant germplasm (Figure 4.2). Appreciable improvements in yield have been noted in several *D. virgifera* tolerant maize hybrids (Branson et al. 1982, 1983), and more recently, visual root ratings have proven to be a more accurate estimate of resistance (Knutson et al. 1999). Tolerance in maize also exists to ear damage by *H. zea* Boddie, (Wiseman and Widstrom 1992, Wiseman et al. 1972), the chinch bug (Painter et al. 1935), the spotted stalk borer (Dabrowski and Nyangiri 1983), *O. nubilalis* (Jarvis et al. 1991), as well as the fall armyworm, *Spodoptera frugiperda* (J. E. Smith), the southwestern corn borer, *Diatraea grandiosella* Dyar, and the sugarcane borer, *Diatraea saccharalis* Fabricus (Kumar and Mihm 1995).

The resistance of several cultivars of alfalfa, *Medicago sativa* L., to a complex of aphids also involves tolerance. The cultivars 'Dawson' and 'KS-10' tolerate damage by the pea aphid, *Acyrthosiphon pisum* (Harris), and the spotted alfalfa aphid, *Therioaphis maculata* (Buckton), due to increased production of dry matter, carotene, and protein (Kehr et al. 1968, Kindler et al. 1971). The newer, improved cultivars 'Lahontan' and 'Lahontan PGL' (polygenic) have tolerance to several *T. maculata* biotypes (Nielson and Olson 1982, Nielson and Kuehl 1982). Tolerance also exists in some high-yielding New Zealand alfalfa cultivars to an aphid complex formed by *A. pisum*, *T. maculata* and the blue alfalfa aphid, *Acyrthrosiphon kondoi* Shinji (Bishop et al. 1982, Turner and Robins 1982). Plants of some alfalfa cultivars can also tolerate the effects of defoliation by larvae of the alfalfa weevil, *Hypera postica* (Gyllenhal) (Showalter et al. 1975). Tolerance to adult feeding by the threecornered alfalfa hopper, *Spississtilus festinus* (Say), was identified in improved alfalfa cultivars by Moellenbeck et al. (1993).

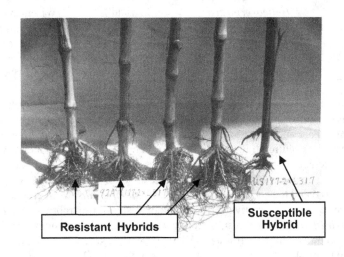

Figure 4.1. Comparative injury by Diabrotica *spp. larvae to roots of a resistant maize hybrid [92A x (187-2 x 317)] (left) and a susceptible maize hybrid, [187-2 x 317] (right). (From Painter 1968)*

Tolerance to arthropod feeding also exists in forage and turf grasses. Nilakhe (1987) identified tolerance to feeding by the spittlebugs *Zulia entreriana* (Berg.) and *Deois flavopicta* (Stal) in the forage grasses *Andropogon gayanus* Kunth, *Bracharia brizantha* (Hochst. ex. A. Rich.), *Bracharia humidicola* (Rendle), *Paspalum guenoarum* Archevaleta, and *Paspalum plicatulum* Michx. Ferrufino and LaPointe (1989) also detected tolerance in several of the same grasses to a related spittlebug, *Zulia colombina* (Lallemand).

An assessment of over fifty warm-season turf grasses yielded cultivars of paspalum, bermudagrass, and zoysiagrass exhibiting tolerance to the two-lined spittlebug, *Prosapia bicincta* (Say) (Shortman et al. 2002). In a similar effort, Crutchfield and Potter (1995) assayed several cultivars of cool-season turf grasses for resistance to feeding by grubs of the Japanese beetle, *Popillia japonica* Newman, and the southern masked chafer, *Cyclocephala lurida* Bland. Although all cultivars tolerated damage, cultivars of creeping bentgrass, *Agrostis stolonifera* L., sustained proportionately lower root mass losses.

Some cultivars of cotton, *Gossypium hirsutum* L., with high lint yields and the pubescent leaf character are tolerant to feeding by the tarnished plant bug, *Lygus lineolaris* (Palisot de Beauvois) (Meredith and Laster 1975, Meredith and Schuster 1979). Sharma and Agarwal (1984) determined that tolerance in cotton cultivars to stem damage from feeding by the spotted bollworm, *Earias vitella* (F.), is due to the production of greater numbers of branches in response to *E. vitella* feeding. Tolerance in cotton to thrips (Thysanoptera: Thripidae) has been identified in *Gossypium barbedense* genotypes with thickened lower epidermal cells (Bowman and McCarty 1997).

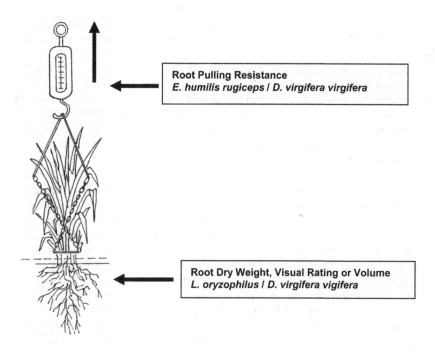

Figure 4.2. Measurements used in maize, rice and sugarcane to assess tolerance Diabrotica virgifera virgifera, Lissorhoptorus oryzophilus *and* Euetheola humilis rugiceps. *(From Knutson et al. 1999, Oliver et al. 1972, Ortman et al. 1968, Robinson et al. 1981, Rogers et al. 1976 and Zuber et al. 1971)*

Tolerance is also a component of the resistance of some cultivars of fruits and vegetables. Tolerance to the twospotted spider mite, *Tetranychus urticae* Koch, exists in cultivars of tomato, tomato, *Lycopersicon esculentum* Mill., (Gilbert et al. 1966) and strawberry, *Fragaria* x *ananassa* Duchesne, (Schuster et al. 1980, Gimenez-Ferrer et al. 1994). Tolerant tomato cultivars have high levels of defoliation but yields that are similar to those with little defoliation. Reductions in both the number and weight of fruit of the *T. urticae*-tolerant strawberry cultivars 'Florida Belle' and 'Sequoia' are lower than that of the susceptible cultivars 'Tioga' and 'Siletz' (Schuster et al. 1980).

Tolerance is an active component of the resistance of cassava, *Manihot esculenta* Crantz, to the cassava mealybug, *Phenacoccus manihoti* Matile-Ferrero (Leru and Tertuliano 1993). Tolerance in some cultivars of okra, *Abelmoschus esculentus* L., to feeding by the jassid, *Amrasca biguttula* (Ishida), was noted by (Teli and Dalaya 1981), and tolerance is a major component of resistance in eggplant, *Solanum melongena* L., to the shoot-and-fruit borer, *Leucinodes orbonalis* Guen. (Mukhopadhyay and Mandal 1994). Bohn et al. (1973) determined that tolerance in cantaloupe, *Cucumis melo cantalupensis* Naudin., to the melon aphid, *Aphis gossypii* Glover, is due to the lack of leaf curling after aphid infestation. Differing levels of tolerance in plants affect the uniformity of the expression of tolerance to leaf curling.

Tolerance to feeding by the flea beetle, *Phyllotreta cruciferae* (Goeze.), was identified in both *Brassica juncea* L. and *Sinapis alba* L. (Brandt and Lamb 1994), as well as in lines of oilseed rape, *Brassica para* L. (Lamb et al. 1993). Cardoza et al. (1999) conducted extensive experiments on the basis of resistance in zucchini squash, *Cucurbita* spp., to the squash silverleaf disorder, caused by feeding of the silverleaf whitefly, *Bemisia argentifolii* Bellows and Perring. Tolerance is the principal mode of silverleaf resistance.

Tolerance is also a major component in the resistance of common beans, *Phaseolus vulgaris* L., to hopperburn, caused by infestation and feeding damage by the potato leafhopper, *Empoasca fabae* (Harris) (Schaafsma et al. 1998) and *Empoasca kraemeri* Ross and Moore (Kornegay and Cardona 1990). Hopperburned plants are stunted, with downwardly curled, chlorotic leaves. These symptoms develop when hopper feeding causes blockages of xylem and phloem (Serrano and Backus 1998).

Tolerance to arthropod attack has been documented as an important factor in commercial forestry in North America. Vigorous, rapidly growing genotypes of trees have been identified that exhibit tolerance to several pest arthropods. These include the tolerance of paper birch, *Betula papyrifera* Marsh., to the bronze birch borer, *Agrilus anxius* Gory (Miller et al. 1991); the tolerance of white spruce, *Picea glauca* (Moench.) Voss., to the white pine weevil, *Pissodes strobi* (Peck) (King et al. 1997); and the tolerance of Douglas fir, *Pseudotsuga menziesii* [Mirb.] Franco, to the western spruce budworm, *Choristoneura occidentalis* Freeman (Clancy et al. 1993).

3. QUANTITATIVE MEASUREMENTS OF TOLERANCE

Several different techniques have been developed to evaluate the plant characteristics most commonly associated with arthropod tolerance. These characteristics include increases in the size and growth rate of plant leaves stems, petioles, roots, and seed or fruit. Tolerance to aphids among small grain plants has been measured based on seedling damage and survival (Wood 1961, Webster and Starks 1984). However, tolerance in sorghum to *S. graminum* is affected by plant maturity, and Doggett et al. (1970) found that yield differences in older, actively tillering sorghum plants are a more accurate measure of *S. graminum* tolerance than seedling survival. Tolerance in rice to damage by *N. lugens* is also more accurately assessed and identified as "field resistance" in tillering vegetative plants than in seedlings (Ho et al. 1982).

Schweissing and Wilde (1979) and Panda and Heinrichs (1983) developed formulae to assess rice and sorghum tolerance to arthropod damage. These formulae are calculated as [(mg dry weight of uninfested plants - mg dry weight of infested plants)/ mg arthropod dry weight]. Morgan et al. (1980) devised a further measure of sorghum tolerance, a functional plant loss index (FPLI) that combines measurements of leaf area loss and visual *S. graminum* damage ratings.

FPLI is defined as:

$$1 - (\text{Area}_C - \text{Area}_I / \text{Area}_C) \times 1 - (\text{Average visual damage rating}) \times 100,$$

where Area C is the leaf area of the uninfested control plants and Area I is the leaf area of the infested treatment plants. When arthropod damage is mild (in this case due to a short test duration) only leaf area is measured, using a functional plant loss (FPL) measurement where:

$$\text{Area}_C - \text{Area}_I / \text{Area}_C \times 100.$$

Panda and Heinrichs (1983) developed a modified FPLI to determine the tolerance of rice to feeding by *N. lugens* that was calculated as:

$$1 - (W_I / W_C) \times 1 - (\text{Damage rating} / 9) \times 100,$$

Where W_I is the dry weight of the infested plant and W_C is the dry weight of the control, uninfested plant.

As indicated by Reese et al. (1994) none of these measurements factor out the potential contributions of antibiosis on arthropod populations used in tolerance experiments. The first attempt to correct this problem was the development of a weight index (WI), by Brammel-Cox et al. (1986) for measuring antibiosis and tolerance in sorghum to the *S. graminum*. WI is calculated as:

$$(W_C - W_T / GB) \times 100,$$

Where W_C and W_T are the dry weights of control and infested (treated) plants, respectively, and GB is the number of *S. gramninum* on the infested plants at the end of the experiment.

However, since WI expresses tolerance on an absolute scale, there is still a need to quantify tolerance on a proportional basis. This is accomplished by dividing the plant weight loss ($W_C - W_T$) by W_T, in order to minimize the variation between sources of germplasm. A tolerance index (TI) developed by Dixon et al. (1990), expresses tissue loss on a proportional loss and includes a correction factor for aphid population size. TI is calculated as:

$$[(W_C - W_T / W_C) / GB] \times 100$$

Reese et al. (1994) presented evidence to demonstrate the improvement in accuracy of TI over WI, in several sorghum genotypes evaluated for *S. gramninum* resistance.

TI also has limitations, as TI values near zero may be obtained due to small differences between the weights of arthropods on infested and control plants, or because of the occurrence of very high arthropod populations on each infested plant. For this reason, Reese et al. (1994) proposed the use of the ratio DWT, a measure of relative or proportional dry weight change, to determine differences in tissue loss among test genotypes. DWT is defined as:

$$[(W_C - W_T) / W_C)] \times 100$$

DWT also has limitations, primarily statistical, because it does not allow for differences in the rate of plant tissue weight changes between plant genotypes. Nevertheless, Reese et al. (1994) compared measurements of TI and DWT to the slope resulting from regressing W_T on W_C, and found good agreement between all three methods. Both DWT and TI are used frequently in plant resistance to arthropods research.

Girma et al. (1999) and Deol et al. (2001) demonstrated how aphid-induced chlorophyll loss in wheat and sorghum can be measured photometrically with a SPAD 502 chlorophyll meter. Girma et al. (1999) developed a SPAD-based chlorophyll loss index, based on DWT. The SPAD Index is calculated as:

$$(SPAD_C - SPAD_T) / SPAD_C,$$

where $SPAD_C$ represents the SPAD meter reading on the control plant and $SPAD_T$ is the SPAD meter reading on the treated (arthropod infested) plant. Sorghum cultivars known to be tolerant to *S. gramninum* biotypes E and I based on TI and/or DWT are also generally tolerant based on SPAD indices (Girma et al. 1999). Similar results have been observed for wheat genotypes tolerant to *S. gramninum* biotype I (Flinn et al. 2001, Boina et al. 2005). A more complete discussion of the SPAD meter technique and its application in generating a SPAD Index rating are detailed in Chapter 6.

Measurements to identify plant material resistant to arthropod root feeding also compensate for the differences between infested and uninfested plants. Zuber et al. (1971) compared the root volume of insecticide treated-and untreated plots of maize inbreds for resistance to *D. virgifera* (Figure 4.2).   Though not defined as an FPLI, Robinson et al. (1981) evaluated the resistance of rice plant roots to feeding by *L. oryzophilus* larvae, using visual root ratings.   Differences between the root volumes of insecticide treated- and untreated-rice plants were also used by Oliver et al. (1972) as a measure of tolerance to *L. oryzophilus* feeding (Figure 4.2). Tseng et al. (1987) developed an array of plant growth measurements to evaluate the tolerance of rice genotypes to *L. oryzophilus* larval feeding.   These include: seedling survival, plant height, the number of leaves, the number of tillers (shoots), root length, root weight, plant weight, and grain weight. Ho et al. (1982) monitored the photosynthetic rate of various rice cultivars and determined that the photosynthetic activity of the tolerant cultivar 'Triveni' was less affected after infestation by *N. lugens* than the susceptible cultivar 'Taichung Native 1.'

Serrano et al. (2000) developed a proportional yield index to measure tolerance in common bean cultivars to hopperburn by *E. kraemeri*. The index is the proportion of the yield of one cultivar or genotype to that of all genotypes tested in a given field experiment. This proportional measure standardizes differences in growing season and location, allowing comparisons among different experiments.   Index values range from 0 to 2.0, with a genotype value of less than 1.0 indicating tolerance.

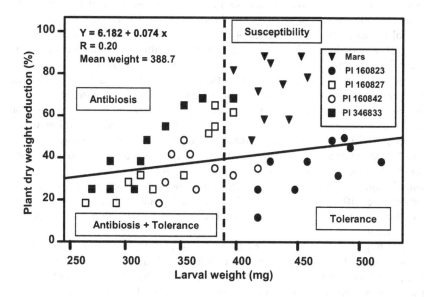

Figure 4.3. Relationship between rice plant biomass reduction (tolerance) and S. frugiperda *larval weight (antibiosis) among plants of four plant introductions (PI) and the susceptible cultivar 'Mars'. (From Lye & Smith 1988, reprinted with permission of the Florida Entomological Society)*

Regression analysis techniques have also been used to study the relationship between tolerance and antibiosis in populations of plant material evaluated for arthropod resistance. Lye and Smith (1988) demonstrated the partitioning of the tolerance and antibiosis resistance components to *S. frugiperda* in rice plant introductions by plotting plant dry weight reduction against larval weight gain (Figure 4.3). The intersection of a line marking the mean maximum larval weight and the regression line forms four quadrants that indicate different combinations of tolerance and antibiosis. The resulting scatter diagram provides an estimate of the different combinations of antibiosis and/or tolerance present in each plant introduction evaluated. A similar regression was used by Panda and Heinrichs (1983) to differentiate tolerance from antibiosis in rice cultivars resistant to *N. lugens* and by Ortega et al. (1980) to delineate tolerance and antibiosis resistance in maize to *S. frugiperda*.

4. TOLERANCE MECHANISMS

Research addressing the mechanisms of plant tolerance has been limited, likely due to the reluctance of plant breeders and producers to use arthropod-resistant crop cultivars that harbor high pest populations. However, recent studies by different research groups have shown the direct involvement of plant photosynthesis, plant hormones, and plant physical structures in the expression of plant tolerance. The tolerance of different wheat genotypes to the *D. noxia* has been shown to involve the ability of resistant plants to withstand or recover from damage to their photosynthetic system. *D. noxia* phloem feeding causes significant reductions in total chlorophyll, carotenoids, and chlorophyll A and B content in susceptible plants (Burd and Elliott 1996, Haile et al. 1999, Riedell and Blackmer 1999, Heng-Moss et al. 2003). These reductions are then manifested as reduced photosynthetic and chlorophyll fluorescence rates, resulting in greatly reduced plant photosynthetic efficiency and biomass production (Burd and Elliott 1996, Haile et al. 1999). Boyko et al. (2005) used molecular techniques to develop a cDNA library of foliage from a resistant wheat cultivar infested with *D. noxia* 1 (see Chapter 11). Numerous DNA sequences encoding photosystem and chlorophyll genes involved in photosynthesis are highly expressed in foliage of resistant plants. This is the first actual evidence that plant genes controlling tolerance are differentially expressed in resistant plants. For a more detailed discussion of more than twenty gene sequences expressed in aphid resistant plants, see Chapter 9.

Depending on the source of germplasm expressing tolerance, plant photosystems are either unaffected by feeding or recover fully in as little as one week after aphids are removed from plants. Experiments conducted by Nagaraj et al. (2002a,b) have shown that *S. graminum* feeding on sorghum plants of from one to four days duration also reduces the chlorophyll content and photosynthetic rate of sorghum leaves. Although chlorophyll content and photosynthesis are correlated, a small decrease in chlorophyll content from one day of infestation can result in a sharp reduction in sorghum photosynthesis. These effects are much more pronounced in susceptible plants than in resistant plants, which recover more rapidly over a 10-day period. Related studies indicate that chlorophyll is lost gradually for up to 10 days as a result of *D. noxia* feeding, but that plants fed on by *S. graminum* loose chlorophyll rapidly until the fourth day of feeding and lose chlorophyll more gradually thereafter (Deol et al. 2001). Overall, *S. graminum*-related chlorophyll loss is greater than that resulting from *D. noxia* feeding.

Maxwell and Painter (1962a,b) determined that the stunting of barley seedlings is a result of the removal of the plant hormone auxin during feeding by *S. graminum*. No auxin occurs in the honeydew of *S. graminum* feeding on *S. graminum*-tolerant barley cultivars and very limited amounts of auxin were obtained from the honeydew of *S. graminum* feeding on tolerant wheat cultivars, compared to that obtained from *S. graminum* feeding on susceptible cultivars. Maxwell and Painter (1962b) proposed that the binding of auxins to proteins or enzyme systems in the stem tissues of barley might prevent *S. graminum* from drinking auxins from tolerant cultivars. Alternatively, tolerance may also be due to the lack of penetration of *S. graminum* feeding stylets into stem tissues of *S. graminum* resistant barley plants.

The role of another plant hormone, abscissic acid (ABA), in the resistance of tomato to the carmine spider mite, *Tetranychus cinnabarinus* Boisd., was evaluated by Gawronska and Kielkiewicz (1999). In uninfested plants, ABA levels are normally higher in leaves of a susceptible cultivar than a tolerant cultivar. In leaflets of mite-damaged plants, leaf ABA content in the tolerant cultivar increases by 37% compared to only a 13% increase in the mite-susceptible cultivar (Figure 4.4A). Mechanical wounding also increases the ABA content of wounded leaves of tolerant plants to a lesser extent (21%), but the ABA content in tissues of susceptible plants actually decreases (Figure 4.4B). The increased ABA content of adjacent, non-infested leaflets in tolerant plants is likely due to intra-plant ABA signals from infested tissues. Synergism between ABA and the defense response elicitor jasmonic acid (see Chapter 9) has also been reported in plant defensive responses to disease infection (Chao et al. 1999). Other plant growth hormones may also be involved in tolerance. Wittmann and Schonbeck (1996) induced tolerance in barley plants to *R. padi*, by foliar application of *Bacillus subtilis*. Induced, *R. padi*-infested plants produced over 100% greater concentrations of indole-3-acetic acid and produced greater grain yields than non-induced plants. A discussion of signaling molecules in Chapter 9 will include how ABA and other elicitor molecules signal the infested plant to also produce defensive allelochemicals in resistant plants.

The tolerance of some bean cultivars to *Empoasca* damage may be the result of tolerance characters causing a behavioral shift in hopper feeding behavior to one that involves less damaging probing behavior. *E. kraemeri* modifies its probing behavior on tolerant plants. The number of short-duration probes causing multiple bean cell wall lacerations decreases, and the number of less damaging, single-cell puncture probes increases (Calderon and Backus 1992). In addition, Serrano and Backus (1998) found that *E. kraemeri* probing may elicit a physical compensatory response in

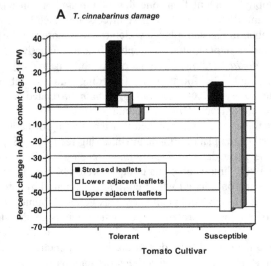

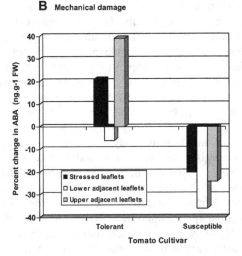

Figure 4.4. Effect of Tetranychus cinnabarinus *feeding (A) and mechanical damage (B) on the percent increase or decrease in abscissic acid (ABA) content of leaflets of tomato cultivars tolerant or susceptible to* T. cinnabarinus. *(From Gawronska & Kielkiewicz, 1999, Reprinted with permission of The Franciszek Górski Institute of Plant Physiology, Polish Academy of Sciences and the Institute of Plant Physiology, Warsaw Agricultural University, under the auspices of the Committee of Physiology, Genetics and Breeding of Plants, Polish Academy of Sciences, Warsaw, Poland)*

tolerant plants.   These plants have reduced overall plant cell damage and corresponding tracheal elements with a larger internal area that promotes increased relative nutrient flow rates.

Other than eventual leaf chlorosis, the differences and similarities between the compensation and recovery of cereal plants tolerant to aphids and bean plants tolerant to leafhoppers have not been analyzed. It will be interesting to learn if future research can elucidate common plant defense pathways used against arthropod attacks, as knowledge increases about the different plant genes expressed in response to arthropod feeding damage.

5. CONCLUSIONS

Tolerance resistance occurs over a wide variety of crop plant families and genera, indicating that it is a widespread, but generally expressed trait among crop genotypes selected for arthropod resistance.   The many examples of tolerance identified indicate that the key to its expression appears to be a plant's ability to maintain normal or enhanced levels of photosynthetic capacity in the presence of arthropod feeding and biomass removal. The identification of over-expressed DNA sequences encoding chlorophyll and photosynthesis genes in *D. noxia* tolerant wheat is a major breakthrough in the understating of plant tolerance to arthropods. These sequences can now be used as candidate gene probes to enable researchers to conduct more accurate and efficient analyses of the genes involved in arthropod tolerance in other crops. Sequences involved in the production of plant hormones such as abscissic acid may be able to be used in a similar manner. Since tolerance is a complex genetic trait (Chapter 8) it will be necessary to identify the gene sequences of several different components, in order to fully understand the contributions of each to the total phenotypic effect identified as plant tolerance to arthropods.

A great deal of tolerance research has been conducted with cereal crop plants, and this has lead to some of the most advanced and refined techniques known for the phenotypic quantification of tolerance. The SPAD chlorophyll meter quantification of sorghum and wheat tolerance of aphid feeding damage for example, allows for the selection of tolerance in a rapid and accurate manner, a development which is greatly accelerating at least the preliminary selection of aphid resistant sorghum genotypes (Girma et al. 1999, Nagaraj et al. 2002a,b). Given the breadth of the occurrence of tolerance, however, there are many opportunities for the use of the SPAD technique with other crops, especially fiber crops, where biomass production is essential.

REFERENCES CITED

Bishop, A. L., P. J. Walters, R. H. Holtkamp, and B. C. Dominiak. 1982. Relationships between *Acyrthosiphon kondoi* and damage in three varieties of alfalfa. J. Econ. Entomol. 75:118–122.

Bohn, G. W., A. N. Kishaba, J. A. Principe, and H. H. Toba. 1973. Tolerance to melon aphid in *Cucumis melo* L. J. Am. Soc. Hort. Sci. 98:37–40.

Boina, D., S. Prabhakar. C. M. Smith, S. Starkey, L. Zhu, E. Boyko, and J. C. Reese. 2005. Categories
    of resistance to greenbug (Homoptera: Aphididae) biotype I in wheats expressing the *Gby* and *Gbz*
    genes. *J. Kansas Entomol.Soc.* (accepted).

Bowman, D.T., and J. C. McCarty. 1997. Thrips (Thysanoptera: Thripidae) tolerance in cotton: Sources
    and heritability. J. Entomol. Sci. 32:460–471.

Boyko, E. V. , C. M. Smith, J. Bruno, T. Vankatappa, Y. Deng, S. R. Starkey, and P. Voothuluru. Aphid
    Avr- and wheat *Pto/Pti*-like sequences modulate aphid – cereal plant interactions. 2005. Science.
    (Submitted.).

Brammel-Cox, P. J., A. G. O. Dixon, J. C. Reese, and T. L. Harvey. 1986. New approaches to the
    identification and development of sorghum germplasm resistant to biotype E greenbug. Proc. 41st
    Ann. Corn & Sorghum Res. Conf., Am. Seed Trade Assn, December 6–7, 1986. Washington, DC.
    41: 1–16.

Brandt, R. N., and R. J. Lamb. 1994. Importance of tolerance and growth rate in the resistance of oilseed
    rapes and mustards to flea beetles, *Phyllotreta cruciferae* (Goeze) (Coleoptera: Chrysomelidae).
    Can. J. Plant Sci. 74:169–176.

Branson, T. F., G. R. Sutter, and J. R. Fisher. 1982. Comparison of a tolerant and susceptible maize
    inbred under artificial infestations of *Diabrotica virgifera virgifera*: yield and adult emergence.
    Environ. Entomol. 11:371–372.

Branson, T. F., V. A. Welch, G. R. Sutter, and J. R. Fisher. 1983. Resistance to larvae of *Diabrotica
    virgifera virgifera* in three experimental maize hybrids. Environ. Entomol. 12:1509–1512.

Burd, J. D., and N. C. Elliott. 1996. Changes in chlorophyll a fluorescence induction kinetics in cereals
    infested with Russian wheat aphid (Homoptera: Aphididae). J. Econ. Entomol. 89:1332–1337.

Byrne, D. H., J. M. Guerrero, A. C. Bellotti, and V. E. Gracen. 1982. Yield and plant growth responses
    of *Mononychellus* mite resistant and susceptible cassava cultivars under protected vs. infested
    conditions. Crop Sci. 22:486–490.

Calderon, J. D., and E. A Backus. 1992. Comparison of the probing behaviors of *Empoasca fabae* and
    E. kraemeri (Homoptera: Cicadellidae) on resistant and susceptible cultivars of common beans. J.
    Econ. Entomol. 85:88–99.

Cardona, C., A. Frei, J. M. Bueno, J. Diaz, H. Gu, and S. Dorn. 2002. Resistance to thrips
    (Thysanoptera: Thripidae) in beans. J. Econ. Entomol. 95: 1066–1073.

Cardoza, Y. J. H. J. MacAuslane, and S. E. Webb. 1999. Mechanisms of resistance to whitefly-induced
    squash silverleaf disorder in zucchini. J. Econ. Entomol. 92:700–707.

Chao, W. S., Y-Q., Gu, V. Pautot, E. A. Bray, and L. L. Walling. 1999. Leucine aminopeptidase RNAs,
    proteins, and activities increase in response to water deficit, salinity, and the wound signals
    systemin, methyl jasmonate, and abscisic acid. Plant Physiol. 129:979–992.

Clancy, K. M., J. K. Itami, and D. P. Huebner. 1993. Douglas-fir nutrients and terpenes: Potential
    factors to western spruce budworm defoliation. Forest Sci. 39:78–94.

Crutchfield, B. A., and D. A. Potter. 1995. Tolerance of cool-season turfgrasses to feeding by Japanese
    beetle and southern masked chafer (Coleoptera: Scarabaeidae) grubs. J. Econ. Entomol. 88:1380–
    1387.

Dabrowski, Z. T., and E. L. Kidiarai. 1983. Resistance of some sorghum lines to the spotted stalk-borer *Chilo partellus* under western Kenya conditions. Insect Sci. Appl. 4:119–126.

Dabrowski, Z. T., and E. O. Nyangiri. 1983. Some field and screenhouse experiments on maize resistance to *Chilo partellus* under western Kenya conditions. Insect Sci. Appl. 4:109–118.

Dahms, R. G. 1948. Comparative tolerance of small grains to greenbugs from Oklahoma and Mississippi. J. Econ. Entomol. 41:825–826.

Das, Y. T. 1976. Cross resistance to stemborers in rice varieties. J. Econ. Entomol. 69:41–46.

Deol, G. S., J. C. Reese, B. S. Gill, G. E. Wilde, and L. E. Campbell. 2001. Comparative chlorophyll losses in susceptible wheat leaves fed upon by Russian wheat aphids or greenbugs (Homoptera: Aphididae). J. Kansas Entomol. Soc. 74:192–198.

Dixon, A. G. O., P. J. Brammel-Cox, J. C. Reese, and T. L. Harvey. 1990. Mechanism of resistance and their interactions in twelve sources of resistance biotype E greenbug (Homoptera: Aphididae). J. Econ. Entomol. 83:2324–240.

Doggett, H., K. J. Starks, and S. A. Eberhart. 1970. Breeding for resistance to the sorghum shootfly. Crop Sci. 10:528–531.

du Toit, F. 1989. Components of resistance in three bread wheat lines to Russian wheat aphid (Homoptera: Aphididae) in South Africa. J. Econ. Entomol. 82:1779–1781.

Ferrufino, A., and S. L. LaPointe. 1989. Host plant resistance in *Brachiaria* grasses to the spittlebug *Zulia colombiana*. Entomol. Exp. Appl. 51:155–162.

Flinn, M. F., C. M. Smith, J. C. Reese, and B. S. Gill. 2001. Categories of resistance to greenbug (Homoptera: Aphididae) biotype I in *Aegilops tauschii* germplasm. J. Econ. Entomol. 94: 558–563.

Gawronska, H., and M. Kielkiewicz. 1999. Effect of the carmine spidermite (Acarida: Tetranychidae) infestation and mechanical injury on the level of ABA in tomato plants. Acta Physiol. Plantarum. 21:297–303.

Gimenez-Ferrer, R. M., W. A. Erb, B. L. Bishop, and J. C. Scheerens. 1994. Host-pest relationships between the two-spotted spider mite (Acari: Tetranychidae) and strawberry cultivars with differing levels of resistance. J. Econ. Entomol. 87:168–175.

Gilbert, J. C., J. T. Chinn, and J. S. Tanaka. 1966. Spider mite tolerance in multiple disease resistant tomatoes. Proc. Am. Soc. Hort. Sci. 89:559–562.

Girma, M., K. D. Kofoid, and J. C. Reese. 1999. Sorghum germplasm tolerant to greenbug (Homoptera: Aphididae)feeding damage as measured by reduced chlorophyll loss. J. Kansas Entomol. Soc. 71:108–115.

Grigarick, A. A. 1984. General control problems with rice invertebrate pests and their control in the United States. Prot. Ecol. 7:105–114.

Haile, F. J., L. G. Higley, X. Z. Ni, and S. S. Quisenberry. 1999. Physiological and growth tolerance in wheat to Russian wheat aphid (Homoptera : Aphididae) injury Environ. Entomol. 28: 787–794.

Havlickova, H. 1997. Differences in level of tolerance to cereal aphids in five winter wheat cultivars. Rostlinna Vyroba. 43:593–596.

Heng-Moss, T. M., X. Ni, T. Macedo, J. P. Markwell, F. P. Baxendale, S. S. Quisenberry, and V. Tolmay. 2003. Comparison of chlorophyll and carotenoid concentrations among Russian wheat aphid (Homoptera: Aphididae) – infested wheat isolines. J. Econ. Entomol. 96:475–481.

Hewitt, G. B. 1980. Tolerance of ten species of *Agropyron* to feeding by *Labops hesperius*. J. Econ. Entomol. 73:779–782.

Ho, D. T., E. A. Heinrichs, and F. Medrano. 1982. Tolerance of the rice variety Triveni to the brown planthopper, *Nilaparvata lugens*. Environ. Entomol. 11:598–602.

Hudon, M., M. S. Chiang, and D. Chez. 1979. Resistance and tolerance of maize inbred lines to the European corn borer *Ostrinia nubilalis* (Hubner) and their maturity in Quebec. Phytoprotection 60:1–22.

Jarvis, J. L., W. A. Russell, J. E. Campbell, and W. D. Guthrie. 1991. Level of resistance in maize to prevent or reduce yield losses by second-generation European corn borers. Maydica. 36:267–273.

Jung-Tsung, W., E. A. Heinrichs, and F. G. Medrano. 1986. Resistance of wild rices, *Oryza* spp., to the brown planthopper, *Nilaparvata lugens* (Homoptera: Delphacidae). Environ. Entomol. 15: 648–653.

Kehr, W. R., G. R. Manglitz, and R. L. Ogden. 1968. Dawson alfalfa, a new variety resistant to aphids and bacterial wilt. Nebr. Agr. Exp. Sta. Bull. 497.23 pp.

Kindler, S. D., W. R. Kehr, and R. L. Ogden. 1971. Influences of pea aphids and spotted alfalfa aphids on the stand, yield, dry matter and chemical composition of resistant and susceptible varieties of alfalfa. J. Econ. Entomol. 64:653–657.

King, J. N., A. D. Yanchuk, G. K. Kiss, and R. I. Alfaro. 1997. Genetic and phenotypic relationships between weevil (*Pissodes strobi*) resistance and height growth in spruce populations of British Columbia. Can. J. For. Res. 27:732–739.

Knutson, R. J., B. E. Hibbard, B. D. Barry, V. A. Smith, and L. L. Darrah. 1999. Comparison of screening techniques for western corn rootworm (Coleoptera: Chrysomelidae) host-plant resistance. J. Econ. Entomol. 92:714–722.

Kornegay, J., and C. Cardona. 1990. Development of an appropriate breeding scheme for tolerance to *Empoasca kraemeri* in common bean. Euphytica. 47:223–231.

Kumar, H., and J. A. Mihm. 1995. Antibiosis and tolerance to fall armyworm, *Spodoptera frugiperda* (J. E. Smith), southwestern corn borer, *Diatraea grandiosella* Dyar and sugarcane borer, *Diatraea saccharalis* Fabricius in selected maize hybrids and varieties. Maydica. 40:245–251.

Lage, J., B. Skovmand, and S. B. Andersen. 2003. Characterization of greenbug (Homoptera: Aphididae) resistance in synthetic hexaploid wheats. J. Econ. Entomol. 96: 1922–1928.

Lamb, R. J., P. Palaniswamy, K. A. Pivnick, and M. A. H. Smith. 1993. A selection of oilseed rape, *Brassica rapa* L., with resistance to flea beetles *Phyllotreta cruciferae* (Goeze) (Coleoptera: Chrysomelidae). Can. Entomol. 125:703–713.

Leru, B., and M. Tertuliano. 1993. Tolerance of different host-plants to the cassava mealybug *Phenacoccus manihoti* Matile - Ferrero (Homoptera, Pseudococcidae). Intl. J. Pest Mgmnt. 39:379–384.

Lye, B. H., and C. M. Smith. 1988. Evaluation of rice cultivars for antibiosis and tolerance resistance to fall armyworm (Lepidoptera: Noctuidae). Florida Entomol. 71:254–261.

Macharia, M., and J. M. Mueke. 1986. Resistance of barley varieties to barley fly *Delia flavibasis* Stein (Diptera:Anthomyiidae). Insect Sci. Applic. 7: 75–79.

Maxwell, F. G., and R. H. Painter. 1962a. Auxin content of extracts of host plants and honeydew of different biotypes of the corn leaf aphid, *Rhopalosiphum maidis* (Fitch). J. Kans. Entomol. Soc. 35:219–233.

Maxwell, F. G., and R. H. Painter. 1962b. Auxins in honeydew of *Toxoptera graminum, Therioaphis maculata*, and *Macrosiphum pisi*, and their relation to degree of tolerance in host plants. Ann. Entomol. Soc. Am. 55:229–233.

Meredith, W. R., Jr., and M. L. Laster. 1975. Agronomic and genetic analysis of tarnished plant bug tolerance in cotton. Crop Sci. 15:535–538.

Meredith, W. R., Jr., and M. F. Schuster. 1979. Tolerance of glabrous and pubescent cottons to tarnished plant bug. Crop Sci. 19:484–488.

Miller, H. R., T. L. Randolph, and F. B. Peairs. 2003. Categories of resistance at four growth stages in three wheats resistant to the Russian wheat aphid (Homoptera: Aphididae). J. Econ. Entomol. 96:673–679.

Miller, R. O., P. D. Bloese, J. W. Hanover, and R. A. Haack. 1991. Paper birch and European white birch vary in growth and resistance to bronze birch borer. J. Amer. Soc. Hort. Sci. 116:580–584.

Mize, T. W., and G. Wilde. 1986. New resistant plant material to the chinch bug (Heteroptera: Lygaeidae) in grain sorghum: Contribution of tolerance and antixenosis as resistance mechanisms. J. Econ. Entomol. 79:42–45.

Moellenbeck, D. J., S. S. Quisenberry, and M. W. Alison, Jr. 1993. Resistance of alfalfa cultivars to the threecornered alfalfa hopper (Homoptera: Membracidae). J. Econ. Entomol. 86:614–620.

Morgan, J., G. Wilde, and D. Johnson. 1980. Greenbug resistance in commercial sorghum hybrids in the seedling stage. J. Econ. Entomol. 73:510–514.

Mukhopadhyay, A., and A. Mandal. 1994. Screening of brinjal *(Solanum melgonea)* for resistance to major insect pests. Indian J. Agric. Sci. 64:798–803.

Nair, N. R., S. S. Nair, and N. Ramabai. 1978. A new high yielding brown planthopper tolerant variety of rice. Oryza. 17:161.

Nagaraj, N., J. C. Reese, M. B. Kirkham, K. Kofoid, L. R. Campbell, and T. M. Loughin. 2002a. Relationship between chlorophyll loss and photosynthetic rate in greenbug (Homoptera: Aphididae) damaged sorghum. J. Kansas Entomol. Soc. 75: 101–109.

Nagaraj, N., J. C. Reese, M. B. Kirkham, K. Kofoid, L. R. Campbell, and T. M. Loughin. 2002b. Effect of greenbug, *Schizaphis graminum* (Rondani) (Homoptera: Aphidiade) biotype K on chlorophyll content and photosynthetic rate of tolerant and susceptible sorghum hybrids. J. Kansas Entomol. Soc. 75: 299–307.

N'Guessan, F. K., and S. S. Quisenberry. 1994. Screening selected rice lines for resistance to the rice water weevil (Coleoptera, Curculionidae). Environ. Entomol. 23:665–675.

Nielson, M. W., and R. O. Kuehl. 1982. Screening efficacy of spotted alfalfa aphid biotypes and genic systems for resistance in alfalfa. Environ. Entomol. 11:989–996.

Nielson, M. W., and D. L. Olson. 1982. Horizontal resistance in 'Lahontan' alfalfa to biotypes of the spotted alfalfa aphid. Environ. Entomol. 11:928–930.

Nilakhe, S. 1987. Evaluation of grasses for resistance to spittlebugs. Pesq. Agropec. Bras., Brasilia. 22:767–783.

Oliver, B. F., J. R. Gifford, and G. B. Trahan. 1972. Studies of the differences in root volume and dry root weight of rice lines where rice water weevil larvae were controlled and not controlled. Ann. Prog. Rpt. Rice Res. Sta., LAES, LSU Agric. Ctr. 64:212–217.

Ortega, A., S. K. Vasal, J. Mihm, and C. Hershey. 1980. Breeding for insect resistance in maize. In: F. G. Maxwell and P. R. Jennings (Eds.), Breeding Plants Resistant to Insects. John Wiley, New York. pp. 372–419.

Ortman, E. E., D. C. Peters, and P. J. Fitzgerald. 1968. Vertical-pull technique for evaluating tolerance of corn root systems to northern and western corn rootworm. J. Econ. Entomol. 61:373–375.

Painter, R. H. 1951. Insect Resistance in Crop Plants. University of Kansas Press, Lawrence. 520 pp.

Painter, R. H. 1968. Crops that Resist Insects Provide a Way to Increase World Food Supply. Kansas Agric. Exp. Sta. Bulletin 520, 22 pp.

Painter, R. H. R. O. Snelling, and A. M. Brunson. 1935. Hybrid vigor and other factors in relation to chinch bug resistance in corn. J. Econ. Entomol. 28:1025–1030.

Panda, N., and E. A. Heinrichs. 1983. Levels of tolerance and antibiosis in rice varieties having moderate resistance to the brown planthopper, *Nilaparvata lugens* (Stal) (Hemiptera: Delphacidae). Environ. Entomol. 12:1204–1214.

Papp, M.and A. Mesterhazy, 1993. Resistance to bird cherry-oat aphid (*Rhopalosiphum padi* L.) in winter wheat varieties. Euphytica. 67: 49–57.

Reese, J. C., J. R. Schwenke, P. S. Lamont, and D. D. Zehr. 1994. Importance and quantification of plant tolerance in crop pest management programs for aphids: greenbug resistance in sorghum. J. Agric. Entomol. 11:255–270.

Riedell, W. E., and T. M. Blackmer. 1999. Leaf reflectance spectra of cereal aphid-damaged wheat. Crop Sci. 39:1835–1840.

Robinson, J. F., C. M. Smith, and G. B. Trahan. 1981. Evaluation of rice lines for rice water weevil resistance. Ann. Prog. Rpt., Rice Res. Sta., LAES, LSU Agric. Ctr. 73:260–272.

Rogers, R. R., W. A. Russell, and J. C. Owens. 1976. Evaluation of a vertical-pull technique in population improvement of maize for corn rootworm tolerance. Crop Sci. 16:591–594.

Schaafsma, A. W., C. Cardona, J. L. Kornegay, A. M. Wylde, and T. E. Michaels. 1998. Resistance of common bean lines to the potato leafhopper (Homoptera: Cicadellidae). J. Econ. Entomol. 91:981–986.

Schuster, D. J., and K. J. Starks. 1973. Greenbugs: Components of host plant resistance in sorghum. J. Econ. Entomol. 66:1131–1134.

Schuster, D. J., J. F. Price, F. G. Martin, C. M. Howard, and E. E. Albregts. 1980. Tolerance of strawberry cultivars to twospotted spider mites in Florida. J. Econ. Entomol. 73:52–54.

Schweissing, F. C., and G. Wilde. 1979. Temperature and plant nutrient effects on resistance of seedling sorghum to the greenbug. J. Econ. Entomol. 72:20–23.

Serrano, M. S., and E. A. Backus. 1998. Differences in cellular abnormalities induced by the probing behaviors of *Emposaca kraemeri* (Homoptera: Cicadellidae) on tolerant and susceptible common beans. J. Econ. Entomol. 91:1481–1491.

Serrano, M. S., E. A. Backus, and C. Cardona. 2000. Comparison of AC electronic monitoring and field data for estimating tolerance to *Empoasca kraemeri* (Homoptera: Cicadellidae) in common bean genotypes. J. Econ. Entomol. 93: 1796 –1809.

Sharma, H. C., and R. A. Agarwal. 1984. Factors imparting resistance to stem damage by *Earias vittella* F. (Lepidoptera: Noctuidae) in some cotton phenotypes. Prot. Ecol. 6:35–42.

Sharma, H. C., and V. F. Lopez. 1993. Survival of *Calocoris angustatus* (Hemiptera: Miridae) nymphs on diverse sorghum genotypes. J. Econ. Entomol. 86:607–613.

Shortman, S. L., S. K. Braman, R. R. Duncan, W. W. Hana, and M. C. Engelke. 2002. Evaluation of turfgrass species and cultivars for potential resistance to twolined spittlebug (Hemiptera: Cercopidae). J. Econ. Entomol. 95:478–486.

Showalter, A. H., R. L. Pienkowski, and D. D. Wolf. 1975. Alfalfa weevil: host response to larval feeding. J. Econ. Entomol. 68:619–621.

Smith, C. M., D. J. Schotzko, R. S. Zemetra, and E. J. Souza. 1992. Categories of resistance in wheat plant introductions resistant to the Russian wheat aphid (Homoptera: Aphididae). J. Econ. Entomol. 85:1480–1484.

Snelling, R. O., and R. G. Dahms. 1937. Resistant varieties of sorghum and corn in relation to chinch bug control in Oklahoma. Okla. Agr. Exp. Sta. Bul. 232.

Snelling, R. O. 1941. Resistance of plants to insect attack. Bot. Rev. 7:543–586.

Stern, V. M., R. Sharma, and C. Summers. 1980. Alfalfa damage from *Acrythosiphon kondoi* and economic threshold studies in southern California. J. Econ. Entomol. 73:145–148.

Strauss, S. Y., and A. A. Agrawal. 1999. The ecology and evolution of plant tolerance to herbivory. Trends Ecol. Evol. 14:179–185.

Teli, V. S., and V. P. Dalaya. 1981. Varietal resistance on okra to *Amrasca biguttula biguttula* (Ishida). Indian J. Agric. Sci. 51:729–731.

Tseng, S. T., C. W. Johnson, A. A. Grigarick, J. N. Rutger, and H. L. Carnahan. 1987. Registration of short stature, early maturing, and water weevil tolerant plant material lines of rice. Crop. Sci. 27:1320–1321.

Turner, J. W., and P. A. Robins. 1982. Aphid resistant lucernes. Queensland Agric. J. 108:153.

van den Berg, J., W. G. Wenzel, and M. C. van der Westhuizen. 1994. Tolerance and recovery resistance of grain sorghum genotypes artificially infested with *Busseola fusca* (Fuller) (Lepidoptera, Noctuidae). Insect Sci. Appl. 15: 61–65.

Velusamy, R., and E. A. Heinrichs. 1986. Tolerance in crop plants to insect pests. Insect Sci. Applic. 7:689–696.

Webster, J. A., and D. R. Porter. 2000. Plant resistance components of two greenbug (Homoptera: Aphididae) resistant wheats. J.Econ. Entomol. 93:1000–1004.

Webster, J. A., and K. J. Starks. 1984. Sources of resistance in barley to two biotypes of the greenbug *Schizaphis graminum* (Rondani), Homoptera: Aphididae. Prot. Ecol. 6:51–55.

Wilde, G. E., and M. R. Tuinstra. 2000. Greenbug (Homoptera : Aphididae) resistance in sorghum: Characterization of KS 97. J. Agric. Urban Entomol. 17:15–19.

Williams, C. T., M. N. Ukwungwu, B. N. Singh, O. Okhidievbie, and J. Nnabo. 1999. Farmer-managed trails in south-east Nigeria to evaluate the rice variety Cisadane and estimate yield losses by the African rice gall midge, *Orseolia oryzivora* Harris & Gagne. Internat. J. Pest Man. 45:117–124.

Wiseman, B. R., and N. W. Widstrom. 1992. Resistance of corn populations to larvae of the corn earworm (Lepidoptera: Noctuidae). J. Econ. Entomol. 85:601–605.

Wiseman, B. R., W. W. McMillian, and N. W. Widstrom. 1972. Tolerance as a mechanism of resistance in corn to the corn earworm. J. Econ. Entomol. 65:835–837.

Wittmann, J., and F. Schonbeck. 1996. Studies of tolerance induction in wheat infested with powdery mildew or aphids. Z. Pflanzenkr. Pflanzensch. 103:300–309.

Wood, E. A., Jr. 1961. Description and results of a new greenhouse technique for evaluating tolerance of small grains to the greenbug. J. Econ. Entomol. 54:303–305.

Zuber, M. S., G. J. Musick, and M. L. Fairchild. 1971. A method of evaluating corn strains for tolerance to the western corn rootworm. J. Econ. Entomol. 64:1514–1518.

CHAPTER 5

LOCATING SOURCES OF RESISTANCE

1. WHY COLLECT AND PRESERVE GERMPLASM ?

There are numerous reasons for plant biologists to collect and preserve germplasm. Some are biological and some are humanitarian. In spite of the past successes of the "Green Revolution" of the 1960s and the current accomplishments of the "Gene Revolution" to increase food productivity, about 800 million people in the world, predominantly in underdeveloped countries, remain undernourished. Most global projections indicate a nearly 50% increase in world population by 2030 (Brown 1994), coupled to an overall decrease in the *per capita* surface area available for food crop production (Iwanaga 1999).

To address this problem, sources of resistance to abiotic and biotic (including arthropods) stresses must be identified, quantified and used to promote crop genetic diversity and crop improvement (Strauss et al. 1988). Clement and Quisenberry (1999) reviewed the existing global genetic resources in arthropod resistant crop plants. This chapter will draw extensively on this rich source of information.

The use of diverse genetic diversity in crops is important, as demonstrated by the vulnerability exhibited in the past by different crops with very narrow genetic bases (Harlan 1972). Most European cultivars of potato, *Solanum tuberosum* L., were destroyed in the pandemics of the potato late blight, *Phytophthora infestans,* resulting in the Irish "Great Famine" of the 1880s. An epidemic of corn leaf blight, caused by *Helimthosporium turcicum* Pass, in the southern United States during the 1970's resulted in losses to production of maize, *Zea mays* L., valued at several million dollars (Smale et al. 1998). The extreme susceptibility of French *Vitus* cultivars to the grape phylloxera, *Daktulosphaira vitifoliae* (Fitch), and the ensuing collapse of the industry in the 1870s (described in Chapter 1) is an additional example of narrow genetic crop diversity. Although not as spectacular, or well documented, genetic vulnerability in crops to invasive species of arthropods has been equally costly. For example, total crop losses in barley, *Hordeum vulgare* L., and wheat, *Triticum aestivum* L., due to damage by the Russian wheat aphid, *Diuraphis noxia* (Mordvilko), in the United States from 1987 to 1993 amounted to greater than $800 million before the introduction of aphid-resistant cultivars (Webster et al. 2000).

In addition, new sources of genetic diversity beneficial to virtually all crops must be re-acquired over time, because of the difficulties involved in carrying out systematic collections of new accessions and the decline in the conditions of germplasm due to deterioration of storage facilities. Although germplasm information is becoming easier to access, there also remain many gaps in the information available on germplasm traits of many crops, especially underutilized 'minor' crops, without a high market demand. Functional global databases are also yet to be developed and coordinated in a uniform manner, because existing information remains to a large extent, scattered throughout the scientific literature. The lack of information about both *ex situ* and *in situ* genetic materials are among the most significant obstacles to the increased use of plant genetic resources for agriculture (FAO 1996).

Issues involving the commercialization of germplasm have also affected the development of crop diversity for several years (Beuselinck and Steiner 1992). Prior to changes in industry that have led to the patenting of plant genetic properties, plant breeders and collaborating scientists engaged in generally free exchange of germplasm towards the purpose of overall genetic crop improvement. In 1988, the research centers of the Consultative Group for International Agricultural Research (CGIAR, see Chapter 1) adopted a policy that prohibited germplasm collections developed from international collaborations to become the property of an individual country. However, many countries enforced the 1993 Convention on Biodiversity that in essence prohibited the export of genetic materials to prevent the loss of their genetic resources. This made it very difficult for gene banks to add new and diverse genetic materials to their collections.

In 2001, the International Treaty on Plant Genetic Resources was adapted by 116 countries, (excluding the United States and Japan) mandating the free exchange of 35 major crop plant species, including maize, wheat and rice, *Oryza sativa* (L.). Some species of forage grasses, peanut, *Arachis hypogaea* L., soybean, *Glycine max* (L.) Merr., and tomato, *Lycopersicon esculentum* Mill., were not included in the agreement. The incentive for the participating countries is that companies using seeds from public gene banks to create cultivars must pay a royalty into a FAO fund to preserve global genetic biodiversity through preservation and maintenance of gene banks (Charles 2001a). For the species included in the treaty a common "multilateral" gene bank of material has been created from which breeders may develop new cultivars for a use fee (Charles 2001b).

There are more than 1,300 international, national and regional germplasm collections worldwide, resulting in combined holdings of more than 6.1 million accessions stored worldwide in *ex situ* collections, including approximately 520,000 accessions stored worldwide in field collections and approximately 35,000 accessions stored in *in vitro* collections (FAO 1996). Many accessions held in these collections are sources of resistance to several significant arthropod pests. However, much of the germplasm in the collections has been under-utilized and remains a vast reservoir of potential pest resistance genes for crop improvement. The U. S. National Plant Germplasm System, for example, contains more than 500,000 accessions. The combined seed stocks of the International Agricultural Research

Centers (see below) also amount to roughly 500,000 accessions. However, data are generally unavailable on how many accessions maintained by gene banks have been used by breeding programs to improve cultivars. Because of this situation, entomologists and plant breeders must strive to obtain and evaluate germplasm collections for arthropod resistance, and develop quantitative data on the frequency of pest resistance occurring in existing and new germplasm.

2. PROCUREMENT OF GERMPLASM

Several steps are necessary to begin a program of evaluating plant germplasm for resistance to arthropods. The differences between these steps are determined by the ease with which each can be accomplished. Normally, the search for resistance begins by evaluating crop cultivars grown in the geographic area where resistance is required. Sources of resistance are not randomly distributed, and this approach does not guarantee a high probability of identifying high levels of durable resistance. Almost routinely, searches for resistance in germplasm grown outside of the required location follow, requiring importation of foreign plant introductions for evaluation.

Resistance may also be obtained from related species of plants; however, use of this material involves interspecific crosses with the crop of interest. Interspecific crosses with arthropod resistance have been produced, such as *Secale x Triticum* crosses which have yielded *Triticale* hybrids resistant to the greenbug, *Schizaphis graminum* Rondani, and *D. noxia* (Nkongolo et al. 1992, Deol et al. 1995).

3. OCCURRENCE OF ARTHROPOD RESISTANCE

For the most part, the greatest diversity of wild crop plant relatives and land races is still found in the areas mapped by N. I. Vavilov, the famous Russian geneticist and botanist. Vavilov noted that diversity in agricultural crops is not equally dispersed, but clustered in certain world geographic regions (Table 5.1). Humans have been an integral element of crop evolution in these different regions. As a result, genetic diversity in domestic crop plant species is distributed in a very different manner than biological diversity.

Arthropod resistance is frequently found in a low frequency among the germplasm evaluated. Results summarized by Heinrichs (1986) and Heinrichs and Quisenberry (1999) for resistance to rice insect pests in Asia and Africa indicate that from 0.1 to 2.6 % of the germplasm evaluated was resistant. The one exception was resistance to the zigzag leafhopper, *Recilia dorsalis* (Motschulsky), where 33 % of 237 accessions were resistant.

A similar frequency of resistance has been identified in evaluations of sorghum and barley germplasm. Porter et al. (1999) found that resistance to *D. noxia* ranged from >1% to 13% of barley accessions evaluated and that resistance to *S. graminum* ranged from >1% to 33%. In a review of arthropod resistance to various sorghum pests, Teetes et al. (1999) found a similar range of resistance (>1% to 11%) in ten different evaluations of germplasm. Clement et al. (1999) summarized the frequency of arthropod resistance in the cool season legumes chickpea, *Cicer arietinum* L., faba bean, *Vicia faba* L., lentil, *Lens culinaris* Medik., and pea, *Pisum sativum* L., and found that the frequency of resistance to various arthropod pests varied from >1% to 23%.

Searches for resistance can be directed to select for sympatric or allopatric resistance. Sympatric resistance evolves in plants in the presence of pest arthropods, and allopatric resistance occurs in plants with no previous evolutionary contact with the pest arthropod (Harris 1975).

Table 5.1. World centers of origin of crop plants as defined by Vavilov (1951)

Geographic Region(s)	Crop plant(s)
Chinese	lettuce, rhubarb, soybean, turnip
Indian	cucumber, mango, oriental cotton, rice
Indo-Malayan	banana, coconut, rice
Central Asiatic	almond, cantaloupe, flax, lentil
Near-Eastern	alfalfa, apple, cabbage, rye
Mediterranean	celery, chickpea, durum wheat, peppermint
Ethiopian (formerly Abyssinian)	castor, coffee, grain sorghum, pearl millet
South Mexican and Central American	lima bean, maize, papaya, upland cotton
South American (Peru-Ecuador-Bolivia)	Egyptian cotton, potato, pumpkin, tomato
Chile	potato
Brazilian-Paraguayan	manioc, peanut, pineapple, rubber tree

Sympatric resistance, resulting from long-term evolutionary contact between the pest arthropod and plant, is often expressed as tolerance. Leppik (1970) proposed that searches for arthropod resistance be conducted in the original home of the arthropod and plant. This sympatric relationship was identified by Flanders et al. (1992) in primitive potato families resistant to the Colorado potato beetle, *Leptinotarsa decemlineata* (Say), and to the potato leafhopper, *Empoasca fabae* (Harris). Sympatric resistance also exists in cotton for resistance to the boll weevil, *Anthonomus grandis* (Boheman), (Farias et al. 1999) and in grapes resistant to *D. vitifoliae* (Fergusson-Kolmes and Dennehy 1993, Martinez-Peniche 1999).

In several cases, however, sources of allopatric arthropod resistance have been obtained outside of the geographic center of origin of the pest (Table 5.2). Examples include resistance in sorghum to the chinch bug, *Blissus leucopterus leucopterus* (Say); resistance in wheat to the wheat stem sawfly, *Cephus cinctus* Norton; resistance in soybeans to the Mexican bean beetle, *Epilachna varivestis* Mulsant; resistance in maize to the European corn borer, *Ostrinia nubilalis* (Hübner), southwestern corn borer, *Diatraea grandiosella* (Dyar), and fall armyworm *Spodoptera frugiperda* (J. E. Smith); resistance in potatoes to *L. decemlineata*; resistance in raspberry, *Rubus phoenicolasius* Maxim., to the raspberry aphid, *Amphorophora rubi* (Kalt.); and resistance in pear, *Pirus communis* L., to the pear psylla, *Psylla pyricola* Foerster. Pimentel (1991) reviewed a similar phenomenon in arthropod biological control, which he referred to as the 'new-association' approach. Here, the selection of parasites and predators originating away from the native home of the pest arthropod greatly improves the success of these organisms when they are introduced into a new environment for biological control. What are the advantages of allopatric resistance? This type of resistance is often polygenic, and if so can be more durable. There often has been no gene-for-gene co-evolutionary progression between arthropod and host plant, and sources of resistance may contain several genes, which offer defense against many kinds of stresses.

Table 5.2. Examples of host plant resistance apparently evolved in the absence of the arthropod [a]

Host	Arthropod(s)
Andropogon sorghum	*Blissus leucopterus leucopterus*
Cocoa	Cocoa capsid
Glycine max	*Epilachna varivestis*
Malus sylvestris	*Rhagoletis pomonella,*
	Conotrachelus nenuphar
	Empoasca fabae
	Eriosoma lanigerum
Oryza glaberrima	*Nephotettix virescens*
Oryza sativa	*Tagosodes orizicolus*
	Nilaparvata lugens
	Spodoptera frugiperda
Pyrus	*Psylla pyricola*
Rubus idaeus	*Aphis rubicola*
	Amphorophora agathonica
Triticum species	*Cephus cinctus*
Zea mays	*Ostrinia nubilalis*
	Spodoptera frugiperda
	Diatraea grandiosella

[a] *Harris (1975), Hudon & Chiang (1991), Jennings & Pineda (1970), Pathak (1977)*

Molecular marker maps of many crops are being continually supplemented with new information about markers linked to chromosome loci of agricultural importance (see Chapter 8). In the case of tomato, rice and soybean for example, genes of interest are frequently inherited as complex, quantitatively inherited traits known as quantitative trait loci (QTLs).

Various QTL analyses have shown that though complex traits are controlled by several loci, much of the genetic variation in a segregating plant population is controlled by a few QTLs with fairly major effects. Tanksley and McCouch (1997) recommended the use of QTLs and other molecular markers to probe exotic germplasm in gene banks for beneficial traits, since this type of germplasm has been shown to contain many new and useful genes (including those for arthropod resistance). As more information accumulates in a given crop molecular map, patterns are likely to emerge about the chromosome locations of key loci linked to useful genes (see Chapter 8 for a discussion of how these patterns are emerging in some arthropod resistance genes in crop plants). When this information is available, progeny from crosses involving exotic germplasm can be assayed in targeted chromosome areas to determine the presence of the gene(s) of interest (Nevo 1998). Molecular analyses of exotic germplasm may also be aided by information about plant geographic distribution. Flanders et al. (1997) surveyed over 1,000 accessions of potato from 92 *Solanum* species for resistance to a complex of arthropods including *E. fabae, L. decemlineata*, the green peach aphid, *Myzus persicae* (Sultzer), the potato aphid, *Macrosiphum euphorbiae* (Thomas), and the potato flea beetle, *Epitrix cucumeris* (Harris). For each arthropod, geographic areas of germplasm origin were identified where a higher frequency of resistance occurred than anticipated. The authors suggest the use of geographic indicators to identify likely starting points in efforts to obtain additional sources of exotic arthropod-resistant germplasm.

4. EXISTING GERMPLASM SYSTEMS

There are several sources from which to obtain crop germplasm with potential arthropod resistance. Agricultural scientists in many countries have developed or have access to a wealth of germplasm. Efforts to obtain germplasm from scientists in foreign countries should be coordinated through agencies within the government of the country of origin. In the United States, this agency is the New Crop Introduction Branch of the U. S. Department of Agriculture. Within the United States there are several national seed collections of various crops (see Table 5.3). With the exception of the base collection at the National Center for Genetic Resources Preservation in Fort Collins, CO and the U. S. Potato Genebank at Sturgeon Bay, WI, most germplasm in the United States is available to agricultural scientists for experimental use.

The Consultative Group for International Agricultural Research, an association of 58 public and private sector members established in 1971, supports 16 international agricultural research centers (Table 5.3) that promote sustainable agricultural development based on the environmentally sound management of natural resources (Plucknett and Horne 1990). The CGIAR centers and their major germplasm holdings and locations include CIAT (beans, cowpeas), Cali, Columbia; CIP (cassava, potato), Lima, Peru; CIMMYT (maize, wheat), Mexico City, Mexico; the International Plant Genetic Resources Institute (IPGRI, formerly IBPGR) (all crops) Rome, Italy; ICARDA (legumes), Allepo, Syria, ICRISAT (millet, sorghum), Patancheru, India; IITA (cassava, cowpeas), Ibadan, Nigeria; and IRRI (rice), Los Banos, Philippines.

(see reviews by Clement and Quisenberry 1999, FAO 1996, Muehlbauer and Kaiser 1994). IPGRI strengthens the conservation and use of all plant genetic resources through research and training. IPGRI efforts have resulted in increased genetic stability and diversity in several crops. Many sources of germplasm also exist in foreign national seed collections and private seed companies (Table 5.3). Within each collection, facilities exist for receiving, processing and storing germplasm accessions. In order to maintain collections, however, borrowers are frequently requested to return samples of seed produced by plants in experimental plantings.

In addition to agencies that distribute germplasm, organizations also exist to provide advice to governmental officials about the status of germplasm. In the United States, the National Plant Genetic Resources Board exists to advise the Secretary of Agriculture and National Association of State Universities and Land Grant Colleges about national germplasm needs. This advisory board also recommends plans to coordinate the collection, maintenance, description, evaluation, and utilization of germplasm between the United States and international organizations such as IPGRI.

Table 5.3. Major world sources of crop germplasm for evaluation of plant resistance to arthropods

Bean, *Phaseolus* species and Cowpea, *Vigna* species
Centro Internacional de Agricultura Tropical (CIAT), Colombia
Instituto Nacional de Investigaciones Agricolas (INIA) Mexico
International Institute of Tropical Agriculture (IITA), Nigeria
University of Cambridge, UK
USDA-ARS Western Regional Plant Introduction Station, Pullman, WA, USA
N. I. Vavilov Institute of Plant Industry, St. Petersburg, Russia

Chickpea, *Cicer arietinum* and Lentil, *Lens culinaris*
Ethiopian Gene Bank, Addis Ababa, Ethiopia
International Center for Agricultural Research in the Dry Areas (ICARDA), Aleppo, Syria
International Crops Research Institute for the Semi-Arid Tropics (ICRISAT), Patancheru, India
Institut für Pflanzengenetik und Kulturpflanzenforshung (IPKG), Gaterslaben, Germany
Institute of Crop Germplasm Resources, Beijing, and People's Republic of China
Laboratorio del Germiplasmo, Bari, Italy
Nordic Gene Bank, Alnarp, Sweden
USDA-ARS Western Regional Plant Introduction Station, Pullman, WA USA
N. I. Vavilov Institute of Plant Industry, St. Petersburg, Russia

Table 5.3. Continued

Cassava, *Manihot esculenta*
Centro Internacional de Agricultura Tropical (CIAT), Colombia
International Institute of Tropical Agriculture (IITA), Nigeria

Maize, *Zea mays*
Instituto Colombiana Agropecuorio, Colombia
Instituto Nacional de Investigaciones Agricolas (INIA), Mexico
International Maize and Wheat Improvement Center (CIMMYT), Mexico
International Center for Insect Physiology and Ecology (ICIPE), Kenya
International Crops Research Institute for the Semi-Arid Tropics (ICRISAT), India
Pioneer Hi-Bred International, Inc., Johnson City, IA USA
Northrup, King & Co, Eden Prairie, MN USA
USDA-ARS North Central Regional Plant Introduction Station, Ames, IA USA

Potato, *Solanum* **species**
Brunswick Genetic Resources Centre, Braunschweig, Germany
Central Columbia Collection, Tibaitata Columbia
Center for Genetic Resources, Wageningen, Netherlands
Chilean Tuberous Solanum Collection, Valdivia, Chile
Collection of Tuberous Solanum of Argentina, Balcarce, Argentina
Commonwealth Potato Collection, Pentlandfield, Scotland
Gross Lusewitz Potato Species Collection, IPKG, Gatersleben, Germany
International Potato Center (CIP), Lima, Peru
USDA-ARS United States Potato Genebank, Sturgeon Bay, WI USA
N. I. Vavilov Institute of Plant Industry, St. Petersburg, Russia

Rice, *Oryza* **species**
Bangaldesh Rice Research Institute
Central Rice Research Institute, India
Centro Nacional de Recursos Geneticos e Biotecnologia, Brazil
Centre de Cooperation Internationale en Recherche Agronomique pour le
     Developement, France
Institute of Crop Resources Research, Jiangsu Academy of Agricultural Sciences,
     China
International Center for Crop Germplasm Resources, Chinese Academy of
     Agricultural Sciences
Institute of Crop Research, Sichuan Academy of Agricultural Sciences, China
International Rice Research Institute (IRRI), Philippines
International Institute of Tropical Agriculture (IITA), Nigeria
Malaysian Agricultural Research and Development Institute
National Board for Plant Genetic Resources, India
National Institute of Agrobiological Resources, Japan

Table 5.3. Continued

Rice Division, National Board for Plant Genetic Resources, Bangkok,
    Thailand
University of Kyushu, Japan
USDA-ARS Genetic Stocks - *Oryza* Collection, Stuttgart, AR USA
USDA-ARS National Small Grains Collection, Aberdeen, ID USA
West African Rice Development Association, Buoke, Ivory Coast

Sorghum, *Sorghum* species
International Crops Research Institute for the Semi-Arid Tropics (ICRISAT),
    Hyderabad, India
Northrup King & Co., Eden Prairie, MN USA
Texas A&M Sorghum Germplasm Collection, Mayaguez, Puerto Rico
USDA-ARS Plant Genetic Resources Conservation Unit, Griffin, GA USA

Barley, *Hordeum vulgare;* Oat, *Avena sativa*; and Rye, *Secale cereale*
International Center for Agricultural Research in the Dry Areas (ICARDA), Syria
International Maize and Wheat Improvement Center (CIMMYT), Mexico
Waite Agricultural Research Institute Barley Collection, Adelaide, Australia
Ohara Institute for Agricultural Biology, Okayama University, Kurashiki, Japan
Agricultural Research Institute, Kromeriz, Czech Republic
Swedish Seed Association, Svalov, Sweden
USDA-ARS National Small Grains Collection, Aberdeen, ID USA

Soybean, *Glycine max*
Asian Vegetable Research and Development Center (AVRDC), Tainan, Taiwan
Australian National Soybean Collection, Canberra
USDA-ARS Soybean Germplasm Collection, Urbana, IL USA
Soybean Germplasm Collection, University of Illinois, Urbana, IL USA

Sugarcane, *Saccharum* species
Indian National Sugarcane Germplasm Collection, Coimbatore, India
USDA-ARS National Germplasm Repository, Miami, FL USA

Sweet Potato, *Ipomoea batatas*
International Potato Center (CIP), Lima, Peru
USDA-ARS Plant Genetic Resources Conservation Unit, Griffin, GA USA

Wheat, *Triticum aestivum*
Australian Wheat Collection, Tamworth, New South Wales, Australia
Plant Breeding Institute, Cambridge, England
Laboratorio del Germoplasmo, Bari, Italy

Table 5.3. Continued

Crop Research and Introduction Center, Ismir, Turkey
Czech Research Institute for Crop Production Gene Bank, Prague, Czech Republic
USDA-ARS National Small Grains Collection, Aberdeen, ID USA
North American Plant Breeders, Brookston, IN USA
Wheat Genetics Resource Center, Kansas State University, Manhattan, KS USA

5. CONDITIONS OF EXISTING GERMPLASM STOCKS

Public gene banks continually maintain germplasm collections and attempt to avoid the genetic vulnerability that results when a widely planted crop is found to be uniformly susceptible to a pest, pathogen or environmental hazard because of its genetic constitution (NAS 1972, Plucknett et al. 1987). The efforts of gene banks involve the collection, preservation and maintenance of germplasm of the major world food crops. Gene banks also work to develop and provide as much genetic diversity as possible to avoid the occurrence of outbreaks of diseases and arthropods in crop plants with similar genes for pest susceptibility. Defining genetic diversity is complicated by the fact that there has never been a global inventory of plant genetic resources. There are current shortages of diversity in pulse crops, root crops, fruits and vegetables, with the exception of potato and tomato. Collections of forage and tree species are especially small. Collections of cereals are comparatively more diverse (FAO 1996). There is also concern by the IPGRI however, that rice and wheat germplasm collections may be especially inadequate in their content of wild species (FAO 1996, Hargrove et al. 1985).

Global germplasm preservation efforts are jeopardized by numerous human activities, including crop plant habitat destruction from slash and burn agricultural practices, population expansion, civil strife and war, and unchecked timber and mining industry (Clement and Quisenberry 1999, FAO 1996, Raven 1983). Deregulation and destruction of forests and bushlands, as well as civil strife and war, have caused large scale genetic erosion in Africa, Asia and Latin America. Hundreds of species of plants with potential medical and agricultural uses are disappearing in many regions of the world, due to the types of human activities described above, further eroding the geographic bases of many crop plants. Finally, the development and introduction of new crop cultivars has in many cases led to the unintended consequence of the replacement and loss of traditional, genetically diverse farmer varieties or land races. The process is reported to be the cause of genetic erosion of crops such as barley, maize, potato, rice, tomato and wheat (FAO 1996). In at least one case, the development of improved crop hybrids for arthropod resistance may have narrowed the genetic base of a crop to permit future yield increases. Jordan et al. (1998) noted that >80% of the Australian sorghum hybrids had some level of resistance to the sorghum midge, *Stenodiplosis sorghicola* (Coquillet), by 1995, but that the shift to midge resistant hybrids was linked to narrowed genetic diversity and heterozygosity in commercial hybrids. If linkage drag (inclusion of genes from the resistance source that have a negative influence on the improved cultivar) is contributing to the reduced diversity, the authors suggest that sorghum breeders broaden the genetic base of Australian sorghum.

6. CONCLUSIONS

There is a continuous global need to preserve existing global crop plant germplasm collections. Additional efforts are now necessary to increase the diversity and amount of collections, to collect new genetic materials to broaden the genetic composition of domestic crop plant species, and to better utilize existing germplasm stocks. Efforts of plant resistance researchers to both preserve existing diversity and to add new sources of genetic diversity are expressly needed. There are many opportunities available for close interdisciplinary research between entomologists, geneticists, plant biologists, and plant breeders to accomplish these goals.

REFERENCES CITED

Beuselinck, P. R., and J. J. Steiner. 1992. A proposed framework for identifying, quantifying and utilizing plant germplasm resources. Field Crops Res. 29:261–272.

Brown, L. R. 1994. State of the World. Norton, New York.

Charles, D. 2001a. Seeds of discontent. Science. 294:772–775.

Charles, D. 2001b. Seed treaty signed: U. S., Japan abstain. Science. 294:1263–1264.

Clement, S. L., and S. S. Quisenberry. (Eds.) 1999. Global Plant Genetic Resources for Insect Resistant Crops. CRC. Press, Boca Raton, FL. 295 pp.

Clement, S. L., M. Cristofaro, S. E. Cowgill, and S. Weigand. 1999. Germplasm resources, insect resistance, and grain legume improvement. In: S. L. Clement and S. S. Quisenberry (Eds.), Global Plant Genetic Resources for Insect Resistant Crops. CRC Press, Boca Raton, FL.

Deol, G. S., G. E. Wilde, and B. S. Gill. 1995. Host plant resistance in some wild wheats to the Russian wheat aphid, *Diuraphis noxia* (Mordvilko) (Homoptera: Ahididae). Plant Breed. 114: 545–546.

FAO (Food and Agriculture Organization of the United Nations). 1996. The state of the world's plant genetic resources for food and agriculture. FAO, Rome.

Flanders, K. L., J. G. Hawkes, E. B. Radcliffe, and F. I. Lauer. 1992. Insect resistance in potatoes: sources, evolutionary relationships, morphological and chemical defenses, and ecogeographical associations. Euphytica. 61:83–111.

Flanders, K. L., E. B. Radcliffe, and J. G. Hawkes. 1997. Geographic distribution of insect resistance in potatoes. Euphytica. 93:201–221.

Farias, F. J. C., M. J. Lukefahr, J. N. DaCosta, and E. C. Freiren. 1999. Behavior of lines from cotton primitive race stocks to attack of the boll weevil. Pesquisa Agropecuaria Brasileira. 34:2235–2240.

Fergusson-Kolmes, L. A., and T. J. Dennehy. 1993. Differences in host utilization by populations of North American grape phylloxera (Homoptera: Phylloxeridae). J. Econ. Entomol. 86:1502–1511.

Hargrove, T. R., V. L. Cabanilla, and W R. Coffman. 1985. Changes in rice breeding in 10 Asian countries: 196–84. Diffusion of genetic materials, breeding objectives, and cytoplasm. IRRI Research Paper Series No. 111. 18 pp.

Harlan, J. R. 1972. Genetics of disaster. J. Environ. Qual. 1:212–215.

Harris, M. K. 1975. Allopatric resistance: Searching for sources of insect resistance for use in agriculture. Environ. Entomol. 4:661–669.

Heinrichs, E. A. 1986. Perspectives and directions for the continued development of insect-resistant rice varieties. Agric., Ecosyst. Environ. 18:9–36.

Heinrichs, E. A., and S. S. Quisenberry. 1999. Germplasm evaluation and utilization for insect resistance in rice. In: S. L. Clement and S. S. Quisenberry (Eds.), Global Plant Genetic Resources for Insect Resistant Crops. CRC Press, Boca Raton, FL. pp. 3–25.

Hudon, M., and M. S. Chiang. 1991. Evaluation of resistance of maize germplasm to the univoltine European corn borer *Ostrinia nubilalis* (Hübner) and relationship with maize maturity in Quebec. Maydica. 36:69–74.

Iwanaga, M. 1999. Foreword. In: S. L. Clement and S. S. Quisenberry (Eds.), Global Plant Genetic Resources for Insect Resistant Crops. CRC Press, Boca Raton, FL.

Jennings, P. R., and A. Pineda. 1970. Effect of resistant rice plants on multiplication of the plant hopper, *Sogatodes oryzicola* (Muir.). Crop Sci. 10:689–690.

Jordan, D. R., Y. Z. Tao, I. D. Godwin, R. G. Henzell, M. Cooper, and C. L. McIntyre. 1998. Loss of genetic diversity associated with selection for resistance to sorghum midge in Australian sorghum. Euphytica. 102:1–7.

Leppick, E. E. 1970. Gene centers of plants as sources of disease resistance. Ann. Rev. Phytopath. 8:324–344.

Martinez-Peniche, R. 1999. Effect of different phylloxera (*Daktulosphaira vitifoliae* Fitch) populations from South France, upon resistance expression of rootstocks 41 B and Aramon x Rupestris Ganzin No. 9. Vitis. 38:167–178.

Muehlbauer, F. J., and W. J. Kaiser. 1994. Using host plant resistance to manage biotic stresses in cool season food legumes . Euphytica. 73:1–10.

National Academy of Sciences. 1972. Genetic Vulnerability of Major Crops. National Academy of Sciences, Washington, DC.

Nevo, E. 1998. Genetic diversity in wild cereals: Regional and local studies and their bearing on conservation *ex situ* and *in situ*. Genet. Res. Crop Evol. 45: 355–370.

Nkongolo K.K., J. S. Quick, F. B. Peairs. 1992. Inheritance of resistance of three Russian triticale lines to the Russian wheat aphid. Crop Sci. 32:689–692.

Pathak, M. D. 1977. Defense of the rice crop against insect pests. Ann. N. Y. Acad. Sci. 287:287–295.

Pimentel, D. 1991. Diversification of biological control strategies in agriculture. Crop Protect. 10:243–254.

Plucknett, D. L., and M. E. Horne. 1990. The consultative group on international agricultural research-goals, accomplishments, and current activities. Food Rev. Internat. 6:67–89.

Plucknett, D. L., N. J. H. Smith, J. T. Williams, and N. M. Anishetty. 1987. Gene Banks and the World's Food. Princeton University Press, Princeton, NJ.

Porter, D. R., D. W. Mornhinweg, and J. A. Webster. 1999. Insect resistance in barley germplasm. In: S. L Clement and S. S. Quisenberry (Eds.), Global Plant Genetic Resources for Insect Resistant Crops. CRC Press, Boca Raton, FL. pp. 51–63.

Raven, P. 1983. The challenge of tropical biology. Bull. Entomol. Soc. Am. 29:5–12.

Smale, M., M. R. Bellon, and P. L. Pingali. 1998. Farmers, gene banks, and crops. In: E. M. L. Smale. (Ed.), Farmers, Gene Banks, and Crop Breeding: Economic Analyses and Diversity in Wheat, Maize, and Rice. Kluwer Academic Publishers, Boston.

Strauss, M. L., J. A. Pinto, and J. I. Cohen. 1988. Quantification of diversity in *ex situ* plant collections. Diversity. 16:30–32.

Tanksley, S. D., and S. R. McCouch. 1997. Seed banks and molecular maps: Unlocking genetic potential from the wild. Science. 277:1063–1066.

Teetes, G. L., G. C. Peterson, K. F. Nwanze, and B. B. Pendelton. 1999. Genetic diversity of sorghum: A source of insect-resistant germplasm. In: S. L. Clement and S. S. Quisenberry (Eds.), Global Plant Genetic Resources for Insect Resistant Crops. CRC Press, Boca Raton, FL. pp. 63–85.

Vavilov, N. I. 1951. The origin, variation, immunity and breeding of cultivated plants. Chronica Botanica. 13:1–364.

Webster, J., P. Treat, L. Morgan, and N. Elliott. 2000. Economic Impact of the Russian Wheat Aphid and Greenbug in the Western United States. 199–94, 199–95, and 199–96. USDA ARS Service Report PSWCRL Rep. 00–001.

CHAPTER 6

TECHNIQUES TO MEASURE RESISTANCE

1. MANIPULATION OF ARTHROPOD POPULATIONS

1.1. Field Populations

In order to determine if insect resistance exists in a diverse group of plant material, it is necessary to manipulate the pest arthropod population, the test plant population, or both. Rarely is a researcher able to simply plant a group of plant material and accurately evaluate the arthropod damage sustained. Without proper planning, the researcher will arrive at the time of bioassay with plants with insufficient arthropod populations to inflict differential damage or arthropod populations available to infest plants that are not at the proper phenological stage of development.

Researchers evaluating germplasm in the early stages of a plant resistance program normally use field populations of pest insects. The final proving ground for plant material found to be resistant is in replicated field tests. However, field evaluations have some inherent problems that may directly affect the search for resistance. Unmanaged arthropod populations may be too low or unevenly distributed in space and/or time to inflict a consistent level of damage. Year to year variation in population levels of the target pest arthropod may also make interpreting the results of field evaluations difficult. Finally, unmanaged field populations may be contaminated with non-target pests that have feeding symptoms similar to those of the target pest.

Selective insecticides can be applied to plants in field plots to eliminate non-target pests and natural enemies of pest arthropod populations, causing a resurgence in population levels (Chelliah and Heinrichs 1980). Placing light traps, pheromone traps, or kairomone traps in the experimental plots can also be used to augment pest insect populations. Finally, researchers may find it useful to make mass collections of indigenous pest insect populations collected from surrounding areas and re-releasing them onto test plants (Sharma et al. 1992).

Managed or supplemented populations will insure a more uniform distribution of insects or mites, but these populations are also subject to mortality by natural enemies such as predators, parasites, or pathogens. For this reason, test plants should be treated with a selective insecticide before and after infestation to eliminate populations of natural enemies (Sharma et al. 1992). In spite of these precautions, some supplemented pest arthropod populations may suffer high mortality from pathogen infection, due to the abnormally high population densities in cages. The most useful insect population is the one that causes sufficient damage for the researcher to observe the maximum differences among plant material evaluated.

Several procedures can be employed to obtain a useful field population of test arthropods. Begin by planting a trap crop consisting of a mixture of susceptible cultivars of differing maturity (Sharma et al. 1988a,b). This mixture should be planted as border rows around the actual test plots and at regular intervals within the plots. Ideally, the trap crop mixture should occur after every experimental row, but this may not be possible, because of field plot space limitations. If large-scale populations of pest arthropods fail to accumulate in the test plots, trap crop rows can be mechanically cut, promoting arthropod movement onto test plants. If the supply of experimental plant material is sufficient, the entire experiment may be planted again in duplicate or triplicate, at different points in time, so that at least one planting will coincide with the peak pest population in the field (Sharma et al. 1997). Chesnokov (1962) recorded some of the first efforts employed to use trap crops for enhancing evaluation of small grain crop resistance to the frit fly, *Oscinella frit* (L.) (Figure 6.1).

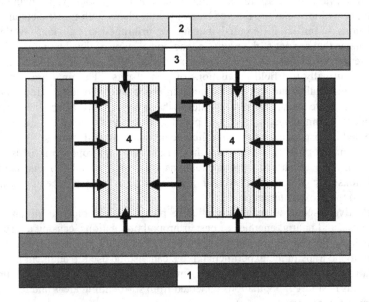

Figure 6.1. The use of trap crops of cultivated and wild grasses to enhance populations of Oscinella frit *in evaluations of wheat cultivars. From Chesnokov 1962. 1) existing cereal crops; (2) wild grasses; (3) thin spaced, early fall - planted winter wheat and rye; (4) thin-spaced late spring - planted wheat test plants.*

1.2. Caged Arthropod Populations

In spite of all of the efforts outlined above, caging arthropods on test plants may be necessary. Cages offer two major advantages to the plant resistance researcher. They limit emigration of the test arthropod from plants being evaluated, and they protect the pest population from predation and parasitism. In greenhouse and field tests, arthropods may be housed in small clip-on cages over small sections of intact plant leaves, stems, or flowers. However, East et al. (1992) found that evaluations of muskmelon, *Cucumis melo* (L.), resistance to the two spotted spider mite, *Tetranychus urticae* Koch, using whole leaves or leaf discs (see Section 2.2) are better correlated to *T. urticae* damage than evaluations using plants in clip cages.

Crafts-Brandner and Chu (1999) evaluated the effect of feeding by arthropods in clip cages on photosynthesis in leaves of muskmelon and cotton, *Gossypium hirsutum* L. Leaves sustaining feeding damage from arthropods in clip cages had increased leaf temperature and chlorophyll content, and decreased incident radiation, CO_2 exchange rate and leaf soluble protein content. Thus, clip cages cause changes in leaf biochemistry that may affect arthropod nutrition, and cage effects should be considered when interpreting results of plant-arthropod bioassays. Deol et al. (1997) used a foam leaf cage (see Figure 6.7) in place of the clip cage to successfully evaluate resistance in sorghum, *Sorghum bicolor* (L.) Moench, to the greenbug, *Schizaphis graminum* (Rondani).

Pathak et al. (1982) developed a parafilm sachet cage to collect honeydew from leafhoppers and planthoppers feeding on resistant and susceptible rice cultivars (Figure 6.2). The parafilm sachet technique has been successfully with the green leafhopper, *Nephotettix virescens* (Distant) (Khan and Saxena 1985b) and the white-backed planthopper, *Sogatella furcifera* (Horvath) (Khan and Saxena 1985c). These cages were modified by Tedders and Wood (1987) for use with aphid pests of pecan. Larger plant tissue can be enclosed in sleeve cages made of polyester organdy cloth (Figure 6.3), nylon cloth (Sharma et al. 1992), or hydroponic plant growth pouches (Figure 6.4).

Special modifications of plant growth conditions, such as the slant board technique, have been used to evaluate legumes for resistance to root feeding insects (Byers and Kendall 1982, Powell et al. 1983, Murray and Clements 1992). However, Byers et al. (1996) noted that growth and survival of the clover root curculio, *Sitona hispidulus* (F.), was greater on plants in plastic 'conetainers' than on roots of plants grown in a slant board pouch. Whole plants can be placed in cages constructed of wood, plexiglass, or metal frames, supporting screened aluminum panels of nylon or saran. The type and number of test plants that must be evaluated dictate cage size and shape. Dimensions vary from small Cornell type cages to large cages that are annually placed over galvanized metal frames to cover entire field experiments (Figure 6.5) (Lambert 1984). Saran screening, a polyester coated nylon material, offers superior resistance to environmental deterioration. Cage openings can be closed using heavy-duty zippers or the Velcro® fiber.

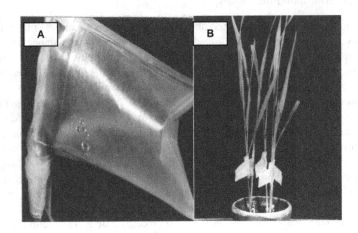

Figure 6.2. A parafilm sachet for collecting honeydew excreted by Nilaparvata lugens while feeding on rice stems. (A) Completed sachet showing honeydew collected from one female feeding for 24 hr (B) Sachets attached to the bases of several individual plants. (From Pathak et al. 1982. Reprinted with permission from J. Econ. Entomol., Vol. 75:194-195. Copyright 1982, Entomological Society of America)

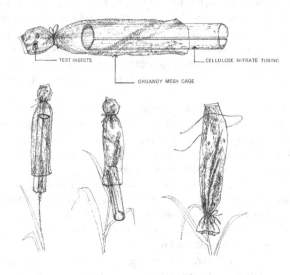

Figure 6.3. Organdy mesh cages to confine Oebalus pugnax adults on rice plant panicles.

Despite their advantages, however, cages also have some inherent disadvantages that must be anticipated and compensated for. Some cages may cause abnormal environmental conditions that can alter plant growth or cause foliar disease outbreaks. Not all plants are affected similarly, and cage effects, if any, must be determined on a case-by-case basis before being used on routinely to evaluate germplasm for resistance. Additional discussions and comparisons are available in Smith et al. (1994) and Sharma et al. (1992).

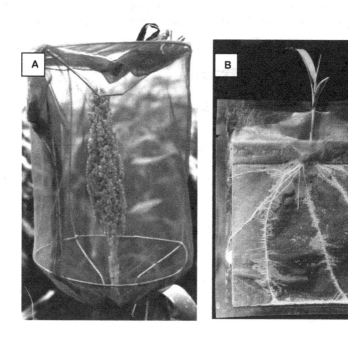

Figure 6.4. (A) Nylon cage to confine insects on sorghum panicle (From Sharma et al. 1992. ICRISAT Info. Bull. 32 (B) Hydroponic plant growth pouch to confine Diabrotica vigifera vigifera *on maize roots. (From Örtman & Branson 1976. Reprinted with permission from J. Econ. Entomol., Vol. 69:380-382. Copyright 1976, Entomological Society of America)*

1.3. Supplementing Populations with Artificially Reared Arthropods

If the pest insect can be mass reared, then insects are available on a year-round basis for evaluating plant material. Scores of artificial diets for rearing insects have been developed. For complete discussions, readers are referred to Anderson and Leppla (1992) and Singh and Moore (1985). The greatest successes in insect mass rearing for plant resistance evaluations have been with foliar and stalk feeding Lepidoptera.

Commercial hybrids of maize, *Zea mays* L., with resistance to the European corn borer, *Ostrinia nubilalis*, Hübner, the southwestern corn borer, *Diatraea grandiosella* (Dyar), and the fall armyworm, *Spodoptera frugiperda* (J. E. Smith), have been produced because of the development and refinement of techniques to handle and rear these insects (Davis and Guthrie 1992, Mihm 1983a,b, c). Mechanical techniques have also been developed to greatly reduce the amount of time required to mix, dispense, and inoculate artificial diets, remove pupae from diet, and harvest insect eggs (Davis 1980a, 1982, Davis et al. 1985, 1990). The net result of these accomplishments has been a quantum increase in the annual amount of plant material that can be accurately evaluated.

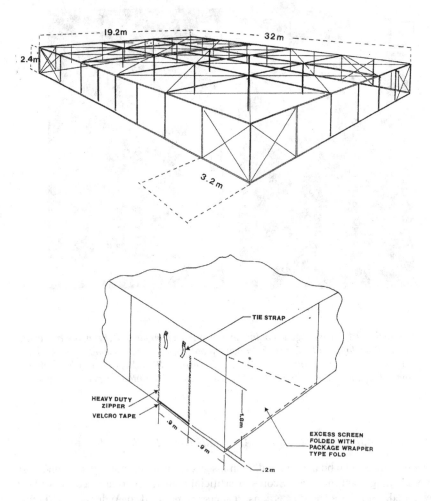

Figure 6.5. Galvanized permanent metal cage frame supporting saran screen cover used to confine insect populations on field plantings of soybeans. Reprinted from Lambert, L. 1984. An improved screen-cage design for use in plant and insect research. Agron. J. 76:168-170, Copyright 1984 American Society of Agronomy, with permission of the American Society of Agronomy.

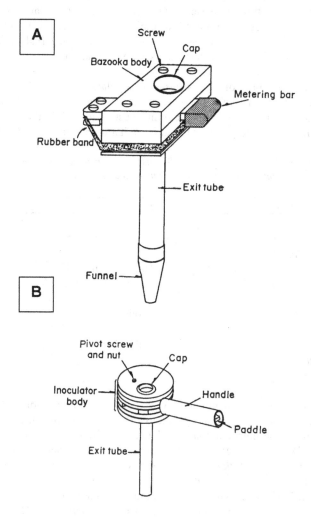

Figure 6.6. Manual dispensers for inoculating plant tissues with immature lepidopterous larvae (A) modified 'Bazooka' inoculator (From Mihm et al. 1978), (B) Davis larval inoculator. (From Davis & Oswalt 1979)

Techniques to infest plants have also been developed and refined. Early methods made use of agar-based suspensions containing corn earworm, *Helicoverpa zea* (Boddie), eggs that were injected with maize silk masses (Widstrom and Burton 1970), or bollworm, *Heliothis virescens* (F.) eggs applied to the fruiting structures of cotton plants (Dilday 1983). The development of a larval plant inoculator (Davis and Oswalt 1979, Mihm et al.

1978, Wiseman et al. 1980) (Figure 6.6) allowed the application of a suspension of immature larvae and fine-mesh sterilized corncob grits onto plant tissues. This inoculation method provided a means for rapid, accurate placement of larvae onto plants (Wiseman and Widstrom 1980, Davis and Williams 1980). The larval inoculator has been used successfully to infest test plants with several species of insects (Table 6.1).

Continued production of test arthropod populations on artificial diet often decreases their genetic diversity (Berenbaum 1986). To avoid these problems, quality control measures in the rearing program must insure that the behavior and metabolism of the laboratory-reared insect is similar to that of wild individuals.

Table 6.1. Insects successfully dispensed unto crop plant cultivars in plant resistance evaluations using the larval inoculator

Crop and Arthropod	References
Oryza sativa	
Spodoptera frugiperda	Pantoja et al. 1986
Sorghum bicolor	
Chilo partellus	Harvey et al. 1985
Schizaphis graminum	Nwanze & Reddy 1991
Zea mays	
Diatraea grandiosella	Davis 1980b
Heliothis virescens	Diawara et al. 1992
Heliothis zea	Mihm 1983a, b, c
Ostrinia nubilalis	
Spodoptera frugiperda	

An effective means of avoiding the development of these problems is to infuse wild individuals into the laboratory colony and to insure that the artificial diet closely resembles the nutritional and allelochemical composition of the host plant. Slansky and Wheeler (1992) and Wheeler et al. (1992) found that rearing the velvetbean caterpillar, *Anticarsia gemmatalis* (Hubner), in the laboratory on artificial diet for several generations did not select for reduced levels of *A. gemmatalis* detoxifying enzymes and that laboratory-reared larvae were suitable for studying the detoxification of plant xenobiotics by *A. gemmatalis*. The responses of both laboratory-reared and field strain (first generation) larvae to artificial diet nutrient levels are also similar. However, field strain larvae may initially adapt to artificial diets poorly.

2. WHEN AND HOW TO EVALUATE TEST PLANTS

2.1. Methods to Differentiate Between Resistance Categories

Different experimental test procedures are necessary to differentiate between the antixenosis, antibiosis, and tolerance categories of plant resistance to arthropods. Much of the effort in a program to develop resistant cultivars involves elimination of susceptible plant materials. Therefore, large-scale evaluations where arthropods are offered a free choice of plant materials, either in field plots or greenhouse experiments are often conducted initially. Materials identified from these tests as potentially resistant are then reevaluated in a smaller group that includes a susceptible control cultivar. To confirm the antixenosis category of resistance, plant materials are planted and infested together within each experimental replication in either the field or greenhouse. Different cultivars under evaluation are often planted in a circular arrangement in greenhouse pots and test insects are released in the center of the test plants. However, Webster and Inayatullah (1988) found that a completely random design in greenhouse flats gave a more accurate estimate of *S. graminum* antixenosis than the circular arrangement. Test arthropod populations are left on plants until the susceptible control cultivars have sustained heavy damage or accumulate large pest populations, at which time plants are evaluated for damage and/or populations (See Chapter 2). By identifying resistant plants in choice tests, the researcher is assured that the potentially resistant material possesses antixenosis.

To identify antibiosis, plant materials are planted, caged, and infested separately (Davis 1985). Test arthropods have no choice but to feed or not feed on plants of each cultivar being evaluated. Antibiosis measurements related to pest survival and development such as those discussed in Chapter 3 are then recorded during the course of the development of pest arthropods on test plants.

The intrinsic rate of increase r_m has been adopted in aphid research as an improved measure of antibiosis. The utility of r_m is based on results of Lewontin (1965) that indicate a small delay in the reproduction of an organism with a high intrinsic rate of increase can reduce net reproduction more than proportionally. When r_m is low, fecundity becomes a critical factor in altering the rate of population growth. Wyatt and White (1977) developed an estimation of r_m for aphids and tetranychid mites, where $r_m = 0.738$ ($\log_e M_d$)/d; and d is time required for a newly emerged aphid (F_1) to produce its first offspring; M_d is the total number of progeny produced by P_1, the mother of F_1; and 0.738 is the mean regression slope of (M_d /d) for four aphid species. When P_1 reproduction begins, the first nymph produced (F_1) is moved to a different leaf of the same plant and caged (Flinn et al. 2001). When aphid F_1 produces its first offspring, d and M_d are determined.

The intrinsic rate of increase has been used to assess antibiosis in wheat, *Triticum aestivum* L., to *S. graminum* (Webster and Porter 2000, Lage et al. 2003, Smith et al. 2003, Boina et al. 2004,) and the Russian wheat aphid, *Diuraphis noxia* (Mordvilko) (Webster et al. 1996, Miller et al. 2003). For *S. graminum* on wheat, r_m values in these studies range from 0.109 (resistant) (Lage et al. 2003) to 0.251 (susceptible) (Boina et al. 2004). For *D. noxia* on wheat, r_m values range from 0.24 (resistant) to 0.29 (susceptible) (Miller et al. 2003). Values for *D. noxia* on barley, *Hordeum vulgare* L., are similar – 0.224 (resistant) and 0.273 (susceptible) (Webster et al. 1996). Ruggle and Gutierrez (1995) used intrinsic rate of increase to explain performance of the spotted alfalfa aphid,

Therioaphis maculata (Buckton), on resistant and susceptible cultivars of alfalfa, *Medicago sativa* L. Both large and small-scale differences in antibiosis in cultivars of tomato, *Lycopersicon esculentum* Mill., resistant to the greenhouse whitefly, *Trialeurodes vaporariorum* (Westwood), were demonstrated by Romanow et al. (1991). r_m values range from 0.066 (resistant) to 0.097 (susceptible). Kocourek et al. (1994) used r_m to assess the growth rates of melon aphid, *Aphis gossypii* Glover, on cultivars of greenhouse cucumber, *Cucurbita sativus* L. The population fluxes of both *A. gossypii* and the green peach aphid, *Myzus persicae* (Sultzer), placed on cultivars of chrysanthemum, *Chrysanthemum leucanthemum* L., were evaluated by Guldemond et al. (1998) using intrinsic rate of increase measurements. Values of r_m in these studies ranged from 0.214 to 0.239 for *M. persicae* and 0.267 to 0.317 for *A. gossypii*.

Antibiosis and antixenosis are not always easily distinguishable from one another. This is especially evident when experiments are conducted with early instars of immature insects (Horber 1980). These two resistance categories may also be difficult to separate, since the death of test arthropods in an antibiosis test may result from either the toxic factor(s) involved in antibiosis or the deterrent factor(s) involved in antixenosis (Kishaba and Manglitz 1965, Renwick 1983).

Entirely different techniques are employed to assess plant tolerance, since it does not involve a plant interaction with arthropod behavior or physiology. Normally, the existence of tolerance is determined by comparing the production of plant biomass (yield) in insect-infested and non-infested plants of the same cultivar. Yield differences between infested and plants can then be used to calculate percent yield loss based on the ratio: yield of infested plants / yield of infested plants.

A tolerance evaluation involves preparing replicated plantings that include the different cultivars being evaluated and a susceptible control cultivar, caging all plants in each replicate, and infesting caged plants in one-half of each replicate with insect populations at or above the economic injury level for that insect. Plants should remain infested until susceptible controls exhibit marked growth reduction or until the pest insect has completed at least one generation of development. Volumetric or plant biomass production measurements (see Chapter 4) are then taken to calculate percent yield loss measurements.

As mentioned in Chapter 4, aphid-induced chlorophyll loss in wheat and sorghum can be measured photometrically with a SPAD 502 chlorophyll meter (Minolta Camera Co., Ltd., Japan), designed to measure chlorophyll A and B (Yadava 1986). A linear relationship between chlorophyll content and SPAD (chlorophyll) unit values has been established by Markwell et al. (1995). To concentrate aphids into an area where chlorophyll loss can be measured, a double-sided adhesive foam leaf cage is placed on the top of a leaf (Figure 6.7). A quantity of aphids sufficient to cover the caged leaf surface area (~0.5 cm diameter) is released into the cage. A small piece of organdy cloth (2.5 x 2.5cm) is then placed on the adhesive cage surface and aphids are allowed to feed for 4 days. Aphids are then removed and differences in chlorophyll content of infested and non-infested leaf tissue are compared on each leaf.

A SPAD-based chlorophyll loss index, based on proportional plant dry tissue weight loss (DWT) (see Chapter 4), is calculated as:

$$(SPAD_C - SPAD_T) / SPAD_C,$$

where $SPAD_C$ represents the SPAD meter reading on the control plant and $SPAD_T$ is the SPAD meter reading on the infested plant (Deol et al. 1997).

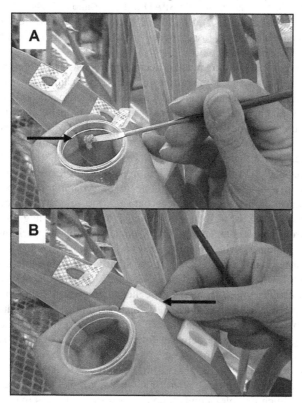

Figure 6.7. A double-sided adhesive foam leaf cage for measuring leaf chlorophyll loss resulting from aphid feeding (A) Cages are placed on the top of a leaf and a quantity of aphids sufficient to cover the caged leaf surface area (~0.5 cm diameter) is released into each cage (B) A small piece of organdy cloth is placed on the adhesive surface of each cage. (Images courtesy of Dr. John Reese, Kansas State University)

Five representative SPAD unit measurements are taken at each leaf cage site and averaged, yielding a mean cage site SPAD unit measurement. These measurements are used to calculate a mean cage site SPAD index. The three cage site SPAD index measurements are then used to calculate a mean plant SPAD index, and each plant

SPAD index value is then used to calculate a mean genotype SPAD index value. Data are subjected to analysis of variance and differences between treatments in mean chlorophyll loss are determined. Percent chlorophyll loss is calculated as: (SPAD index value x 100).

Sorghum cultivars known to be tolerant to *S. graminum* biotypes E and I and wheat genotypes tolerant to biotype I also have a significantly lower chlorophyll loss than susceptible cultivars based on SPAD indices (Girma et al. 1999, Flinn et al. 2001, Nagaraj et al. 2002). Deol et al. (2001) used SPAD index measurements to show that *S. graminum*-induced chlorophyll losses occur more quickly on susceptible wheat leaves than those caused by *D. noxia*.

Very few studies have compared the accuracy of measuring tolerance by the SPAD index to other tolerance measurements. For the majority of such studies conducted, SPAD indices are equivalent to those involving plant proportional dry weight (DWT) or height reductions. Girma et al. (1999) compared the SPAD index ratings of sorghum genotypes resistant to *S. graminum* to DWT and tolerance (TI) measurements. All three measurements provide similar results in identifying tolerance in the resistant genotypes. Results of Flinn et al. (2001) showed no difference between SPAD index and DWT ratings for the tolerance of an *Aegilops tauschii* accession and a wheat breeding line to *S. graminum* biotype I. Both techniques produce significant differences between the tolerant control line and susceptible control line, as well as between the tolerant breeding line and the susceptible control. Boina et al. (2005) found a similar trend in comparisons of DWT and SPAD index ratings of wheat plants containing different genes for *S. graminum* biotype I resistance. Plants with both genes are significantly more tolerant than the susceptible control, whether measured by DWT or SPAD index. Lage et al. (2003) examined the tolerance of several synthetic hexaploid wheat genotypes to   *S. graminum* biotype E, using SPAD index and proportional plant height reductions. In these experiments, plant height reduction measurements revealed differences in tolerance among genotypes and between tolerant genotypes and the susceptible control. However, SPAD index measurements did not.

Discussions in Chapters 2, 3, and 4 illustrate how resistance is frequently mediated by multiple categories, often involving multiple plant chemical and physical factors. Such complex combinations of resistance factors may complicate decisions about which resistance factors are of the greatest importance to study. A plant resistance index (PRI) was developed by Inayatullah et al. (1990) to combine normalized mean values for antibiosis, antixenosis and tolerance into a single index value and facilitate comparisons between genotypes. Normalized values for each category are computed on a zero-to-one scale by dividing the mean value of each genotype evaluated by the maximum mean value of all entries. Normalized tolerance values are calculated as percent reductions of plant biomass relative to a non-infested control. Normalized indices for antibiosis (X), antixenosis (Y) and tolerance (Z) are used to calculate PRI in the formula: $PRI = 1/(XYZ)$. Webster and Porter (2000) used a PRI to compare and contrast two different *S. graminum* resistant wheat cultivars, each expressing antibiosis, antixenosis and tolerance. Lage et al.

(2003) used a PRI to illustrate major differences in the resistance to *S. graminum* between three synthetic hexaploid wheat genotypes.

Although the PRI has utility, even decisions about the effects of a single resistance category may involve many possibilities. For instance, when assessing antibiosis effects, is it more important to measure larval mortality, larval weight, pupation, or to quantify a predictive allelochemical or biophysical factor? Principal component analyses (PCA) have been used by several researchers to develop internal comparisons as a way of addressing this question. Indices of resistance in spruce, *Picea* spp., to the white pine weevil, *Pissodes strobi* (Peck), developed by Alfaro et al. (1996) and Tomlin and Borden (1997), have identified tree and arthropod characters that best describe resistance, as well as important tree habitat factors that facilitate resistance. PCA have also been used to identify the key arthropod or plant biological measurements that most accurately describe and predict arthropod resistance in alfalfa to potato leafhopper, *Empoasca fabae* (Harris) (Shockley et al. 2002), resistance of cowpea, *Vigna unguiculata* (L.) Walp., to the legume pod borer, *Maruca testulalis* Grey (Oghiakhe and Odulaja 1993), potato resistance to the Colorado potato beetle *Leptinotarsa decemlineata* (Say), (Horton et al. 1997), and resistance of sugarcane, a complex hybrid of *Saccharum* species, to *D. saccharalis* (White 1993).

Standardized evaluation methods exist for the evaluation of the different categories of arthropod resistance in alfalfa (Nielson 1974, Hill and Newton 1972, Simonet et al. 1978, Sorensen 1974); apple, *Malus* spp. (Wearing 1998, Wearing and Colhoun 1999); common bean, *Phaseolus vulgaris* L. (Beebe et al. 1993, Impe and Hance 1993, Kornegay and Cardona 1999, Schaafsma et al. 1998); cassava, *Manihot esculenta* Crantz (Schoonhoven 1974); cotton (Benedict 1983, Dilday 1983, George et al. 1983, Jenkins et al. 1983, Leigh 1983, Maredia et al. 1994, Schuster 1983, Tugwell 1983); cowpea (Oghiakhe et al. 1995); groundnut, *Apios americana* Medik., (Wightman et al. 1990); rice (Heinrichs et al. 1985, Kalode et al. 1989); maize (Davis et al. 1989, 1992, Guthrie et al. 1960, 1978, Hudon and Chiang 1991, Kaster et al. 1991, Mihm 1983a,b,c); muskmelon, *Cucumis melo* (L.) (Simmons and McCreight 1996); pearl millet, *Pennisetum glaucum* (L.) R. Br., (Sharma and Sullivan 2000, Sharma and Youm 1999), sorghum (Johnson and Teetes 1979, Soto 1972, Sharma et al. 1992, Starks and Burton 1977); sugarcane (Agarwal et al. 1971, Martin et al. 1975, White et al. 2001); turfgrass (Busey and Zaenkar 1992) and wheat (Berzonsky et al. 2003, Webster and Smith 1983). For further reading see the reviews of Davis (1985), Sharma et al. (1992), Smith et al. (1994) and Tingey (1986).

2.1.1. Seedlings

Greenhouse experiments allow the researcher to make large-scale evaluations of seedlings in a relatively short period of time. This technique is commonly used to evaluate plant material for resistance to leaf and stem feeding insects, and has proven to be beneficial in eliminating large numbers of susceptible plants (Table 6.2). However, some plant material resistant as a seedling may be susceptible in later growth stages, necessitating field verification of seedling resistance.

2.1.2. Mature plants

If insect damage occurs in the later vegetative stages or in the reproductive stages of plant development, field or greenhouse tests should be conducted with mature plants, regardless of the increased amounts of time space and labor necessary to grow plants to the age of evaluation. A common initial procedure is to identify several potential

Table 6.2. Crop plants evaluated as seedlings for resistance to arthropod pests

Crop Plant and Arthropod(s)	Reference(s)
Brachiaria spp.	
Aeneolamia varia	Cardona et al. 1999
Gossypium hirsutum	
Lygus hesperus	Leigh 1983
Tetranychus urticae	Schuster 1983
Medicago sativa	
Acyrthosiphon pisum	Nielson 1974
Empoasca fabae	Sorensen & Horber 1974
Hypera postica	Sorensen 1974
Philaenus spumarius	Hill & Newton 1972
Sitona hispidulus	Byers et al. 1996
Therioaphis maculata	Nielson 1974
Melilotus/Trifolium	
Acyrthosiphon pisum	Zheng et al. 1994
Halotydeus destructor	Marshall et al. 1977
Hypera meles	Smith et al. 1975
Sitona cylindricollis	Gross et al. 1964, Murray 1996
Therioaphis trifolii	Gorz et al. 1979
Oryza sativa	
Mythimna separata	Heinrichs et al. 1985
Nephotettix cincticeps	
Nephotettix nigropictus	
Nephotettix virescens	
Nilaparvata lugens	
Nymphula depunctalis	
Orseolia oryzae	
Sogatella furcifera	
Sorghum biocolor	
Atherigona soccata	Nwanze et al. 1992
Chilo partellus	Nwanze & Reddy 1991
Schizaphis graminum	Starks & Burton 1977
Lepidopterous larvae	Soto 1972
Triticum aestivum	
Diuraphis noxia	du Toit 1987, Webster et al. 1987
Oulema melanopus	Webster & Smith 1983
Schizaphis graminum	Starks & Burton 1977
Sipha flava	Merkle & Starks 1985

sources of resistant in preliminary bioaasays, and to evaluate several of these genotypes simultaneously (Figure 6.8). Rufener et al. (1987) described a laboratory-greenhouse procedure that evaluates plants of soybean, *Glycine max* (L.) Merr., in the vegetative stage of development for resistance to feeding by larvae of Mexican bean beetle, *Epilachna varivestis* Mulsant. Identification of resistant plants before pollination allows crosses involving these plants to be made in the same growing season and reduces the amount of time required to develop beetle-resistant cultivars. In field studies, planting dates should be adjusted to coincide with the expected time of peak insect abundance. If necessary, two or three separate plantings should be made over time, in order to have one planting that best coincides with the arthropod population peak.

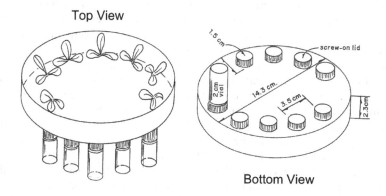

Figure 6.8. A clear plastic bioassay chamber used to measure resistance in alfalfa cuttings to Empoasca fabae. *(From Roof et al. 1976, Reprinted with permission from Environ. Entomol., Vol. 5:295-301. Copyright 1976, the Entomolgical Society of America)*

2.2. Altered Plant Tissues

Once resistance has been positively identified, several methods can be employed to alter the configuration of plant tissues, in order to conduct in-depth determinations of the factors that mediate resistance. The effects of physical structures involved in resistance, such as trichomes, can be removed to determine the effects of these structures (if any) on arthropod behavior (Gibson 1976, Khan et al. 1986). Harman et al. (1996) developed a bioassay to evaluate the effect of anacardic acids produced by leaf trichomes of geranium, *Pelargonium x hortorum* (L.), in resistance to the two-spotted spider mite, *Tetranychus urticae* Koch. Intact resistant plants are temporarily rendered susceptible by removal of acids with water rinses, and the role of acids in resistance can be evaluated as they are regenerated by trichomes.

Tissues can then be altered physically by drying and grinding, followed by removal of extractable phytochemicals by solvent extraction. However, this technique should be used with caution, as Muzika et al. (1990) found large-scale differences in the amount of monoterpenes extracted from conifer needles when comparisons were made between solvent extraction, steam distillation, and liquid CO_2 extraction. Extracts can then be assayed for deterrence by applying them onto inert substrates such as discs of filter paper, polyurethane foam (Ascher and Nemny 1978), glass fibers (Adams and Bernays 1978, Stadler and Hanson 1976) and cellulose nitrate membrane filters (Bristow et al. 1979, Doss and Shanks 1986). The concentrations of allelochemicals in this type of assay substrate may not always be the biological equivalent of those occurring in living plant tissues, as Woodhead (1983) found that 75 to 80% of the chemicals applied to glass fiber discs were located along the disc periphery. Extracts can also be evaluated for their effects on arthropod growth and metabolism by adding them in solution to an inert substrate such as cellulose. After removal of solvent, the extract-amended cellulose "cake" is incorporated into the artificial diet mixture for infestation and bioassay (Chan et al. 1978).

Biochemical interactions may occur between allelochemicals obtained from plants and components in the artificial diet, masking the effects of the extract or allelochemical (Reese 1983). In addition, many artificial diets are super-optimal, and subtle allelochemical effects may be masked (Rose et al. 1988). Un-extracted leaf powders (Smith and Fischer 1983, Quisenberry et al. 1988) and homogenized fresh plant parts (Wiseman et al. 1986) can also be added to diets, but their effects may be proportionately reduced by dilution in the diet. For this reason, it is important to assay allelochemicals in diets that closely resemble the nutritional content of the pest insect's host plant, at concentrations in which the allelochemical occurs in fresh plant tissue. Diawara et al. (1991) showed that the effects of dried florets from sorghum genotypes resistant to *S. frugiperda* were greater when larvae were fed artificial diet containing florets but lacking pinto bean (a protein source), compared to diets containing pinto bean.

Several different systems have been developed for the collection of volatile allelochemicals released from intact tissues of growing plants (Heath and Manukian 1992, 1994, Loughrin et al. 1990). Volatiles emitted by plants can be collected directly from the air by absorption onto porous adsorbent polymers as Porapak Q® (ethyl-vinylbenzone-divinyl benzene copolymer), Tennax (2,6-diphenyl-p-phenylene oxide), or activated carbon. Plant volatiles from air pulled downward in the collection chamber are passed through a collection point and trapped on the adsorbent. After collection, the adsorbent is removed and the volatiles are eluted with various non-polar organic solvents. The extract is then concentrated and analyzed with gas chromatography/mass spectroscopy (GC/MS). Dilutions equivalent to in-plant concentration are then applied to an inert material and bioassayed for attraction or repellency in an olfactometer (See Section 3.2.3).

Allelochemicals may also be collected from intact plant root systems using various types of adsorbent resins (Tang 1986). Collection of allelochemicals is achieved by cycling nutrient solutions through the plant root system several times to accumulate allelochemicals produced by plant roots.   Extraction of these allelochemicals from the adsorbent resins is similar to that for volatile allelochemicals. Quantification of allelochemicals from aqueous solutions is often accomplished using high-pressure liquid chromatography (HPLC).

Containers used to evaluate allelochemicals should be small, easy to handle, and afford the researcher the ability to view arthropod behavior during the bioassay. Common test containers include polystyrene plastic insect rearing cups, trays with preformed wells for diet and insects, and plastic or glass petri dishes with quadrant divisions.

Though they lack humidity control, paper cartons, with ventilated inserts in the lids, may also be suitable for use. For an in depth discussion of insect feeding bioassays, see the review of Lewis and van Emden (1986).

Variation in maize callus tissue exhibits levels of resistance to *S. frugiperda*, *D. grandiosella*, and *H. zea* similar to whole plant foliage (Williams and Davis 1985, Williams et al. 1985, 1987a, 1987b).   Callus tissues from *S. frugiperda*-resistant cultivars of Bermuda grass, *Cynodon dactylon* (L.), also exhibit a resistance reaction similar to whole plant foliage (Croughan and Quisenberry 1989).

3. MEASUREMENTS OF RESISTANCE

3.1. Plant Measurements

3.1.1. Direct Arthropod Feeding Injury

Measurements of insect damage to plants are usually more useful than measurements of insect growth or population development on plants, because reduced plant damage and the corresponding increases in yield or quality are the ultimate goals of most crop improvement programs. Often, measurements of yield reduction indicate direct insect feeding injury in plants. Soft x-ray photography has been used to determine the effect of insect infestation on cottonseed quality (George et al. 1983) and on maize root growth (Villani and Gould 1986). Tissue damage in plants can also be determined by measuring the incidence of tissue necrosis, fruit abscission, stem damage, or grain damage (Figure 6.9). The severity of virus-related stunting, yellowing, or curling can indicate resistance to the virus vector or resistance to the virus itself. Measurements of the cosmetic grade of fruits and vegetables can also be used to measure the effect of insect damage on the aesthetic value of produce. Insect defoliation to plants is routinely determined by rating scales that make use of visual estimates of plant damage based on percentages or numerical ratings (Figure 6.10).

Several such scales are used to evaluate foliar damage by important insect pests of maize (Table 6.3), rice (Table 6.4), and sorghum (Table 6.5). Bohn et al. (1999) demonstrated a high correlation between maize yield reduction and stalk damage ratings in evaluations of maize for resistance to *O. nubilalis*.

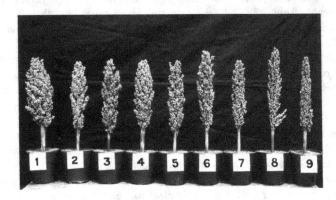

*Figure 6.9. Visual damage scale used to rate sorghum for resistance (1) and susceptibility (9)
to* Contarinia sorghicola. *(From Sharma et al. 1992, Reprinted with permission of ICRISAT)*

*Table 6.3. Rating scale used to evaluate maize and sorghum genotypes for leaf feeding
resistance to Chilo, Diatraea, Helicoverpa, Heliothis, Ostrinia, and Spodoptera [a]*

Resistance Level	Rating	Description
Highly Resistant	1	No damage or few pinholes
Resistant	2	Few shot holes on a few leaves
	3	Shot holes on several leaves
Intermediately Resistant	4	Shot holes on several leaves, few long lesions
	5	Several leaves with long lesions
	6	Several leaves with lesions <2.5 cm
	7	Long lesions common on 1/2 of leaves
Susceptible	8	Long lesions common on 1/2-2/3 of leaves
	9	Most leaves with long lesions

[a] *From Davis et al. (1989, 1992), Guthrie et al. (1960, 1978) and Mihm (1983a,b)*

Photometric leaf area meters have commonly been used to assess differences in
arthropod defoliation to different plant cultivars for many years (Kogan and Goeden
1969). Fladung and Ritter (1991) developed an inexpensive, accurate means of
assessing difference in defoliation using a hand held scanner and a personal
computer. The scanner records defoliation data and scanner software transfers the
defoliation image to the computer. Digitized image data are then quantified and
analyzed. This technology has been used to assess leaf - feeding deterrence to the
armyworm *Spodoptera litura* (F.), and damage by western flower thrips,
Frankliniella occidentalis (Pergande), to cucumber (Escoubas et al. 1993, Mollema et al.
1992).

Indirect feeding injury measurements related to plant growth such as photosynthetic, transpiration, and respiratory rates can also be recorded, although they are farther removed from reductions in plant biomass yield. Insect feeding injury can also be simulated by mechanical defoliation. However, plants respond very differently to artificial defoliation than to actual arthropod tissue removal.

Table 6.4. Rating scales used to evaluate rice for resistance to common insect pests [a]

Score	C. suppressalis (% dead heart)	O. oryzae (% galls)	Leafhopper/ planthopper damage	Hydrellia philippina damage
0	0	0	None	No lesions
1	1-10	<1	Slight	Pinhead lesions
3	11-25	1-5	Leaves 1 & 2 yellow	Lesions ~1 cm long
5	26-40	6-15	Plants stunted; >1/2 leaves yellow	Lesions >1 cm, but on < ½ total leaf
7	41-60	16-50	> ½ plants dead; ½ severely stunted & wilted	Lesions on ~ ½ leaf
9	61-100	51-100	All plants dead	Large lesions on ½ leaf, leaf broken

[a] *From Heinrichs et al. (1985)*

Table 6.5. Rating scales used to measure insect resistance in sorghum [a]

Damage Score	Schizaphis graminum damage [b]	Contarinia sorghicola damage (percent)
0	---------------	0
1	No red spots on leaves	1-10
2	Red spots on leaves	11-20
3	Part of 1 leaf dead	21-30
4	One leaf dead	31-40
5	Two leaves dead	41-50
6	Four leaves dead	51-60
7	Six leaves dead	61-70
8	Eight leaves dead	71-80
9	Entire plant dead	81-90
10	---------------	91-100

[a] *From Johnson and Teetes (1979)*
[b] *Estimates of* S. graminum *population size should also be taken if possible*

Figure 6.10. A visual rating scale for measuring resistance in maize to Spodoptera
frugiperda. Ratings of 4 (resistant) [A] and 9 (susceptible) [B], 7 days after infestation;
Ratings of 5 (resistant) [C] and 9 (susceptible) [D], 14 days after infestation. (From Davis et
al. 1992, Reprinted with permission of the Mississippi Agricultural and Forestry Experiment
Station)

The relationship between artificial and natural defoliation should be closely scrutinized before accepting and utilizing results based exclusively on artificial defoliation.

3.1.2. Correlations of Plant Factors to Arthropod Resistance

A thorough knowledge of the actual cause of resistance may not be essential for the development of resistant cultivars. However, this information may target specific phenotypic and genotypic characters that can be monitored during the breeding and selection process (see Chapter 8). If resistance is chemically or morphologically based, concentrations of allelochemicals or the density and/or size of morphological structures present in tissues of resistant cultivars can be measured, allowing a more rapid determination of potentially resistant plant materials.

If plant allelochemical and morphological factors are evaluated separately from arthropod damage, experimental variation due to variation in the test arthropod is avoided. Separate experiments can then be designed to study the effect of a specific resistance factor on the arthropod in replicated field or greenhouse experiments. However, the demonstration of allelochemical or morphological differences between resistant and susceptible plants does not always conclusively demonstrate that these factors are involved in conditioning resistance.

Some of the physical and allelochemical resistance factors described in Chapters 2, 3 and 4 have been used to monitor for arthropod resistance in different crop plants. Quantitative differences in the trichome density, length or type in wheat and alfalfa have been used to select for resistance to feeding by larvae of the cereal leaf beetle, *Oulema melanopus* L., and the alfalfa weevil, *Hypera postica* Gyllenhal (Hoxie et al. 1975, Kitch et al. 1985).

Differences in the physical leaf and stalk characteristics of maize have also been correlated with arthropod resistance. Leaf toughness, as measured by an instron leaf penetrometer is correlated to maize resistance to the *O. nubilalis* larval survival (Bergvinson et al. 1994). Maize resistance to *O. nubilalis* may also be correlated to physical stalk strength, as measured by resistance penetrometer (Abedon and Tracy 1996, Zuber and Grogan 1961). The resistance of sugarcane to *D. saccharalis* is also correlated to stalk strength using stalk resistance penetrometery (Martin et al. 1975). Kolb et al. (1998) also used penetrometry to correlate resistance of ponderosa pine, *Pinus ponderosa* var. *scopularum* Englm., to various species of *Dendroctonus* spp.

The vertical maize root pull technique described in Chapter 4 was used to successfully identify initial sources of resistance to the western corn rootworm, *Diabrotica virgifera virgifera* LeConte. However, more recent results obtained by Knutson et al. (1999) indicate that visual root damage measurements originally developed by Hills and Peters (1971) are a more reliable and concise measurement of *D. virgifera* resistance.

These examples withstanding, it is very important to understand the inter-relationships of all potential resistance factors before placing total confidence in a single physical factor for determining or predicting resistance. Navon et al. (1991)

demonstrated that resistance in pubescent cotton to *Heliothis armigera* remained even after shaving trichomes from the resistant cultivars. In fact, the resistance is likely due to antibiotic effects of allelochemicals from pigment glands in the leaves of resistant cultivars, and not in leaf trichomes.

Concentrations of organic acids have been shown to predict arthropod resistance in plants. Robinson et al. (1982) were one of the first groups to do so, developing an accurate, efficient thin layer chromatography technique to identify maize lines with high concentrations of 6-methoxybenzoxazolinone (MBOA) for resistance to the European corn borer. Gérard et al. (1993) analyzed leaves of genotypes of pear, *Pirus communis* L., resistant to pear psylla, *Cacopsylla pyricola* Forester, using HPLC and GC/MS to demonstrate a correlation between resistance and 16-hydroxyhexadecanoic acid. The fluorescence of roots of lettuce, *Lactuca sativa* L., resistant to the lettuce root aphid, *Pemphigus bursarius* L., is due to higher quantities of isochlorogenic acid and was shown to be a predictor of aphid resistance (Cole 1987). Rutherford (1998) used near-infrared spectroscopy of sugarcane chlorogenates and flavonoids to reliably predict resistance to the stalk borer *Eldana saccharina* Walker. Leaf surface isoflavones of subterranean clover, *Trifolium subterraneum* (L.), have been proposed by Wang et al. (1999) as a simple method of determining relative clover resistance to the redlegged earth mite, *Halotydeus destructor* Tucker.

Allelochemical 'profiling' has been described as a predictive means of selecting for resistance to several insects. Ekman et al. (1973) was one of the first to do so, after developing a rapid method for determining wheat resistance to the sunn pest, *Eurygaster integriceps* Put., based on degradation rates of endosperm starch. Ellsbury et al. (1992) evaluated numerous accessions of white clover, *Trifolium repens* L., for resistance to *H. postica* and suggested that resistance can be improved by selection of clover with increased cyanogenic glycoside levels. Discriminant analyses of phenotypic reaction and gas chromatography data was used by Brennan et al. (1992) to demonstrate that resistance in blackcurrant, *Ribes lacustre* (Pers.) Poir., to the gall mite, *Cecidophyopsis ribis* Westw., is highly correlated to bud terpenoid content. Brignolas et al. (1998) used principal component analyses to show that the phloem phenolic content of Norway spruce is a strong predictor of spruce resistance to the bark beetle, *Ips typographus* L., and its associated fungus, *Ceratocystis polonica*. Allelochemical profiles have been used by plant breeders to improve cultivars for insect resistance in cotton (Hedin et al. 1983, 1991), maize (Hedin et al. 1984), strawberry, *Fragaria* x *ananassa* Duchesne (Hamilton-Kemp et al. 1988), and tomato (Quiros et. al. 1977, Andersson et al. 1980).

As previously mentioned for physical factors, however, reliance on correlations of a single allelochemical factor as the sole predictor of resistance may be unwise, especially without confirming field data of resistant plant performance. The antibiotic effects of DIMBOA and 6-MBOA have been extensively demonstrated with pest lepidoptera of maize and with resistance of cereals to the rose grain aphid, *Metopolophium dirhodum* (Walker) (see Chapter 3). However, DIMBOA and DIBOA concentrations in barley, wheat and rye, rye, *Secale cereale* L., have only been correlated to aphid resistance (Argandona 1981, Barria et al. 1992, Thackray et al. 1990) and have not been shown to play a role in aphid resistance under field conditions. The same is true of the leaf content of the indole alkaloid gramine, which has been correlated with barley resistance to the bird cherry oat aphid, *Rhopalosiphum padi* (L.). Moharramipour et al. (1997) and Ahman et al. (2000) found no correlation between field *R. padi* population levels and gramine content, indicating that gramine does not confer barley resistance to *R. padi*.

3.2. Arthropod Measurements

Measurements of arthropod population development and behavior provide important supplemental information to plant measurements of resistance. These measurements are of primary importance in determining the existence of antibiosis, antixenosis, and/or tolerance.

3.2.1. Sampling Arthropod Populations

Arthropod populations should be sampled at the plant site where damage occurs, during the phenological development of the plant when the pest is normally present, and at a time of day when the arthropod is normally active on the plant. Populations of non-mobile arthropod may be estimated visually, but this method is subject to error because of the variations in canopy size between different cultivars of crop plants. Shaking or beating plant foliage to dislodge larger, visible arthropods onto a ground cloth for counting or collection is a more accurate method of population estimation. More mobile or active arthropods can be better sampled with a sweep net, a mobile vacuum collection machine (D-Vac), or by anesthetizing caged insects with carbon dioxide. The vacuum collection method is usually less damaging than the sweep net method, but it is most effective in collecting light-bodied arthropods. Pitfall traps, sticky traps, light traps, and pheromone traps may also be used to measure insect population density, but these measurements are, at best, only indirect, since they are competing with the test plants for attractiveness.

3.2.2. Measurements of Arthropod Growth

Arthropod development can be monitored to determine if antibiotic and antixenotic effects are exhibited by arthropods confined to the foliage of resistant cultivars. Several measures of arthropod metabolic efficiency can also be determined, using various nutritional indices (Waldbauer 1968).

The consumption Index (CI) is calculated as:
    [weight of food eaten/ (mean weight of larvae during testing/ test duration)].
Approximate digestibility (AD) is calculated as:
    [(weight of food eaten - weight of feces)/ weight of food eaten].
The efficiency of conversion of ingested food (ECI) is calculated as:
    (weight gain of larvae/weight of food eaten).

The efficiency of conversion of digested food (ECD) is calculated as:
[weight gain of larvae / (weight of food eaten – weight of feces)].

The CI indicates whether antixenotic properties are present in the resistant cultivar, because consumption depends on positive gustatory stimulation in the early stages of feeding. AD, ECI, and ECD are all indicators of potential antibiotic effects of a resistant cultivar, because each measures metabolic processes that affect insect nutritional physiology. Nutritional index measurements have been shown to give the greatest precision when insects consume at least 80% of the available food offered in the experiments (Schmidt and Reese 1986). Bowers et al. (1991) found that nutritional indices were affected by plant leaf age and arthropod test cohort size (see Chapter 7 for additional discussion of other arthropod and plant variables). Nutritional indices have been used to quantify resistance mechanisms in cotton (Mulrooney et al. 1985, Montandon et al. 1986, and Shaver et al. 1970) and soybean (Reynolds et al. 1984). For a more in-depth discussion of the use of nutritional indices, see Reese (1978).

Nutritional index values are difficult to determine for feeding by Heteroptera and Homoptera. However, Khan and Saxena (1984a) developed a technique to monitor feeding by the green rice leafhopper, *Nephotettix cincticeps* (Uhler), after uptake by rice plants of a safranine dye that is higly selective of lignin and translocated in xylem tissues. Red honeydew excreted by hoppers feeding on treated resistant seedlings is indicative of xylem feeding. Hoppers fed treated susceptible seedlings excrete clear honeydew, indicating phloem feeding. In related research, Pathak and Heinrichs (1982) developed a technique to monitor feeding of the brown planthopper, *Nilaparvata lugens* Stål, on rice using filter paper treated with bromocresol green. Where drops of honeydew containing amino acids contact the treated filter paper, a colorometric reaction occurs, and dark blue spots appear. The area of the spots is strongly correlated to the weight of actual honeydew produced during feeding on both resistant and susceptible cultivars. Habbi et al. (1993) developed a miniaturized chamber slide bioassay system to evaluate the effects of plant allelochemicals on survival of *E. fabae*. The system was used to effectively evaluate allelochemicals for toxicity and deterrence to *S. graminum* (Formusoh et al. 1997). The review of Chansigaud and Strebler (1986) offers an extensive discussion of many of the methods used to study the effects of allelochemicals on the feeding behavior of Homoptera.

3.2.3. Measurements of Arthropod Behavior

Several techniques used to quantify arthropod behavior may also be of use in the determination of plant antixenosis. The olfactory responses of arthropods to volatile allelochemical stimuli have been observed since the beginning of the 20th century, when Barrows (1907) and McIndoo (1926) developed the first olfactometers. Simple two-choice (Y-tube) olfactometers or multi-choice devices have been used to measure the response of several species of insects to volatile plant allelochemicals. More complex olfactometers such as a low speed wind tunnel (Visser 1976) and an olfactometer room (Payne et al. 1976) have also been used to measure insect olfactory responses to plant volatiles. Information about the olfactory responses of pest arthropods to plant volatiles provides plant breeders an additional means to select cultivars with reduced levels of plant attractants or increased levels of plant repellents. For additional in-depth discussions of olfactometers, see comments in reviews by Finch (1986), Visser (1986) and Smith et al. (1994).

Arthropod behavior can also be measured electrophysiologically, to indicate the effects of the resistant plant cultivar on arthropod olfaction. Electroantennograms (EAGs) are measures of the response of olfactory receptors commonly located on antennal sensilla to plant olfactory stimuli. Perception of attractant stimuli elicits hyperpolarizing nerve potentials, while perception of repellent stimuli elicits hypopolarizing nerve potentials. These electrical potentials are transduced by a clamping preamplifier and placed in computer storage or displayed on either an oscilloscope or a strip chart recorder.

Electroantennograms are routinely obtained by splitting the outlet from a gas chromatograph column between the chromatograph detector and the arthropod antenna. Tentative assignments of the effects of individual components of plant aroma on olfactory perception can be made by comparing the retention times of plant volatiles separated chromatographically with the EAG response.

The response of arthropod contact chemoreceptors to various allelochemicals can also be determined electrophysiologically. The allelochemical stimulus is applied as a vapor or as a solution and the resulting stimulant or deterrent nerve potentials are recorded and observed. Frazier and Hanson (1986) provide a detailed explanation and discussion of the use of electrophysiological recording techniques to monitor arthropod chemosensory responses.

Electroretinograms (ERGs) measure arthropod visual response spectra and can provide information concerning arthropod perception of both monochromatic light and color. Techniques for determining the ERGs of several different insects were developed by Agee (1977). Electroretinogram data may be useful to plant breeders by providing them data that will enable the development and production of plant cultivars with pigmentation patterns outside of or abnormal to an arthropod's normal color perception range (see Chapter 2).

An electronic feeding monitor, originally developed by McLean and Kinsey (1966) and McLean and Weigt (1968), has been acknowledged as the "single most important advance in the development of specialized techniques for studying homopteran feeding behaviors" (Walker 2000). The device, now more commonly known as the electronic penetration graph (EPG) (Tjallingii 1985), involves passing a small electrical current across a test insect and plant, both of which are wired to a tape recording device. Recorded data are transferred to a personal computer, digitized and analyzed, and may also be displayed on a strip chart recorder. Feeding activity is detected when the insect feeding stylets penetrate the plant tissue at various depths, causing a change in the electrical conductance by the plant tissues (Figure 6.11).

These changes are converted electronically and displayed as feeding waveforms (Figure 6.12). Differences in the type of waveform produced during insect feeding indicate the frequency of feeding and differences in feeding on either plant xylem or phloem. Both AC and DC systems are used to generate EPG data. Reese et al. (2000) compared differences and similarities between the two systems. All aspects of recording, storage, and analyzing feeding data measured by EPG monitoring are described and discussed in detail by Walker and Backus (2000).

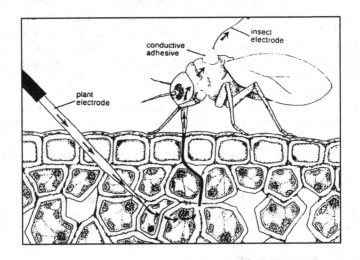

Figure 6.11. Pathway of electrical current at the insect-plant interface used to create an electronic penetration graph monitor. Plant electrode placement for illustrative purposes only. (From Walker (2000. Reprinted with permission from Thomas Say Publications in Entomology, Copyright 2000, Entomological Society of America)

Several studies have been documented correlations between the EPG waveforms of various species of aphids to specific probing behaviors during feeding (Backus 1994). Reese et al. (1994) reviewed the previous uses of EPG in investigations of plant resistance to aphids and noted that resistance to several aphid species is related to a longer period of time required by aphids to contact phloem tissues and to the reduced ingestion of phloem contents.

EPG monitoring has been used to identify the plant phloem as the site of resistance in melon to *A. gossypii* (Klingler et al. 1998), and in tomato cultivars resistant to *T. vaporariorum* (Lei et al. 1999). Van Helden and Tjallingii (1993) found that the phloem is the site of resistance factors in lettuce cultivars resistant to the lettuce aphid, *Nasonovia ribisnigri* (Mosley). Phloem sap collected from *N. ribisnigri*-resistant plants is less preferred for aphid feeding than sap from susceptible plants (van Helden et al. 1995). These phloem factors have thus far eluded detection, however, as experiments by van Helden et al. (1994) failed to demonstrate differences in the chemical composition of the phloem sap collected from stylectomized aphids feeding on resistant and susceptible lettuce cultivars.

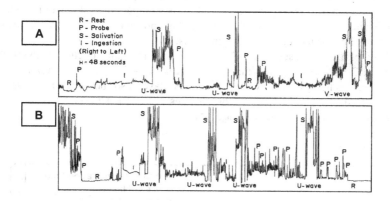

Figure 6.12. Electronic waveforms recorded during feeding of Sogatella frucifera *on (A) susceptible rice plants and (B) resistant rice plants using an electronic penetration graph monitor. (From Khan & Saxena 1984b. Reprinted with permission from J. Econ. Entomol., Vol. 77:1479-1482. Copyright 1984, the Entomolgical Society of America)*

EPG monitoring has also revealed differences in the feeding responses of planthoppers on resistant and susceptible rice cultivars (Khan and Saxena 1984b, Velusamy and Heinrichs 1986) (Figure 6.12).

Calderon and Backus (1992) developed EPG data by monitoring the feeding of *E. fabae* and *Empoasca krameri* Ross & Moore, on hopper-resistant or susceptible common bean cultivars. The time spent probing is no different for either *Empoasca* species on resistant and susceptible cultivars, but the type of *Empoasca* feeding does differ. On the tolerant cultivars, hoppers decrease the duration of feeding and the number of multiple-cell laceration probes. Serrano et al. (2000) used principal component analyses to identify the major components of *E. krameri* feeding, to develop a stylet penetration index (SPI), computed as: $SPI = 1 - [(r + 1) - p]$.

The first principal component, (p) is the pulsating laceration score. The second principal component, (r) is the cell rupturing score. The third principal component, (I) is the lancing ingestion score. SPI values greater than 1.0 indicate that damaging *E. krameri* feeding occurs, leading to yield-reducing leaf 'hopper burn'. Values of less than 1.0 indicate *E. krameri* feeding is much less damaging and hopper burn does not occur.

Serrano et al. (2000) compared the SPI values of different bean genotypes to their field-derived yield index values for *E. krameri* tolerance (Figure 6.13). SPI and yield index values are closely correlated among two tolerant genotypes and a susceptible control, but unrelated for two moderately tolerant genotypes. SPI provides valuable information on the types of hopper feeding on different bean genotypes, and is less sensitive to environmental fluctuation than field screening. SPI may also significantly reduce the time required to screen genotypes. However, SPI does not provide yield data about genotypes and is quite sensitive to changes in *E. krameri* probing behavior. Serrano et al. (2000) suggest that SPI may be better utilized in the later generations of bean breeding programs to detect *E. krameri* tolerance among advanced generation genotypes.

Caillaud et al. (1995) used a related technique, discriminant function analysis, to delineate EPG differences in feeding by the cereal leaf aphid, *Sitobion avenae* (F.), among wheat genotypes. EPG-based indices have also been used to assess differences in feeding by *M. persicae* on potatoes (Holbrook and Reeves 1979) and sugar beet, *Beta vulgaris* L. (Haniotakis and Lange 1974).

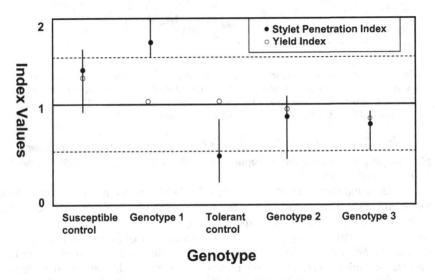

Figure 6.13. Comparison of mean stylet penetration index values and mean yield index values for five genotypes of Phaseolus vulgaris *evaluated for tolerance to* Empoasca kraemeri. *(From Serrano et al. 2000. Reprinted with permission from J. Econ. Entomol., Vol. 93:1796-1809. Copyright 2000, Entomological Society of America)*

4. CONCLUSIONS

The techniques and devices used to record arthropod activity on plants and the resulting plant damage are as varied as the particular combination of pest arthropod and host plant being investigated. Techniques such as mass seedling evaluation and mechanical inoculation of plants with artificially- reared arthropods have led to great increases in the speed and efficiency of plant resistance evaluations, and have decreased the time required to release an arthropod resistant cultivar. The adoption of standardized damage rating scales for many key pest species, in particular foliar feeding lepidopteran larvae, has also contributed to the rapid, accurate development of arthropod resistant cultivars. As a result, the development of resistant cultivars of several crop plants have been developed and placed into integrated pest management systems with precision and confidence. The adoption of various plant resistance indices to evaluate antixenosis, antibiosis and tolerance has added a new dimension to plant resistance bioassays and increased confidence in the results of germplasm assessments. These measurements have increased the degree of accuracy involved in defining the levels of arthropod resistance in germplasm and greatly improved the researcher's ability to accurately assess the contribution of each category of resistance.

Although much progress has been made in developing accurate and efficient techniques to access plant resistance to arthropods, new and improved techniques are still needed. A common problem is the lack of a constant supply of test arthropods with which to conduct research. This problem usually stems from the lack of an artificial diet with which to produce arthropods or the need to improve procedures to rear and handle the test arthropod in question. The rearing of the major pest species of foliar-feeding Lepidoptera has become a largely commercial operation. However, this is not the case for many pest species of Coleoptera, Hemiptera and Homoptera. There is much research remaining to be conducted to develop rearing methods for these insects.

In laboratory research, there is still a need for an improved knowledge of plant nutrient composition, in order to design artificial diets that more accurately reflect the nutrient composition of an arthropod's host plant. This understanding is critical to the success of determining the true contributions of plant allelochemicals to arthropod resistance. Many of the artificial diets currently used for laboratory culture of Lepidoptera are super-optimal in nutrient concentration, masking the effects of allelochemicals that may have subtle effects on the expression of plant resistance. There is a growing number of new micro-analytical techniques to determine and quantify allelochemical resistance factors in plants. Prudent use of these techniques, tied to a thorough knowledge of how resistance is mediated by the allelochemical, should contribute to faster, more accurate resistance assays.

Can the "perfect" cage be designed for plant-arthropod evaluations? There remains a need to develop cage materials that allow even better light transmission and airflow while maintaining sufficiently small pore sizes to contain arthropods. Improved arthropod infestation techniques and devices that consume less time and minimize handling damage to arthropods, such as the larval Lepidoptera inoculator, will also be very useful. Finally, the development and refinement of standardized rating scales to determine arthropod damage or defoliation in crops other than those described and discussed previously (maize, sorghum, rice, wheat) would greatly facilitate the development of arthropod resistant cultivars in several additional crop plant species.

Much of the progress in plant resistance to arthropods has been due to conventional plant evaluation (phenotyping) and plant breeding techniques.

Chapters 8, 9, and 10 include extensive descriptions of the use of molecular plant technologies that have been adapted to develop arthropod resistant plants. However, problems inherent to the development of arthropod resistance (non-target arthropod susceptibility, development of virulent biotypes, and poor agronomic qualities in resistant parent material) will likely follow the development of resistant cultivars, whether by conventional or transgenic means. New techniques to accurately identify plant resistance to arthropods must be a dominant element of future crop cultivar improvement research efforts.

REFERENCES CITED

Abedon, B. G., and W. F. Tracy. 1996. *Crongrassl* of maize (*Zea mays* L.) delays development of adult plant resistance to common rust (*Puccinia sorghi* Schw.) and European corn borer (*Ostrinia nubilalis* Hübner). J. Hered. 87:219-223.

Adams, C. M., and E. A. Bernays. 1978. The effect of combinations of deterrents on the feeding behavior of *Locusta migratoria*. Entomol. Exp. Appl. 23:101-109.

Agee, H. R. 1977. Instrumentation and techniques for measuring the quality of insect vision with the electroretinogram. USDA-ARS-S-162. 13 pp.

Agarwal, R. A., J. P. Singh, and C. B. Tiwari. 1971. Technique for screening of sugarcane varieties resistant to top borer, *Scirpophaga nivella* F. Entomophaga. 16:209-220.

Ahman, I., S. Tuvesson, and M. Johansson. 2000. Does indole alkaloid gramine confer resistance in barley to aphid *Rhopalosiphum padi*? J. Chem. Ecol. 26:233-255.

Alfaro, R. I., F. He, G. Kiss, J. King, and A. Yanchuk. 1996. Resistance of white spruce to white pine weevil: development of a resistance index. Forest Ecol. Management. 81:51–62.

Anderson, T. E., and N. C. Leppla. Eds. 1992. Advances in Insect Rearing for Research and Pest Management, Oxford & IBH Publishing Co., New Delhi.

Andersson, B. A., R. T. Holman, L. Lundgren, and G. Stenhagen. 1980. Capillary gas chromatograms of leaf volatiles. A possible aid to breeders for pest and disease resistance. J. Agric. Food Chem. 28:985–989.

Argandona, V. H., H. M. Niemeyer, and L. J. Corcuera. 1981. Effect of content and distribution of hydroxamic acids in wheat on infestation by the aphid *Schizaphis graminum*. Phytochem. 20:637–76.

Ascher, K. R. S., and N. E. Nemny. 1978. Use of foam polyurethane as a carrier for phagostimulant assays with *Spodoptera littoralis* larvae. Entomol. Exp. Appl. 24:546–548.

Backus, E. A. 1994. History, development, and applications of the AC electronic monitoring system for insect feeding. In: M. M. Ellsbury, E. A. Backus, and D. L. Ullman (Eds.), History, Development, and Application of AC Electronic Insect Feeding Monitors. Thomas Say Publications in Entomology. Entomological Society of America, Lanham, MD, pp. 1–51.

Barria, B. N., S. V. Copaja, and H. M. Niemeyer. 1992. Occurrence of DIBOA in wild *Hordeum* species and its relation to aphid resistance. Phytochem. 31:89–91.

Barrows, W. M. 1907. The reactions of the pomace fly, *Drosophila ampelophila* Loew., to odorous substances. J. Exp. Zool. 4:515–537.

Beebe, S., C. Cardona, O. Díaz, F. Rodríguez, E. Maníca, and S. Ajquejay. 1993. Development of common bean (*Phaseolus vulgaris* L.) lines resistant to the pod weevil, *Apion godmani* Wagner, in Central America. Euphytica. 69:83–88.

Benedict, J. H. 1983. Methods of evaluating cotton for resistance to the boll weevil. In: Host Plant Resistance Research Methods for Insects, Diseases, Nematodes and Spider Mites in Cotton. Southern Cooperative Series Bulletin 280. pp. 19–26.

Berenbaum, M. 1986. Postingestive effects of phytochemicals on insects: On paracelsus and plant products. In: J. R. Miller and T. A. Miller (Eds.), Insect-Plant Interactions. Springer-Verlag , New York. pp. 121–153.

Bergvinson, D. J., J. T. Arnason, R. I. Hamilton, J. A. Mihm, and D. C. Jewell. 1994. Determining leaf toughness and its role in maize resistance to the European corn borer (Lepidoptera: Pyralidae). J. Econ. Entomol. 87:1743–1748.

Berzonsky, W. A., H. Dong, S. D. Haley, M. O. Harris, R. J. Lamb, R. I. H. McKenzie, H. W. Ohm, F. L. Patterson, F. B. Peairs, D. R. Porter, R. H. Ratcliffe, and T. G. Shanower. 2003. Breeding wheat for resistance to insects. Plant Breed. Rev. 22:221–296.

Bohn, M., R. C. Kreps, D. Klein, and A. E. Melchinger. 1999. Damage and grain yield losses caused by European corn borer (Lepidoptera: Pyralidae) in early maturing European maize hybrids. J. Econ. Entomol. 92:723–731.

Boina, D., S. Prabhakar, C. M. Smith, S. Starkey, L. Zhu, E. V. Boyko, and J. C. Reese. 2005. Categories of resistance to greenbug (Homoptera: Aphididae) biotype I in the wheat genes *Gby* and *Gbz*. J. Kansas Entomol. Soc.

Bowers, M. D., N. E. Stamp, and E. D. Frajer. 1991. Factors affecting the calculation of nutritional indices for foliage-fed insects: an experimental approach. Entomol. Exp. Appl. 61:101–116.

Brennan, R. M., G. W. Robertson, J. W. McNicol, L. Fyfee, and J. E. Hall. 1992. The use of metabolic profiling in the identification of gall mite (*Cecidophyopsis ribis* Westw.) - resistant blackcurrant (*Ribes nigrum* L.) genotypes. Ann. Appl. Biol. 121:503–509.

Brignolas, F., F. Lieutier, D. Sauvard, E. Christiansen, and A. A. Berryman. 1998. Phenolic predictors for Norway spruce resistance to the bark beetle *Ips typographus* (Coleoptera: Scolytidae) and an associated fungus, *Ceratocystis polonica*. Can. J. For. Res. 28:720–728.

Bristow, P. R., R. P. Doss, and R. L. Campbell. 1979. A membrane filter bioassay for studying phagostimulatory materials in leaf extracts. Ann. Entomol. Soc. Am. 72:16–18.

Busey, P., and E. I. Zaenker. 1992. Resistance bioassay from Southern chinch bug (Heteroptera: Lygaeidae) excreta. J. Econ. Entomol. 85:2032–2038.

Byers, R. A., and W. A. Kendall. 1982. Effects of plant genotypes and root nodulation on growth and survival of *Sitona* spp. larvae. Environ. Entomol. 11:440–443.

Byers, R. A., W. A. Kendall, R. N. Peaden, and D. W. Viands. 1996. Field and laboratory selection of *Medicago* plant introductions for resistance to the clover root curculio (Coleoptera: Curculionidae). J. Econ. Entomol. 89:1033–1039.

Caillaud, C. M., J. S. Pierre, B. Chaubet, and J. P. Dipietro. 1995. Analysis of wheat resistance to the cereal aphid *Sitobion avenae* using electrical penetration graphs and flow charts combined with correspondence analysis. Entomol. Exp. Appl. 75:9–18.

Calderon, J. D., and E. A. Backus. 1992. Comparison of the probing behaviors of *Empoasca fabae* and *E. kraemeri* (Homoptera : Cicadellidae) on resistant and susceptible cultivars of common beans. J. Econ. Entomol. 85: 88–99.

Cardona, C., J. W. Miles, and G. Sotelo. 1999. An improved methodology for massive screening of *Brachiaria* spp. genotypes for resistance to *Aeneolamia varia* (Homoptera: Cercopidae). J. Econ. Entomol. 92:490–496.

Chan, B. G., A. C. Waiss, Jr., W. L. Stanley, and A. E. Goodban. 1978. A rapid diet preparation method for antibiotic phytochemical bioassay. J. Econ. Entomol. 71: 366–368.

Chansigaud, J., and G. Strebler. 1986. Methods for investigating the mode of action of natural or synthetic substances on the feeding behaviour of Homoptera. Agronomie. 6:845–856.

Chelliah, S., and E. A. Heinrichs. 1980. Factors affecting insecticide-induced resurgence of the brown planthopper, *Nilaparvata lugens* on rice. Environ. Entomol. 9:773–777.

Chesnokov, P. G. 1962. Methods of Investigating Plant Resistance to Pests. National Science Foundation, Washington, DC. 107 pp.

Cole, R. A. 1987. Intensity of radicle fluorescence as related to the resistance of seedlings of lettuce to the lettuce root aphid and carrot to the carrot fly. Ann. Appl. Biol. 111:629–639

Crafts-Brandnera, S. J., and C.-C. Chu. 1999. Insect clip cages rapidly alter photosynthetic traits of leaves. Crop Sci. 39:1896–1899.

Croughan, S. S., and S. S. Quisenberry. 1989. Enhancement of fall armyworm (Lepidoptera: Noctuidae) resistance in bermudagrass through cell culture. J. Econ. Entomol. 82:236–237.

Davis, F. M. 1980a. A larval dispenser-capper machine for mass rearing the southwestern corn borer. J. Econ. Entomol. 73:692–693.

Davis, F. M. 1980b. Fall armyworm resistance programs. Fla.Entomol. 63:420–433.

Davis, F. M. 1982. Mechanically removing southwestern corn borer pupae from plastic rearing cups. J. Econ. Entomol. 75:393–395.

Davis, F. M. 1985. Entomological techniques and methodologies used in research programmes on plant resistance to insects. Insect Sci. Applic. 6: 391–400.

Davis, F. M., and W. D. Guthrie. 1992. Rearing Lepidoptera for plant resistance research. In: T. E. Anderson and N. C. Leppla (Eds.), Advances in Insect Rearing for Research and Pest Management, Oxford & IBH Publishing Co., New Delhi. pp. 211–228.

Davis, F. M., S. Malone, T. G. Oswalt, and W. C. Jordan. 1990. Medium-sized lepidopterous rearing system using multicellular rearing trays. J. Econ. Entomol. 83:1535–1540.

Davis, F. M., S. S. Ng, and W. P. Williams. 1992. Visual rating scales for screening whorl-stage corn for resistance to fall armyworm. Miss. Agric. Agric. For. Exp. Sta. Technical Bull. 186. pp. 1–9.

Davis, F. M., and T. G. Oswalt. 1979. Hand inoculator for dispensing lepidopterous larvae. USDA/ARS AAT-S-9/October, 5 pp.

Davis, F. M., T. G. Oswalt, and S. S. Ng. 1985. Improved oviposition and egg collection system for the fall armyworm (Lepidoptera: Noctuidae). J. Econ. Entomol. 78: 725–9.

Davis, F. M., and W. P. Williams. 1980. Southwestern corn borer: Comparison of techniques for infesting corn for plant resistance studies. J. Econ. Entomol. 73: 704–706.

Davis, F. M., W. P. Williams, and B. R. Wiseman 1989. Methods used to screen maize for and to determine mechanisms of resistance to the southwestern corn borer and fall armyworm. In: Toward Insect Resistant Maize for the Third World: Proceedings of the International Symposium on Methodologies for Developing Host Plant Resistance to Maize Insects. Mexico, D. F.:CIMMYT.

Deol, G. S., J. C. Reese, and B. S. Gill. 1997. A rapid nondestructive technique for assessing chlorophyll loss from greenbug (Homoptera: Aphididae) feeding damage on sorghum leaves. J. Kansas Entomol. Soc. 70:305–312.

Deol, G. S., J. C. Reese, B. S. Gill, G. E. Wilde, and L. R. Campbell. 2001. Comparative chlorophyll losses in susceptible wheat leaves fed upon by Russian wheat aphids or greenbugs (Homoptera: Aphididae). J. Kansas Entomol. Soc.74: 192–198.

Diawara, M. M., B. R. Wiseman, and D. J. Isenhour. 1991. Bioassay for screening plant accessions for resistance to fall armyworm (Lepidoptera: Noctuidae) using artificial diets. J. Entomol. Sci. 26: 367–374.

Diawara, M. M., B. R. Wiseman, D. J. Isenhour, and N. S. Hill. 1992. Sorghum resistance to whorl feeding by larvae of the fall armyworm (Lepidoptera: Noctuidae). J. Agric. Entomol. 9: 41–53.

Dilday, R. H. 1983. Methods of screening cotton for resistance to *Heliothis* spp. In: Host Plant Resistance Research Methods for Insects, Diseases, Nematodes and Spider Mites in Cotton. Southern Cooperative Series Bulletin 280. pp. 26–36.

Doss, R. P., and C. H. Shanks, Jr. 1986. Use of membrane filters as a substrate in insect feeding bioassays. Bull. Entomol. Soc. Am. 32:248–249.

du Toit, F. 1987. Resistance in wheat (*Triticum aestivum*) to *Diuraphis noxia* (Hemiptera: Aphididae). Cer. Res. Comm. 15:175–179.

East, D. A., J. V. Edelson, E. L. Cox, and M. K. Harris. 1992. Evaluation of screening methods and search for resistance in muskmelon, *Cucmis melo* (L.), to the two spotted spider mite, *Tetranychus urticae* Koch. Crop Protection. 11:39–44.

Ekman, N. V., N. A. Vilkova, and I. D. Shapiro. 1973. Express-method for determining resistance levels of cereals to *Eurygaster integriceps* Put. according to disintegration rates of caryopsis starch. Proc. All-Union Inst. For. Plant Prot. 37:176–180.

Ellsbury, M. M., G. A. Pederson, and T. E. Fairbrother. 1992. Resistance to foliar- feeding Hyperine weevils (Coleoptera: Curculionidae) in cyanogenic white clover. J. Econ. Entomol. 85:2467–2472.

Escoubas, P., L. Lajide, and J. Mitzutani. 1993. An improved leaf-disk antifeedant method bioassay and its application for the screening of Hokkaido plants. Entomol. Exp. Appl. 66:99–107.

Finch, S. 1986. Assessing host-plant finding in insects. In: J. R. Miller and T. A. Miller (Eds.), Insect-Plant Interactions. Springer-Verlag, New York. pp. 23–64.

Fladung, M., and E. Ritter. 1991. Plant leaf area measurements by personal computers. J. Agron. Crop Sci. 166:69–70.

Flinn, M. F., C. M. Smith, J. C. Reese, and B. S. Gill. 2001. Categories of resistance to greenbug (Homoptera: Aphididae) biotype I in *Aegilops tauschii* germplasm. J. Econ. Entomol. 94:558–563.

Formusoh, E. S., J. C. Reese, and G. Bradfisch. 1997. A miniaturized bioassay system for screening compounds deleterious to greenbugs (Homoptera: Aphididae) on artificial diets. J. Kansas Entomol. Soc. 70:323–328.

Frazier, J. L., and F. E. Hanson. 1986. Electrophysiological recording and analysis of insect chemosensory responses. In: J. R. Miller and T. A. Miller (Eds.), Insect-Plant Interactions. Springer-Verlag, New York. pp. 285–330.

George, B. W., F. D. Wilson, and R. L. Wilson. 1983. Methods of evaluating cotton for resistance to pink bollworm, cotton leaf perforator, and lygus bugs. In: Host Plant Resistance Research Methods for Insects, Diseases, Nematodes and Spider Mites in Cotton. Southern Cooperative Series Bulletin 280. pp. 41–45.

Gérard, H. C., W. F. Fett, R. A. Moreau, S. F. Osman, and R. L. Miller. 1993. Chemical and enzymatic investigation of the leaf cuticle of pear genotypes differing in resistance to pear psylla. J. Agric. Food Chem. 41:2437–2441.

Gibson, R. W. 1976. Glandular hairs on *Solanum polyadenium* lessen damage by the Colorado potato beetle. Ann. Appl. Biol. 82:147–150.

Girma, M., K. D. Kofoid, and J. C. Reese. 1999. Sorghum germplasm tolerant to greenbug (Homoptera: Aphididae) feeding damage as measured by reduced chlorophyll loss. J. Kansas Entomol. Soc. 71: 108–115.

Gorz, H. J., G. R. Manglitz, and F. A. Haskins. 1979. Selection for yellow clover aphid and pea aphid resistance in red clover. Crop Sci. 19:257–260.

Gross, A. T. H., and G. A. Stevenson. 1964. Resistance in *Melilotus* species to the sweetclover weevil (*Sitona cylindricollis*). Can. J. Plant Sci. 44:487–488.

Guldemond, J. A., W. J. van den Brink, and E. den Belder. 1998. Methods of assessing population increase in aphids and the effect of growth stage of the host plant on population growth rates. Entomol. Exp. Appl. 86:161–173.

Guthrie, W. D., F. F. Dicke, and C. R. Neiswander. 1960. Leaf and sheath feeding resistance to the European corn borer in eight inbred lines of dent corn. Ohio Agric. Exp. Stn. Res. Bull. 860. 38 pp.

Guthrie, W. D., W. A. Russell, G. L. Reed, A. R. Halbauer, and D. F. Cox. 1978. Methods of evaluating maize for sheath-collar-feeding resistance to the European corn borer. Maydica. 23:45–54.

Habibi, J., E. A. Backus, and T. H. Czapla. 1993. Plant lectins affect survival of the potato leafhopper (Homoptera: Cicadellidae). J. Econ. Entomol. 86:945–951.

Hamilton-Kemp, R. A. Anderson, J. G. Rodriguez, J. H. Longhrin, and C. G. Patterson. 1988. Strawberry foliage headspace vapor components at periods of susceptibility and resistance to *Tetranychus urticae* Koch. J. Chem. Ecol. 14:789–791.

Haniotakis, G. E., and W. H. Lange. 1974. Beet yellows virus resistance in sugar beets: mechanism of resistance. J. Econ. Entomol. 67:25–28.

Harman, J., P. Paul, R. Craig, D. Cox-Foster, J. Medford, and R. O. Mumma. 1996. Development of a mite bioassay to evaluate plant resistance and its use in determining regeneration of spider mite resistance. Entomol. Exp. Appl. 81:301–305.

Harvey, T. L., H. L. Hackerott, T. J. Martin, and W. D. Stegmeier. 1985. Mechanical insect dispenser for infesting plants with greenbugs (Homoptera: Aphididae). J. Econ. Entomol.78:489–492.

Heath, R. R., and A. Manukian. 1992. Development and evaluation of systems to collect volatile semiochemicals from insects and plants using a charcoal-infused medium for air purification. J. Chem. Ecol. 18:1209–1226.

Heath, R. R., and A. Manukian. 1994. An automated system for use in collecting volatile chemicals released from plants. J. Chem. Ecol. 20:593–608.

Hedin, P. A., J. N. Jenkins, D. H. Collum, W. H. White, W. L. Parrott, and M. W. MacGowan. 1983. Cyanidin-3-α-glucoside, a newly recognized basis for resistance in cotton to the tobacco budworm, *Heliothis virescens* (Fab.) (Lepidoptera: Noctuidae). Experientia. 39:799–801.

Hedin, P. A., F. M. Davis, W. P. Williams, and M. L. Salin. 1984. Possible factors of leaf-feeding resistance in corn to the southeastern corn borer. J. Agric. Food Chem. 32:262–267.

Hedin, P. A., W. L. Parrott, and J. N. Jenkins. 1991. Effects of cotton plant allelochemicals and nutrients on behavior and development of tobacco budworm. J. Chem. Ecol. 17:1107–1121.

Heinrichs, E. A., F. G. Medrano, and H. R. Rapusas. 1985. Genetic Evaluation for Insect Resistance in Rice. International Rice Research Institute, Los Banos, Laguna, Philippines. 356 pp.

Hill, R. R., Jr., and R. C. Newton. 1972. A method for mass screening alfalfa for meadow spittle bug resistance in the greenhouse during the winter. J. Econ. Entomol. 65:621–623.

Hills, T. M., and D. C. Peters. 1971. A method of evaluating post planting insecticide treatments for control of western corn rootworm larvae. J. Econ. Entomol. 64:764–765.

Holbrook, F.R., and A. F. Reeves. 1979. Indexing selections from the Maine potato breeding program for resistance to the green peach aphid. Research in Life Sciences. 26:1–5.

Horber, E. 1980. Types and classification of resistance. In: F. G. Maxwell and P. R. Jennings (Eds.), Breeding Plants Resistant to Insects, John Wiley, New York, pp. 15–21.

Horton, D. R., R. L. Chauvin, T. Hinojosa, D. Larson, C. Murphy, and K. D. Biever. 1997. Mechanisms of resistance to Colorado potato beetle in several potato lines and correlation with defoliation. Entomol. Exp. Appl. 82:239–246.

Hoxie, R. P., S. G. Wellso, and J. A. Webster. 1975. Cereal leaf beetle response to wheat trichome length and density. Environ. Entomol. 4:365–370.

Hudon, M., and M. S. Chiang. 1991. Evaluation of resistance of maize germplasm to the univoltine European corn borer *Ostrinia nubilalis* (Hübner) and relationship with maize maturity in Quebec. Maydica. 36:69–74.

Impe, G. van, and T. Hance. 1993. Une technique d'evaluation de la sensibilité varietale au tetranyque tisserand, *Tetranuchus urticae* Koch (Acarina: Tetranychidae), Application au haricot, au concombre, a la tomate et au fraisier. Agronomie. 13:739–749.

Inayatullah, C., J. A. Webster, and W. S. Fargo. 1990. Index for measuring plant resistance to insects. Southwest Entomol. 109:146–152.

Jenkins, J. N., W. L. Parrott, and J. C. McCarty. 1983. Breeding cotton for resistance to the tobacco budworm: Techniques to achieve uniform field infestations. In: Host Plant Resistance Research Methods for Insects, Diseases, Nematodes and Spider Mites in Cotton. Southern Cooperative Series Bulletin 280. pp. 36–41.

Johnson, J. W., and G. L. Teetes. 1979. Breeding for arthropod resistance in sorghum. In: M. K. Harris (Ed.), Biology and Breeding for Resistance to Arthropods and Pathogens on Agricultural Plants. TAMU Publ. MP-1451. pp. 168–180.

Kalode, M. B., J. S. Bentur, and T. E. Srinivasan. 1989. Screening and breeding rice for stem borer resistance. In: K. F. Nwanze (Ed.), International Workshop on Sorghum Stem Borers. International Crops Research Institute for the Semi-Arid Tropics (ICRISAT), Patancheru, Andhra Pradesh, India. pp. 153–157.

Kaster, L. V., M. A. Carson, M. E. Meehan, and R. Sisco. 1991. Rapid method of evaluating maize for sheath-collar feeding resistance to the European corn borer (Lepidoptera: Pyralidae). J. Econ. Entomol. 84:324–327.

Khan, Z. R., and R. C. Saxena. 1984a. Techniques for demonstrating phloem or xylem feeding by leafhoppers (Homoptera: Cicadellidae) and planthoppers (Homoptera: Delphacidae) in rice plant. J. Econ. Entomol. 77:550–552.

Khan, Z. R., and R. C. Saxena. 1984b. Electronically recorded waveforms associated with the feeding behavior of *Sogatella furcifera* (Homoptera: Delphacidae) on susceptible and resistant rice varieties. J. Econ. Entomol. 77:1479–1482.

Khan, Z. R., and R. C. Saxena. 1985a. Mode of feeding and growth of *Nephotettix virescens* (Homoptera: Cicadellidae) on selected resistant and susceptible rice varieties. J. Econ. Entomol. 78:583–587.

Khan, Z. R., and R. C. Saxena. 1985b. Behavioral and physiological responses of *Sogatella furcifera* (Homoptera: Delphacidae) to selected resistant and susceptible rice varieties. J. Econ. Entomol. 78:1280–1285.

Khan, Z. R., and R. C. Saxena. 1985c. Behavior and biology of *Nephotettix virescens* (Homoptera: Cicadellidae) on tungro-infected rice plants: epidemiology implications. Environ. Entomol. 14:297–303.

Khan, Z. R., J. T. Ward, and D. M. Norris. 1986. Role of trichomes in soybean resistance to cabbage looper, *Trichoplusia ni*. Entomol. Exp. Appl. 42:109–117.

Kishaba, A. N., and G. R. Manglitz. 1965. Non-preference as a mechanism of sweetclover and alfalfa resistance to the sweetclover aphid and spotted alfalfa aphid. J. Econ. Entomol. 58: 566–569.

Kitch, L. W., R. E. Shade, W. E. Nyquist, and J. D. Axtell. 1985. Inheritance of density of erect glandular trichomes in the genus *Medicago*. Crop Sci. 25:607–611

Klingler, J., G. Powell, G. A. Thompson, and R. Isaacs. 1998. Phloem specific aphid resistance in *Cucumis melo* line AR5: effects on feeding behaviour and performance of *Aphis gossypii*. Entomol. Exp. Appl. 86:79–88.

Knutson, R. J., B. E. Hibbard, B. D. Barry, V. A. Smith, and L. L. Darrah. 1999. Comparison of screening techniques for western corn rootworm (Coleoptera: Chrysomelidae) host-plant resistance. J. Econ. Entomol. 92:714–722.

Kocourek, F., J. Havelka, J. Berankova, and V. Jarosik. 1994. Effect of temperature on development rate and intrinsic rate of increase of *Aphis gossypii* reared on greenhouse cucumbers. Entomol. Exp. Appl. 71:59–64.

Kogan, M., and R. D. Goeden. 1969. A photometric technique for quantitative evaluation of feeding preferences of phytophagous insects. Ann. Entomol. Am. 62:319–322.

Kolb, T. E., K. M. Holmberg, M. R. Wagner, and J. E. Stone. 1998. Regulation of ponderosa pine foliar physiology and insect resistance mechanisms by basal area treatments. Tree Physiol. 18:375–381.

Kornegay, J., and C. Cardona. 1999. Development of an appropriate breeding scheme for tolerance to *Empoasca kraemeri* in common bean. Euphytica. 47:223–231.

Lage, J., B. Skovmand, and S. B. Andersen. 2003. Characterization of greenbug (Homoptera: Aphididae) resistance in synthetic hexaploid wheats. J. Econ. Entomol. 96:1922–1928.

Lambert, L. 1984. An improved screen-cage design for use in plant and insect research. Agron. J. 76:168–170.

Lei, H., J. C. Van Lenteren, and W. F. Tjallingii. 1999. Analysis of resistance in tomato and sweet pepper against the greenhouse whitefly using electrically monitored and visually observed probing and feeding behaviour. Entomol. Exp. Appl. 92:299–309.

Leigh, T. F. 1983. Research methods for cotton resistance to spider mites. In: Host Plant Resistance Research Methods for Insects, Diseases, Nematodes and Spider Mites in Cotton. Southern Cooperative Series Bulletin 280. pp. 56–58.

Lewis, A. C., and H. F. van Emden. 1986. Assays for Insect Feeding. In: J. R. Miller and T. A. Miller (Eds.), Insect-Plant Interactions. Springer-Verlag, New York. pp. 95–119.

Lewontin, R. C. 1965. Selection for colonizing ability. In: H. G. Baker and G. L. Stebbins (Eds.), The Genetics of Colonizing Species, Academic Press, New York, pp. 79–94.

Loughrin, J. H., T. R. Hamilton-Kemp, R. A. Anderson, and D. F. Hildebrand. 1990. Volatiles from flowers of *Nicotania sylvestris, N. otophora,* and *Malus x domesetica:* headspace components and day/night changes in their relative concentrations. Phytochem. 29:2473–2477.

Maredia, K. M., N. P. Tugwell, B. A. Waddle, and F. M. Bourland. 1994. Technique for screening cotton germplasm for resistance to tarnished plant bug, *Lygus lineolaris* (Palisot de Beauvois). Southwest. Entomol. 19:63–70.

Markwell, J., J. C. Osterman, and J. L. Mitchell. 1995. Calibration of the Minolta SPAD - 502 leaf chlorophyll meter. Photosyn. Res. 46:467–472.

Marshall, S. L., T. J. Ridsdill-Smith, and R. A. Prestidge. 1997. Resistance of seedling white clover cultivars to redlegged earth mite *Halotydeus destructor*. Proc. Fiftieth New Zealand Plant Protect. Conf., pp. 56–60.

Martin, F. A., C. A. Richard, and S. D. Hensley. 1975. Host resistance to *Diatraea saccharalis* (F.): Relationship of sugarcane internode hardness to larval damage. Environ. Entomol. 4:687–688.

McIndoo, N. E. 1926. An insect olfactometer. J. Econ. Entomol. 12:545–571.

McLean, D. L., and M. G. Kinsey. 1964. A technique for electronically recording aphid feeding and salivation. Nature. 202:1358–1359.

McLean, D. L., and W. A. Weigt. 1968. An electronic measurement system to record aphid salivation and ingestion. Ann. Entomol. Soc. Am. 61:180–185.

Merkle, O. G., and K. J. Starks. 1985. Resistance of wheat to the yellow sugarcane aphid (Homoptera:Aphididae). J. Econ. Entomol. 78:127–128.

Mihm, J. A. 1983a. Techniques for efficient mass rearing and infestation in screening for host plant resistance to corn earworm, *Heliothis zea*. Centro Internacional de Mejoramiento de Maiz y Trigo. El Batan, Mexico. 16 pp.

Mihm, J. A. 1983b. Efficient mass-rearing and infestation techniques to screen for host plant resistance to fall armyworm, *Spodoptera frugiperda*. Centro Internacional de Mejoramiento de Maiz y Trigo CIMMYT, El Batan, Mexico. 16 pp.

Mihm, J. A. 1983c. Efficient mass rearing and infestation techniques to screen for host plant resistance to maize stem borers, *Diatraea* sp. Centro Internacional de Mejoramiento de Maiz y Trigo. El Batan, Mexico. 23 pp.

Mihm, J. A., F. B. Peairs, and A. Ortega. 1978. New procedures for efficient mass production and artificial infestation with lepidopterous pests of maize. CIMMYT Review. International Wheat and Maize Improvement Center, El Batan, Mexico, 138 pp.

Miller, H. R., T. L. Randolph, and F. B. Peairs. 2003. Categories of resistance at four growth stages in three wheats resistant to the Russian wheat aphid (Homoptera: Aphididae). J. Econ. Entomol. 96:673–679.

Moharramipour, S., H. Tsumki, K. Sato, and H. Yoshida. 1997. Mapping resistance to cereal aphids in barley. Theor. Appl. Genet. 94:592–596.

Mollema, C., F. van Dijken, K. Reinink, and R. Jansen. 1992. An automatic and accurate evaluation of thrips-damage. Image-analysis: a new tool in breeding for resistance. In: S. B. J. Menken, J. H. Visser, and P. Harrewijn. (Eds.), Proc. 8th International Symposium on Insect-Plant Relationships, Kluwer Academic Publ., Dordrecht, The Netherlands. pp. 1–2.

Montandon, R., R. D. Stipanovic, H. J. Williams, W. L. Sterling, and S. B. Vinson. 1986. Nutritional indices and excretion of gossypol by *Alabama argillacae* (Hübner) and *Heliothis virescens* (F.) (Lepidoptera: Noctuidae) fed glanded and glandless cotyledonary cotton leaves. J. Econ. Entomol. 80:32–36.

Mulrooney, J. E., W. L. Parrott, and J. N. Jenkins. 1985. Nutritional indices of second-instar tobacco budworm larvae (Lepidoptera: Noctuidae) fed different cotton strains. J. Econ. Entomol. 78:757–761.

Murray, P. J. 1996. Evaluation of a range of varieties of white clover for resistance to feeding by weevils of the genus *Sitona*. Plant Var. Seeds. 9:9–14.

Murray, P. J., and R. O. Clements. 1992. A technique for assessing damage to roots of white clover caused by root feeding insects. Ann. Appl. Biol. 121:715–719.

Muzika, R. M., C. L. Campbell, J. W. Hanover, and A. L. Smith. 1990. Comparison of techniques for extracting volatile compounds from conifer needles. J. Chem. Ecol. 16:2713–2722.

Nagaraj, N., J. C. Reese, M. B. Kirkham, K. Kofoid, L. R. Campbell, and T. M. Loughin. 2002. Effect of greenbug, *Schizaphis graminum* (Rondani) (Homoptera: Aphidiade) biotype K on chlorophyll content and photosynthetic rate of tolerant and susceptible sorghum hybrids. J. Kansas Entomol. Soc. 75: 299–307.

Navon, A., V. Melamed-Madjar, M. Zur, and E. Ben-Moshe. 1991. Effect of cotton cultivars on feeding of *Heliothis armigera* and *Spodoptera littoralis* larvae and on oviposition of *Bemisia tabaci*. Agric., Ecosyst. Environ. 35:73–80.

Nwanze, K. F. Y. V. R. Reddy, S. L. Taneja , H. C. Sharma, and B. L. Agrawal . 1991. Evaluating sorghum genotypes for multiple insect resistance. Insect Sci. Appl. 12:183–188.

Nielson, M. W. 1974. Evaluating spotted alfalfa aphid resistance. In: K. K. Barnes (Ed.), Standard Tests to Characterize Pest Resistance in Alfalfa Varieties. U. S. Dept. Agric. NC–19:19–20.

Nwanze, K. F., and Y. V. R. Reddy. 1991. A rapid method for screening sorghum for resistance to *Chilo partellus* (Swinhoe) (Lepidoptera: Pyralidae). J. Agric. Entomol. 8:41–44.

Nwanze, K. F., R. J. Pring, P. S. Sree, D. R. Butler, Y. V. R. Reddy, and P. Soman. 1992. Resistance in sorghum to the shoot fly, *Atherigona soccata* - epicuticular wax and wetness of the central whorl leaf of young seedlings. Ann. Appl. Biol. 120:373–382.

Oghiakhe, S., L. E. N. Jackai, and W. A. Makanjuola. 1995. A rapid visual field screening technique for resistance of cowpea (*Vigna unguiculata)* to the legume pod borer, *Maruca testulalis* (Lepidoptera: Pyralidae). Bull. Entomol. Res. 82:507–512.

Oghiakhe, S., and A. Odulaja. 1993. A multivariate analysis of growth and development parameters of the legume pod borer, *Maruca testulalis* on variably resistant cowpea cultivars. Entomol. Exp. Appl. 66:275–282.

Ortman, E. E., and T. F. Branson. 1976. Growth pouches for studies of host plant resistance to larvae of corn rootworms. J. Econ. Entomol. 69:380–382.

Pantoja, A., C. M. Smith, and J. F. Robinson. 1986. Evaluation of rice plant material for resistance to the fall armyworm (Lepidoptera: Noctuidae). J. Econ. Entomol. 79:1319–1323.

Pathak, P. K., and E. A. Heinrichs. 1982. Bromocresol green indicator for measuring feeding activity of *Nilaparvata lugens* on rice varieties. Philipp. Ent. 5:195–198.

Pathak, P. K., R. C. Saxena, and E. A. Heinrichs. 1982. Parafilm sachet for measuring honeydew excretion by *Nilaparvata lugens* on rice. J. Econ. Entomol. 75: 194–195.

Payne, T. L., E. R. Hart, L. J. Edson, F. A. McCarty, P. M. Billings, and J. E. Coster. 1976. Olfactometer for assay of behavioral chemicals for the southern pine beetle. *Dendroctonus frontalis* (Coleoptera: Scolytidae). J. Chem. Ecol. 2:411–419.

Powell, G. S., W. V. Campbell, W. A. Cope, and D. S. Chamblee. 1983. Ladino clover resistance to the clover root curculio (Coleoptera: Curculionidae). J. Econ. Entomol. 76:264–268.

Quiros, C. F., M. A. Stevens, C. M. Rick, and M. L. Kok-Yokomi. 1977. Resistance in tomato to the pink form of the potato aphid (*Macrosiphum euphorbiae* Thomas): The role of anatomy, epidermal hairs, and foliage composition. J. Am. Soc. Hort. Sci. 102:166–177.

Quisenberry, S. S., P. Caballero, and C. M. Smith. 1988. Influence of bermudagrass leaf extracts on development of fall armyworm (Lepidoptera: Noctuidae) larvae. J. Econ. Entomol. 81:910–913.

Reese, J. C. 1978. Chronic effects of plant allelochemicals on insect nutritional physiology. Entomol. Exp. Appl. 24:625–631.

Reese, J. C. 1983. Nutrient-allelochemical interactions in host plant resistance. In: P. A. Hedin (Ed.), Plant Resistance to Insects. ACS Symposium Series No. 208. American Chemical Society, Washington DC, pp. 231–243.

Reese, J. C., D. C. Margolies, E. A. Backus, S. Noyes, P. Bramel-Cox, and A. G. O. Dixon. 1994. Characterization of aphid host plant resistance and feeding behavior through use of a computerized insect feeding monitor. In: M. M. Ellsbury, E. A. Backus, and D. L. Ullman (Eds.), History, Development, and Application of AC Electronic Insect Feeding Monitors. Thomas Say Publications in Entomology. Entomological Society of America, Lanham, MD, pp. 52–72.

Reese, J. C., W. F. Tjallingii, M. van Helden, and E. Prado. 2000. Waveform comparisons among AC and DC electronic monitoring systems for aphid (Homoptera: Aphididae) feeding behavior. In: G. P. Walker and E. A. Backus (Eds.), Principles and Applications of Electronic Monitoring and Other Techniques in the Study of Homopteran Feeding Behavior. Thomas Say Publications in Entomology: Proceedings. Entomological Society of America, Lanham, MD, pp. 70–101.

Renwick, J. A. A. 1983. Nonpreference mechanisms: Plant characteristics influencing insect behavior. In: P. A. Hedin (Ed.), Plant Resistance to Insects. ACS Symposium Series No. 208. American Chemical Society, Washington DC, pp. 199–213.

Reynolds, G. W., C. M. Smith, and K. M. Kester. 1984. Reductions in consumption, utilization, and growth rate of soybean looper (Lepidoptera: Noctuidae) larvae fed foliage of soybean genotype PI 227687. J. Econ. Entomol. 77: 1371–1375.

Robinson, J. F., A. Klun, W. D. Guthrie, and T. A. Brindley. 1982. European corn borer leaf feeding resistance: A simplified technique for determining relative differences in concentrations of 6-methoxy-benzoxazolinone (Lepidoptera: Pyralidae). J. Kansas Entomol. Soc. 55:297–301.

Romanow, L. R., O. M. B. de Ponti, and C. Mollema. 1991. Resistance in tomato to the greenhouse whitefly: analysis of population dynamics. Entomol. Exp. Appl. 60:247–259.

Roof, M. E., E. Horber, and E. L Sorensen. 1976. Evaluating alfalfa cuttings for resistance to the potato leafhopper. Environ. Entomol. 5:295–301.

Rose, R. L., T. C. Sparks, and C. M. Smith. 1988. Insecticide toxicity to the soybean looper and the velvetbean caterpillar (Lepidoptera: Noctuidae) as influenced by feeding on resistant soybean (PI227687) leaves and coumestrol. J. Econ. Entomol. 81:1288–1294.

Rufener, II. G. K., R. B. Hammond, R. L. Cooper, and S. K. St. Martin. 1987. Larval antibiosis screening technique for Mexican bean beetle resistance in soybean. Crop Sci. 27:598–600.

Ruggle, P., and A. P. Gutierrez. 1995. Use of life tables to assess host plant resistance in alfalfa to Therioaphis maculata (Homoptera: Aphididae): Hypothesis for maintenance of resistance. Environ. Entomol. 24:313–325.

Rutherford, R. S. 1998. Prediction of resistance in sugarcane to stalk borer Eldana saccharina by near-infrared spectroscopy on crude budscale extracts: involvement of chlorogenates and flavonoids. J. Chem. Ecol. 24:1447–1463.

Schaafsma, A. W., C. Cardona, J. L. Kornegay, A. M. Wylde, and T. E. Michaels. 1998. Resistance of common bean lines to the potato leafhopper (Homoptera: Cicadellidae). J. Econ. Entomol. 91:981–986.

Schmidt, D., and J. C. Reese. 1986. Sources of error in nutritional index studies of insects on artificial diet. J. Insect Physiol. 32: 193–198.

Schoonhoven, A. V. 1974. Resistance to thrips damage in cassava. J. Econ. Entomol. 76:728–730.

Schuster, M. F. 1983. Screening cotton for resistance to spider mites. In: Host Plant Resistance Research Methods for Insects, Diseases, Nematodes and Spider Mites in Cotton. Southern Cooperative Series Bulletin 280. pp. 54–56.

Serrano, M. S., E. A. Backus, and C. Cardona. 2000. Comparison of AC electronic monitoring and field data for estimating tolerance to *Empoasca kraemeri* (Homoptera: Cicadellidae) in common bean genotypes. J. Econ. Entomol, Vol. 93:1796–1809.

Sharma , H. C., Y. O. Doumbia, and N. Y. Diorisso. 1992. A headcage technique to resistance to mirid head bug, *Eurystylus immaculatus* Odh. in west Africa. Insect Sci. Applic. 13:417–427.

Sharma, H. C., F. Singh, and K. F. Nwanze (Eds.). 1997. Plant Resistance to Insects in Sorghum. International Crops Research Institute for the Semi-Arid Tropics (ICRISAT), Patancheru, Andhra Pradesh, India. 216 pp.

Sharma, H. C., and D. J. Sullivan. 2000. Screening for plant resistance to Oriental armyworm, *Mythimna separata* (Lepidoptera: Noctuidae) in pearl millet, *Pennisetum glaucum*. J. Agric. Urban Entomol. 17:125–134.

Sharma, H. C., S. L. Taneja, K. Leuschner, and F. Nwanze. 1992. Techniques to screen sorghums resistance to insects. Information Bulletin No. 32. International Crops Research Institute for the Semi-Arid Tropics (ICRISAT), Patancheru, Andhra Pradesh, India. 48 pp.

Sharma, H. C., P. Vidyasagar, and K. Leuschner. 1988a. Field screening for resistance to sorghum midge (Diptera: Cecidomyiidae). J. Econ. Entomol. 81:327–334.

Sharma, H. C., P. Vidyasagar, and K. Leuschner. 1988b. No-choice cage technique to screen for resistance to sorghum midge (Diptera: Cecidomyiidae). J. Econ. Entomol. 81:415–422.

Sharma, H. C., and O. Youm. 1999. Integrated pest management in pearl millet with special reference to host plant resistance to insects. In: I. S. Khairwal, K. N. Raj, D. J. Andrews, and G. Harinarayana (Eds.), Pearl Millet Improvement. Oxford and IBH, New Delhi. pp. 381–425.

Shaver, T. N., M. J. Lukefahr, and J. A. Garcia. 1970. Food utilization, ingestion, and growth of larvae of the bollworm and tobacco budworm on diets containing gossypol. J. Econ. Entomol. 63:1544–1546.

Shockley, F. W., E. A. Backus, M. R. Ellersieck, D. W. Johnson, and M. McCaslin. 2002. Glandular-haired alfalfa resistance to potato leafhopper (Homoptera: Cicadellidae) and hopperburn: development of resistance indices. J. Econ. Entomol. 95:437–447.

Simonet, D. E., R. L. Pienkowski, D. G. Martinez, and R. D. Blakeslee. 1978. Laboratory and field evaluation of sampling techniques for the nymphal stages to the potato leafhopper on alfalfa. J. Econ. Entomol. 71:840–842.

Simmons, A. M., and J. D. McCreight. 1996. Evaluation of melon for resistance to *Bemisia argentifolii* (Homoptera: Aleyrodidae). J. Econ. Entomol. 89:1663–1668.

Singh, P., and R. F. Moore. (Eds.) 1985. Handbook of Insect Rearing. Vol. I. Elsevier Science
    Publishing Co. New York. 481 pp.

Slansky, F. Jr., and G. S. Wheeler. 1992. Feeding and growth responses of laboratory and field strains
    of velvetbean caterpillars (Lepidoptera: Noctuidae) to food nutrient level and allelochemicals. J.
    Econ. Entomol. 85:1717–1730.

Smith, C. M., and N. F. Fischer. 1983. Chemical factors of an insect resistant soybean genotype
    affecting growth and survival of the soybean looper. Entomol. Exp. Appl. 33:343–345.

Smith, C. M., Z. R. Khan, and M. D. Pathak. 1994. Techniques for Evaluating Insect Resistance in
    Crop Plants. Lewis Publ. Co. 320 pp.

Smith, C. M., H. N. Pitre, and W. E. Knight. 1975. Evaluation of crimson clover for resistance to leaf
    feeding by the adult clover head weevil. Crop Sci. 15:257–258.

Smith, C. M., H. Havlickova, S. Starkey, B. S. Gill, and V. Holubec. 2003. Identification of *Aegilops*
    germplasm with multiple aphid resistance. *Euphytica.* 135:265–273.

Sorensen, E. L. 1974. Evaluating pea aphid resistance. In: D. K. Barnes (Ed.), Standard Tests to
    Characterize Pest Resistance in Alfalfa Varieties. USDA NC-19:18–19.

Sorensen, E. L., and E. Horber. 1974. Selecting alfalfa seedlings to resist the potato leafhopper. Crop
    Sci. 14: 85–86.

Soto, P. E. 1972. Mass rearing of the sorghum shootfly and screening for host plant resistance under
    greenhouse conditions. In: M. G. Jotwani and W. R. Young (Eds.), Control of Sorghum Shootfly.
    Oxford Press, New Delhi, 324 pp.

Stadler, E., and F. E. Hanson. 1976. Influence of induction of host preference on chemoreception of
    Manduca sexta: behavioral and electrophysiological studies. Symp. Biol. Hung. 16:267–273.

Starks, K. J., and R. L. Burton. 1977. Greenbugs: determining biotypes, culturing, and screening for
    plant resistance. U. S. Dept. Agric. ARS Tech. Bull. 1556: 12 pp.

Tang, C. 1986. Continuous trapping techniques for the study of allelochemicals from higher plants. In:
    A. R. Putnam and C. Tang (Eds.), The Science of Allelochemistry. John Wiley, New York, pp.
    113–131.

Tedders, W. L., and B. W. Wood. 1987. Field studies of three species of aphids on pecan: An improved
    cage for collecting honeydew and glucose-equivalents contained in honeydew. J. Entomol. Sci.
    22:23–28.

Thackray, D.J., S. D. Wratten, P. J. Edwards, and H. M. Niemeyer. 1990. Resistance to the aphids
    Sitobion avenae and *Rhopalosiphum padi* in gramineae in relation to hydroxamic acid levels. Ann.
    Appl. Biol. 116: 573–582.

Tingey, W. M. 1986. Techniques for evaluating plant resistance to insects. In: J. A. Miller and T. A.
    Miller (Eds.), Plant Insect Interactions. Springer Verlag, New York, pp. 251–284.

Tjallingii, W. F. 1985. Electrical nature of recorded signals during penetration by aphids. Entomol.
    Exp. Appl. 38:177–186.

Tomlin, E. S., and J. H. Borden. 1997. Multicomponent index for evaluating resistance by Sitka spruce
    to the white pine weevil (Coleoptera: Curculionidae). J. Econ. Entomol. 90:704–714.

Tugwell, Jr., N. P. 1983. Methods: evaluating cotton for resistance to plant bugs. In Host Plant Resistance Research Methods for Insects, Diseases, Nematodes and Spider Mites in Cotton. Southern Cooperative Series Bulletin 280. pp. 46–53.

Van Helden, M., H. P. N. F. van Heest, T. A. van Beek, and W. F. Tjallingii. 1995. Development of a bioassay to test phloem sap samples from lettuce for resistance to *Nasonovia ribisnigri* (Homoptera, Aphididae). J. Chem. Ecol. 21:761–774.

Van Helden, M., and W. F. Tjallingii. 1993. Tissue localisation of lettuce resistance to the aphid *Nasonovia ribisnigri* using electrical penetration graphs. Entomol. Exp. Appl. 68:269–278.

Van Helden, M., W. F. Tjallingii, and T. A. van Beek. 1994. Phloem sap collection from lettuce (*Lactuca sativa* L.): chemical comparison among collection methods. J. Chem. Ecol. 20:3191–3206.

Velusamy, R., and E. A. Heinrichs. 1986. Electronic monitoring of feeding behavior of *Nilaparvata lugens* (Homoptera: Delphacidae) on resistant and susceptible rice cultivars. Environ. Entomol. 15:678–682.

Villani, M. G., and F. Gould. 1986. Use of radiographs for movement analysis of the corn wireworm, *Melanotus communis* (Coleoptera: Elateridae). Environ. Entomol. 15:462–464.

Visser, J. H. 1976. The design of a low speed wind tunnel as an instrument for the study of olfactory orientation in the Colorado beetle (*Leptinotarsa decemlineata*). Entomol. Exp. Appl. 20:275–288.

Visser, J. H. 1986. Host odor perception in phytophagus insects. Ann. Rev. Entomol. 31: 121–144.

Waldbauer, G. P. 1968. The consumption and utilization of food by insects. Adv. Insect Physiol. 5:229–288.

Walker, G. P. 2000. A beginner's guide to electronic monitoring of Homopteran probing behavior. In: G. P. Walker and E. A. Backus (Eds.), Principles and Applications of Electronic Monitoring and Other Techniques in the Study of Homopteran Feeding Behavior. Thomas Say Publications in Entomology: Proceedings. Entomological Society of America, Lanham, MD, pp. 14–40.

Walker, G. P., and E. A. Backus (Eds.). 2000. Principles and Applications of Electronic Monitoring and Other Techniques in the Study of Homopteran Feeding Behavior. Thomas Say Publications in Entomology: Proceedings. Entomological Society of America, Lanham, MD.

Wang, S. F., T. J. Ridsdill-Smith, and E. L. Ghisalberti. 1999. Levels of isoflavonoids as indicators of resistance of subterranean clover trifoliates to redlegged earth mite *Halotydeus destructor*. J. Chem. Ecol. 25:795–803.

Wearing, C. H. 1998. Bioassays for measuring ovipositional and larval preferences of leafhoppers (Lepidoptera: Tortricidae) for different cultivars of apple. New Zealand J. Crop Hort. Sci. 26:269–278.

Wearing, C. H., and K. Calhoun. 1999. Bioassays for measuring the resistance of different apple cultivars to the development of leafrollers (Lepidoptera: Tortricidae). New Zealand J. Crop Hort. Sci. 27:91–99.

Webster, J. A., and C. Inayatullah. 1988. Assessment of experimental designs for greenbug (Homoptera: Aphididae) antixenosis tests. J. Econ. Entomol. 81:1246–1250.

Webster, J. A., and D. R. Porter. 2000. Plant resistance components of two greenbug (Homoptera: Aphididae) resistant wheats. J. Econ. Entomol. 93:1000–1004.

Webster, J. A., D. R. Porter, J. D. Burd, and D. W. Mornhinweg. 1996. Effect of growth stage of resistant and susceptible barley on the Russian wheat aphid, *Diuraphis noxia* (Homoptera: Aphididae). J. Agric. Entomol. 13: 283–291.

Webster, J. A., and D. H. Smith, Jr. 1983. Developing small grains resistant to the cereal leaf beetle. USDA Tech. Bull. No. 1673. 10 pp.

Webster, J. A., K. J. Starks, and R. L. Burton. 1987. Plant resistance studies with *Diuraphis noxia* (Homoptera: Aphididae), a new United States wheat pest. J. Econ. Entomol. 80: 944–949.

Wheeler, G. S., F. Slansky Jr., and S. J. Yu. 1992. Laboratory colonization has not reduced constitutive or induced polysubstrate monooxygenase activity in velvetbean caterpillars. J. Chem. Ecol. 18:1313–1325.

White, W. H. 1993. Cluster analysis for assessing sugarcane borer resistance in sugarcane line trials. Field Crops Res. 33:159–1168.

White, W. H. J. D. Miller, S. B. Milligan, D. M. Burner, and B. L. Legendre. 2001. Inheritance of sugarcane borer resistance in sugarcane derived from two separate measures of insect damage. Crop Sci. 41:1706–1710.

Widstrom, N. W., and R. L. Burton. 1970. Artificial infestation of corn with suspensions of corn earworm eggs. J. Econ. Entomol. 63:443–446.

Wightman, J. A., K. M. Dick, G. V. Ranga Rao, T. G. Shanower, and C. G. Gold. 1990. Pests of groundnut in the semi-arid tropics. In: S. R. Singh (Ed.), Insect Pests of Legumes. Longman and Sons, New York. pp. 243–322.

Williams, W. P., and F. M. Davis. 1985. Southwestern corn borer larval growth on corn callus and its relationship with leaf feeding resistance. Crop Sci. 25: 317–319.

Williams, W. P., P. M. Buckley, and F. M. Davis. 1985. Larval growth and behavior of the fall armyworm (Lepidoptera: Noctuidae) on callus initiated from susceptible and resistant corn hybrids. J. Econ. Entomol. 78:951–954.

Williams, W. P., P. M. Buckley, and F. M. Davis. 1987a. Tissue culture and its use in investigations of insect resistance in maize. Agric. Ecosyst. Environ. 18:185–190.

Williams, W. P., P. M. Buckley, and F. M. Davis. 1987b. Feeding response of corn earworm (Lepidoptera: Noctuidae) to callus and extracts of corn in the laboratory. Environ. Entomol. 16:532–534.

Wiseman, B. R., and N. W. Widstrom. 1980. Comparison of methods of infesting whorl-stage corn with fall armyworm larvae. J. Econ. Entomol. 73:440–442.

Wiseman, B. R., F. M. Davis, and J. E. Campbell. 1980. Mechanical infestation device used in fall armyworm plant resistance programs. Fla. Entomol. 63: 425–428.

Wiseman, B. R., H. N. Pitre, S. L. Fales, and R. R. Duncan. 1986. Biological effects of developing sorghum panicles in a meridic diet on fall armyworm (Lepidoptera: Noctuidae) development. J. Econ. Entomol. 79:1637–1640.

Woodhead, S. 1983. Distribution of chemicals in glass fibre discs used in insect bioassays. Entomol. Exp. Appl. 34:119–120.

Wyatt, I. J., and P. F. White. 1977. Simple estimation of intrinsic increase rates for aphids and tetranychid mites. J. Appl. Ecol. 14: 757–766.

Yadava, U. L. 1986. A rapid and nondestructive method to determine chlorophyll in intact leaves. HortScience. 21:1449–1450.

Zheng, F., G. A. Pederson, F. M. Davis, and M. M. Ellsbury. 1994. Modalities of resistance of N-2 red clover germplasm to pea aphid (Homoptera: Aphididae). J. Agric. Entomol. 11:349–359.

Zuber, M. S., and C. O. Grogan. 1961. A new technique for measuring stalk strength in corn. Crop Sci. 1:378–380.

CHAPTER 7

FACTORS AFFECTING THE EXPRESSION OF PLANT RESISTANCE TO ARTHROPODS

Variation in the test arthropod, the plant material to be evaluated, and in the environment all affect the expression of plant resistance to arthropods. The ability of each variable to influence the outcome of plant resistance evaluations in the laboratory, the greenhouse, or field experiments, should be determined before drawing conclusions about relative plant resistance or susceptibility. Interactions of different combinations of these three types of variables can also confound the experimental error involved in plant resistance evaluations. Therefore, it is important that these sources of error be understood and limited as much as possible before beginning large-scale evaluations of plant genotypes.

Environmentally induced changes in plant development have significant effects on the expression of arthropod resistance in plants, since plant growth is determined by existing environmental conditions. Severe environmental perturbations, such as those brought on by drought or flood conditions, cause wholesale changes in temperature and/or soil conditions. These changes then are expressed as stresses to plant metabolism and growth that, in turn, can affect the expression of plant resistance. Human-induced changes in plant growth environments, such as air pollutants have been shown to have stress effects on plant growth and metabolism. Examples of these various types of stresses and their relation to arthropods and other plant variables will be discussed at length later in this chapter. For a complete in-depth discussion of the effects of additional different stresses on plants in relation to arthropod susceptibility, readers are referred to Heinrichs (1988).

1. PLANT VARIABLES

1.1. Plant Density

The density of plant foliage affects the expression of resistance to arthropods. Webster et al. (1978) determined that oat, *Avena sativa* (L.), planted at low seeding rates have lower populations of the cereal leaf beetle, *Oulema melanopus* (L.), than those planted at intermediate and high seeding rates. Results of experiments involving density of wheat, *Triticum aestivum* L., plants and resistance to the wheat stem sawfly, *Cephus cinctus* (Norton), reveal a similar relationship (Miller et al. 1993). Fly-resistant cultivars become susceptible as stand densities increase. In long-term field experiments, Kolb et al. (1998) observed that ponderosa pine trees, *Pinus ponderosa* var. *scopulorum* Englm., growing in dense stands have greater susceptibility to insects than trees growing in thin stands. Susceptible trees have a more dilute phloem concentration, reduced resin production and flow, and reduced foliar toughness compared to resistant trees growing in thin stands. Similarly, Alghali (1984) determined that as spacing of rice, *Oryza sativa* (L.), plants increases, plants produce greater numbers of tillers (stems) and that populations of the stalk-eyed fly, *Diopsis thoracica* (West), and stem damage caused by fly infestation also increase. An overall theme in these examples is that plants growing at reduced density have better access to and utilization of available soil moisture and nutrients, enabling them to more efficiently produce higher quantities of chemical and physical defenses.

Fery and Cuthbert (1972) also observed that damage to southern pea, *Pisum sativum* var. *macrocarpon* L., by the cowpea curculio, *Chalcodermus aeneus* Boheman, increases in proportion to changes in plant density. In this instance, a 15-fold increase in plant density led to a 300% increase in pod damage. A similar relationship exists between plant density in tomato, *Lycopersicon esculentum* Mill., and damage by larvae of the tomato fruit worm, *Helicoverpa zea* (Boddie) (Fery and Cuthbert 1974). Fery and Cuthbert (1973) also demonstrated that significant differences in the percentage of *H. zea* damaged fruits among tomato accessions are nonexistent when the data are adjusted for vine size. Harris et al. (1987) identified large differences in the number of onion fly, *Delia antiqua* (Meigen), eggs among different breeding lines of onion, *Allium cepa* L. However, the use of covariance analysis indicated that there are no differences between lines when data are adjusted for plant size. Results from experiments on the relationship between plant density in cotton, *Gossypium hirsutum* L., condensed tannin concentration (see Chapter 3) and resistance to the two spotted spider mite, *Tetranychus urticae* (Koch), support this trend, as there are no significant differences in tannin content between large plant density (129,000 plants ha^{-1}) and small plant density (30, 000 plants ha^{-1}) (Lege et al 1993). The above examples serve to illustrate that investigators should standardize the densities of plants in evaluations of resistance and should express arthropod damage or infestations in proportion to plant density or biomass.

1.2. Plant Height

Differences in plant height also affect the expression of resistance to arthropods. Smith and Robinson (1983) noted a relationship between cultivar height in rice, *Oryza sativa* (L.), and infestation by the least skipper, *Ancyloxypha numitor* (F.). Short stature cultivars are heavily fed upon. Tingey and Leigh (1974) detected a similar height preference of the plant bug, *Lygus hesperus* Knight, for oviposition among cotton cultivars in field experiments. When adjusted for plant height, plant bug oviposition is greatest on short cultivars. de Jager et al. (1995) noted the opposite trend in a relationship between cultivar height in chrysanthemum, *Chrysanthemum leucanthemum* L., and resistance to the thrips species *Frankliniella occidnetalis* (Pergande). Tall cultivars with large amounts of foliage are presumed to invest less in herbivore defense than shorter cultivars and are more susceptible to thrips damage.

1.3. Plant Tissue Age

The resistance of plant tissues to arthropod damage varies markedly during the life of the plant and is age-specific for many crop plant species. In many plant - arthropod interactions where fruiting structures are infested, resistance factors occur later in plant phenological development or accumulate in older plant tissues as arthropod defenses. In several graminaceous crops, resistance is not manifested until later stages of plant development (Figure 7.1).

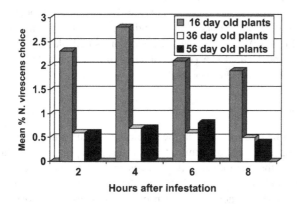

Figure 7.1. Effect of rice plant age (16, 36 and 56 day old plants) on the antixenosis to Nephotettix virescens. *(From Rapusas & Heinrichs 1987. Reprinted with permission from* Environ. Entomol., Vol. 16:106-110. Copyright 1987, Entomological Society of America)

A general, similar trend exists in the resistance of numerous crop plant, pasture plant and tree plant species to their respective pest arthropods. These include (in order of plant taxonomic relationship) resistance in barley, *Hordeum vulgare* L., to the bird-cherry oat aphid, *Rhopalosiphum padi* Fitch, (Leather and Dixon 1981); resistance in maize, *Zea mays* L., to *H. zea* (Wann and Hills 1966); rice resistance to the brown planthopper, *Nilaparvata lugens* Stal, and the green leafhopper, *Nephotettix virescens* (Distant) (Velusamy et al. 1986, Rapusas and Heinrichs 1987); pasture grass resistance to the grass aphid, *Metopolophium festucae cerealium* (Dent and Wratten 1986); tomato resistance to the Colorado potato beetle, *Leptinotarsa decemlineata* (Say) (Sinden et al. 1978); resistance in celery, *Apium graveolens* Linn., to the beet armyworm, *Spodoptera exigua* (Hubner) (Diawara et al. 1994); resistance soybean, *Glycine max* (L.) Merr., in to a lepidopteran complex (Rowan et al. 1993) and the resistance of balsam fir, *Abies balsamea* (L.) Mill., to the spruce budworm, *Choristoneura fumiferana* (Clemens) (Bauce et al. 1997). The involvement of various allelochemical defenses in several of these interactions has been discussed previously in Chapters 2 and 3.

Resistance may also occur early in plant phenological development to circumvent the development of arthropod pests that synchronize their development to coincide with the abundance of young vegetative plant tissues. For example, resistance in sorghum, *Sorghum bicolor* (L.) Moench, to the corn leaf aphid, *Rhopalosiphum maidis* (Fitch), and the migratory locust, *Locusta migratoria migratoroides* (R. & F.) is greater in younger, smaller plants (Woodhead and Bernays 1977, Fisk 1978). Locust resistance in young sorghum plants is related to higher release rates of cyanide (Woodhead and Bernays 1977) and higher concentrations of *p*-hydroxybenzaldehyde in leaf surface waxes (Woodhead 1982). Resistance in maize is greater in younger plants to several foliar feeding Lepidoptera, including *Chilo partellus* Swinhoe (Kumar and Asino 1993); *H. zea* (Wiseman and Snook 1995); the European corn borer, *Ostrinia nubilalis* (Hubner) (Klun and Robinson 1969) and the fall armyworm, *Spodoptera frugiperda* (J. E. Smith) (Videla et al. 1992). In addition, small plants of some maize inbred lines contain more DIMBOA (see Chapters 2 and 3) than taller, more mature plants (Guthrie et al. 1986, Klun and Robinson 1969).

1.4. Plant Phenology

Resistance may also be due to a shift in the phenology of the plant that disrupts the normal synchronous relationship between the arthropod pest and its host. An excellent example of the phenomenon is the resistance of sorghum cultivars to the sorghum midge, *Stenodiplosis sorghicola* (Couquillet), documented by Diarisso et al. (1998). Spikelets of resistant genotypes open during darkness and close during the day, when midge oviposition commences. Spikelets of susceptible sorghum genotypes flower during the day, coinciding with and permitting midge oviposition. A related phenological trend, occurring seasonally rather than daily, is the delayed or accelerated development of different species of trees to create asynchrony to arthropod pest development. Resistance to the white pine weevil, *Pissodes strobi* (Peck), occurs on trees of Sitka spruce, *Picea sitchensis* (Bong.) Carr., that produce a new "flush" of spring growth before ambient temperatures are high enough to stimulate weevil oviposition (Hulme et al. 1995). Quiring (1994) detected the opposite type of synchrony in trees of white spruce, *Picea glauca* (Moench) Voss, resistant to oviposition and larval damage by the spruce bud moth, *Zeiraphera canadensis* Mutt. & Free. The most resistant trees produce the latest flush of new growth, which delays exposure to moth oviposition and ensuing larval feeding and development.

1.5. Plant Tissue Type

With a few exceptions, the foliage of younger, more succulent plants is more preferred by arthropods over older foliage in plant - arthropod interaction studies and arthropod plant resistance experiments. In the canopy of the soybean plant, younger more succulent leaves near the plant apex are preferred for feeding by *H. zea* and *T. urticae* than lower leaves (McWilliams and Beland 1977, Rodriguez et al. 1983). The top two fully expanded leaves of a soybean cultivar resistant to the soybean looper, *Pseudoplusia includens* (Walker), are less resistant than other primary leaves further down the stem of the plant (Reynolds and Smith 1985) (Figure 7.2). Nault et al. (1992) confirmed this trend in foliage of soybean cultivars resistant to *H. zea* and the velvetbean caterpillar, *Anticarsia gemmatalis* Hubner. A similar relationship exists in the foliage of *Melilotus infesta* Guss., a sweet clover species resistant to feeding by the sweet clover weevil, *Sitona cylindricollis* (F.) (Beland et al. 1970). The same pattern exists in cultivars of cucumber, *Cucurbita sativus* L., resistant to *F. occidentalis* (de Kogel 1997a) and in clones of cottonwood, *Populus* spp., resistant to *C. scripta* (Bingaman and Hart 1993). In the cases of cottonwood and soybean resistance, there are some indications that resistance-mediating allelochemicals accumulate in mature plant tissues and deter arthropods from feeding (Rose et al.1988, Bingaman and Hart 1992). One of the few exceptions is in the resistance of sweet pepper, *Capsicum baccatum* L., to the greenhouse whitefly, *Trialeurodes vaporariorum* (Westwood), where the upper leaves of a resistant pepper cultivar are more resistant to whitefly feeding than older, more mature leaves (Laska et al. 1986).

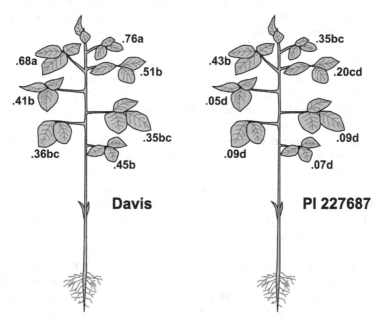

Figure 7.2. Differential mean growth rates [ln (final wt. - initial wt.)24/hrs duration of assay] of Pseudoplusia includens *larvae feeding on trifoliates of susceptible ('Davis') and resistant (PI227687) soybean cultivars. (From Smith 1985. Reprinted with permission from Insect Sci. Applic. Vol. 6:243-248. Copyright 1985, ICIPE Science Press)*

1.6. Infection of Plant Tissues By Diseases

Plant resistance to arthropods may also be enhanced by the immune response of plants to invasion by infectious diseases. Fungal endophytes have been investigated thoroughly in this regards, and reports associating their occurrence in grasses to enhanced arthropod resistance were first published during the 1980s, when *Acremonium lolii* was shown to be involved in the resistance of perennial ryegrass, *Lolium perenne* L., to the Argentine stem weevil, *Listronotus bonariensis* (Kuschel) and several *Crambus* species of sod webworm (Funk et al. 1983, Gaynor and Hunt 1983). Since then, over 40 species of arthropods (Table 7.1) have been shown to exhibit adverse antibiosis or antixenosis effects from exposure to or feeding on perennial ryegrass or tall fescue, *Festuca arundinaceae* Schreb., infected with *Acremonium coenophialum* (Clement et al. 1994, Popay and Rowan 1994). Several pest arthropods are adversely affected by *Acremonium* infection in multiple grass species.

The effects of endophyte-infected plants, when deleterious, are manifested by as antixenosis, antibiosis, or both, and there is good experimental evidence for an allelochemical basis for these effects. Permine, a pyrrolizidine alkaloid from ryegrass, deters *L. bonariensis* feeding on endophyte-infected perennial ryegrass and occurs in many species of grass infected with endophytes (Rowan and Gaynor 1986). Loline alkaloids from ryegrass are antibiotic to *L. bonariensis* (Prestidge and Gaynor 1988). Ergot alkaloids infecting fescue mediate antibiotic effects in *R. padi* (Dahlman et al. 1991), *S. frugiperda* (J. E. Smith) (Clay and Cheplick 1989) and the Russian wheat aphid, *Diuraphis noxia* Mordvilko (Clement et al. 1997). Both permine and loline alkaloids convey antibiosis to the greenbug, *Schizaphis graminum* Rondani, in endophyte-infected grasses (Breen 1992, Siegel et al. 1990).

In some instances, endophyte infection is of no competitive advantage the plant. Certain combinations of endophyte-infected tall fescue and perennial ryegrass have no detrimental effects on the grain aphid, *Sitbion avenae* (Fabricius), the strawberry aphid, *Sitobion fragariae* (Walker) and the rose grain aphid, *Metopolophium dirhodum* (Walker). Some endophytes are ineffective against chewing arthropods as well, such as the southern armyworm, *Spodoptera eridania* (Stoll) and the true armyworm, *Pseudaletia unipunctata* (Haworth) (Johnson et al. 1985, Eichenseer and Dahlman 1993). The effects of different species of grasses infected with *Acremonium* to the same arthropod also vary widely. Breen (1993) demonstrated that the development of *S. eridania* fed *F. arundinaceae* infected with *A. coenophialum* is unaffected. However, the development of *S. eridania* fed perennial ryegrass infected with *Acremonium lolii was* adversely affected. *Festuca rubra* L. subsp. *commutate* Gaud., infected with *Acremonium* have severely reduced development.

The identification of endophyte-mediated arthropod resistance in turf grasses has resulted in the development and marketing of perennial ryegrass and fescue cultivars with resistance to many turf grass arthropod pests. However, the full potential for endophyte-assisted resistance is yet to be realized. Cultivars of Kentucky bluegrass *Poa pratensis* L., have not yet been developed.

Table 7.1. Field crop insect pests exhibiting antibiosis (Ab) or antixenosis (Ax) effects after exposure to or feeding on Acremonium-infected grasses

Coleoptera	Category	References
Chaetocnema pulicaria	Ab	Kirfman et al.1986
Costelytra zealandica	Ab	Popay et al. 1993
Cyclocephala lurida	Ab	Potter et al. 1992
Heteronychus arator	Ax	Prestidge & Ball 1993
Listronotus bonariensis	Ab, Ax	Prestidge & Gallagher 1988
Sphenophorus parvulus	Ab, Ax	Johnson et al. 1985
Sphenophorus venatus	Ab	Murphy et al. 1993
S. inaequalis, S. minimus	Ab	Johnson et al. 1985, Murphy et al. 1993
Hemiptera		
Blissus leucopterus hirtus	Ab, Ax	Mathias et al. 1990
Draculacephala spp.	Ab	Muegge et al. 1991
Endrica inimical	Ab	Kirfman et al.1986
Agallia constricta	Ab	Kirfman et al.1986
Exitianus exitiosus	Ab	Muegge et al. 1991
Graminella nigrifons	Ab	Muegge et al. 1991
Prosapia bicincta	Ab	Muegge et al. 1991
Homoptera		
Balanoccus poae	Ab	Pearson et al. 1988
Rhopalosiphum padi	Ab, Ax	Eichenseer & Dahlman 1992
Rhopalosiphum maidis	Ax	Buckley et al. 1991
Schizaphis graminum	Ab, Ax	Breen 1992
Diuraphis noxia	Ab	Clement et al. 1992
Sipha flava	Ax	Funk et al. 1993
Aploneura lentisci	Ab	Schmidt 1993
Rhopalomyzus poae	Ab	Schmidt 1993
Lepidoptera		
Crambus spp.	Ab, Ax	Murphy et al. 1993
Ostrinina nubilalis	Ab, Ax	Riedell et al. 1991
Spodoptera frugiperda	Ab, Ax	Breen 1993
Spodoptera eridania	Ab	Ahmad et al. 1987
Mythimna convecta	Ab, Ax	Quigley et al. 1993
Agrotis infusa	Ax	Quigley et al. 1993
Graphania mutans	Ab, Ax	Dymock et al. 1988
Orthoptera		
Teleogryllus commodus	Ax	Quigley et al. 1993
Acheta domestica	Ax	Ahmad et al. 1986

Artificial infection of Kentucky bluegrass with *Acremonium* endophytes has proven partially successful (Leuchtmann 1992), but highly specific associations of the plant and endophyte are yet to be identified (Leuchtmann and Clay 1993). If such combinations can be developed, the use of endophyte-associated arthropod resistant cultivars will increase significantly.

The use of endophyte-induced resistance in arthropod management of cereal grain crops is problematic, because of potential toxic effects to humans from endophyte-produced alkaloids in foliage and seeds of such crops. For more in-depth reading about endophyte-enhanced resistance to arthropods, readers are referred to reviews by Breen (1994), Clement et al. (1994), and Popay and Rowan (1994).

1.7. Evaluation of Excised and Intact Plant Tissues

Various investigators have conducted research to determine if removing tissues from plants for evaluation has an effect on the expression of resistance. Sams et al. (1975) compared the results of evaluations of *Solanum* plant material for resistance to the green peach aphid *Myzus persicae* Sulzer, using excised leaflet bioassays in the laboratory and aphid population counts on plants in the field. Correlations between the ratings of plant material in the two evaluations were highly significant, suggesting the use of excised leaf assays as a means of rapid assessment of plant material for aphid resistance. Similarly, Raina et al. (1980) found no differences in the amount of feeding of Mexican bean beetles, *Epilachna varivestis* Mulsant, on excised and intact leaves from plants of common bean, *Phaseolus vulgaris* L.

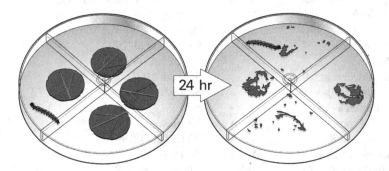

Figure 7.3. Plastic petri dish with divided base for evaluating soybean leaf discs for resistance to Helicoverpa zea. *Left and right quadrants - resistant cultivar, Top and bottom quadrants - susceptible cultivar. (From Smith 1985. Reprinted with permission from Insect Sci. Applic. Vol. 6:243-248. Copyright 1985, ICIPE Science Press)*

East et al. (1992) compared the resistance of muskmelon plant introductions to *T. urticae*, using excised whole leaves, floating leaf discs and intact leaves with clip cages. Variation in experiments using both excised whole leaves and floating leaf discs was significantly lower than that for leaf-clip cage assays, indicating that cut leaf tissues can be used to accurately assess mite development.

In contrast, excision of plant tissues may affect the involvement of allelochemicals in the expression of plant resistance to arthropods. Thomas et al. (1966) detected significantly greater survival of nymphs of the spotted alfalfa aphid, *Therioaphis maculata* (Buckton), on excised trifoliates of several clones of alfalfa, *Medicago sativa* L., than on intact trifoliates. Although there is no indication that allelochemicals are involved in alfalfa resistance to *T. maculata,* it is possible that the concentration of such factors could decrease after excision. Leaves of cassava, *Manihot esculenta* Crantz, normally unpalatable and resistant to the locust, *Zonocercus variegatis* (L.), are readily eaten within one hour after excision, due to a drastic decline in the amount of cyanide produced by intact leaves (Bernays et al. 1977).

1.8. Pre-assay Damage to Tissues

The expression of plant resistance to arthropods is also affected by previous exposure to various stimuli, and as will be discussed in Chapter 9, prior wounding by arthropod or mechanical means induces increased resistance of many crop plants to arthropod damage.

2. ARTHROPOD VARIABLES

Variables within test arthropods must also be compensated for, so that arthropod in laboratory or greenhouse experiments behavior reflects their responses in field populations. Arthropod age, gender, and the effects of pre-assay conditioning are important variables to be defined before proceeding with the evaluation of experimental plant genotypes. The peak activity period of the test arthropod and the level of arthropod infestation on test plants should also be determined, in order to develop reliable, long-term bioassay protocols.

2.1. Arthropod Age

The age of the test arthropod is directly proportional to the amount of plant biomass it will ingest during the evaluation of plant material. Therefore, the arthropod age that most accurately exhibits differences between resistant and susceptible cultivars or genotypes is the appropriate one to use. *E. varivestis* adults feed equally on the leaves of resistant and susceptible soybean cultivars 3 days after eclosion, but at 14 and 35 days after eclosion, differences in the amount of foliage consumed is significant (Smith et al. 1979) (Figure 7.4). Though differences are significant at 35 days, the amount of beetle feeding is only one-half that of 14 day-old beetles.

At the sensory level, arthropod age also affects the outcome of a bioassay. Blaney and Simmonds (1994) demonstrated this trend, using electrophysiological recordings of tarsal sensilla of the turnip root fly, *Delia floralis* (Fallen). Fly tarsal sensillar stimulation to sucrose and potassium chloride are initially high, but decline during the life of the fly. Interestingly, flies not responding to potassium chloride do not respond to either sucrose or the fly modifying allelochemical sinigrin. Rees

(1970) noted the same age-related trend in decline of the labellar sensilla of *Psila terraenovae* (Robineau-Desvoidy).

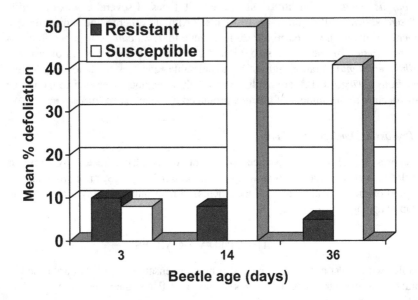

Figure 7.4. Effect of Epilachna varvestis *age on defoliation of resistant and susceptible soybean cultivars. (from Smith et al. 1979. Adapted with permission from J. Econ. Entomol., Vol. 72:374-377. Copyright 1979, Entomological Society of America)*

2.2. Arthropod Gender

Gender-based differences in the behavior of test arthropods also affect the outcome of plant resistance evaluations. Phytophagous female arthropods often consume more foliage than males, due to the high dietary protein requirements of egg production. In addition to differential feeding, mating activity may interfere with female feeding. Schalk and Stoner (1976) determined that female *Leptinotarsa decemlineata* consume significantly more foliage of both resistant and susceptible tomato cultivars than males (Figure 7.5). Cook (1988) noted that female rice water weevils, *Lissorhoptrus oryzophilus* Kuschel, fed rice cultivars with varying levels of weevil resistance consume more foliage than males. Similar results were noted by Smith et al. (1979) in evaluations of *E. varivestis* feeding on soybean cultivars. Although sorting individuals by gender may be time consuming, this practice can contribute to greatly improving the accuracy of experimental results.

Gender-based responses to plant allelochemicals also exist. The attraction of clover head weevils, *Hypera meles* (F.), to *Trifolium* spp. flower bud volatiles is greater in females than in males (Smith et al. 1976).   Similarly, the attraction of ponderosa pine cone beetles, *Conopthorus ponderosae* Hopkins, to pine cone resin is also greater in females than in males (Kinzer et al 1972).

2.3. Density and Duration of Arthropod Infestation Level

Determining the proper arthropod infestation density and duration on test plants is necessary in order to avoid over- or under estimation of the resistance or susceptibility of test plants. The starting point and usually the minimum level of arthropod infestation, is one that causes measurable, economically significant damage to plants. In order to rapidly eliminate large amounts of susceptible plant material in resistance screening assays, larger than normal arthropod infestation levels may be placed on plants. In field evaluations, monitoring the increase of pest natural populations in field plot border rows (Chapter 6), can aid the investigator in determining when and if supplemental infestations should be placed on plants at the same time when natural populations reach their peak. Whether in the greenhouse or field, high-density infestations constitute a calculated risk, however, and may mask the appearance of low-level resistance in some plant material.

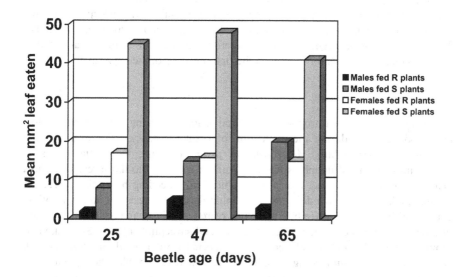

Figure 7.5. Consumption of resistant (R) and susceptible (S) tomato foliage by Leptinotarsa decemlineata *adults. (From Schalk & Stoner 1976. Adapted with permission from J. Am. Soc. Hort. Sci., Vol.101:74-76. Copyright 1976, American Society for Horticultural Science)*

Results of Bosque-Perez and Schotzko (2000) further illustrate arthropod density effects. Plants of the wheat cultivar 'Stephens', which is susceptible to *D. noxia*, are unable to support large *D. noxia* populations when infested at a rate of 20 aphids per plant, compared to a lower initial infestation of 5 aphids per plant. Although the high infestation rate may yield susceptibility data more rapidly, the lower infestation rate more accurately represents field infestation conditions, and has fewer adverse effects on plant growth and survival. Plants infested at a lower rate remain alive for a longer period of time and support aphid populations for comparison to resistant genotypes. Results of Burd and Burton (1992) support this concept, but suggest that infestation duration also warrants consideration by investigators. Their results indicate that when *D. noxia* are removed from wheat plants, the previously infested plants recover. Thus, *D. noxia* infestation duration may be more important than infestation density.

However, it is typical and anticipated for plant damage and arthropod population development to differ on resistant and susceptible genotypes, regardless of infestation duration. Annan et al. (1995) noted this to be the case for cowpea, *Vigna unguiculata* (L.) Walp., resistant to the cowpea aphid, *Aphis craccivora* Koch. When resistant and susceptible plants are infested with aphids for from 3 to 28 days, resulting *A. craccivora* populations and cowpea pod production are greater on susceptible plants than on resistant plants at all durations of infestation. Schotzko and Smith (1991b) employed spatial and geostatistical analyses of *D. noxia* population densities on aphid-resistant and susceptible plants to reveal unique aphid distribution patterns during population development and dispersal. On susceptible plants, dispersal from plants is delayed, but on resistant plants, aphids disperse more rapidly. These results serve as a reminder that equal arthropod densities cannot be assumed to persist in plant bioassays for more than the first few days of the experiment.

2.4. Pre-assay Conditioning

The plants in which arthropods are reared prior to testing directly affect test arthropod behavior during the evaluation of plant resistance. For example, the responses of pea aphids, *Acyrthosiphon pisum* (Harris), to alfalfa plants is significantly enhanced in aphids reared on alfalfa compared to those reared on faba bean, *Vicia faba* L. (Girousse et al. 1999) (Figure 7.6). Feeding by aphids reared on alfalfa result in distinctive differences in aphid damage to resistant and susceptible control cultivars, whereas feeding by aphids reared on faba bean results in few if any differences in aphid damage between resistant and susceptible plants. Schotzko and Smith (1991a) and Worrall and Scott (1991) detected similar trends in studies of *D. noxia* and *S. graminum* feeding on aphid-resistant and susceptible wheat plants. The conditioning effects of pre-test host plant had significant effects on both aphids in antibiosis experiments. In antibiosis experiments with the same aphids, Robinson (1993) and Schweissing and Wilde (1979a) demonstrated that the experimental control plant has a greater effect on the aphid that the pre-assay host plant.

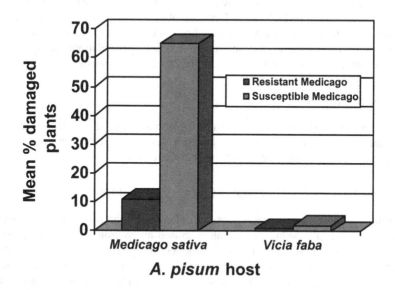

Figure 7.6. Responses of Acyrthosiphon pisum reared on Medicago sativa or Vicia faba plants to aphid resistant and susceptible alfalfa cultivars. (From Girousse et al. 1998. Adapted with permission from Phytoprotection. 79:139-148. Copyright 1998, The Quebec Society for the Protection of Plants).

In short-term behavioral assays, the conditions at which test arthropods are held also influence their behavior. Saxena (1967) observed that the attraction of the red cotton bug, *Dysdercus koenigii* (F.), to cotton seed extracts increased in proportion to the amount of time that arthropods were starved prior to testing. A similar relationship exists in the response of starved desert locusts, *Schistocerca gregaria* Forskal, to grass odors (Moorehouse 1971). Wiseman and McMillian (1980) found that *H. zea* larvae fed on the foliage of several different crop plants exhibit different feeding responses to maize silk extracts, indicating the need for standardization of the diet used to rear arthropods for resistance assays.

2.5. Arthropod Activity Period

Both diel and diurnal arthropod activity patterns affect the accuracy of measurements of plant resistance. Warner and Richter (1974) monitored the annual abundance of adult alfalfa weevils, *Hypera postica* (Gyllenhal), during day and night hours. Weevils are present in greater numbers on plants in the fall, winter, and spring during the day, but are more prevalent on plants at night during the warm summer months. Boiteau et al. (1979) detected a diel population fluctuation in the bean leaf beetle, *Cerotoma trifurcata* (Forster), in a field planting of soybean. Beetle abundance is lowest from 11:00 am to 1:00 pm and greatest after 3:00 pm. These examples stress the need to monitor arthropod populations from field evaluations at times when test arthropods are the most abundant and to avoid sampling populations during periods when arthropod activity patterns or habitat preferences are changing.

2.6. Arthropod Biotypes

Changes in the physiology of the pest arthropod may result in the development of resistance-breaking biotypes that alter the expression of resistance significantly. A complete discussion of biotypes is presented in Chapter 11.

3.   ENVIRONMENTAL VARIABLES

In addition to the variation inherent in both plant and arthropod, the evaluation of plant resistance is also subject to variation caused by environmental effects. Such variations include those resulting from changes in lighting, temperature, relative humidity, soil nutrient conditions, and agrochemicals commonly found in contact with the crop plant. The following discussion is intended as a general guide for illustrative purposes. For more detailed discussions, readers are referred to reviews of Tingey and Singh (1980) and Sharma et al. (2001).

3.1. Light Duration, Quantity and Quality

The duration, quantity and quality of light have each been shown to affect the expression of plant resistance to arthropods. Resistance to the tobacco hornworm, *Manduca sexta* (L.), in the wild tomato, *Lycopersicon hirsutum* Humb. and Bonpl., increases in plants grown under long daylengths (Kennedy et al. 1981), and the same plants produce higher quantities of the hornworm toxin 2 - tridecanone (Chapter 3). Strawberry plant resistance to *T. urticae* rapidly increases in plants grown under long daylength, compared to plants grown under short daylength (Patterson et al. 1994). Normally susceptible plants grown in natural daylengths gradually develop moderate resistance over the length of the growing season. The resistance effects of supplemental daylength are expressed in a short-day sensitive cultivar as well as a day-neutral cultivar, but are readily reversible under short daylengths. In at least one example, reduced daylength increase arthropod resistance. Plants of *Lycopersicon hirsutum* have high levels of arthropod resistance and possess both glandular and non-glandular trichomes. Trichome production is directly affected by daylength, and plants grown in short days produce a greater density of glandular trichomes and a reduced density of non-glandular trichomes than plants grown in long days (Snyder et al. 1998). *Bemisia argentifolii* Bellows and Perring, fed plants grown in short day conditions exhibit antibiosis effects, presumably because of greater densities of *L.*

hirsutum glandular trichomes. Resistance in sorghum to the midge *Stenodiplosis sorghicola*, is also more readily expressed in short day growing conditions, as midge susceptibility of plants increases under long daylength (Sharma et al. 2003).

In at least one instance, resistance can apparently be photosynthetically over-stimulated so that the resistance factors normally produced by the plant cease to function. Resistance to *T. ni* in soybean is completely negated by exposure of plants to continuous illumination, but is regained after growth of plants in a 16 hour daylength for 14 days (Khan et al. 1986).

The intensity of the light received by plants also tempers the expression of resistance in several plant-arthropod interactions (Table 7.2). The negative effect of reduced light intensity on the solid stem character of wheat cultivar resistance to the wheat stem sawfly, *Cephus cinctus* (Norton), has been demonstrated by several investigators (Roberts and Tyrell 1961, DePauw and Read 1982, Holmes 1984). Tomato plants grown under long-day, low light intensity greenhouse conditions have reduced allelochemical defenses and are more susceptible to *M. sexta* and *T. urticae* (Nihoul 1993, Jansen and Stamp 1997).

Table 7.2. Decreased plant resistance to arthropods as a result of reduced light intensity

Plant	Arthropod	Reference
Beta vulgaris	*Myzus persicae*	Lowe 1967
Chrysanthemum	*Frankliniella occidentalis*	de Kogel et al. 1997b
Glycine max	*Epilachna varivestis*	Elden & Kenworthy 1995
Lycopersicon	*Manduca sexta*	Jansen & Stamp 1997
esculentum	*Tetranychus urticae*	Nihoul 1993
Medicago sativa	*Hypera postica*	Shade et al. 1975
Rhododendron	*Stephanitis pyroides*	Trumbule & Denno 1995
Sorghum bicolor	*Atherigona soccata*	Woodhead 1981
	Chilo partellus	
Triticum aestivum	*Cephus cinctus*	Roberts & Tyrell 1961
		DePauw & Read 1982
		Holmes 1984
Vigna unguiculata	*Aphis craccivora*	Nkansah-Poku & Hodgson 1995
Zea mays	*Ostrinia nubilalis*	Manuwoto & Scriber 1985a

Nihoul (1993) documented how trichomes of tomato plants growing in low light intensity trap fewer *T. urticae* than plants growing in normal, higher intensity lighting. Reduced light intensity also decreases exudate production by glandular trichomes on stems of alfalfa cultivars resistant to *H. postica* (Shade et al. 1975). Woodhead (1981) noted that the lower light intensity resulting from atmospheric cloudiness reduces the production of phenolics involved in arthropod resistance in sorghum. Resistance in chrysanthemum cultivars to *F. occidentalis* is suppressed under high intensity summer lighting conditions (de Kogel 1997), but when plants are grown under low (reduced) light intensity, resistance increases (Figure 7.7). Changes in light intensity have no effect on susceptible cultivars. The expression of resistance in azalea, *Azalea indica* L., cowpea, maize, sugar beet, *Beta vulgaris* L., and soybean is also diminished by reduced light intensity (Lowe 1967, Manuwoto and Scriber 1985a, Reynolds and Smith 1985, Elden and Kenworthy 1995, Trumbule and Denno 1995).

The results of Franca and Tingey (1994) are a notable exception to the preceding examples of how reduced light intensity negatively affects arthropod resistance. Plants of the wild potato, *Solanum berthaultii*, are resistant to *L. decemlineata*, and resistance is mediated by exudates of glandular trichomes that trap *L. decemlineata* immatures (Neal et al. 1989) (See Figure 2.8). Growing *S. berthaultii* plants in shaded conditions results in reduced trichome density, exudate production and trichome phenolic oxidation activity, yet the level of *L. decemlineata* resistance is unaffected. In spite of reduced trichome metabolism under shaded illumination, the threshold level of density and exudation are presumably sufficient to provide an adequate level of *L. decemlineata* resistance.

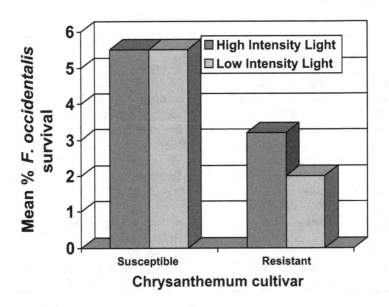

Figure 7.7. Effect of reduced light intensity on chrysanthemum cultivars and levels of antibiosis resistance (survival) to female Frankliniella occidentalis. Reprinted from de Kogel, W. J., M. van der Hoek, M. T. A. Dik, B. Gebala, F. R. van Dijken, and C. Mollema. 1997b. Seasonal variation in resistance of chrysanthemum cultivars to Frankliniella occidentalis (Thysanoptera: Thripidae). Euphytica. 94:283-288, Copyright 1997 Kluwer Academic Publishers, with kind permission of Springer Science and Business Media.

In addition to light duration and intensity, the quality of light received by plants also affects the expression of resistance. Both ultraviolet (UV) and shortwave ultraviolet (UVB) radiation enhances resistance. The antibiotic effect of psoralen, a linear furanocoumarin from celery, *Apium graveolens* (L.), to *S. exigua* is enhanced when larvae are exposed to UV light (Trumble et al. 1991). Foliage from plants of rough lemon, *Citrus jambhiri* Lush, also contain furanocoumarins, and *C. jambhiri* plants gown under enhanced UVB intensity have increased resistance to *T. ni* larvae (McCloud and Berenbaum 1994). Rice plants grown under UVB irradiation produce higher levels of phenolics than non-irradiated plants and inhibit growth of the rice leaffolder, *Marasmia patnalis* Bradley, more than non-irradiated plants (Caasi-Lit 1998).

3.2. Temperature

Plant resistance to arthropods may not be expressed at abnormally low or high temperatures. Numerous studies have established that high temperature diminishes the expression of different genes controlling resistance in wheat to different biotypes of the Hessian fly, *Mayetiola destructor* Say (Sosa and Foster 1976, Sosa 1979, Tyler and Hatchett 1983, Ratanatham and Gallun 1986). For example, the resistance of the *H19* gene is fully effective at 19° C but losses all effectiveness against *M. destructor* at 23° C and 26° C, while the *H27* gene is effective at all three temperatures (Ohm et al. 1997) (Figure 7.8).

The opposite trend exists in the expression of sorghum resistance to different biotypes of *S. graminum*. Initially, biotype B antixenosis present at 32° C was shown to be lost at 21° C (Wood and Starks 1972, Webster and Starks 1987). Other studies since have documented that antibiosis, antixenosis and tolerance to *S. graminum* biotypes C, E, and I are lost at low (15-21° C) temperatures but fully functional at high temperatures (28-30° C) (Schweissing and Wilde 1979b, Harvey et al. 1994, Thindwa and Teetes 1994). *S. sorghicola* resistance in sorghum is also diminished at temperatures lower than those normally observed in field growing conditions (Sharma et al. 2003). Conversely, tolerance to *S. graminum* biotype C in barley, oat and rye is greater at lower temperatures (Schweissing and Wilde 1978).

The expression of arthropod resistance in alfalfa is also altered at low temperatures. Schalk et al. (1969) found evidence of a breakdown of resistance in some alfalfa clones to *T. maculata* at temperatures 10 to 15° C below those in normal field growing conditions. Karner and Manglitz (1985) and Johnson et al. (1980) observed similar results for alfalfa resistance to *A. pisum* and *H. postica*.

The effects of temperature on the allelochemical bases of resistance have been evaluated less extensively, but temperature effects have been documented. Walters et al. (1991) studied the allelochemistry of resistance in an inbred line of geranium, *Pelargonium x hortorum* (L.), to *T. urticae*, conferred by glandular trichome exudates containing anacardic acid.

Resistance is lost at 25° C compared to 15° C, presumably because of a shift to a larger composition of short chain acids in trichome exudates, reducing exudate viscosity, and causing exudates to flow from the trichome tip at high temperatures. Although several studies have been conducted, a significant effect of temperature on

hydroxamic acid content of cereal plants in relation to arthropod resistance has not
been established (Gianoli and Niemeyer 1996, 1997).

Given the loss of resistance to various pest arthropods in different crops at
temperature extremes, temperatures more closely related to field conditions would
seem better choices for greenhouse of laboratory resistance bioassays, and several
studies have documented this to be the case. Kindler and Staples (1970a) noted that
differences between susceptible alfalfa clones and clones resistant to *T. maculata*
were more apparent at fluctuating temperatures similar to natural field growing
conditions, than at a constant temperature equivalent to the mean of the fluctuating
regime. van de Klashorst and Tingey (1979) reported similar results. Potato resistance
to *E. fabae* at fluctuating temperature is more similar to resistance ratings of field-
grown plants than ratings of plants grown at constant temperatures. Harvey et al.
(1994) observed the same trend with *S. graminum* biotype I on resistant sorghum
genotypes. Resistance lost at a constant low temperature is regained at an alternating
high and low temperature regime.

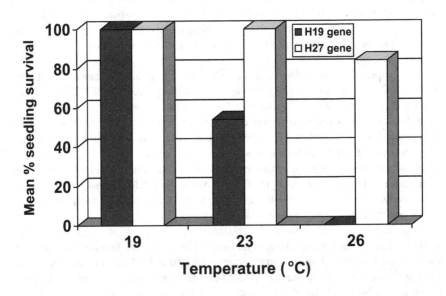

Figure 7.8. *Affect of temperature change on expression of the H19 and H27 genes in wheat*
for resistance to Mayetiola destructor. *(From Ohm 1997. Adapted with permission from*
Crop Sci. 37:113-115. Copyright 1997, Crop Science Society of America)

Breen (1992) observed a functional temperature 'window' at which antixenosis
resistance to *S. graminum* in *Acremonium* endophyte-infected ryegrass was most
fully expressed. Antixenosis, endophyte concentration and concentration of the
alkaloid arthropod deterrent peramine were each reduced in plants grown at 7°C and
28°C, compared to plants grown at 14°C and 21°C.

3.3. Soil Nutrients

The macronutrients nitrogen, phosphorous, potassium, calcium and sulfur are essential for plant growth. Variations in the amounts of these nutrients applied to the medium in which plants are grown prior to evaluation can significantly affect the expression of plant resistance (Tingey and Singh 1980, Fageria and Scriber 2001). Nitrogen (N) is directly linked to plant and arthropod growth. Increases in soil N content generally increase plant vegetative tissue mass, supplying arthropods with more tissue of increased nutritional composition and digestibility than plants of lower N content (Table 7.3). Increasing the amount of N fertilization decreases *S. frugiperda* resistance in maize (Wiseman et al. 1973), grasses (Chang et al. 1985), and peanuts (Leuck and Hammons 1974). Similar effects occur when greater than normal amounts of N are applied to maize resistant to *S. eridania* (Manuwoto and Scriber 1985b), alfalfa resistant to *T. maculata* (Kindler and Staples 1970b), ryegrass resistant to *L. bonariensis* (Gaynor and Hunt 1983) and cowpea resistant to *A. craccivora* (Annan et al. 1997). Barbour et al. (1991) found that increasing the rate of N, P, and K fertilization by approximately 10-fold, greatly reduced the glandular trichome-based resistance in *L. hirsutum* f. *glabratum* to *M. sexta* and *L. decemlineata*.

However, in some instances, increased amounts of N fertilization have been shown to increase resistance. Such is the case of resistance in pearl millet, *Pennisetum glaucum* (L.) R. Br., to *S. frugiperda* (Leuck 1972). Riedell et al. (1996) reported that N applications increase the root growth of maize inbred lines exhibiting tolerance to the western corn rootworm, *Diabrotica virgifera virgifera* LeConte. Inducible resistance in Alaskan birch, *Betula resinifera* L., to the spear-marked black moth, *Rheumaptera hastate,* is directly tied to increased N fertilization (Bryant et al. 1993), while constitutively expressed antibiosis and tolerance resistance in cassava to *P. manihoti* are directly proportional to N availability (Le Ru et al. 1994). Archer et al. (1990) evaluated sorghum cultivars of varying N use efficiency for resistance to the Banks grass mite, *Oligonychus pratensis* (Banks), and quantified lower mite densities on N-use inefficient cultivars. Thus, in some instances, the efficiency of N use or N metabolism may contribute to plant resistance as much as the direct actions of arthropod resistance genes.

Increasing the amount of potassium (K) and phosphorous (P) in the soil medium also increases arthropod resistance. The level of *T. maculata* resistance in alfalfa (Kindler and Staples 1970b) and of *S. graminum* resistance in sorghum (Schweissing and Wilde 1979b) increase in plants receiving supplemental K. Annan et al. (1997) observed that supplemental applications of K to cowpea cultivars increase their resistance to *A. craccivora*. Salim and Saxena (1991) observed similar results for resistance in rice to the whitebacked planthopper, *Sogatella furcifera* (Horvath). Increased K application negatively affects *S. furcifera* behavior (antixenosis) and survival (antibiosis). Resistance to *T. maculata* in alfalfa and *S. frugiperda* in pearl millet also increase after additional phosphorous applications (Kindler and Staples 1970b, Leuck 1972). The foliar P concentration in soybean genotypes with *E. varivestis* resistance is significantly lower than in susceptible genotypes (Elden and Kenworthy 1994). However, this trend may also be due to efficiency of soybean P utilization.

The silica content of plants may adversely affect maize and rice pest arthropod survival (see Chapters 2 and 3). Salim and Saxena (1992) noted that silica supplements increase *S. furcifera* resistance in a normally resistant rice cultivar, while iron supplements actually decrease resistance by promoting increased *S. furcifera* population growth. The addition of aluminum to rice plant soils has no effect on resistance to *S. furcifera*, but does provide a small degree of pseudoresistance to *S. furcifera* in a normally susceptible rice cultivar. The results of these studies indicate the need for careful standardization of the quantity and quality of soil amendments used to grow plants for arthropod resistance evaluation. The adoption of such protocols will allow accurate within year and year-to-year comparisons of experimental results from different locations.

Table 7.3. Decreased plant resistance to arthropods as a result of applications of increased amounts of soil nitrogen

Plant	Arthropod	Reference
Arachis glabrata	Spodoptera frugiperda	Leuck & Hammons 1974
Betula resinifera	Rheumaptra hastale	Bryant et al. 1993
Graminae spp.	S. frugiperda	Chang et al. 1985
Lolium	Listronotus bonariensis	Gaynor & Hunt 1983
Lycopersicon hirsutum	Leptinotarsa decemlineata	Barbour et al. 1991
	Manduca sexta	
Manihot esculenta	Phenacoccus manihoti	Le Ru et al. 1994
Medicago	Therioaphis maculata	Kindler & Staples 1970b
Zea mays	S. frugiperda	Manuwoto & Scriber 1985b
	Spodoptera eridania	Wiseman et al. 1973

It is equally important to understand differences in nutrient use efficiencies in plant. Interactions between soil nutrients, resistance-mediating allelochemicals, and environmental variations (see below) constitute an extremely complicated set of variables to untangle in the process of developing an understanding of the molecular and/or biochemical mechanisms controlling arthropod resistance (Hermes and Mattson 1992, Hammerschmidt and Schultz 1996). For additional discussions of the involvement of soil nutrients in plant-arthropod interactions, readers are directed to the review of Fageria and Scriber (2001).

3.4. Soil Moisture

The effects of plant moisture deficiency on arthropod resistance have been researched in only a few instances. Moisture deficiency accelerates leaf protein break down, increasing the N content of phloem sap, increases starch hydrolysis and increases phloem sucrose content. As N and sucrose content increases, aphid development and survival increase (Kennedy and Booth 1959). Oswald and Brewer (1997) noted that *D. noxia* susceptibility in barley cultivars increases in water deficient plants. *D. noxia* survival in general is better on moisture deficient plants in hot, dry conditions than in cooler more humid conditions. An accurate assessment of resistance may be masked in plants receiving excessive amounts of soil moisture. Extreme moisture deficiency causes phloem sap viscosity to increase, due to decreased cell turgor. This increased viscosity complicates aphid ingestion of plant sap. Wearing (1972) demonstrated that at moderate wilting, plants of Brussels sprouts, *Brassica oleracea* var. *gemmifera* Zenker., support the same number of *M. persicae* as turgid plants, but during severe water deficient, *M. persicae* development is restricted.

Jenkins et al. (1997) assessed the resistance of soybean cultivars to *E. varivestis* in moisture deficient and moisture excess growing conditions. Adult and larval mortality on resistant cultivars increase in moisture deficits but resistance is diminished or lost, depending on the cultivar assessed, in vegetative stage plants growing in excess moisture conditions. In at least one instance of allelochemical-based resistance, moisture deficiency has no adverse affect on resistance. Pelletier (1990) examined *S. berthaultii* plants grown in moisture deficits and found no differences in the numbers of resistance-mediating glandular trichomes on plant terminal leaflets. Leaflets are smaller, but have a greater density of trichomes than plants grown in normal moisture conditions. As with soil nutrients, greenhouse soil moisture conditions to as much of an extent as possible, should resemble field moisture conditions, so that greenhouse data most accurately reflects measurement of actual arthropod resistance in field-grown plants.

3.5. Agrochemicals

Several insecticides, herbicides, and plant growth regulators adversely affect phytophagous arthropods (Tingey and Singh 1980, Kogan and Paxton 1983). Gall and Dogger (1967) demonstrated how 2, 4-D herbicide causes the wheat cultivar 'Selkirk', (normally susceptible to *C. cinctus*) to become resistant. Treatment of plants of the resistant cultivar 'Rescue' increases resistance only slightly. Application of growth retardant phosfon (2, 4-dichloro-benzyltributyl phosphonium chloride) decreases survival and reproduction of *T. urticae* on the normally mite-susceptible chrysanthemum cultivar 'Golden Princess Anne', but has little effect on the resistant cultivar '#4 Golden Princess Anne' (Worthing 1969).

The plant growth regulators (2-chloroethyl)-trimethylammonium chloride (CCC) and N, N-dimethyl-piperidinium chloride (PIX) increase resistance in a *S. graminum* susceptible sorghum cultivar, but neither compound enhances the resistance of a resistant cultivar (Dreyer et al. 1984). The application of PIX increases the content of cotton plant gossypol, (see Chapters 2 and 3) but has no beneficial effect on cotton resistance to *H. virescens* larvae (Mulrooney et al. 1985).

3.6. Relative Humidity

Relative humidity is an especially important factor in determining plant resistance to arthropod pests of stored grain. Russell (1966) found that the rate of development of the rice weevil, *Sitophilus oryzae* (L.), increases considerably on weevil-resistant sorghum cultivars when cultures are maintained in humidity 20% less than normal. Rogers and Mills (1974) found that sorghum resistance to the maize weevil, *Sitophilus zeamais* (Motschulsky), is stable in one resistant cultivar regardless of humidity, but decreases in a moderately resistant cultivar 'Redlan' as humidity increases.

3.7. Atmospheric Fluctuations

Changes in atmospheric components of the plant growth environment may alter plant metabolism and the performance of arthropods on those plants. Ozone is a major air pollutant in the United States and ozone-related injury to crop plants is a significant loss factor (Hughes 1987). Lin et al. (1990) evaluated the susceptibility of a soybean cultivar to *E. varivestis* in plants wounded by *P. includens* to induce resistance, followed by subsequent exposure to high or low concentrations of ozone. At low ozone concentrations, induced resistance was neutralized and *E. varivestis* showed no preference between non-induced plants in normal air and induced plants in ozone. At high ozone concentrations however, the *P. includens*-induced resistance disappeared and *E. varivestis* fed more readily on ozone treated plants than on control plants in normal air. Chappelka et al. (1988) noted similar ozone-related effects on non- induced foliage of other soybean cultivars for *E. varivestis* susceptibility. Trumble et al. (1987) found that tomato plants treated with high concentrations of atmospheric ozone had greatly increased susceptibility to the tomato fruit worm, *Keferia lycopersicella* (Walsingham). Excess amounts of atmospheric sulfur dioxide have also been shown to alter the metabolism of common bean plants and increase their susceptibility to arthropod attack (Hughes et al. 1981, 1982).

Evidence also exists to demonstrate how a more recent shift in atmospheric conditions related to global climate change, increased CO_2 concentration, also affects plant arthropod interactions. Coviella et al. (2002) subjected cotton plants to varying CO_2 concentrations and demonstrated how plants treated with high levels of CO_2 have proportionally reduced N content and as a result, increased resistance to *S. exigua* larval damage. As with previous examples of soil variables however, this is only one example of the effects of increased CO_2 concentration on arthropod performance. Many other experiments are required over a broad plant taxonomic range and varying levels of CO_2 concentration to fully understand the impact of altered CO_2 concentrations on plant-arthropod interactions and plant resistance.

4. CONCLUSIONS

The factors discussed in this chapter are examples of the fact that plant resistance to arthropods is a relative and highly variable phenomenon. As such, resistance is dependent on several interacting factors involving the test arthropod, the test plant, and test environment. The differences cited here in each of these variables indicate the need to adopt standardized temperature, lighting, humidity, soil moisture and soil nutrient conditions in laboratory and greenhouse bioassays that reflect those of field growing conditions. If this can be accomplished, results of these experiments can be accurately compared to field and greenhouse data collected at different times or locations.

Variation in the expression of plant resistance also indicates that the condition of both plant tissues and test arthropods must be defined as accurately as possible, in order to minimize the variations and related sources of error of these organisms in resistance assays. In field experiments, useful arthropod resistance is that which remains stable over several years in a broad range of environmental conditions. For this reason, year-to-year environmental variation may prove useful in documenting the stability and operating range of arthropod resistance.

REFERENCES CITED

Ahmad, S., J. M. Johnson-Cicalese, W. K. Dickson, and C. R. Funk. 1986. Endophyte-enhanced resistance in perennial ryegrass to the bluegrass billbug, *Sphenophorus parvulus*. Entomol. Exp. Appl. 41:3–10.

Ahmad, S., S. Govindarajan, J. M. Johnson-Cicalese, and C. R. Funk. 1987. Association of a fungal endophyte in perennial ryegrass with antibiosis to larvae of the southern armyworm, *Spodoptera eridania*. Entomol. Exp. Appl. 45:287–294.

Alghali, A. M. 1984. Effect of plant spacing on the infestation levels of rice by the stalk-eyed borer, *Diopsis thoracica* West (Diptera: Diopsidae). Trop. Agric. (Trinidad). 61:74–75.

Annan, I. B., K. Ampong-Nyarko, W. M. Tingey, and G. A. Schaefers. 1997. Interactions of fertilizer, cultivar selection, and infestation by cowpea aphid (Aphididae) on growth and yield of cowpeas. Intl. J. Pest Manage. 43:307–312.

Annan, I. B., G. A. Schaefers, and W. M. Tingey. 1995. Influence of duration of infestation by cowpea aphid (Aphididae) on growth and yield of cowpea resistant and susceptible cultivars. Crop Protection. 14:533–538.

Archer, T. L., A. B. Onken, E. D. Bynum, Jr. and G. C. Peterson. 1990. Banks grass mite (*Oligonychis partensis*) abundance on sorghum cultivars with different levels of nitrogen use and metabolism efficiency. Exp. Appl. Acarol. 9:177–182.

Barbour, J. D., R. R. Farrar, and G. G. Kennedy. 1991. Interaction of fertilizer regime with host-plant resistance in tomato. Entomol. Exp. Appl. 60:289–300.

Bauce, E., M. Crepin, and N. Carisey. 1997. Spruce budworm growth, development and food utiliaztion on young and old balsam fir trees. Oecologia. 97:499–507.

Beland, G. L., W. R. Akeson, and G. R. Manglitz. 1970. Influence of plant maturity and plant part on nitrate content of the sweet clover weevil-resistant species *Melilotus infesta*. J. Econ. Entomol. 63:1037–1039.

Bernays, E. A., R. F. Chapman, E. M. Leather, A. R. McCaffery, and W. W. D. Modder. 1977. The relationship of *Zonocerus variegatus* (L.) (Acridoidea: Pyrgomorphidae) with cassava (*Manihot esculenta*). Bull Entomol. Res. 67:391–404.

Bingaman, B. R., and E. R. Hart. 1993. Clonal and leaf age variation in *Populus* phenolic glycosides, implications for host selection by *Chrysomela scripta* (Coleoptera, Chrysomelidae). Environ. Entomol. 22:397–403.

Blaney, W. M., and M. S. J. Simmonds. 1994. Effect of age on the responsiveness of peripheral chemosensory sensilla of the turnip root fly (*Delia floris*). Entomol. Exp. Appl. 70:253–262.

Boiteau, G., J. R. Bradley, and J. W. Van Duyn. 1979. Bean leaf beetle: Diurnal population fluctuations. Environ. Entomol. 8:615–618.

Bosque-Perez, N. A., and D. J. Schotzko. 2000. Wheat genotype, early plant growth stage and infestation density effects on Russian wheat aphid (Homoptera: Aphididae) population increase and plant damage. J. Entomol. Sci. 35:22–38.

Breen, J. P. 1992. Temperature and seasonal effects on expression of *Acremonium* endophyte-enhanced resistance to *Schizaphis graminum* (Homoptera: Aphididae. Environ. Entomol. 21:68–74.

Breen, J. P. 1993. Enhanced resistance to fall armyworm (Lepidoptera: Noctuidae) in *Acremonium* endophyte-infected turfgrasses. J. Econ. Entomol. 86:621–629.

Breen, J. P. 1994. *Acremonium* endophyte interactions with enhanced plant resistance to insects. Ann. Rev. Entomol. 39:401–423.

Bryant, J. P., P. B. Reichardt, T. P. Clausen, and R. A. Werner. 1993. Effects of mineral nutrition on delayed inducible resistance in Alaska paper birch. Ecology. 74:2072–2084.

Buckley, R. J., P. M. Halisky, and J. P. Breen. 1991. Variation in feeding deterrence of the corn leaf aphid related to *Acremonium* endophytes in grasses. Phytopathol. 81:120–125.

Burd, J. D., and R. L. Burton. 1992. Characterization of plant damage caused by Russian wheat aphid (Homoptera: Aphididae). J. Econ. Entomol. 85:2017–2022.

Caasi-Lit, M. T. 1998. Effects of ultraviolet-B irradiated rice plants on the growth and development of the rice leaffolder, *Marasmia patnalis* Bradley. Philipp. Ent. 12:179–193.

Chappelka, A. H., M. E. Kraemer, T. Mebrahtu, M. Rangappa, and P. S. Benepal. 1988. Effects of ozone on soybean resistance to the Mexican bean beetle (*Epilachna varivestis* Mulsant). Environ. Exp. Bot. 28:53–60.

Chang, N. T., B. R. Wiseman, R. E. Lynch, and D. H. Habeck. 1985. Resistance to fall armyworm: Influence of nitrogen fertilizer application on nonpreference and antibiosis in selected grasses. J. Agric. Entomol. 2:137–146.

Clay, K., and G. P. Cheplick. 1989. Effect of ergot alkaloids from fungal endophyte-infected grasses on fall armyworm (*Spodoptera frugiperda*). J. Chem. Ecol. 15:169–182.

Clement, S. L., W. L. Kaiser, and H. Eichenseer. 1994. *Acermonium* endophytes in germplasms of major grasses and their utilization for insect resistance. In: C. W. Bacon and J. F. White, Jr. (Eds.), Biotechnology of Endophytic Fungi of Grasses. CRC Press, Boca Raton. pp. 185–199.

Clement, S. L., D. G. Leister, A. D. Wilson, and K. S. Pike. 1992. Behavior and performance of *Diuraphis noxia* (Homoptera: Aphididae) on fungal endophyte-infected and uninfected perennial ryegrass J. Econ. Entomol. 85:583–588.

Clement, S. L., A. D. Wilson, D. G. Leister, and C. M. Davitt. 1997. Fungal endophytes of wild barley and their effects on *Diuraphis noxia* population development. Entomo. Appl. Exp. 82:275–281.

Colwell A. Cook. 1987. Categories of Resistance in Rice to the Rice Water Weevil, *Lissorhoptrus oryzophilus* Kuschel. M.S. Thesis. Louisiana State University. 87 pages.

Coviella, C. E., R. D. Stipanovic, and J. T. Trumble. 2002. Plant allocation to defensive compounds: interactions between elevated CO_2 and nitrogen in transgenic cotton plants. J. Exp. Bot. 53:323–332.

Dahlman, D. L., H. Eichenseer, and M. R. Siegel. 1991. Chemical perspectives on endophyte-grass interactions and their implication in herbivory. In: P. Barbosa, V. Kirchik, and C. G. Jones (Eds.), Multi-trophic Level Interactions Among Microorganisms, Plants and Insects. John Wiley, New York. pp. 227–252.

de Jager, C. M., R. P. T. Butot, P. G. L. Klinkhamer, T. J. Dejong, K. Wolff, and E. Vandermeijden . 1995. Genetic variation in chrysanthemum for resistance to *Frankliniella occidentalis*. Entomol. Exp. Appl. 77:277–287.

de Kogel, W. J., A. Balkemaboomstra, M. Vanderhoek, S. Zijlstra, and C. Mollema. 1997a. Resistance to western flower thrips in greenhouse cucumber: Effect of leaf position and plant age on thrips reproduction. Euphytica. 94:63–67.

de Kogel, W. J., M. van der Hoek, M. T. A. Dik, B. Gebala, F. R. van Dijken, and C. Mollema. 1997b. Seasonal variation in resistance of chrysanthemum cultivars to *Frankliniella occidentalis* (Thysanoptera: Thripidae). Euphytica. 94:283–288.

Dent, D. R., and S. D. Wratten. 1986. The host-plant relationships of apterous virginoparae of the grass aphid *Metopolophium festucae cerealium*. Ann. Appl. Biol. 108:567–576.

DePauw, R. M., and D. W. L. Read. 1982. The effect of nitrogen and phosphorous on the expression of stem solidness in Canuck wheat at four locations in southwestern Saskatchewan. Can. J. Plant Sci. 62:593–598.

Diarisso, N. Y., B. B. Pendleton, G. L. Teetes, G. C. Peterson, and R. M. Anderson. 1998. Spikelet flowering time: Cause of sorghum resistance to sorghum midge (Diptera : Cecidomyiidae). J. Econ. Entomol. 91:1464–1470.

Diawara, M. M., J. T. Trumble, C. F. Quiros, K. K. White, and C. Adams. 1994. Plant age and seasonal variations in genotypic resistance of celery to beet armyworm (Lepidoptera, Noctuidae). J. Econ. Entomol. 87: 514–522.

Dreyer, D. L., B. C. Campbell, and K. C. Jones. 1984. Effect of bioregulator-treated sorghum on greenbug fecundity and feeding behavior: implications for host-plant resistance. Phytochem. 23:1593–1596.

Dymock, J. J., D. D. Rowan, and I. R. McGee. 1988. Effect of endophyte-produced mycotoxins on Argentine stem weevil and the cutworm, *Graphiana mutans*. In: P. P. Stahle (Ed.), Proc. Fifth Aust. Conf. Grasslands Invert. Ecol. D&D Printing, Victoria, Australia, 35.

East, D.A., J. V. Edelson, E. L. Cox, and M. K. Harris. 1992. Evaluation of screening methods and search for resistance in muskmelon, *Cucumis melo* L., to the twospotted spider mite, *Tetranychus urticae* Koch. Crop Protect. 11:39–44.

Eichenseer, H., and D. L. Dahlman. 1992. Antibiotic and deterrent qualities of endophyte-infected tall fescue to two aphid species (Homoptera: Aphididae). Environ. Entomol. 21:1046–1051.

Eichenseer, H., and D. L. Dahlman. 1993. Survival and development of the true armyworm *Pseudaletia unipunctata* (Haworth) (Lepidoptera: Noctuidae), on endophyte-infected and endophyte-free tall fescue. J. Entomol. Sci. 28:462–467.

Elden, T. C., and W. J. Kenworthy. 1994. Foliar nutrient concentrations of insect susceptible and resistant soybean germplasm. Crop Sci. 34:695–699.

Elden, T. C., and W. J. Kenworthy. 1995. Physiological responses of an insect-resistant soybean line to light and nutrient stress. J. Econ. Entomol. 88:430–436.

Fageria, N. K., and J. M. Scriber. 2001. The role of essential nutrients and minerals in insect resistance in crop plants. In: T. N. Ananthakrishnan (Ed.), Insect and Plant Defense Dynamics. Science Publishers, Enfield, NH (USA). pp. 23–53.

Fery, R. L., and F. P. Cuthbert, Jr. 1972. Association of plant density, cowpea curculio damage and *Choanephora* pod rot in southern peas. J. Amer. Soc. Hort. Sci. 97:800–802.

Fery, R. L., and F. P. Cuthbert, Jr. 1973. Factors affecting evaluation of fruitworm resistance in the tomato. J. Amer. Soc. Hort. Sci. 98:457–459.

Fery, R. L., and F. P. Cuthbert, Jr. 1974. Effect of plant density on fruitworm damage in the tomato. Hort. Sci. 9:140–141.

Fisk, J. 1978. Resistance of *Sorghum bicolor* to *Rhopalosiphum maidis* and *Peregrinus maidis* as affected by differences in the growth stage of the host. Entomol. Expl. Appl. 23:227–236.

Franca, F. H., and W. M. Tingey. 1994. Influence of light level on performance of the Colorado potato beetle on *Solanum tuberosum* L. and on resistance expression in *S. berthaultii* Hawkes. J. Amer. Soc. Hort. Sci. 119:915–919.

Funk, C. R., P. M. Halisky, M. C. Johnson, M. R. Siegel, A. V. Stewart, S. Amhad, R. H. Hurley, and I. C. Harvey. 1983. An endophytic fungus and resistance to sod webworms: association in *Lolium perenne* L. Bio/Technology 1:189–191.

Funk, C. R., R. H. White, and J. P. Breen. 1993. Importance of *Acremonium* in turfgrass breeding and management. Agric. Ecosyst. Environ. 44:215–232.

Gall, A., and J. R. Dogger. 1967. Effect of 2, 4-D on the wheat stem sawfly. J. Econ. Entomol. 60:75–77.

Gaynor, D. L., and W. F. Hunt. 1983. The relationship between nitrogen supply, endophytic fungus, and argentine stem weevil resistance in ryegrasses. Proc. N. Z. Grassland Assn. 44:257–263.

Gianoli, E., and H. M. Niemeyer. 1996. Environmental effects on the induction of wheat chemical defences by aphid infestation. Oecologia. 107:549–552.

Gianoli, E., and H. M. Niemeyer. 1997. Environmental effects on the accumulation of hydroxamic acids in wheat seedlings: The importance of plant growth rate. J. Chem. Ecol. 23:543–551.

Girousse, C., R. Bournville, and I. Badehhausser. 1999. Evaluation of alfalfa resistance to the pea aphid, *Acyrthosiphon pisum* [Hompotera: Aphididae] Methodological aspects to improve a standardized seedling test. Phytoprotection. 79:139–148.

Guthrie, W. D., Wilson, R. L., Coats, J. R., J. C. Robbins, C. T. Tseng, J. L. Jarvis, and W. A Russell. 1986. European corn borer (Lepidoptera: Pyralidae) Leaf-feeding resistance and DIMBOA content in inbred lines of dent maize grown under field versus greenhouse conditions. J. Econ. Entomol. 79: 1492–1496.

Hammerschmidt, R., and J. C. Schultz. 1996. Multiple defenses and signals in plant defense against pathogens and herbivores. In: J. T. Romero, J. A Saunders, and P. Barbosa (Eds.), Phytochemical Diversity and Redundancy in Ecological Interactions. Plenum Press, NY. pp. 121–154.

Harris, M. O., J. R. Miller, and O. M. B. de Ponti. 1987. Mechanisms of resistance to onion fly egg-laying. Entomol. Exp. Appl. 43:279–286.

Harvey, T. L., G. E. Wilde, K. D. Kofoid, and P. J. Bramel-Cox. 1994. Temperature effects on resistance to greenbug (Homoptera: Aphididae) biotype I in sorghum. J. Econ. Entomol. 87:500–503.

Heinrichs, E.A. (Ed.). 1988. Plant Stress-Insect Interactions. John Wiley, New York. 600 pp.

Hermes, D. A., and W. J. Mattson. 1992. The dilemma of plants: To grow or to defend. Quart. Rev. Biol. 67:283–355.

Holmes, N. D. 1984. The effect of light on the resistance of hard red springs wheats to the wheat stem sawfly, *Cephus cinctus* (Hymenoptera: Cephidae). Can. Entomol. 116:677–684.

Hughes, P. R. 1987. Insect populations on host plants subjected to air pollution. In: E. A. Heinrichs (Ed.), Plant Stress-Insect Interactions. Wiley, New York. pp. 249–319.

Hughes, P. R., J. E. Potter and L. H. Weinstein. 1981. Effects of air pollutants on plant-insect interactions: reactions of the Mexican bean beetle to SO_2-fumigated pinto beans. Environ. Entomol. 10:741–744.

Hughes, P. R., J. E. Potter and L. H. Weinstein. 1982. Effects of air pollution on plant-insect interactions: increased susceptibility of greenhouse-grown soybeans to the Mexican bean beetle after exposure to SO_2. Environ. Entomol. 11:173–176.

Hulme, M. A. 1995. Resistance by translocated sitka spruce to damage by *Pissodes strobi* (Coleoptera: Curculionidae) related to tree phenology. J. Econ. Entomol. 88:1525–1530.

Jansen, M. P. T., and N. E. Stamp. 1997. Effects of light availability on host plant chemistry and the consequences for behavior and growth of an insect herbivore. Entomol. Exp. Appl. 82:319–333.

Jenkins, E. B., R.B. Hammond, S. K. St. Martin, and R. L. Cooper. 1997. Effect of soil moisture and soybean growth stage on resistance to Mexican bean beetle (Coleoptera:Coccinellidae). J. Econ. Entomol. 90:697–703.

Johnson, K. J. R., E. L. Sorensen, and E. K. Horber. 1980. Effect of temperature and glandular-haired *Medicago* species on development of alfalfa weevil larvae. Crop Sci. 20:–631–633.

Johnson, M. C., D. L. Dahlman, M. R. Siegel, L. P. Bush, G. C. M. Latch, D. A. Potter, and D. R. Varney. 1985. Insect feeding deterrents in endophyte-infected tall fescue, Appl. Environ. Microbiol. 49: 568–571.

Karner, M. A., and G. R. Manglitz. 1985. Effects of temperature and alfalfa cultivar on pea aphid (Homoptera: Aphididae) fecundity and feeding activity of convergent lady beetle (Coleoptera: Coccinelidae). J. Kansas Entomol. Soc. 58:131–136.

Kennedy, G. G., R. T. Yamamoto, M. B. Dimock, W. G. Williams, and J. Bordner. 1981. Effect of daylength and light intensity on 2-tridecanone levels and resistance in *Lycopersicon hirsutum* f. *glabratum* to *Manduca sexta*. J. Chem. Ecol. 7:707–716

Kennedy, J. S., and C. O. Booth. 1959. Responses of *Aphis fabae* Scop. to water shortage in hostplants in the field. Entomol. Exp. Appl. 2:1–11.

Khan, Z. R., D. M. Norris, H. S. Chiang, N. E. Weiss and S. S. Oosterwyk. 1986. Light-induced susceptibility in soybean to cabbage looper, *Trichoplusia ni* (Lepidoptera: Noctuidae). Environ. Entomol. 15:803–808.

Kirfman, G. W., R. L. Brandenburg, and G. B. Garner. 1986. Realtionship between insect abundance and endophyte infestation level in tall fescue in Missouri. J. Kansas Entomol. Soc. 59:552–554.

Kindler, S. D., and R. Staples. 1970a. The influence of fluctuating and constant temperatures, photoperiod, and soil moisture on the resistance of alfalfa to the spotted alfalfa aphid. J. Econ. Entomol. 63:1198–1201.

Kindler, S. D., and R. Staples. 1970b. Nutrients and the reaction of two alfalfa clones to the spotted alfalfa aphid. J. Econ. Entomol. 63:938–40.

Kinzer, H. G., B. J. Ridgill, and J. M. Reeves. 1972. Response of walking *Conophthorus ponderosae* to volatile attractants. J. Econ. Entomol. 65:726–729.

Klun, J. A., and J. F. Robinson. 1969. Concentration of two 1,4-benzoxazinones in dent corn at various stages of development of the plant and its relations to resistance of the host plant to the European corn borer. J. Econ. Entomol. 62:214–220.

Kogan, M., and J. Paxton. 1983. Natural inducers of plant resistance to insects. In: P. A. Hedin (Ed.), Plant Resistance to Insects. Am. Chem. Soc. Symp. Series 208, American Chemical Society, Washington, DC. pp. 153–171.

Kolb, T. E., K. M. Holmberg, W. R. Wagner, and J. E. Stone. 1998. Regulation of ponderosa pine foliar physiology and insect resistance mechanisms by basal area treatments. Tree Physiol. 18:375–381.

Kumar, H., and G. O. Asino. 1993. Resistance of maize to *Chilo partellus* (Lepidoptera, Pyralidae) - effect of plant phenology. J. Econ. Entomol. 86:969–973.

Laska, P., J. Betlach, and M. Havrankova. 1986. Variable resistance in sweet pepper, *Capsicum annuum*, to glasshouse whitefly, *Trialeurodes vaporariorum* (Homoptera, Aleyrodidae). Acta Entomol. Bohemoslov. 83:347–353.

Leather, S. R., and A. F. G. Dixon. 1981. The effect of cereal growth stage and feeding site on the reproductive activity of the bird-cherry aphid, *Rhopalosiphum padi*. Ann. Biol. 97:135–141.

Lege, K. E., C. W. Smith, and J. T. Cothren. 1993. Planting date and plant density effects on condensed tannin concentration in cotton. Crop Sci. 33:320–324.

Le Ru, B., J. P. Diangana, and N. Beringer. 1994. Effects of nitrogen and calcium on the level of resistance of cassava to the mealybug *P. manihoti*. Insect Sci. Appl. 15:87–96.

Leuck, D. B. 1972. Induced fall armyworm resistance in pearl millet. J. Econ. Entomol. 65:1608–1611.

Leuck, D. B., and R. O. Hammons. 1974. Nutrients and growth media: Influence on expression of resistance to the fall armyworm in the peanut. J. Econ. Entomol. 67:564.

Leuchtmann, A. 1992. Systematics, distribution and host specificity of grass endophyte. Natural Toxin. 1:150–162.

Leuchtmann, A., and C. Clay. 1993. Nonreciprocal compatibility between *Epichoe typhina* and four grasses. Mycologia. 85:157–163.

Lin. H., M. Kogan, and A. G. Endress. 1990. Influence of ozone on induced resistance in soybean to the Mexican bean beetle (Coleoptera: Coccinellidae). Environ. Entomol. 19:854–858.

Lowe, H. J. B. 1967. Interspecific differences in the biology of aphids (Homoptera: Aphididae) on leaves of *Vicia faba*. II. Growth and excretion. Entomol. Exp. Appl. 10:413–420.

Manuwoto, S., and J. M. Scriber. 1985a. Neonate larval survival of European corn borers, *Ostrinia nubilalis*, on high and low DIMBOA genotypes of maize: effects of light intensity and degree of insect inbreeding. Agric., Ecosyst. Environ. 14:221–236.

Manuwoto, S., and J. M. Scriber. 1985b. Differential effects of nitrogen fertilization of three corn genotypes on biomass and nitrogen utilization by the southern armyworm, *Spodoptera eridania*. Agric., Ecosyst. Environ. 14:25–40.

Mathias, J. K., R. H. Ratcliffe, and J. L. Hellman. 1990. Association of an endophytic fungus in perennial ryegrass and resistance to the hairy chinch bug (Hemiptera: Lygaeidae). J. Econ. Entomol. 83:1640–1646.

McCloud, E. S., and M. R. Berenbaum. 1994. Stratospheric ozone depletion and plant-insect interactions: Effects of UVB radiation on foliage quality of *Citrus jambhiri* for *Trichoplusia ni*. J. Chem. Ecol. 20:525–539.

McWilliams, J. M., and G. L. Beland. 1977. Bollworm: Effect of soybean leaf age and pod maturity on development in the laboratory. Ann. Entomol. Soc. Am. 70:214–216.

Miller, R. H., S. El Masri, and K. Al Jundi. 1993. Plant density and wheat stem sawfly (Hymenoptera: Cephidae) resistance in Syrian wheats. Bull. Ent. Res. 83:95–102.

Moorehouse, J. E. 1971. Experimental analysis of the locomotor behavior of *Schistocerca gregaria* induced by odor. J. Insect Physiol. 17:913–920.

Muegge, M. A., S. S. Quisenberry, G. E. Bates, and R. E. Joost. 1991. Influence of *Acremonium* infection and pesticide use on seasonal abundance of leafhoppers and froghoppers (Homoptera: Cicadellidae; Cercopidae) in tall fescue. Environ. Entomol. 20:1531–1536.

Mulrooney, J. E., P. A. Hedin, W. L. Parrott, and J. N. Jenkins. 1985. Effects of PIX, a plant growth regulator, on allelochemical content of cotton and growth of tobacco budworm larvae (Lepidoptera: Noctuidae). J. Econ. Entomol. 78:1100–1104.

Murphy, J. A., S. Sun, and L. L. Betts. 1993. Endophyte–enhanced resistance to billbug (Coleoptera: Curculionidae), sod webworm (Lepidoptera: Pyralidae) and white grub (Coleoptera: Scarabaeidae) in tall fescue. Environ. Entomol. 22:699–703.

Nault, B. A., J. N. All, and H. R. Boerma. 1992. Resistance in vegetative and reproductive stages of a soybean breeding line to three defoliating pests (Lepidoptera: Noctuidae). J. Econ. Entomol. 85:1507–1515.

Neal, J. J., J. C. Steffens, and W. M. Tingey. 1989. Glandular trichomes of *Solanum berthaultii* and resistance to the Colorado potato beetle. Entomol.Axp. Appl. 51:133–140.

Nebeker, T. E. and J. D. Hodges. 1983. Influence of forestry practices on host-susceptibility to bark beetles. Z. Angew. Entomol. 96: 194–208.

Nihoul, P. 1993. Do light intensity, temperature and photoperiod affect the entrapment of mites on glandular hairs of cultivated tomatoes? Exp. Appl. Acarol. 17: 709–718.

Nkansah-Poku, J., and C. J. Hodgson. 1995. Interaction between aphid resistant cowpea cultivars, three clones of cowpea aphid, and the effect of two light intensity regimes on this interaction. Intl. J. Pest Manage. 41:161–165.

Ohm, H. W., R. H. Ratcliffe, F. L. Patterson, and S. E. Cambron. 1997. Resistance to Hessian fly conditioned by genes *H19* and proposed gene *H27* of durum wheat line PI422297. Crop Sci. 37:113–115.

Oswald, C. J., and M. J. Brewer. 1997. Aphid-barley interactions mediated by water stress and barley resistance to the Russian wheat aphid (Homoptera: Aphididae). Environ. Entomol. 26:591–602.

Patterson, C. G., D. D. Archbold, J. G. Rodriguez, and T. R. Hamilton-Kemp. 1994. Daylength and resistance of strawberry foliage to the two spotted spider mite. HortSci. 29:1329–1331.

Pearson, W. D. 1988. The poasture mealy bug, *Balanoccus poae* (Maskell) in Canterbury: a prelikminary report. In: P. P. Stahle (Ed.), Proc. Fifth Aust. Conf. Grasslands Invert. Ecol. D&D Printing, Victoria, Australia. 297.

Pelletier, Y. 1990. The effect of water stress and leaflet size on the density of trichomes and the resistance to Colorado potato beetle larvae (*Leptinotarsa decemlineata* [Say]) in *Solanum berthaultii* Hawkes. Can. Entomol. 122:1141–11147.

Popay, A. J., and D. R. Rowan. 1994. Endophytic fungi as mediators of plant-insect in teractions. In E. A. Bernays (Ed.), Insect-Plant Interactions Vol. V. CRC Press, Boca Raton, pp. 83– 103.

Popay, A. J., R. A. Mainland, and C. J. Saunders. 1993. The effect of endophytes in fescue grass on growth and survival of third instar grass grub larvae. In: D. E. Hume, G. C. M. Hatch, and H. S.

Easton (Eds.), Proc. Second Intl. Symp. *Acremonium* /Grass Interactions, AgResearch, Grasslands Resaerch Centre, Palmerston North, New Zealand, 170.

Potter, D. A., C. G. Patterson, and C. T. Redmond. 1992. Influence of turfgarss species and tall fescue endophyte on feeding ecology of Japanese beetle and southern masked chafer grubs (Coleoptera: Scarabaeidae). J. Econ. Entomol. 85: 900–909.

Prestidge, R. A., and O. J.-P. Ball. 1993. The role of endophytes in alleviating plant biotic stress in New Zealand. In: D. E. Hume, G. C. M. Hatch, and H. S. Easton (Eds.), Proc. Second Intl. Symp. *Acremonium*/Grass Interactions: Plenary Papers, AgResearch, Grasslands Research Centre, Palmerston North, New Zealand, 141.

Prestidge, R. A., and R. T. Gallagher. 1988. Endophyte fungus confers resistance to ryegrass: Argentine stem weevil larval studies. Ecol. Entomol. 13:429–435.

Quigley, P., X. Li, G. McDonald, and A. Noske. 1993. Effects of *Acremonium lolii* on mixed pastures and associated insect pests in southeastern Australia. In: D. E. Hume, G. C. M. Hatch, and H. S. Easton (Eds.), Proc. Second Intl. Symp. *Acremonium*/Grass Interactions, AgResearch, Grasslands Resaerch Centre, Palmerston North, New Zealand, 177 pp.

Quiring, D. 1994. Influences of inter-tree variation in time of budburst of white spruce on herbivory and the behaviour and survivorship of *Zeiraphera canadensis*. Ecol. Entomol. 19:17–25.

Raina, A. K., P. S. Benepal, and A. Q. Sheikh. 1980. Effects of excised and intact leaf methods, leaf size, and plant age on Mexican bean beetle feeding. Entomol. Exp. Appl. 27:303–306

Rapusas, H. R., and E. A. Heinrichs. 1987. Plant age effect on resistance of rice 'IR36' to the green leafhopper, *Nephotettix virescens (*Distant) and rice tungro virus. Environ. Entomol. 16:106–110.

Ratanatham, S., and R. L. Gallun. 1986. Resistance to Hessian fly (Diptera:Cecidomyiidae) in wheat as affected by temperature and larval density. Environ. Entomol. 15:305–310.

Rees, C. J. C. 1970. Age dependency of response in an insect chemoreceptor sensillum. Nature. 227:740–742.

Riedell, W. E., R. E. Kieckhefer, R. J. Petroski, and R. G. Powell. 1991. Naturally-occurring and synthetic loline alkaloid derivatives: insect feeding behavior modification and toxicity. J. Entomol. Sci. 26:122–129.

Riedell, W. E., T. E. Shumacher, and P. D. Everson. 1996. Nitrogen fertilizer management to improve crop tolerance to corn rootworm larva feeding damage. Agron. J. 88:27–32.

Reynolds, G. W., and C. M. Smith. 1985. Effects of leaf position, leaf wounding, and plant age of two soybean genotypes on soybean looper (Lepidoptera: Noctuidae) growth. Environ. Entomol. 14: 475–478.

Roberts, D. W. A., and C. Tyrrell. 1961. Sawfly resistance in wheat. IV. Some effects of light intensity on resistance. Can. J. Plant Sci. 41:457–465.

Robinson, J. 1993. Conditioning host plant affects antixenosis and antibiosis to Russian wheat aphid (Homoptera: Aphididae). J. Econ. Entomol. 86:602–606.

Rodriguez, J. G., D. A. Reicosky, and C. G. Patterson. 1983. Soybean and mite interactions: Effects of cultivar and plant growth stage. J. Kansas Entomol. Soc. 56:320–326.

Rogers, R. R., and R. B. Mills. 1974. Reactions of sorghum varieties to maize weevil infestation under relative humidities. J. Econ. Entomol. 67:692.

Rose, R. L., T. C. Sparks, and C. M. Smith. 1988. Insecticide toxicity to larvae of *Pseudoplusia includens* (Walker) and *Anticarsia gemmatalis* (Hubner) (Lepidoptera) as influenced by feeding on resistant soybean (PI227687) leaves and coumestrol. J. Econ. Entomol. 81:1288–1294.

Rowan, G. B., H. R. Boerma, J. N. All and J. W. Todd. 1993. Soybean maturity effect on expression of resistance to lepidopterous insects. Crop Sci. 33:433–436.

Rowan, D. D., and D. L. Gaynor. 1986. Isolation of feeding deterrents against Argentine stem weevil from ryegrass infected with the endophyte *Acremonium lolii*. J. Chem. Ecol. 12:647–648.

Russell, M. P. 1966. Effects of four sorghum varieties on the longevity of the lesser rice weevil, *Sitophilus oryzae* (L.). J. Stored Prod. Res. 2:75–79.

Salim, M., and R. C. Saxena. 1991. Nutritional stresses and varietal resistance in rice - Effects on whitebacked planthopper. Crop Sci. 31:797–805.

Salim, M., and R. C. Saxena. 1992. Iron, silica, and aluminum stresses and varietal resistance in rice - Effects on whitebacked planthopper. Crop Sci. 32:212–219.

Sams, D. W., F. I. Lauer, and E. B. Radcliffe. 1975. Excised leaflet test for evaluating resistance to green peach aphid in tuber-bearing *Solanum* plant material. J. Econ. Entomol. 68:607–609.

Saxena, K. N. 1967. Some factors governing olfactory and gustatory responses in insects. In T. Hayashi (Ed.), Olfaction and Taste II. Pergamon Press, Oxford. pp. 799–820.

Schalk, J. M., S. D. Kindler, and G. D. Manglitz. 1969. Temperature and preference of the spotted alfalfa aphid for resistant and susceptible alfalfa plants. J. Econ. Entomol. 62:1000–1003.

Schalk, J. M. and A. K. Stoner. 1976. A bioassay differentiates resistance to the Colorado potato beetle on tomatoes. J. Am. Soc. Hort. Sci. 101:74–76.

Schmidt, D. 1993. Effects of *Acremonium uncinatum* and a *Phialophora*-like endophyte on vigour, insect and disease resistance of meadow fescue. In: D. E. Hume, G. C. M. Hatch, and H. S. Easton (Eds.), Proc. Second Intl. Symp. *Acremonium*/Grass Interactions, AgResearch, Grasslands Resaerch Centre, Palmerston North, New Zealand, 177.

Schweissing, F. C., and G. Wilde. 1978. Temperature influence on greenbug resistance of crops in the seedling stage. Environ. Entomol. 7:831–834.

Schwessing, F. C., and G. Wilde. 1979a. Predisposition and nonpreference of greenbug for certain host cultivars. Environ. Entomol. 8:1070–1072.

Schwessing, F. C., and G. Wilde. 1979b. Temperature and plant nutrient effects on resistance of seedling sorghum to the greenbug. J. Econ. Entomol. 72:20–23.

Shade, R. E., T. E. Thompson, and W. R. Campbell. 1975. An alfalfa weevil larval resistance mechanism detected in *Medicago*. J. Econ. Entomol. 68:399–404.

Sharma, H. C., B. U. Singh, and R. Ortiz. 2001. Host Plant Resistance to Insects: Measurement, Mechanisms and Insect-Plant Environment Interactions. In: T. N. Ananthakrishnan (Ed.), Insect and Plant Defense Dynamics. Science Publishers, Enfield, NH (USA). pp. 143–159.

Sharma, H. C., G. Venkateswarulu, and A. Sharma. 2003. Environmental factors influence the expression of resistance to sorghum midge, *Stenodiplosis sorghicola*. Euphytica. 130: 365–375.

Schotzko, D. J., and C. M. Smith. 1991a. Effects of preconditioning host plants on population development of Russian wheat aphids (Homoptera: Aphididae). J. Econ. Entomol. 84:1083–1087.

Schotzko, D. J., and Smith, C. M. 1991b. Effects of host plant on the between-plant spatial distribution of the Russian wheat aphid (Homoptera: Aphididae). J. Econ. Entomol. 84:1725–1734.

Siegel, M. R., G. C. M. Latch, L. P. Bush, N. F. Fannin, D. D. Rowwan, B. A. Tapper, C. W. Bacon and M. C. Johnosn. 1990. Fungal endophyte-infected grasses: alkaloid accumulation and aphid response. J. Chem. Ecol. 16:3301–3315.

Sinden, S. L., J. M. Schalk, and A. K. Stoner. 1978. Effects of day length and maturity of tomato plants on tomatine content and resistance to the Colorado potato beetle. J. Am. Soc. Hort. Sci. 103:596–599.

Smith, C. M. 1985. Expression, mechanisms, and chemistry of resistance in soybean, *Glycine max* L. (Merr.) to the soybean looper, *Pseudoplusia includens* (Walker). Insect Sci. Appl. 6:243–248.

Smith, C. M., and J. F Robinson. 1983. Effect of rice cultivar height on infestation by the least skipper, *Ancyloxypha numitor* (F.) (Lepidoptera: Hesperiidae). Environ. Entomol. 12:967–969.

Smith, C. M., J. L. Frazier, and W. E. Knight. 1976. Attraction of clover head weevil, *Hypera meles*, to flower bud volatiles of several species of *Trifolium*. J. Insect Physiol. 22:1517–1521.

Smith, C. M., R. F. Wilson, and C. A. Brim. 1979. Feeding behavior of Mexican bean beetle on leaf extracts of resistant and susceptible soybean genotypes. J. Econ. Entomol. 72:374–377.

Snyder, J. C., A. M. Simmons, R. R. Thacker 1998. Attractancy and ovipositional response of adult *Bemisia argentifolii* (Homoptera: Aleyrodidae) to type IV trichome density on leaves of *Lycopersicon hirsutum* grown in three day-length regimes. J. Entomol. Sci. 33:270–281.

Sosa, O., Jr. 1979. Hessian fly: Resistance of wheat as affected by temperature and duration of exposure. Environ. Entomol. 8:280–281.

Sosa, O., Jr., and J. E. Foster. 1976. Temperature and the expression of resistance in wheat to the Hessian fly. Environ. Entomol. 5:333–336.

Thindwa, H. P., and G. L. Teetes. 1994. Effect of temperature and photoperiod on sorghum resistance to biotype C and E greenbug (Homoptera: Aphididae). J. Econ. Entomol. 87:1366–1372.

Thomas. J. G., E. L. Sorenson, and R. H. Painter. 1966. Attached vs. excised trifoliate for evaluation of resistance in alfalfa to the spotted alfalfa aphid. J. Econ. Entomol. 59:444–448.

Tingey, W. M., and T. F. Leigh. 1974. Height preference of *Lygus* bugs for oviposition on caged cotton plants. Environ. Entomol. 3:350–351.

Tingey, W. M., and S. R. Singh. 1980. Environmental factors influencing the magnitude and expression of resistance. In: F. G. Maxwell and P. R. Jennings (Eds.), Breeding Plants Resistant to Insects. John Wiley, New York. pp. 89–113.

Trumble, J. T., J. D. Hare, R. C. Musselman, and P. M. McCool. 1987. Ozone-induced changes in host plant suitability: Interactions of *Keferia lycopersicella* and *Lycopersicon esculentum*. J. Chem Ecol. 13:203–218.

Trumble, J. T., W. J. Moar, M. J. Brewer, and W. G. Carson. 1991. Impact of UV radiation on activity of linear furanocoumarins and *Bacillus thuringiensis* var. *kurstaki* against *Spodoptera exigua:* implications for tritrophic interactions. J. Chem. Ecol.17:973–987.

Trumbule, B. B. and R. F. Denno. 1995. Light intensity, host-plant irrigation and habitat related mortality as determinants of the abundance of azalea lacebug (Heteroptera: Tingidae). Environ. Entomol. 24:898–908.

Tyler, J. M., and J. H. Hatchett. 1983. Temperature influence on expression of resistance to Hessian fly (Diptera: Cecidomyiidae) in wheat derived from *Triticum tauschii*. J. Econ. Entomol. 76:323–326.

van de Klashorst, G., and W. M. Tingey. 1979. Effect of seedling age, environmental temperature, and foliar total glycoalkaloids on resistance of five *Solanum* genotypes to the potato leafhopper. Environ. Entomol. 8:690–693.

Velusamy, R., E. A. Heinrichs, and F. G. Medrano. 1986. Greenhouse techniques to identify field resistance to the brown planthopper, *Nilaparvata lugens* (Stal) (Homoptera: Delphacidae, in rice cultivars). Crop Protection. 5:328–333.

Videla, G. W., F. M. Davis, W. P. Williams, and S. S. Ng. 1992. Fall armyworm (Lepidoptera: Noctuidae) larval growth and survivorship on susceptible and resistant corn at different vegetative growth stages. J. Econ. Entomol. 85:2486–2491.

Walters, D. S., J. Harman, R. Craig, and R. O. Mumma. 1991. Effect of temperature on glandular trichome exudate composition and pest resistance in geraniums. Entomol. Exp. Appl. 60:61–69.

Wann, E. V., and W. A. Hills. 1966. Earworm resistance in sweet corn at two stages of ear development. Proc. Am. Soc. Hortic. Sci. 89:491–496.

Warner, R. W., and P. O. Richter. 1974. Alfalfa weevil: Diel activity cycle of adults in Oregon. Environ. Entomol. 3:939–945.

Wearing, C. H. 1972. Responses of *Myzus persicae* and *Brevicoryne brassicae* to leaf age and water stress in Brussels sprouts grown in pots. Entomol. Exp. Appl. 15:61–80.

Webster, J. A., and K. J. Starks. 1987. Fecundity of *Schizaphis graminum* and *Diuraphis noxia* (Homoptera: Aphididae) at three temperature regimes. J. Kans. Entomol. Soc. 60:580–582.

Webster, J. A., D. H. Smith, Jr., and S. H. Gage. 1978. Cereal leaf beetle (Coleoptera: Chrysomelidae): influence of seeding rate of oats on populations. Great Lakes Entomol. 11:117–120.

Wiseman, B. R., D. B. Leuck, and W. W. McMillian. 1973. Effects of fertilizers on resistance of Antigua corn to fall armyworm and corn earworm. Fla. Entomol. 56:1–7.

Wiseman, B. R., and W. W. McMillian. 1980. Feeding preferences of *Heliothis zea* larvae preconditioned to several host crops. J. Ga. Entomol. Soc. 15:449–453.

Wiseman, B. R., and M. E. Snook. 1995. Effect of corn silk age on flavone content and development of corn earworm (Lepidoptera: Noctuidae) larvae. J. Econ. Entomol. 88:1795–1800.

Wood, E. A., Jr., and K. J. Starks. 1972. Effect of temperature and host plant interaction on the biology of three biotypes of the greenbug. Environ. Entomol. 1:230–234.

Woodhead, S. 1981. Environmental and biotic factors affecting the phenolic content of different cultivars of *Sorghum bicolor.* J. Chem. Ecol. 7:1035–1047.

Woodhead, S. 1982. P-hydroxybenzaldehyde in the surface wax of sorghum: its importance in seedling resistance to acridids. Entomol. Exp. Appl. 31:296–302.

Woodhead, S., and E. A. Bernays. 1977. Changes in release rates of cyanide in relation to palatability of sorghum to insects. Nature. 270:235–236.

Worrall, W. D., and R. A. Scott. 1991. Differential reactions of Russian wheat aphid to various small grain host plants. Crop Sci. 31:312–314.

Worthing, C. R. 1969. Use of growth retardants on chrysanthemums: effect on pest populations. J. Sci. Food Agric. 20:394–397.

CHAPTER 8

INHERITANCE OF ARTHROPOD RESISTANCE

1. GENERAL

The genetics of arthropod resistance have been studied since the early 20th century, when Harlan (1916) demonstrated that resistance to the leaf blister mite, *Eriophyes gossypii* Banks, was a heritable trait in cotton, *Gossypium hirsutum* L. Since then, the genetics and inheritance of genes controlling arthropod resistance in many different crop plant species have been documented in several reviews (Gatehouse et al. 1994, Khush and Brar 1991, Singh 1986).

If arthropod resistance genes are to be utilized in cultivar improvement, it is essential to determine the genetics and mode of inheritance of the resistance genes. This is especially important if arthropod biotypes exist on the crop plant under investigation. Knowledge of the genetics of resistance also provides information to plant breeders about the degree of difficulty involved in tracking the resistance gene in breeding efforts to incorporate the gene into improved cultivars.

Entomologists and plant breeders have frequently used a number of the phenotypic evaluation techniques described in Chapter 6 to determine the genetics of resistance by evaluating segregating progeny from crosses between resistant and susceptible parents. The core idea of these evaluations is to observe as many segregating progeny as possible in order to obtain accurate estimates of the mode of inheritance and action of the resistance gene or genes.

Diallel crosses involving several resistant and susceptible parents are also used in inheritance studies. The general combining ability of a cultivar to transmit resistance is determined from the average resistance levels of the F_1 and F_2 plants in all crosses involving that cultivar. Specific combining ability is a measure of the amount of resistance transferred by a cultivar in a single cross with only one other parent. An additional standard measurement of the genetic expression of resistance involves determination of the heritability, or variation observed in the progeny of a cross. Variation among progeny may be related to several possible factors. Additive effects of several different alleles from genes in resistant plants may contribute to variation. Epistatic effects of alleles may contribute to variation. Finally, simple dominant and recessive effects may contribute to variation in resistance.

2. BREEDING METHODS USED TO DEVELOP ARTHROPOD RESISTANCE

Several methods have been utilized to breed arthropod resistant crop plants (Dahms 1972). Mass selection, pure line selection, and recurrent selection are often used to incorporate arthropod resistance genes into crop plants. These methods can be used in both cross-and self-pollinated plants.   In self-pollinated crops, backcross breeding, bulk breeding, and pedigree breeding are also used to incorporate arthropod resistance genes into agronomically desirable cultivars.

Mass selection involves selecting individual resistant plants after each generation (cycle of breeding), combining their seed, and growing this seed in the following generation as an aggregate group of plants. The objective of this breeding method is to select several sources of resistance in each of several cycles of selection. The largest improvements in resistance are usually made in the initial selection, followed by two to five additional cycles of selection. Mass selection has been effectively used to increase potato resistance to the potato leafhopper, *Empoasca fabae* (Harris) (Sanford and Ladd 1983).

Line breeding and pure line selection are forms of mass selection that both involve the selection of individual resistant plants that are advanced separately. In each cycle of selection, resistant selections are self-fertilized. If resistance is sought in a cross-pollinated crop, individual selections are inter-planted in a later selection cycle to form a composite cross of all selected plants.

Recurrent selection is used to concentrate arthropod resistance genes dispersed among several different sources. In each cycle, resistant plants are selected among the progeny produced by a previous mating of resistant individuals, and the mean level of arthropod resistance is increased. Recurrent selection allows the production of an arthropod resistant cultivar with the minimum amount of inbreeding and the introduction of resistance to additional insects from different sources in later selection cycles.   Recurrent selection has been used to increase resistance to arthropod pests in several different crops (Table 8.1).

Hybridization allows more freedom in the selection of resistance sources, since widely divergent genotypes can be combined to obtain a higher level of genetic diversity. The procedure involves selection of F_2 plants for high levels of resistance, the selection of individual resistant plants within selected F_3 families, and finally selection of resistance in entire primarily F_4 families, with some individual plant selections for resistance. It is important to select for high levels of arthropod resistance and a large genetic variance in the F_2 or F_3 generations, because later cycles of selection usually involve selection for improved agronomic characteristics.

Pedigree breeding involves selection of individual plants in segregating populations on the basis of arthropod resistance and pedigree. Initially, a hybrid is created, and all F_1 seed are saved and replanted. The best F_2 plants are selected and their seed are planted as F_3 families. In the F_3 generation, 25-50 seeds of the resistant families are selected. In the F_4 generation, a sample of each selected resistant F_3 family (seed of 50-100 plants) is planted and selection for resistance is made within families. In the F_5 generation, samples of selected F_4 resistant families (seed of 100-500 plants) are planted, evaluated for resistance, and preliminary yield tests are conducted to eliminate resistant families with poor yields. In later generations, selections are made for families with superior resistance, yield, and other agronomic characters. The advantage of pedigree is that a great deal of susceptible plant material is eliminated early in the breeding program, allowing detailed evaluation of selected resistant plants over a period of several years.

Table 8.1 Arthropod resistance in crop plants developed through the use of recurrent selection breeding

Crop	Arthropod(s)	References
Brassica campestrus	*Hyadaphis erysimi*	Barnes & Cuthbert 1975
Gossypium hirsutum	*Anthonomous grandis*	Bird 1982
	Heliothis virescens	
	Pseudatomoscelis seriatus	
	Lygus lineolaris	
Ipomoea batatas	*Cylas formicarius elegantulus*	Jones & Cuthbert 1973, Jones et al. 1976
Medicago sativa	*Hypera postica*	Hanson et al. 1972
	Therioaphis maculata	Graham et al. 1965
Solanum tuberosum	*Empoasca fabae*	Sanford & Ladd 1987
Zea mays	*Helicoverpa zea*	Widstrom et al. 1982, Butron et al. 2002
	Ostrinia nubilalis	Klenke et al. 1986
	Sitophilus zeamais	Widstrom 1989
	Spodoptera frugiperda	Widstrom et al. 1992a

The major disadvantages of pedigree breeding are that it is limited to use in self-pollinated crops and that only a limited number of entries can be processed due to the extra time required for planting, harvesting, and data acquisition on each entry. Pedigree breeding has been used for increasing the levels of resistance in rice, *Oryza sativa* (L.), to the green leafhopper, *Nephotettix virescens* (Distant), the brown planthopper, *Nilaparvata lugens* Stal, and the rice gall midge, *Orseolia oryzae* (Wood-Mason) (Khush 1980). The bulk breeding method is also used to incorporate arthropod resistance into self-pollinated crops. Bulk breeding is similar to the pedigree breeding method, but selection normally does not occur until the F_5 generation.

Backcross breeding involves the uses of recurring backcrosses to one of the parents (recurrent parent) of a hybrid, accompanied by selection for arthropod resistance. The non-recurrent parent is a source of resistance with a higher level of resistance than that used in the previous backcross. Backcross breeding can be used as a rapid way to incorporate arthropod resistance into agronomically desirable cultivars that are susceptible to insects. After each cross, selections are made for agronomically desirable resistant plants. High-yielding cultivars of rice and soybean, *Glycine max* (L.) Merr., with arthropod resistance have been created using backcross breeding (Khush 1978, Smith and Brim 1979). Backcross breeding has not generally proven useful in breeding maize, *Zea mays* L. For example, reciprocal translocation studies have shown that at least 12 genes are involved in *O. nubilalis* resistance (Scott et al. 1966, Onukogu et al. 1978). For this reason, maize hybrid development has been conducted using other breeding methods.

3. COMBINING PLANT PHENOTYPE AND GENOTYPE TO MAP RESISTANCE GENES

Molecular genetic techniques began to be used in determining the genetics of arthropod resistance in the early 1990s, with the development of high-density genetic maps of several major crop plants. Presently, such maps exist for barley, *Hordeum vulgare* L., maize, potato, *Solanum tuberosum* L., rye, *Secale cereale* L., sorghum, *Sorghum bicolor* (L.) Moench, soybean, tomato, *Lycopersicon esculentum* Mill., and wheat, *Triticum aestivum* L. (Paterson et al. 1991, Song et al. 2004, Hernández et al. 2001, Korzun et al. 2001, Boyko et al. 2002, Sharopova et al. 2002, Somers et al. 2004). Molecular markers in many of these crops are linked to genes expressing resistance to several major arthropod pests (see review by Yencho et al. 2000).

In the pre-molecular age of plant resistance to insects, phenotypic evaluations identified a source of resistance or a resistant progeny from a cross between a resistant and susceptible plant. Genetic mapping techniques are now being used to combine a plant's phenotype with its genotype, which is determined after the amplification or hybridization of plant DNA with a molecular marker of known chromosome location. Both data sets are then combined using computer software programs that generate estimates of the genetic linkage between the gene of interest and a specific molecular marker. Molecular markers that are linked to single major genes for resistance have been identified, as well as markers linked to groups of loci controlling the expression of quantitative resistance, known as quantitative trait loci (QTL). The marker-assisted selection (MAS) of plants based on genotype, before phenotypic resistance or susceptibility is expressed, is beginning to be implemented in many plant improvement programs.

3.1 Gene Mapping

Linkage between a resistance gene and a molecular marker may vary greatly. The two may be completely linked, where no crossing over occurs between the resistance gene and the marker during meiosis. The gene and marker are always linked together from one generation to another. The resistance gene and a molecular marker may be incompletely linked, and crossing over may occur between the gene and the marker during meiosis. The resistance gene and a molecular marker may have no linkage, due to being located on different chromosomes or far apart on the same chromosome.

Estimates of the recombination between the resistance gene and a linked marker are measured as the recombination frequency (RF). RF values are measured among segregating F_2 plants, $F_{2:3}$ families or recombinant inbred lines by matching the phenotype and genotype of each progeny and subjecting the paired data to Mapmaker (Lander et al. 1987), an interactive computer software for constructing genetic linkage maps of markers segregating in experimental crosses and for mapping resistance genes using those linkage maps. Mapmaker simultaneously estimates all recombination fractions for markers segregating as dominant, recessive and co-dominant traits in backcross$_1$ backcrosses, F_2 and F_3 intercrosses, and recombinant inbred lines. The linkage between QTLs and marker loci is determined by the way distribution patterns for the resistance characters are linked with the segregation of the resistance genes and the molecular marker at each locus. Mapmaker/QTL (Lincoln et al. 1993) allows mapping of genes controlling polygenic quantitative traits in F_2 intercrosses and backcross$_1$ backcrosses relative to the constructed genetic linkage map.

In order to estimate the recombination frequency between a resistance gene and a molecular marker, researchers often analyze between 100 and 200 F_2 plants or plants from 100 to 200 $F_{2:3}$ families with known arthropod phenotypic resistance or susceptibility. DNA is collected from the resistant and susceptible parent plants producing the progeny, as well as from each F_2 plant or from tissues of several plants in an F_3 family. Different DNA markers from a variety of chromosome locations are screened to identify those producing polymorphisms between the DNA of parent plants. If parent DNA banding pattern polymorphisms exist, the loci of a marker is said to be informative of the resistance gene location. Two DNA samples, one from several highly resistant and one from several highly susceptible plants, are then amplified with the informative marker. If the parent polymorphisms are apparent in the bulked segregant DNA samples, the marker is referred to as a putatively linked marker, and DNA of all F_2 plants or F_3 families is evaluated with it. At this point, Mapmaker is used to correlate the phenotype and genotype of the plants in each population and to calculate the recombination frequency. Recombination frequencies underestimate true genetic distances due to double or multiple crossovers. For this reason, MapMaker contains mapping functions such as the Haldane (1919) function and the Kosambi function (1944) to correct the underestimation.

There are several advantages to adapting molecular marker technology. Some types of molecular markers behave in a co-dominant manner to detect heterozygotes in segregating populations of progeny from crosses between resistant and susceptible plants. Many morphological markers behave in a dominant or recessive manner and do not detect heterozygotes (Staub et al. 1996). Molecular markers are unaffected by the types of environmental affects discussed in Chapter 7, and are phenotype-neutral. In addition, the allelic variation detected by molecular markers in natural plant populations is considerably greater than that detected by morphological markers. Finally, morphological markers may interact epistatically, and molecular markers do not, which greatly increases the number of markers screened in a single population.

3.2 Types of Molecular Markers

Several types of molecular markers have been used to determine the locations of arthropod resistance genes. They include restriction fragment length polymorphism (RFLP) markers, sequence tagged site (STS) markers, random amplified polymorphic (RAPD) markers, amplified fragment length polymorphism (AFLP) markers, and simple sequence repeats (SSRs) or microsatellite markers (Powell et al. 1996, Staub et al. 1996, Jones et al. 1997).

RFLP markers detect differences between genotype DNA when restriction enzymes from bacteria cut genomic DNA at specific nucleotide binding sites and yield variable sizes of DNA fragments. Restriction enzymes are often called restriction endonucleases because they cut within the DNA molecule at only a particular nucleotide sequence, and different restriction enzymes cut DNA into pieces of different sizes. The digested DNA is size-separated by electrophoreses, normally in an agarose or polyacrylamide gel. Because of the negative charge of DNA phosphate groups, digested DNA fragments migrate toward the positive electrode (anode) when the gel is subjected to a direct electrical current, and smaller DNA fragments migrate farther in the gel. The digested DNA is then transferred to a nylon membrane via Southern blotting (Sambrook et al. 2001) and the membrane is probed with a radioactive or fluorescent labeled, single-stranded DNA molecule (probe) of known chromosome location. The membrane bound DNA is denatured by heat, and some probe sequences bind to complementary sites in the DNA restriction digest. Excess probe is removed by a series of washes and the dried membrane is exposed to x-ray film, which is photographically developed into an autoradiogram.

Complementary binding between the probe DNA and membrane DNA provides information about the possible location of a resistant gene. The differences in DNA banding patterns on an autoradiogram of a Southern blot indicate the presence of one or more restriction sites in a sequence. The sequence containing a restriction site is one allele, while the corresponding sequence missing the restriction site is the other allele. When restriction sites in the vicinity of a gene are compared between two genotypes, one genotype may have the site, while the other does not. If differences exist, they are referred to as polymorphisms between the two genotypes. A hypothetical series of DNA polymorphisms resulting from DNA hybridization with a labeled DNA marker of known chromosome location is illustrated in Figure 8.1.

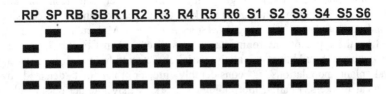

Figure 8.1. Hypothetical polymorphisms resulting from hybridization of DNA from a restriction digest with a labeled DNA marker of known chromosome location showing linkage between the marker and a resistance gene. RP - resistant parent, SP - susceptible parent. RB - resistant bulk, SB - susceptible bulk, R1-R5 - homozygous resistant lines, S1-S5 - homozygous susceptible lines, R6, S6 - heterozygous resistant & susceptible lines, respectively

RFLP probes allow very fine mapping of loci linked to resistance genes, because hundreds of mapped RFLP loci have been mapped in several crop genomes. RFLP molecular markers also have several of the advantages indicated above over morphological markers. These include the ability to behave in a co-dominant manner and detect heterozygotes, where morphological markers do not detect heterozygotes. RFLP markers detect greater allelic variation in natural plant populations than morphological markers and are unaffected by environmental effects. The disadvantages of RFLP linkage analysis include the additional time required to complete an analysis (7 to 10 days) and the use of radioactive isotopes. Nevertheless, RFLP analyses have been used effectively and continue to be used to map arthropod resistance gene loci in several crops (Yencho et al. 2000), as indicated in Table 8.2.

The polymerase chain reaction (PCR), a revolutionary technique developed by Mullis (1990), involves the *in vitro* enzymatic amplification of specific DNA sequences present between two convergent oligonucleotide primers that hybridize to opposite DNA strands. PCR is conducted in a thermal cycler device, in which PCR primers are reacted with template DNA, and the amplification products are electrophoresed to identify primers (markers) yielding polymorphic banding patterns in comparisons of DNA between resistant and susceptible plants. Primers used in PCR are in many instances from known chromosome locations, enabling the researcher to localize the resistance gene location.   Compared to RFLP hybridization, PCR reactions are approximately 40 to 50 times faster and do not normally require the use of radioactive materials.   Initially there were fewer PCR primers than RFLP markers, however the number of PCR markers in many genomic crop maps is growing rapidly.

Several types of PCR primers have been used to identify pathogen and pest resistance genes in plants (Botha and Venter 2000).   Among the first were sequence tagged site (STS) markers, which were sequenced from the ends of RFLP sequences linked to resistance genes. STS markers have been used to map genes for arthropod resistance in barley (Nieto-Lopez and Blake 1994) and rice (Katiyar et al. 2001).   Random amplified polymorphic DNA (RAPD) PCR primers are short random DNA sequences approximately 10 nucelotides long that amplify homologous genomic DNA sequences during the PCR process.   Differences in the DNA sequences of arthropod resistant and susceptible plants result in differential primer binding sites, and allow the visualization of polymorphisms between DNA in resistant and susceptible plants. Nevertheless, arthropod resistance genes in apple, *Malus* spp., rice and wheat have been identified using RAPD primers (Dweikat et al. 1994, 1997, Nair et al. 1995, 1996, Roche et al. 1997, Venter and Botha 2000, Selvi et al. 2002, Jena et al. 2003). Reproducibility of RAPD markers has proven to be difficult in many studies. In addition, DNA banding patterns based on RAPD primer amplification do not reveal heterozygotes or chromosome location.   To overcome this problem, RAPD-generated DNA polymorphic bands can be sequenced and this information used to design location-specific sequence characterized amplified regions (SCARs) (Hernandez et al. 1999). SCARs have been used to identify and map genes for *Oryza* resistance to *O. oryzae* (Sardesai et al. 2001) and *N. lugens* (Renganayaki et al. 2002).

Table 8.2. RFLP loci linked to arthropod resistance genes in different crop plants

Crop Plant	Arthropod	Resistance Gene	Linkage Distance	Reference
Oryza sativa	*Orseolia oryzae* biotype 1	*Gm2* *Gm2*	0.7-2.0 cM >1cM	Mohan et al. 1994
	O. oryzae biotypes 1 & 2			Rajyashri et al. 1998
	O. oryzae biotypes 1-4	*Gm-6t*	1.0-2.3 cM	Katiyar et al. 2001
	Nephotettix nigropictus	*Grh1*	4.4-5.9 cM	Kurata et al. 1994
	N. nigropictus	*Grh1*	0 cM	Tamura et al. 1999
	Nilaparvata lugens	--	3.7 cM	Ishii et al. 1994
	N. lugens	*Bph1*	1.7 cM	Murata et al. 1997
	N. lugens	*bph2*	3.5 cM	Murata et al. 1998
Triticum aestivum	*Mayetiola destructor*	*H13* *H23* *H24*	8.0-35.0 cM 6.9-15.6 cM 5.9-12.9 cM	Gill et al. 1987 Ma et al. 1993
	Schizaphis graminum	*Gb5*	centromeric	Dubcovsky et al. 1998
		Gby	7.5-11.3 cM	Boyko et al. 2004
Vigna radiata	*Callosobruchus chinensis*	*Rac1*	0 cM	Myers et al. 1996
Vigna unguiculata	*Aphis craccivora*	*Bruc*	3.6-27.1 cM	Young et al. 1992

Amplified fragment length polymorphisms (AFLPs) are based on the selective PCR amplification of restriction enzyme-digested DNA fragments, as in RFLP analysis (Vos et al. 1995). The DNA generated in each amplification contains molecular markers of random origin, but the process results in a much larger number of amplified DNA bands in one amplification. Sample DNA is digested with different restriction enzymes and restriction enzyme adaptors are then annealed to the restriction products. Restriction digests are then pre-selected by PCR amplification with general restriction enzymes attached to unique oligonucleotide primers and the pre-selected PCR products are then selectively amplified using specific 3" oligonucleotide primers. Amplified fragments are then denatured using polyacrylamide gel gelectrophoresis and then exposed to X-ray film to view the AFLP polymorphic banding patterns. AFLP markers produce many DNA polymorhphisms and they have been used successfully to identify and map arthropod resistance genes in apple, rice and wheat (Cevik and King 2002, Murai et al. 2001, Sardesai et al. 2001, Weng and Lazar 2002, Sharma et al. 2003).

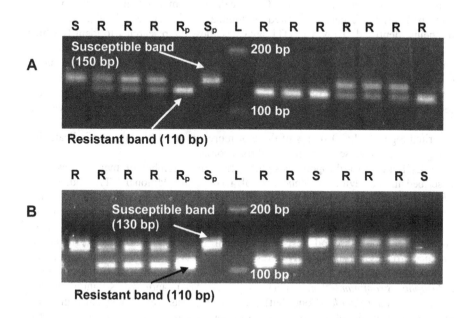

Figure 8.2. DNA fragments amplified from leaves of F₂ progeny from a cross between the wheat cultivar Stanton (S. graminum biotype I susceptible, [Sₚ]) and TA4152La4 (S. graminum biotype I resistant, [Rₚ]). R - resistant phenotype, S - susceptible phenotype. A - microsatellite marker gdm46; B - microsatellite marker gdm67. Amplification products were electrophoresed in 3% agarose gels stained with ethidium bromide Reprinted from Zhu, L. C., C. M. Smith, A. Fritz, E. V. Boyko, and M. B. Flinn. 2004. Genetic analysis and molecular mapping of a wheat gene conferring tolerance to the greenbug (Schizaphis graminum Rondani). Theor. Appl. Genet. 109:289-293, Copyright 2004 Springer Verlag, with kind permission of Springer Science and Business Media.

Microsatellite PCR primers, also called simple sequence repeats (SSRs) are tandem arrays of 2 to 5 base repeat units (particularly dinucleotide repeats) that are widely distributed in eukaryotic DNA. Several advantages have contributed to the success of microsatellite primers. These include their ability to detect single loci and their specific chromosome localization. Microsatellite primers generate high levels of polymorphism and detect patterns of co-dominant inheritance in populations of segregating progeny (Figure 8.2). Some instances of semi-codominance have also been identified in wheat during the process of mapping genes for *D. noxia* resistance (Liu et al. 2001). Microsatellite markers are proving to be very useful in many different crops for the identification of arthropod resistance genes (Yencho et al. 2000). Microsatellite primers have been used successfully to identify arthropod resistance genes in rice and wheat (Liu et al. 2001, 2002, Miller et al. 2001, Malik et al. 2003, Cook et al. 2004, Zhu et al. 2004).

As described previously, the linkage between quantitative trait loci (QTL) and molecular marker loci is determined by the way phenotypic resistance is linked with the segregation of several minor genes for resistance and the molecular markers at multiple loci or genomic regions. QTLs cannot be mapped in the same way as arthropod resistance genes inherited as dominant traits, since individual genes are not being identified. However, QTL analyses allow the researcher to identify which loci from a group contribute the most significantly to explaining the phenotypic variation for a biochemical or biophysical character mediating arthropod resistance (Table 8.3). A key component of QTL analysis is the calculation of a LOD [logarithm of the odds (to base 10)] score, a statistical estimate of the most likely recombination frequency between two loci. LOD scores determine whether two loci are likely to be near one another on a chromosome and are therefore likely to be inherited together. LOD scores of three or more are generally taken to indicate that two gene loci are close together on a chromosome.

QTLs linked to arthropod resistance genes in a number of major crops were described in the review of Yencho et al. (2000) and additional QTL studies have occurred since then. In soybean, QTLs for both antibiosis and antixenosis resistance to *H. zea* have been identified (Rector et al. 1998, 1999, 2000). Terry et al. (1999) used QTL analyses to demonstrate that recombinants from crosses between unrelated susceptible and arthropod resistant soybean lines transgressively segregate to produce progeny varying greatly in resistance to *H. zea*. Among cereal grain crops, QTL analyses have been used to document resistance in sorghum to the greenbug, *Schizaphis graminum* Rondani (Katsar et al. 2002) and the sorghum midge, *Stenodiplosis sorghicola* (Coquillett) (Tao et al. 2003), as well as resistance in wheat and barley to different species of aphids (Moharramipour et al. 1997, Castro et al. 2004). QTLs have been utilized to locate minor genes in rice contributing to resistance to both *N. lugens* and the green rice leafhopper, *Nephotettix cincticeps* (Uhler). In either case, progeny from doubled haploid populations, from indica rice x japonica rice crosses, or *Oryza sativa* x *O. officinalis* (wild rice) crosses were surveyed for QTLs linked to resistance (Huang et al. 1997, Fukuta et al. 1998, Huang et al. 2001). Alam and Cohen (1998) found two QTLs predominantly associated with *N. lugens* resistance. One QTL is tied to antixenosis and a second is linked to the expression of tolerance. In a different population, Xu et al. (2002) found that a main effect QTL for *N. lugens* resistance maps to the vicinity of a major rice gene controlling leaf and stem pubescence, suggesting that this QTL may explain *N. lugens* antixenosis.

One of the first QTL studies of arthropod resistance was that of Bonierbale et al. (1994) who mapped RFLP loci on progeny of crosses between cultivated potato, and the wild potato *Solanum berthaultii* for genes controlling glandular trichome-based resistance to *L. decemlineata*. Several QTLs explain various components of resistance based on potato leaf trichome type, density and exudate. Yencho et al. (1996) mapped *L. decemlineata* resistance genes in reciprocal backcrosses between *S. tuberosum* and *S. berthaultii* using RFLP markers to identify QTLs linked to reduced *L. decemlineata* oviposition or feeding. In general, most, but not all QTLs linked to *L. decemlineata* resistance are similar to those associated with glandular trichomes. Additional RFLP marker studies of the foliar glycoalkaloids solanidine and solasodine produced in *S. tuberosum* and *S. berthaultii* backcross progeny identified QTLs explaining approximately 20% of the variation in the occurrence of these compounds. QTLs have also been identified for tomato resistance to the leaf miner, *Liriomyza trifolii* (Burgess) (Moreira et al. 1999).

Table 8.3 Significant QTL loci in a single factor ANOVA for silk maysin (antibiotic to H. zea *larvae) in a F₂ maize population resulting from the cross SC (Sweet Corn) 102 (high maysin content) x hybrid B31857 (low maysin content)*

QTL locus	Maize chromosome bin[a]	Significance (P)	R^2 (%)[b]	Parent contributing high value allele
npi286 (near p1)[c]	1.03	0.0001	25.6	SC102
csu3 (near p1 locus)	1.05	0.0001	17.9	SC102
umc67	1.06	0.0001	7.0	SC102
a1	3.09	0.0001	15.7	B31857
umc66a (near c2)	4.07	0.007	2.8	B31857
umc105	9.02	0.01	2.3	SC102

[a] *Davis et al. 1999;* [b] *percent phenotypic variance explained;* [c] *LOD scores for p1 locus (21.4) and c2 locus (20.0) determined by Davis et al. 1999. [From Guo et al. (2001), Reprinted with permission from J. Econ. Entomol., Vol. 94:564-571. Copyright 2001, Entomological Society of America]*

Linkage analyses have also identified QTLs responsible for resistance of maize to four species of pest Lepidoptera.   QTLs affecting stem-boring resistance to *O. nubilalis* have been identified in genomic regions of maize chromosomes 2, 3, 5 and 9 (Shon et al. 1993, Cardinal et al. 2001, Jampatong et al. 2002, Krakowsky et al. 2002). Comparisons of QTL linked to stem boring and leaf feeding resistance have also shown that there are few common elements in the genetic control of the two types of *O. nubilalis* resistance (Jampatong et al. 2002).   However, QTL alleles on maize chromosomes 2, 5, 7 and 9 play major roles in resistance to stem boring by *O. nubilalis,* the southwestern corn borer, *Diatrea grandiosella* Dyar, and the sugarcane borer, *Diatrea saccharalis* (Fabricius) (Bohn et al. 1996, 1997, Groh et al. 1998, Khairallah et al. 1998, Cardinal et al. 2001).

As described by Guo et al. (2001) in Table 8.3, progress has also been made in identifying QTLs in maize linked to maysin, a glycosyl flavone that control antibiosis to *H. zea* larvae. Byrne et al. (1996) determined silk maysin concentration and RFLP polymorphic genotypes at flavonoid pathway loci in a segregating F₂ population from a cross between high- and low maysin content parents. The maize chromosome 1 locus *p1*, which activates transcription of parts of the flavonoid pathway, explains 58% of the variance for maysin content. A second QTL on chromosome 9 that is dominant for low maysin levels and interacts epistatically with *p1* is *rem (recessive enhancer of maysin)1*. When a functional *p1* allele is present, *rem1* nearly doubles the maysin concentration. Byrne et al. (1997) generated multiple locus models to demonstrate that the *p1* locus is a highly significant factor in explaining variation in both *H. zea* larval weight reduction and increased silk maysin concentration. Additional loci on chromosomes 1 and 9 also explain significant variation for *H. zea* larval weight and maysin concentration.

Both maysin and apimaysin are closely related glycosyl flavones (Chapter 2). Lee et al. (1998) evaluated over 300 F_2 progeny from a cross between a high maysin, low apimaysin content parent and a moderate maysin, moderate apimaysin content parent for QTLs explaining maysin and apimaysin synthesis. A QTL for maysin on maize chromosome 9 was concluded to be *rem1,* and explained 55% of the variance for maysin synthesis. A QTL for apimaysin from the *pr1* region of chromosome 5 was concluded to be *pr1,* and explained 64% of the variance for apimaysin synthesis. Neither QTL affects the other, indicating that syntheses of maysin and apimaysin occur independently. However, *rem1* accounts for only 14.1% of the *H. zea* antibiosis and *pr1* accounted for 14.7% of the antibiosis, suggesting that other antibiotic compounds may contribute to *H. zea* antibiosis. Chlorogenic acid, a maize phenylproponoid metabolite with an adjacent hydroxyl ring structure similar to maysin, has also been implicated in *H. zea* resistance (Duffey and Stout 1996). Bushman et al. (2002) detected a QTL in maize silks corresponding to the *p1* locus that increases both chlorogenic acid and total flavone content. Chlorogenic acid accumulation is probably due to the *p1* induction of chlorogenic acid synthesis and the induction of flavonoid genes to increase phenylproponoid pathway substrate availability (Bushman et al. 2002).

Cardinal et al. (2001) conducted numerous mapping experiments to determine QTLs for maize resistance to leaf feeding and stem boring Lepidoptera, and to compare results from maize grown in temperate and tropical environments. Resistance to stem boring by *O. nubilalis* in temperate maize and to *Diatrea* species in tropical maize appears to be controlled by QTLs on chromosomes 2, 3, 5, 7 and 9. However, genes controlling the synthesis of DIMBOA, the maize leaf organic acid with antibiotic effects during *O. nubilalis* leaf feeding (Chapter 2), occur on maize chromosome 4, in an area associated with QTLs exerting major effects on *O. nubilalis* leaf feeding resistance Cardinal et al. (2001). As described above, the major QTLs for maysin and apimaysin that control resistance to *H. zea* silk feeding occur on chromosomes 1, 5 and 9. The relationship, if any, between QTLs for allelochemicals controlling *H. zea* leaf feeding resistance and *O. nubilalis* stem-boring resistance on chromosomes 5 and 9 is unknown.

3.3 Molecular Marker-Assisted Selection

Marker-assisted selection (MAS) of plants for pest and pathogen resistance has been in development for several years as a means of accelerating the accuracy and rate at which a resistance gene can be bred into an improved crop cultivar. MAS of a major gene in barley for resistance to the cereal cyst nematode, *Heterodera avenae* Woll., can be accomplished approximately 30 times faster and for ~75% less the cost of phenotype evaluations (Kretshmer et al. 1997). Similar cost and labor savings have been documented for the use of a microsatellite marker linked to a dominant gene controlling resistance to the soybean cyst nematode, *Heterodera glycines* Ichinohe (Mudge et al. 1997) and the marker is used in many public breeding programs. Marker-assisted selection has also been used to pyramid major genes in rice for resistance to bacterial blight, *Xanthomonas oryzae,* and for resistance to the rice blast fungus pathogen *Magnaporthe grisea* (Hittalmani et al. 2000, Toenniessen et al. 2003).

In contrast to the markers linked to resistance genes inherited as simple, dominant traits, the improvement of polygenic traits (QTLs) through the use of MAS is difficult, due to the number of genes involved and their interactions (epistatic effects). Both factors are difficult to assess. Resolving these problems often involves multiple field tests across several different environments. However, such experiments often display significant QTL-environment interactions (see below). Narvel et al. (2001) used microsatellite markers to identify soybean QTLs for resistance to foliar feeding Lepidoptera, to determine the degree to which different QTLs have been transferred into soybean cultivars over a 30-year period. Very few resistant genotypes possess multiple QTLs from different soybean linkage groups, and MAS was suggested as a means of introgressing minor gene QTLs linked to soybean foliar pest resistance into elite soybean germplasm.

Several studies of QTLs linked to Lepidoptera resistance in maize underscore the problems involved with the use of QTLs in MAS. Groh et al. (1998) compared MAS and phenotypic selection for maize leaf-feeding resistance to *D. grandiosella* and *D. saccharalis* using two recombinant inbred line populations. The relative efficiency of both methods was similar, suggesting that phenotypic selection is more favorable, due to lower costs. Bohn et al. (2001) used biometric validation methods to compare MAS and phenotypic selection for leaf feeding resistance to *D. grandiosella* and *D. saccharalis* in a population of tropical maize inbred lines. Both estimates determined that MAS improved the efficiency of selection by only 4%, indicating that MAS is less efficient than phenotypic evaluation.

In related studies of leaf feeding resistance to *D. grandiosella*, Willcox et al. (2002) also determined that QTL-MAS and conventional selection methods are equivalent in improving resistance. Although the cost of MAS alone is approximately 90% less than the cost of conventional selection, the accurate identification of QTL position and the cost to generate initial data for use in the MAS process of QTL analysis makes conventional selection more cost effective. In order to make MAS cost effective for foliar feeding resistance to *Diatrea,* the costs of MAS must greatly decrease and additional QTLs explaining large portions of the variance for resistance need to be identified. Finally, QTLs must also be expressed over a broad range of environments if they are to be used in large-scale MAS programs. As indicated previously, environmental variation affects the utility of QTLs for resistance to both *Diatrea* and *Ostrinia* (Groh et al. 1998, Jampatong et al. 2002), additionally reducing their reliability in MAS.

4. INHERITANCE OF ARTHROPOD RESISTANCE IN MAJOR CROPS

The genetics and inheritance of arthropod resistance in food, fiber, and forage crops has been extensively documented in numerous reviews (Khush 1977, Singh 1986, Khush and Brar 1991, Gatehouse et al. 1994). Pertinent examples of progress in breeding arthropod resistant crop plants are described in the following sections, to provide specific information about the known arthropod resistance genes summarized in Tables 8.4, 8.5, 8.6, 8.7, and 8.8.

4.1 Cotton

Diallel analysis indicates that additive gene effects account for approximately 90% of the total genetic variance in cotton for resistance to the tobacco budworm and gossypol gland number (Wilson and Lee 1971, Wilson and Smith 1977). Wilson and George (1979) evaluated the combining ability of resistance in cotton to seed damage by *P. gossypiella* in a group of cultivars and breeding lines. In two lines, general combining ability, the average performance of a breeding line, was greatly increased for resistance to seed damage. When one of these lines was crossed with cultivars lacking extra floral nectaries, resistance was inherited due to dominant or epistatic effects (Wilson and George 1983). The gene action contributing to resistance in progeny of this cross is additive, and only a few genes condition resistance.

4.2 Legumes

Single dominant genes in alfalfa, *Medicago sativa* L., and sweet clover, *Melilotus infesta* Guss., control resistance to the pea aphid, *Acyrthosiphon pisum* (Harris) Glover and Stanford 1966) and the sweetclover aphid, *Therioaphis riehmi* (Borner) Manglitz and Gorz 1968), respectively. Resistance in alfalfa to the spotted alfalfa aphid, *Therioaphis maculata* (Buckton) is controlled by several genes (Glover and Melton 1966), indicating that resistance is quantitative (Table 8.4).

Diallel analyses have been conducted to determine the combining ability of alfalfa for *E. fabae* resistance. Significant effects for both general and specific combining ability indicate that *E. fabae* resistance in alfalfa can be increased (Soper et al. 1984, Elden et al. 1986). Diallel analyses detected highly significant combining ability effects for resistant clones, suggesting that selective breeding can increase *E. fabae* resistance. Elden and Elgin (1987) increased the level of *E. fabae* resistance in five alfalfa populations, using both recurrent selection and individual plant selection.

The level of resistance in red clover, *Trifolium pratense* L., to *A. pisum*, and the yellow clover aphid, *Therioaphis trifolii* (Monell), has also been increased by recurrent selection (Gorz et al. 1979). Five cycles of selection for *T. trifolii* resistance and three cycles of selection for *A. pisum* resistance were employed to develop the synthetic resistant cultivar 'N-2'.

Resistance in soybean to defoliating insects is multigenic. Heritability estimates for resistance to the cabbage looper, *Trichoplusia ni* (Hubner) (Luedders and Dickerson 1977) and *E. varivestis* (Sisson et al. 1976) suggest quantitative inheritance (Table 8.4).

Table 8.4 Arthropod resistance genes in fiber, fruit, forage, legume and vegetable crops

Crop Plant	Arthropod Pests	Gene [a] / Action [b]	References
Cucumis sativa	Diabrotica undecimpunctata	Bi	DaCosta & Jones 1971
	Tetranychus urticae	Q	De Ponti 1979
Glycine max	Epilachna varivestis	2-3 (Q)	Sisson et al. 1976
	Trichoplusia ni		Luedders & Dickerson 1977
	Pseudoplusia includens		Kilen et al. 1977
Helianthus spp.	Homoeosoma electellum	R	Johnson & Beard 1977
Lactuca virosa	Nasonovia ribis nigri	La1	Eenink et al. 1982b
Lycopersicon pennellii	Tetranychus evansi	r	Resende et al. 2002
Malus domestica	Dysaphis plantaginea	R	Alston & Briggs 1970
	Seppaphis devecta	R	Alston & Briggs 1968
Medicago sativa	Therioaphis maculata	Q	Glover & Melton 1966
Melilotus infesta	Therioaphis riehmi	Sca	Manglitz & Gorz 1968
Pennisetum glaucum	Spodoptera frugiperda Helicoverpa zea	Tr	Burton et al. 1977
Phaseoulus lunataus	Empoasca krameri	r /(A)	Park et al. 1994 Lyman and Cardona 1982
Phaseolus vulgaris	Zabrotes subfasciatus	Arc	Romero Andreas et al. 1986, Kornegay et al. 1993
Prunus persicae	Myzus persicae	Rm1	Monet & Massonie 1994
Solanum sparsipilum	Phtorimaea operculella	Q	Ortiz et al. 1990
Solanum tuberosum	Myzus persicae	2/(Q)	Mehlenbacher et al. 1984
Vigna radiata	Callosobruchus chinensis	R	Tomaka et al. 1992
Vigna unguiculata	Aphis craccivora	R	Redden et al. 1983, 1984
	Callosobruchus maculatus	R & r /(A)	Fatunla & Badaru 1983
	Chalcodermus aeneus	2 /(A)	Fery & Cuthbert 1975
Zea mays	O. nubilalis (brood 1)	5-6/(A)	Scott & Guthrie 1967
	O. nubilalis (brood 2)	7/(A)	Chiang & Hudson 1973
	D. u. howardi	r	Sifuentes & Painter 1964

[a] *Capital - dominant, lowercase - recessive;* [b] *A - additive, Q - quantitative*

F_2 plants from a cross between parents resistant and susceptible to the soybean looper, *Pseudoplusia includens* (Walker), also exhibit partial dominance or a quantitative inheritance action (Kilen et al. 1977).

Resistance in lima bean, *Phaseolus lunatus* L., to the leafhopper, *Empoasca kraemeri* Ross and Moore, is due to a quantitative effect of several genes and is inherited as a recessive trait (Lyman and Cardona 1982). Both additive and dominant gene effects are responsible for *E. kraemeri* resistance in cultivars of the common bean, *Phaseolus vulgaris* L. (Kornegay and Temple 1986). Inheritance of hooked bean trichomes, an *Empoasca* resistance mechanism discussed in Chapter 3, is complex and controlled by additive, dominant and epistatic gene actions (Park et al. 1994). There is also evidence for transgressive segregation (levels of resistance greater than that of the resistant parent) in some progenies from crosses between resistant and susceptible bean cultivars. Resistance in wild stains of *Phaseolus vulgaris* to *Zabrotes subfasciatus* (Boheman) is controlled by the toxic seed protein arcelin (Chapter 3). The presence of arcelin is inherited as a dominant trait (Romero Andreas et al. 1986, Kornegay et al. 1993). Interestingly, resistance to the bean weevil, *Acanthoscelides obtectus* (Say), is also derived from a wild *Phaseolus* accession, but in this case, resistance is inherited as the complementary effect of two separate recessive genes (Kornegay and Cardona 1991).

Resistance in mungbean, *Vigna radiata* (L.) R.Wilczek, to the Azuki bean weevil, *Callosobruchus chinensis* L., and the cowpea weevil, *Callosobruchus maculatus* (F.) is derived from the wild mungbean, *Vigna radiata* var. *sublobata*, and is inherited as a simple dominant trait (Tomaka et al. 1992). Several researchers have investigated the genetics and inheritance of arthropod resistance in cowpea, *Vigna unguiculata* (L.) Walp. Bata et al. (1987) studied segregating F_2 and F_3 progeny from crosses between resistant and susceptible cowpeas, and determined that resistance to the cowpea aphid, *Aphis craccivora* Koch, is inherited as a monogenic dominant trait. Conversely, resistance to *C. maculatus* infesting cowpeas is conferred in a complex inheritance pattern by a combination of major and minor genes expressed as a recessive trait (Redden et al. 1983, 1984) and is controlled by both additive and dominance effects (Fatunla and Badaru 1983). Cowpea resistance to the cowpea curculio, *Chalcodermus aeneus* Boheman, is also additive, and controlled by one pair of genes (Fery and Cuthbert 1975).

4.3 Fruits and Vegetables

Resistance in fruit to several species of aphids is controlled by the action of single dominant genes. In apple, these include genes for resistance to the rosy apple aphid, *Dysaphis plantaginea* (Alston and Briggs 1970) and the rosy leaf-curling aphid, *Dysaphis devecta* (Alston and Briggs 1968). In raspberry, *Rubus phoenicolasius* Maxim., two dominant genes - one from Europe and one from North America - are expressed for resistance to the raspberry aphid, *Amphorophora idaei* Borner, (Daubeny 1966, Jones et al. 2000). Monet and Massonie (1994) identified a single gene in peach, *Prunus persica* (L.), (*Rm1*) for resistance to *M. persicae* (Table 8.4). The interaction of each of these genes to aphid biotype development is discussed in Chapter 11.

Scott (1977) successively self pollinated cultivars of carrot, *Daucus carota* L., for three generations to increase the level of resistance to a complex of the western plant bug, *Lygus hesperus* Knight, and *Lygus elisus* van Duzee. *Lygus* mortality increased from 24% on plants of the S_1 generation to 85% on plants of the S_3 generation.

de Ponti (1979) increased resistance in cucumber to *T. urticae* in cucumber, *Cucurbita sativus* L., by crossing several moderately resistant cultivars, indicating a polygenic inheritance of resistance. A single gene controls the inheritance of cucurbitacin, an arthropod feeding deterrent in cucumbers. However, two or three gene pairs control resistance to the spotted cucumber beetle, *Diabrotica undecimpunctata howardi* Barber. Factors other than low cucurbitacin content in seedlings also condition the expression of resistance (Sharma and Hall 1971).

Resistance in lettuce, *Lactuca sativa* L., to the leaf aphid, *Nasonovia ribis nigri* was transferred from *Lactuca virosa* to *Lactuca sativa* by interspecific crossing (Eenink et al. 1982a). Resistance is monogenic and inherited as a dominant trait (Eenink et al. 1982b).

Cuthbert and Jones (1972) determined that recurrent selection increases arthropod resistance in sweet potato, *Ipomoea batatas* [(L.) Lam.]. In a random interbreeding population of sweet potato genotypes, four cycles of selection increased the incidence of resistance to a complex of soil insects consisting of the grub, *Plectris aliena* Chapin; the southern potato wireworm, *Conoderus falli* Lane; the banded cucumber beetle, *Diabrotica balteata* LeConte; *D. howardi*; the elongate flea beetle, *Systena elongata* (F.); and *Systena frontalis* (F.).

Progeny from crosses between cultivated potato and different *Solanum* species with varying densities of glandular trichomes exhibit heritabilities for resistance to *M. persicae* ranging from 50 to 60 %. Resistance is expressed as a partially dominant trait (Sams et al. 1976). One dominant gene controls glandular trichome-mediated resistance to *M. persicae* in *Solanum tarijense* and *S. berthaultii*, but in *S. phureja* x *S. berthaultii* crosses, two genes are involved in the expression of resistance (Gibson 1979). Mehlenbacher et al. (1983, 1984) conducted heritability studies on the density of lobed type A trichomes and simple type B trichomes on *S. berthaultii* foliage. *M. persicae* resistance is related to a complex interaction between trichome density and droplet size, and for this reason is considered a quantitatively inherited trait. Resistance to the potato tuber moth, *Phtorimaea operculella* (Zeller) has been derived from *S. sparsipilum* and is controlled by a small number of major genes (Ortiz et al. 1990). Segregation also occurs for resistance to the Colorado potato beetle, *Leptinotarsa decemlineata* Say, in the progeny of crosses involving *S. tuberosum* and *S. berthaultii* (Wright 1985).

Acylsugars present in *Lycopersicon pennellii* (Correll), are responsible for high levels of resistance to the spider mite, *Tetranychus evansi* Baker & Pritchard. Resende et al. (2002) examined progeny from segregating populations of the cross *L. esculentum* x *L. pennellii* for correlations between leaflet acylsugar contents and levels of mite repellence. The high acylsugar content in *L. pennellii* is inherited as a recessive trait and is explained by the action of a single major locus.

Recurrent selection has been used to improve potato resistance to *E. fabae.* Seven cycles of selection resulted in major reductions in *E. fabae* damage in potato populations selected for resistance (Sanford 1997). Using the populations created by Sanford and Ladd (1987), Sanford et al. (1992) found that concentrations of the glycoalkaloids solanine and chaconine are highly correlated to increased *E. fabae* resistance.

Theurer et al. (1982) used mass selection techniques to isolate and identify genotypes of sugar beet, *Beta vulgaris* L., with resistance to the sugarbeet root maggot, *Tetanops myopaeformis* Roder. The rate of change in resistance remained unchanged for five cycles of selection, indicating that higher levels of resistance remain to be obtained.

Johnson and Beard (1977) evaluated the inheritance of the phytomelanin (achene) layer (Chapter 3) in different *Helianthus* species resistant to the sunflower moth, *Homoeosoma electellum* (Hulst.). Segregation in progeny from a cross of a susceptible cultivar and a resistant *Helianthus* species indicates that resistance is inherited as a dominant trait.

4.4 Maize

Many different maize genotyes have been identified that possess genes for resistance to *O. nubilalis*. Early on, several genes were found to condition resistance to *O. nubilalis* (Scott and Guthrie 1967, Chiang and Hudson 1973). Jennings et al. (1974) conducted a diallel analysis of two sets of maize inbred lines to study the inheritance of resistance to first and second brood (generation) *O. nubilalis*. Different genes condition resistance to both first and second broods, but some genes condition resistance to both broods. Reciprocal translocation studies also demonstrated that only two or three of the twelve chromosome arms in maize are similar for both first and second brood *O. nubilalis* resistance (Onukogu et al. 1978). Warnock et al. (1998) demonstrasted that the resistance of sweet corn to *O. nubilalis* ear damage is inherited as a complex trait, involving multiple genes, and that gene action involves epistatic as well as additive-dominance effects.

Recurrent selection has been used to increase the levels of resistance in maize to *O. nubilalis* (Klenke et al. 1986, Russell et al. 1979). Borer damage ratings decrease significantly for both broods after only four cycles of selection. Variability for *O. nubilalis* resistance in each plant population also decreases with each successive cycle of selection. Lamb et al. (1994) used estimates of the general combining ability (GCA) and specific combining ability (SCA) of different sources of maize to identify optimal sources of resistance to leaf and sheath collar damage by *O. nubilalis.*

Recurrent selection has also been used to significantly increase the level of resistance in maize breeding populations to *H. zea* and *S. frugiperda* (Widstrom et al. 1982, 1992b, Butron et al. 2002) and to the maize weevil, *Sitophilus zeamais* Motschulsky (Widstrom 1989). GCA has been determined to be more descriptive of *H. zea* and *S. zeamais* resistance than the dominance or epistatic effects represented by SCA (Widstrom 1972, Widstrom et al. 1972, 1992a, Widstrom and McMillian 1973). Both GCA and

SCA have been used to explain significant amounts of the variation in different maize popualtions for resistance to *S. frugiperda* and D. grandiosella (Williams et al. 1995, 1998). Test crossing, a procedure to identify resistance levels in maize breeding lines in a recurrent selection program has also been used to increase the level of maize resistance to *H. zea*.

Resistance in maize to defoliation by *H. zea* is also quantitatively inherited (Widstrom and Hamm 1969). Wiseman and Bondari (1992, 1995) used different maize populations segregating for *H. zea* resistance, expressed by the antibiotic effects of maize silks, to determine the inheritance of resistance. In some populations, the additive-dominant model of genetic variance does not explain resistance. Thus, it appears that more than one pair of genes controls maize silk resistance to *H. zea*, and some of these genes interact in a non-allelic manner. The inheritance of both maize ear and stalk resistance to the stem borer, *Sesamia nonagrioides* (LeFebvre), is also quantitatively inherited (Cartea et al. 1999, 2001). In this interaction, however, additive, dominant, and epistatic effects control gene action. Both additive and dominant effects explain much of the variation in the resistance of maize to the corn leaf aphid, *Rhopalosiphum maidis* (Fitch), (Bing and Guthrie 1991) and to the spotted stem borer, *Chilo partellus* (Swinhoe) (Pathak 1991).

4.5 Rice

Rice plant materials are frequently developed for arthropod resistance evaluations using single crosses made to create F_1 hybrid plants. These plants are used to make top-crosses (an F_1 arthropod resistant-hybrid crossed to a commonly-grown cultivar) or double crosses (F_1 hybrid x F_1 hybrid). Seeds from these crosses are planted and plants evaluated for both arthropod and disease resistance. F_2 plants possessing both traits (about 25% of the original F_1 population) are then placed in pedigree nurseries that are evaluated separately for arthropod and disease resistance. These plants are evaluated as F_3 and F_4 families for resistance and as F_5 and F_6 families for agronomic desirability and yield (Khush 1980). Using this type of selection and breeding scheme, many genes have been identified in rice for resistance to *N. lugens*, the green rice leafhopper, *Nephotettix cincticeps* Unler, *N. virescens, O. oryzae* the whitebacked planthopper, *Sogatella furcifera* Horvath; the zigzag leafhopper, *Recilia dorsalis* (Motschulsky) and the yellow stem borer, *Tryporyza incertulas* (Walker).

Four genes condition resistance in rice to *N. cincticeps* (Table 8.5). *Grh* (green rice leafhopper) *1,* from the cultivar Norin-PL2 (Kobayashi et al. 1980) is located on rice chromosome 5 (Tamura et al. 1999). *Grh2* from rice cultivar DV 85, is located on rice chromosome 11 (Wang et al. 2003). *Grh3* is located on chromosome 6, and *Grh4* is located on rice chromosome 3 (Saka et al. 1997, Fukuta et al. 1998, Yazawa et al. 1998). QTLs identified by Fujita et al. (2003) coincide with each of the four genes.

Seven genes control the expression of resistance in rice to *N. virescens* (Chelliah 1986). *Glh* (green leafhopper)*1, Glh2, Glh3, Glh5, Glh6,* and *Glh7* are inherited as dominant traits, while *glh4* and *glh8* are inherited as recessive traits. *Glh1, Glh2,* and *Glh3* were identified in the cultivars 'Pankari 203', 'ASD7', and 'IR8' respectively (Athwal and Pathak 1971), and *glh4* and *Glh5* were identified in Ptb18 and ASD8, respectively, (Siwi and Khush 1977). *Glh6* and *Glh7* were identified in the cultivars 'TAPL 796' and 'Maddai Karuppan', respectively by Rezaul Kamin and Pathak (1982). *glh8* occurs in the cultivar DV85 (Pathak and Khan 1994). QTLs for *N. virescens* resistance on chromosomes 3 and 11 are very near *Grh2* and *Grh4* (see above), and near isogenic lines containing both *Grh2* and *Grh4* express resistance to *N. virescens* (Wang et al. 2004).

Thirteen *Bph* (brown planthopper) genes control resistance in rice to *N. lugens*. *Bph1, 3, 6, 9, 10,* and *13*) are inherited as dominant traits. *bph2, 4, 5, 7, 8, 11* and *12* are inherited as recessive traits. Several major QTL loci for *N. lugnes* resistance are described in Section 3.2. Athwal and Pathak (1971) were the first to note *Bph1* and *bph2* in the cultivars 'Mudgo' and 'ASD 7', respectively. *Bph3* and *bph4* were found in the cultivars 'Rathu Heenati' and 'Babawee', respectively (Lakshminarayana and Khush 1977). *Bph1* and *bph2* segregate independently of *Bph3* and *bph4*, and both *Bph3* + *bph4* and *Bph1* + *bph2* are closely linked (Ikeda and Kaneda 1981). *bph5* in ARC10550, *Bph6* in Swarnalatha, and *bph7* in T12 were identified by Kabir and Khush (1988). *Bph8* in Thai Col. 15 and *Bph9* in Balamavee were identified by Nemamoto et al. (1989). Kawaguchi et al. (2001) localized two additional recessive genes - *bph11* and *bph12*. The wild rice donors *Oryza australiensis* and *O. officinalis* are the source of *Bph10* and *Bph13*, respectively (Ishii et al. 1994, Renganayaki et al. 2002). *Bph1, bph2, Bph9* and *Bph10* have been mapped to a 25cM block on rice chromosome 12 (Ishiii et al. 1994, Hirabayashi and Ogawa 1995, Murata et al. 2000, Muri et al. 2001). *bph11* and *Bph13* map to chromosome 3 (Kawaguchi et al. 2001, Renganayaki et al. 2002), and *bph4* and *bph12* map to chromosomes 6 and 4 respectively (Hirabayashi et al. 1999, Kawaguchi et al. 2001).

Resistance to *O. oryzae* was first related to the effects of several dominant genes (Sastry and Prakasa Rao 1973) and monogenic recessive gene (Sahu and Sahu 1989). Chaudhary et al. (1986) designated two of these as *Gm* (gall midge) 1 and *Gm2*. Tomar and Prasad (1992) used allelism analyses tests to determine the identy of additional *Gm* genes in several rice cultivars in India. Several possess a single dominant gene allelic to *Gm1* and several possess a single dominant gene allelic to *Gm2*. Both genes separate independently of one another. Since then, extensive research has documented six additional dominant genes *(Gm4, 5, 6, 7, 8* and *9)* and a recessive gene *(gm3)* (Srivastava et al. 1993, Yang et al. 1997, Kumar et al. 1998, 2000a,b, Katiyar et al. 2001, Shrivastava et al. 2003). For additional discussions of *Gm* gene-*O. oryzae* biotype interactions, see Chapter 11).

Both dominant and recessive genes also control the inheritance of *S. furcifera* resistance in rice. *Wbph* (white-backed planthopper) *1*, in the cultivar 'N22', and *Wbph2*, in the cultivar 'ARC10239', are inherited as dominant traits (Angeles et al. 1981, Sidhu et al. 1979). *Wbph3*, in the cultivar 'ADR 52', and *wbph4*, in the cultivar 'Podiwi-A8', are inherited as dominant and recessive traits, respectively (Hernandez and Khush 1981). *Whph5* is a dominant trait for resistance in the cultivar 'N-Daing Marie' (Wu and Khush 1985). Two cultivars from Pakistan have resistance to *S. furcifera* governed by *Whph1* and a recessive gene that segregate independently of one another (Nair et al. 1982). Gupta and Shukla (1986) determined that a recessive gene and an unidentified gene control resistance in the breeding line 274-A/TN1. Rapusas and

Heinrichs (1985) also noted recessive gene effects in the expression of *S. furcifera* resistance. Three different dominant genes for resistance to *R. dorsalis* (*Zlh1*, *Zlh2*, and *Zlh3*) exist in 'Rathu Heenati', 'Ptb21', and 'Ptb 33' (Angeles et al. 1986).

Table 8.5. Oryza genes expressing resistance to arthropod pests

Arthropod	Resistance Genes	References
Nephotettix cincticeps	*Grh1,2,3,4*	Kobayashi et al. 1980, Saka et al. 1997, Fukuta et al. 1998, Yazawa et al. 1998, Tamura et al. 1999, Wang et al. 2003
Nephotettix virescens	*Glh1,2,3,5,6,7* *glh4, 8*	Athwal & Pathak 1971, Rezaul Kamin & Pathak 1982, Pathak & Khan 1994, Siwi & Khush 1977, Wang et al. 2004
Nilaparvata lugens	*Bph1,3,6,9,10,13* *bph2,4,5,7,8,11,12*	Athwal & Pathak 1971, Ikeda & Kaneda 1981, Ishii et al. 1994, Kabir & Khush 1988, Kawaguchi et al. 2001, Lakashminarayana & Khush 1977, Renganayaki et al. 2002
Orseolia oryzae	*Gm1,2,4,5,6,7,8,9* *gm3*	Chaudhary et al. 1986, Katiyar et al. 2001, Kumar et al. 1998, 2000a,b, Sahu & Sahu 1989, Sastry & Praska Rao 1973, Satyanarayanaiah & Reddi 1972, Shrivastava et al. 2003, Srivastava et al. 1993, Tomar and Prasad 1992, Yang et al. 1997
Recilia dorsalis	*Zlh1,2,3*	Angeles et al. 1986
Sogatella furcifera	*Wbph1,2,3,5* *wbph4*	Angeles et al. 1981, Hernandez & Khush 1981, Sidhu et al. 1979, Wu & Khush 1985

Resistance in rice to *T. incertulas* is polygenic and exists in many different rice breeding lines developed from crosses involving traditional cultivars with moderate resistance and breeding lines with higher levels of resistance (Khush 1984). A modified breeding strategy using a male-sterile female parent, recurrent selection, and pedigree selection have been used to develop composite cultivars with increased levels of *T. incertulas* resistance (Chaudhary et al. 1981). Although RAPD markers are linked to *T. incertulas* resistance genes (Selvi et al. 2002), the chromosome location of these genes is not known.

4.6 Graminaceous Crop Plants – Barley, Rye and Wheat

Several interspecific crosses combined with backcrossing, have been employed to transfer arthropod resistance into bread wheat from its graminaceous relatives barley, rye, *Aegilops tauschii* Coss. (the wheat D genome chromosome donor), or *Aegilops speltoides* Tausch. (the wheat B genome chromosome donor). Knowledge of the genetics of arthropod resistance in wheat has been facilitated by the use of the aneuploid condition of *Triticum* species, involving a change in the chromosome number from the normal 2N euploid chrosome number. Changes may involve a single chromosome arm or more than one chromosome. Hexaploid wheat may have several aneuploid configurations, including nullisomic (2N=40), monosmoic (2N=41), disomic (2N=42), trisomic (2n=43), tetrasomic (2N=44), and many combinations of these conditions involving multiple chromosome ams (Joppa 1987). Monosomic wheats have been used to determine the chromosome location of genes from controlling resistance to the *S. graminum, D. noxia*, and *M. destructor* (Gallun and Pattertson 1977, Hollenhorst and Joppa 1983, Schroeder-Teeter et al. 1994, Zhu et al. 2004), disomic (normal chromosome complement), and trisomic (extra chromosome complement) plants with one or two chromosomes less than normal.

The identification, selection and development of wheat germplasm for resistance to *M. destructor* began in the United States in the early 1900s. Presently, 29 genes from rye, wheat, durum wheat, *Triticum turgidum* var. durum L., *A. tauschii or Aegilops truncialis* L. control resistance to *M. destructor*. All but *h4* are inherited as dominant or partially dominant traits (Table 8.6). *H1, H2, H3, H4, H5, H7, H8* and *H12* are derived from wheat. *H6, H9, H10, H11, H14, H15, H16, H17, H18, H19, H20, H28* and *H29* are derived from durum wheat. *H21 and H25* were identified in rye, while *H22, H23, H24* and *H26* are derived from *Ae. tauschii* (Berzonsky et al. 2003). *H30* is a single dominant gene from *Ae. truncialis* (Martin-Sanchez et al. 2003). *H3, H6* and *H9* occur on wheat chromosome 5A, with *H3* linked to *H6* and *H9*. *H5* is inherited independently of *H9* (Gallun and Patterson 1977, Stebbins et al. 1982). Ohm et al. (1995) used monosomic analyses to determine that *H10* and *H12* are also located on wheat chromosome 5A. The deployment of these resistance genes in response to *M. destructor* biotypes is discussed in Chapter 11.

Eleven genes expressing resistance to *S. graminum* have been characterized in *Ae. speltoides, Ae. tauschii*, rye, tall wheatgrass *Agropyron elongatum* (Host.) Beauv., and wheat (Table 8.6). Resistance to biotypes A, B, and C, derived from the wheat line CI 17609 'Amigo' was introgressed from rye (Sebesta and Wood 1978) as the dominant gene, *Gb2*, on wheat chromosome 1A (Hollenhorst and Joppa 1983). The dominant genes *Gb3* and *Gb4* on chromosome 7D originate from *Ae. tauschii* and were bred into wheat germplasm CI17895 'Largo' and CI17959, respectively (Harvey et al. 1980, Hollenhorst and Joppa 1983, Weng and Lazar 2002). The dominant gene *Gb5*, located on wheat chromosome 7A in CI 17882, was most likely introgressed from *Ae. speltoides* (Hollenhorst and Joppa 1983, Tyler et al. 1987). The dominant *Gb6* gene, identified in wheat germplasm GRS 1201, (Porter et al. 1994) is located in the 1AL.1RS wheat-rye translocation. *Gbx* and *Gbz*, from *Ae. tauschii*, are allelic or tightly linked to *Gb3* (Zhu et al. 2004). *Gbx* and *Gbz* are inherited as single dominant genes, and were located in the distal 18% region of the long arm of wheat chromosome 7D by using wheat aneuploid and deletion lines (Zhu et al. 2004). *Gby*, thought to be from wheat, is linked to loci on wheat chromosome 7A (Boyko et al. 2004). *Rsg1a* has been shown to confer inducible *S. graminum* resistance in barley that is triggered by recognition of feeding by an avirulent *S. graminum* biotype (Carver et al. 1988, Hays et al. 1999).

Table 8.6. Graminae genes expressing resistance to arthropod pests

Arthropod	Resistance Genes	References
Aceria tosichilla	*Cm1, 2, 3, 4*	Chen et al. 1996, Cox et al. 1999, Malik et al. 2003, Schlegel & Kynast 1987, Thomas & Conner 1986, Whelan & Hart 1988, Whelan & Thomas 1989
Diuraphis noxia	*Dn1, 2, 4, 5, 6, 7, 8, 9, x, dn3, Rdn1, Rdn2*	du Toit 1987, 1988, 1989, Harvey & Martin 1990, Liu et al. 2001, 2002, Ma et al. 1998, Marais & du Toit 1993, Marais et al. 1994, 1998, Mornhinweg et al. 1995, 2002, Nkongolo et al. 1989, 1991a,b, Saidi & Quick 1996, Zhang et al. 1998
Schizaphis graminum	*Gb2, 3, 5, 6, x, y, z, Rsg1a*	Boyko et al. 2004, Harvey et al. 1980, Hays et al. 1999, Hollenhorst & Joppa 1983, Livers & Harvey 1969, Joppa et al. 1980, Porter et al. 1994, Sebesta & Wood 1978, Tyler et al.1987, Weng & Lazar 2002, Zhu et al. 2004
Mayetiola destructor	*H1, 2, 3, 5, 6, 7, 8, 9, 10, 11, 12, 13, 14, 15, 16, 17, 18, 19, 20, 21, 22, 23, 24, 25, 26, 28, 29, 30, h4*	Maas et al. 1987, Martin-Sanchez et al. 2003, Stebbins et al. 1982, 1983, Ratcliffe & Hatchett 1997

Twelve genes from barley, rye or wheat confer *D. noxia* resistance (Table 8.6). The resistance genes *Dn1* and *Dn2* were identified in South Africa in *T. aestivum* accessions PI137739 (*Dn1*) and PI 262660 (*Dn2*), from Iran and Azerbaijan, respectively (du Toit 1987, 1988, 1989) and are inherited as dominant traits. The recessive gene *dn3* is present in an *Ae. tauschii* parent in an amphiploid wheat derived from crosses between *Ae. tauschii* and durum wheat (Nkongolo et al. 1991a). The dominant gene *Dn5* was identified in Bulgarian wheat accession PI 294994 (du Toit 1987) and characterized by Saidi and Quick (1996) and Zhang et al. (1998). The dominant *Dn4* and *Dn6*, originated from Russian (PI372129) and Iranian (PI243781) wheat, respectively (Nkongolo et al. 1989, 1991b, Saidi and Quick 1996). *Dn7*, a rye gene, was transferred to chromosome 1RS of the 1RS•1BL translocation in the wheat cultivar 'Gamtoos' (Marais and du Toit 1993, Marais et al. 1994, 1998). *Dn8* and *Dn9* are co-expressed with *Dn5* in PI 294994 (Liu et al. 2001). *Dnx*, a wheat accession from Afghanistan PI220127 is inherited as a dominant trait (Harvey and Martin 1990, Liu et al. 2001). *Dn1, Dn2, Dn5, Dn6, Dn8* and *Dnx* are located on the short arm of wheat chromosome 7D (Liu et al. 2001, 2002). *Dn4* is located on wheat chromosome 1DS and *Dn9* is located on chromosome 1DL (Liu et al. 2001). Mornhinweg et al. (1995) found *D. noxia* biotype 1 resistance in the barley line STARS-9301B from PI573080, to be controlled by dominant alleles at two loci. An incompletely dominant allele pair at the *Rdn1* locus and a dominant allele pair at the *Rdn2* locus confer a high level of resistance. Two dominant genes also control *D. noxia* biotype 1 resistance in the barley line STARS-9577B from PI591617 (Mornhinweg et al. 2002). One gene expresses high-level resistance and the second expresses an intermediate level of resistance. The relationship between the dominant genes in this germplasm and *Rdn2* has not been established.

The *Cmc1* gene for resistance to the wheat curl mite, *Aceria tosichella* Keifer, a vector of wheat streak mosaic virus, was transferred from *Ae. tauschii* to the short arm of wheat chromosome 6D (Thomas and Conner 1986, Whelan and Thomas 1989). *Cmc2* was transferred to wheat chromosome 6D from tall wheatgrass (Whelan and Hart 1988). An unnamed gene has also been transferred to wheat chromosome 6A from *Haynaldia villosa* (L.) Schur, as a T6AL·6VS translocation (Chen et al. 1996). Malik et al. (2003) used molecular markers to characterize *Cm* gene(s) transferred to wheat germplasm KS96WGRC40 (Cox et al. 1999) from both *Ae. tauschii* and rye (Schlegel and Kynast 1987). The rye-derived resistance gene, designated *Cmc3*, is present on wheat–rye translocation T1AL·1RS. The *Ae. tauschii*-derived resistance gene, *Cmc4*, segregates independently of *Cmc1*, and is also located on the short arm of wheat chromosome 6D.

4.7 Sorghum

Sorghum resistance to the sorghum shootfly, *Atherigona soccata* Rond., is controlled by additive polygenic effects (Rana et al. 1981) and is expressed as a partially dominant trait at low to moderate levels of *A. soccata* infestation. The additive resistance component increases at high *A. soccata* populations, but the dominance component is unaffected (Borikar and Chopde 1980) (Table 8.7). Leaf trichomes also play a role in the *A. soccata* resistance of some sorghum cultivars (See Chapter 2). Gibson and Maiti (1983) evaluated *A. soccata* resistance in the progeny of crosses between pubescent and glaborous sorghum cultivars, and observed resistance to be expressed as a recessive trait conditioned by a single gene. Pathak and Olela (1983) detected similar additive polygenic effects for resistance in sorghum to the stalk borer, *Chilo partellus* (Swinhoe). Pathak (1990) reviewed the genetics of sorghum tolerance resistance to *C. partellus* and determined that both additive and non-additive gene effects are involved. Additive gene effects govern resistance to leaf feeding and stem boring, but resistance to the dead heart condition is controlled by non-additive gene actions.

Sorghum resistance to *S. graminum* biotype C was first detected in tunis grass, *Sorghum virgatum* (Hack.) Stapf. (Hackerott et al. 1969). These genes were expressed in the sorghum genotype SA7536 and a single gene inherited as an incompletely dominant trait was shown to control resistance (Weibel et al. 1972). Porter et al. (1982) reported the detection of biotype E resistance from the sorghum cultivar 'Capbam', which was used as a source of resistance to develop 'Tx2783' by Peterson et al. (1984). The source of resistance in Capbam is unknown. Resistance to biotype I was first reported in Johnsongrass, *Sorghum halepense* (L.) Pers. (Harvey et al. 1991) which proved difficult to use for breeding.

Table 8.7. Genes from Sorghum bicolor *and* Sorghum virgatum *expressing resistance to arthropod pests*

Genes	Inheritance/ Gene Action	Arthropod	Origin of R Alleles	Refererences
Polygenic	Partial dominant/ Additive	*Atherigona soccata*	*S. bicolor*	Borikar & Chopde 1980, Gibson & Maiti 1983
Few major	Partial dominant/ Additive & non-additive	*Calocoris angustatus*	*S. bicolor*	Sharma et al. 2000
Polygenic	Partial dominant/ Additive & non-additive	*Chilo partellus*	*S. bicolor*	Pathak & Olela 1983
Bigenic	Recessive/?	*Eurystylus oldi*	*S. bicolor*	Aladele & Ezeaku 2003
Gb, Ssg 1, 2, 3, 4, 5, 6, 7, 8, 9	Incompletely dominant/additive	*Schizaphis graminum*	*S. bicolor* *S. virgatum*	Katsar et al. 2002, Weibel et al. 1972, Puterka & Peters 1995, Tuinstra et al. 2001
Bigenic	Recessive/ Additive & non-additive	*Stenodiplosis sorghicola*	*Sorghum bicolor*	Agarwal & Abraham 1985, Boozaya-Angoon et al. 1984, Rossetto & Igue 1983, Widstrom et al. 1984

Kofoid et al. (1991) identified biotype I resistance in the commercial hybrid 'Cargill 607E', and Andrews et al. (1993) confirmed this result and found the accessions PI550607 and PI550610 from Russia to be equally resistant. The same three sources express low-level resistance to biotype K (Harvey et al. 1997). Cargill 607E is the only currently available biotype I-resistant sorghum currently produced in the Southern Plains (Porter et al. 1997).

Results of more recent research have shown that several QTL in linkage groups of *S. graminum*-resistant sorghum germplasm are associated with RFLP loci in regions syntenous to the locations of the *Dn* genes discussed above (Katsar et al. 2002, Nagaraj et al. 2005). Resistance to biotypes C, E, F and I is inherited as an incompletely dominant trait controlled by a few major genes (Weibel et al. 1972, Puterka and Peters 1995, Tuinstra et al. 2001).

Resistance to *S. sorghicola* is inherited as a recessive trait and is controlled at two or more loci (Boozaya-Angoon et al. 1984, Rossetto and Igue 1983). Resistance is inherited from both additive and non-additive gene effects (Widstrom et al. 1984, Agarwal and Abraham 1985). Resistance in sorghum to the head bug *Eurystylus oldi* Poppius, is controlled by a pair of recessive genes (Aladele and Ezeaku 2003), while sorghum resistance to another sorghum head bug, *Calocoris angustatus* Leithiery, is inherited as a partially dominant trait controlled by both additive and nonadditive gene action (Sharma et al. 2000).

4.8 Arthropod Resistance Gene Clusters

In three instances, genes for resistance to arthropod pests occur in clusters or closely located groups. In rice, the *Bph1, bph2, Bph9* and *Bph10* genes for *N. lugens* resistance have been mapped to a 25cM block on rice chromosome 12 (Muri et al. 2001). The *N. lugens* resistance genes *bph11* and *Bph13* map to chromosome 3 (Kawaguchi et al. 2001, Renganayaki et al. 2002).

In wheat, *Dn1, Dn2, Dn5, Dn6,* and *Dnx,* located on the short arm of *T. aestivum* chromosome 7D (Liu et al. 2001, 2002), are likely allelic, or a cluster of completely linked resistance genes, as there is no segregation for susceptibility among progeny from crosses involving plants containing each of the six genes in all possible combinations (Liu et al. 2005).

As mentioned previously, maize QTL alleles on chromosomes 2, 5, 7 and 9 play major roles in resistance to stem boring by *O. nubilalis, D. grandiosella* and *D. saccharalis* (McMullen and Simcox 1996, Bohn et al. 1996, 1997, Groh et al. 1998, Khairallah et al. 1998, Cardinal et al. 2001). The major QTLs for production of silk maysin and apimaysin that control resistance to *H. zea* feeding occur on two of these same chromosomes; chromosomes 5 and 9. The relationship between the QTLs for each of these two different types of resistance on chromosomes 5 and 9 has not been investigated. However, genes controlling the synthesis of DIMBOA, the maize leaf organic acid with antibiotic effects during *O. nubilalis* leaf feeding, occur on maize chromosome 4 (Cardinal et al. 2001).

5. FACTORS INFLUENCING INHERITANCE STUDIES

Plant phenological and genetic factors, environmental factors (as described in Chapter 7), and human physical resource limitations all influence the outcome of inheritance studies of plant resistance to arthropods. Genetic self incompatibility and inbreeding depression in some cross-pollinated crops may also rapidly reduce the vigor of plant material rapidly in a breeding program and make this material difficult to transfer to donor parent plants.

Additive, recessive, dominant, semi-dominant or epistatic gene action and the number of gene loci for a given resistance factor, as described above in various examples, also govern the rate of progress in breeding and inheritance research. Since the resistance of a cultivar may change with maturity, the progeny of crosses involving arthropod resistant and susceptible cultivars of different maturities may be difficult to evaluate for resistance in only one evaluation, and may require several evaluations to compensate for the range in progeny maturation. In these and other cases, the amount of seed required to adequately determine the inheritance of and the number of genes involved in resistance is usually substantial, consuming large quantities of time and space. In order to control this problem, it is necessary to grow sizeable plant populations to analyze inheritance, have controlled uniform infestations of tests arthropods (see also Chapter 7), and to design experiments that allow for the separate study of resistance mechanisms that require destructive sampling.

6. STRATEGIES FOR DEPLOYING ARTHROPOD RESISTANCE GENES

Strategies relying on the effects of either major or minor genes can be utilized to deploy arthropod resistant cultivars into various arthropod pest management systems. Horizontal or polygenic resistance (sometimes previously referred to as field resistance) utilizes a number of sources, each with a minor resistance gene that are mixed and allowed to interbreed. Horizontal resistance is generally considered more stable than vertical resistance and is not readily overcome by resistance-breaking arthropod biotypes (see Chapter 11). Horizontal resistance is usually obtained by simple selection for resistance, by random mating to obtain new gene combinations, and by recurrent selection, to increase the frequency of resistance genes. Potential problems involved with horizontal resistance involve the lack of out-crossing in self-pollinated species of plants, and the need for heavy uniform arthropod infestations to identify resistance. The use of male sterile lines in breeding sorghum and rice cultivars with arthropod resistance has solved part of the out-crossing problem.

Vertical or monogenic resistance relies on the effects of a single major gene that results in a high level of resistance against certain segments of the pest arthropod population. This resistance has been presumed to be less stable than horizontal resistance because it can be overcome by the development of pest virulence. Nevertheless, several options are available in plant breeding programs attempting to employ vertical resistance genes in crop plant improvement. One option is to release one major gene, use it until it becomes ineffective, and make additional sequential releases of other major genes. Sequential cultivar release has been used for the deployment of genes in rice with *N. lugens* resistance (Khush 1979). A second option is to pyramid two or more major genes in one cultivar. Though more time consuming, gene pyramiding increases the longevity of resistance genes, and has been used successfully to protect wheat plantings against stem rust in Australia and in rice cultivars with *N. lugens* resistance (Khush 1984).

Heinrichs (1986) and Khush (1984) describe different gene deployment schemes used in rice cultivars with *N. lugens* resistance. Sequential release of cultivars controlled by monogenic resistance sources has been the strategy of necessity in use for several years. The cultivars currently in use in various areas of world rice production have resistance based on the 13 *N. lugens* resistance genes discussed previously. Efforts to pyramid the effects of one or more genes have been limited.

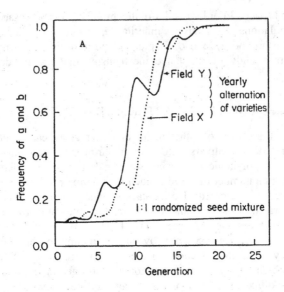

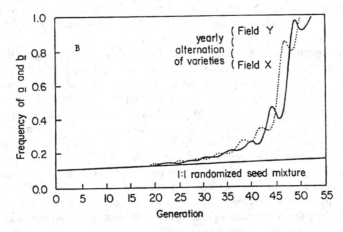

Figure 8.3. Effect of planting a two-gene pyramided Hessian fly, Mayetiola destructor *(Say), resistant wheat cultivar in yearly alternation with a susceptible cultivar in comparison to a 1:1 mixture of resistant and susceptible plants grown annually. Initial M. destructor a & b allele frequencies 0.10 (A) and 0.04 (B) (From Gould 1986a. Adapted with permission from Environ. Entomol., Vol. 15:11-23. Copyright 1986, Entomological Society of America)*

The *Bph3* gene in the cultivar IR64 has remained effective in Asian rice production since 1980 (Alam and Cohen 1998). In Indonesia, gene rotation was used to curb the development of *N. lugens* biotypes in the 1980s (Oka 1983). Cultivars with one set of genes were grown in the wet season production period and cultivars with a different set of genes were grown in the dry season production period.

Porter et al. (2000) compared the effect of deploying the *Gb2, Gb3,* and *Gb6 S. graminum* resistance genes in wheat either sequentially or as two gene pyramids (*Gb2/Gb3, Gb2/Gb,* or *Gb3/Gb6*) against *S. graminum* bioytpes E, F, G, H, and I. Pyramiding provides no additional protection over that provided by single, sequentially-released resistance genes.

A third option is the development and deployment of crop multilines (cultivars composed of different combinations of major and minor resistance genes) for protection against insects. Khush (1980) described a procedure to develop rice multilines with resistance to bacterial blight, grassy stunt disease (vectored by *N. lugens*), *N. lugens,* and *N. virescens.* Rice cultivars produced in Asia have various combinations of these genes. For *N. lugens* specifically, Nemato and Yokoo (1994) noted that *N. lugens* colonies selected on mixtures of resistant lines were more virulent than those selected on lines containing a single resistance gene, and that mixtures of resistant lines delayed the onset of virulence but did not prevent it. Multiline production is also used in the United States, where wheat multilines with resistance to *M. destructor* are grown in different geographic wheat producing areas.

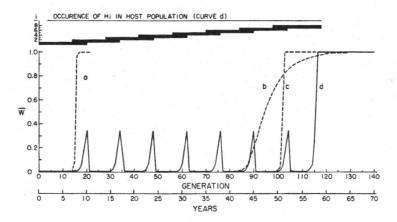

Figure 8.4. Effect of wheat genes with resistance to Mayetiola destructor *on relative mean* M. destructor *fitness (W), (a) genes released simultaneously in one cultivar, (b) genes released simultaneously in eight different cultivars, (c) genes released sequentially in separate cultivars for nine years each, (d) genes released sequentially in separate cultivars for 10 years each. Hi = each of the eight different resistance alleles in (d). (From Cox & Hatchett 1986. Adapted with permission from Environ. Entomol., Vol.15:24-31. Copyright 1986, Entomological Society of America)*

Mixed plantings of *S. sorghicola*-resistant and susceptible sorghum hybrids have also been shown to reduce *S. sorghicola*-related sorghum yield losses (Teetes et al. 1994). Conversely, mixtures of wheat cultivars with resistance to *S. graminum* or to the wheat stem sawfly, *Cephus cinctus* Norton, have no advantage over the use of cultivars containing individual resistance genes (Bush et al. 1991, Weiss et al. 1990).

Gould (1986a) used simulation models to determine how long an arthropod pest would require to adapt to two antibiotic plant resistance factors, when they are deployed sequentially, deployed as a cultivar multiline, or pyramided in a single cultivar. Because of the variations inherent to different arthropod pests and cropping systems, no one strategy is equally durable (long-lasting) over a wide variety of crop management systems. Model results indicate that sequentially released genes and two gene multilines are weaker in resistance than pyramided cultivars, but their resistance is more durable. Planting a percentage of the total crop in an arthropod susceptible cultivar (similar to the refuge concept of transgenic crop management discussed in Chapter 10) further enhances the durability of a pyramided cultivar. More specific simulation model data, based on the interaction of *M. destructor* with resistance in winter wheat (Gould 1986b), confirmed the general predictions about the use of a combination of cultivars with two pyramided resistance genes and some totally susceptible plants (Figure 8.3). Depending on the initial allele frequencies of *M. destructor* populations, resistance was predicted to last from 90 to 400 fly generations (45-200 years).

In contrast, simulation modeling of eight different sequentially released wheat genes involved in *M. destructor* resistance has generated somewhat different results on the lack of durability of pyramided gene cultivars, compared to the sequential release of cultivar release (Figure 8.4) (Cox and Hatchett 1986). In these simulations, the sequential release of one gene every four years maintains resistance approximately 10 times longer than if the resistance of all eight genes were pyramided into a single cultivar and released simultaneously. Additional disadvantages of the release of a cultivar with several pyramided genes are the long-term effort required for cultivar development and the risk involved in the placement of all known resistance genes into a single cultivar.

REFERENCES CITED

Agarwal, B. L. and C. V. Abraham. 1985. Breeding sorghum for resistance to shootfly and midge. In. Proceedings of the International Sorghum Entomology Workshop, July 15-21, 1984, Texas A&M University, College Station, TX. pp. 371-384.

Andrews, D. J., P. J. Bramel-Cox, and G. E. Wilde. 1993. New sources of resistance to greenbug biotype I in sorghum. Crop Sci. 33:198-199.

Aladele, S. E., and I. E. Ezeaku. 2003. Inheritance of resistance to head bug (*Eurystylus oldi*) in grain sorghum (*Sorghum bicolor*). Afr. J. Biotechnol. 2:202-205.

Alam, S. N., and M. B. Cohen. 1998. Detection and analysis of QTLs for resistance to the brown planthopper, *Nilapavata lugens*, in a doubled-haploid rice population. Theor. Appl. Genet. 97:1370-1379.

Alston, F. H., and J. B. Briggs. 1968. Resistance to *Sappaphis devecta* (Wld.) in apple. Euphytica 17:468-472.

Alston, F. H., and J. B. Briggs. 1970. Inheritance of hypersensitivity to rosy apple aphid, *Dysaphis plantaginea* in apple. Can. J. Genet. Cytol. 12:257-258.

Angeles, E. R., G. S. Khush, and E. A. Heinrichs. 1981. New genes for resistance to whitebacked planthopper in rice. Crop Sci. 21:47-50.

Angeles, E. R., G. S. Khush, and E. A. Heinrichs. 1986. Inheritance of resistance to planthoppers and leafhoppers in rice. In Rice Genetics. Proc. Intl. Rice Genetics Symp. May 27-31, 1985. Island Publ. Co., Manila. pp. 537-549.

Athwal, D. S., and M. D. Pathak. 1971. Genetics of resistance to rice insects. In. Rice Breeding, IRRI, Los Banos, Philippines. pp. 375-386.

Barnes, W. C., and F. P. Cuthbert, Jr. 1975. Breeding turnips for resistance to the turnip aphid. Hort. Science 10:59-60.

Bata, H. D., B. B. Singh, S. R. Singh, and T. A. O. Ladeinde. 1987. Inheritance of resistance to aphid in cowpea. Crop Sci. 27:892-894.

Berzonsky, W. A., H. Ding, S. D. Haley, M. O. Harris, R. J. Lamb, R. I. H. McKenzie, H. W. Ohm, F. L. Patterson, F. B. Peairs, D. R. Porter, R. H. Ratcliffe, and T. G. Shanower. 2003. Breeding wheat for resistance to insects. Plant Breed. Rev. 22:221-296.

Bing, J. W., and W. D. Guthrie. 1991. Generation mean analysis for resistance in maize to the corn leaf aphid (Homoptera: Aphididae). J. Econ. Entomol. 84:1080-1082.

Bird, L. S. 1982. Multi-adversity (diseases, insects and stresses) resistance (MAR) in cotton. Plt. Dis. 66:173-176.

Bohn, M, M. M. Khairallah, D. Gonzalez-deLeon, D. A. Hoisington, H. F. Utz, J. A. Deutsch, D. C. Jewell, J. A. Mihm, and A. E. Melchinger. 1996. QTL mapping in tropical maize. 1. Genomic regions affecting leaf feeding resistance to sugarcane borer and other traits. Crop Sci. 36:1352-1361.

Bohn, M., M. M. Khairallah, C. Jiang, D. Gonzalez-deLeon, D. A. Hoisington, H. F. Utz, J. A. Deutsch, D. C. Jewell, J. A. Mihm, and A. E. Melchinger. 1997. QTL mapping in tropical maize. 2. Comparison of genomic regions for resistance to Diatraea spp. Crop Sci. 37:1892-1902.

Bohn, M., S. Groh, M. M. Khairallah, D. A. Hoisington, H. F. Utz, and A. E. Melchinger. 2001. Re-evaluation of the prospects of marker-assisted selection for improving insect resistance against Diatraea spp. in tropical maize by cross validation and independent validation. Theor. Appl. Genet. 103:1059-1067.

Bonierbale, M. W., R. L. Plaisted, O. Pineda, and S. D. Tanksley. 1994. QTL analysis of trichome -mediated insect resistance in potato. Theor. Appl. Genet. 87:973-987.

Boozaya-Angoon, D., K. J. Starks, D. E. Weibel, and G. L. Teetes. 1984. Inheritance of resistance in sorghum, Sorghum bicolor, to the sorghum midge, Contarinia sorghicola (Diptera: Cecidomyiidae). Environ. Entomol. 13:1531-1534.

Borikar, S. T., and P. R. Chopde. 1980. Inheritance of shoot fly resistance under three levels of infestation in sorghum. Maydica. 25:175-183.

Botha, A.M., and E. Venter. 2000. Molcular marker technology linked to pest and pathogen resistance in wheat breeding. S. Afr. J. Sci. 96:233-240.

Boyko, E. V., R. Kalendar, V. Korzun, A. Korol, A. Schulman, and B. S. Gill. 2002. A high density genetic map of Aegilops tauschii includes genes, retro-transposons, and microsatellites which provide unique insight into cereal chromosome structure and function. Plant Mol. Biol. 48:767-790.

Boyko, E. V., S. R. Starkey, and C. M. Smith. 2004. Molecular genetic mapping of *Gby*, a new greenbug resistance gene in bread wheat. Theor. Appl. Genet. 109:1230-1236.

Burton, G. W., W. W. Hanna, J. C. Johnson, Jr., D. B. Leuck, W. G. Monson, J. B. Powell, H. D. Wells, and N. W. Widstrom. 1977. Pleiotropic effects of the tr trichomeless gene in pearl millet transpiration, forage quality and pest resistance. Crop Sci. 17:613-616.

Bush, L., J. E. Slosser, W. D. Worrall, and N. V. Horner. 1991. Potential of wheat cultivar mixtures for greenbug (Homoptera: Aphididae) management. J. Econ. Entomol. 84:1619-1624.

Bushman, B. S., M. E. Snook, J. P. Gerke, S. J. Szalma, M. A. Berhow, K. E. Houchins, and M. D. McMullen. 2002. Two loci exert major effects on chlorogenic acid synthesis in maize silks. Crop Sci. 42:1669-1678.

Butron, A., N. W. Widstrom, M. E. Snook, and B. R. Wiseman. 2002. Recurrent selection for corn earworm (Lepidoptera: Noctuidae) resistance in three closely related corn southern synthetics. J. Econ. Entomol. 95:458-462.

Byrne, P. F., M. D. McMullen, M. E. Snook, T. A. Musket, J. M. Theuri, N. W. Widstrom, B. R. Wiseman, and E. H. Coe. 1996. Quantitative trait loci and metabolic pathways: Genetic control of the concentration of maysin, a corn earworm resistance factor, in maize silks. Proc. Natl. Acad. Sci. USA. 93:8820-8825.

Byrne, P. F., M. D. McMullen, B. R. Wiseman, M. E. Snook, T. A. Musket, J. M. Theuri, N. W. Widstrom, and E. H. Coe. 1997. Identification of maize chromosome regions associated with antibiosis to corn earworm (Lepidoptera: Noctuidae) larvae. J. Econ. Entomol. 90:1039-1045.

Cardinal, A. J., M. Lee, N. Sharopova, W. L. Woodman-Clikeman, and M. J. Long. 2001. Genetic mapping and analysis of quantitative tratit loci for resistance to stalk tunneling by the European corn borer in maize. Crop Sci. 41: 835-845.

Cartea, M. E., R. A. Malvar, A. Butrón, M. I. Vales, and A. Ordás. 1999. Inheritance of antibiosis to *Sesamia nonagrioides* (Lepidoptera: Noctuidae) in maize. J. Econ. Entomol. 92:994-998.

Cartea, M. E., R. A. Malvar, M. I. Vales, A. Butrón, and A. Ordás. 2001. Inheritance of resistance to ear damage caused by *Sesamia nonagrioides* (Lepidoptera: Noctuidae) in maize J. Econ. Entomol. 94:277-283.

Carver, B. F., G. H. Morgan, L. H. Edwards, and J. A. Webster. 1988. Registration of four pairs of greenbug-resistant vs. susceptible near-isolines of winter barley germplasms. Crop Sci. 28:1034-1035.

Castro, A. M., A. Vasicek, C. Ellerbrook, D. O. Gimenez, E. Tocho, M. S. Tacaliti, A. Clua, and J. W. Snape. 2004. Mapping quantitative trait loci in wheat for resistance against greenbug and Russian wheat aphid. Plant Breeding. 123:361-365.

Cevik, V., and G. L. King. 2002. High-resolution genetic analysis of the *Sd*-1 aphid resistance locus in *Malus* spp. Theor. Appl. Genet. 105:346-354.

Chaudhary, B. P., P. S. Srivastava, M. N. Shrivastava, and G. S. Khush. 1986. Inheritance of resistance to gall midge in some cultivars of rice. In. Rice Genetics. International Rice Research Institute, Manila, Philippines. pp. 523-528.

Chaudhary, R. C., E. A. Heinrichs, and G. S. Khush. 1981. Increasing the level of stem borer resistance through male sterile facilitated recurrent selection in rice. Rice Genet. Newslttr. 8:7-8.

Chelliah, S. 1986. Genetics of resistance in rice to planthoppers and leafhoppers. In. Rice Genetics. Proc. Intl. Rice Genetics Symp. May 27-31, 1985. Island Publ. Co., Manila. pp. 513-522.

Chen, Q., R. L. Conner, and A. Laroche. 1996. Molecular characterization of *Haynaldia villosa* chromatin in wheat lines carrying resistance to wheat curl mite colonization. Theor. Appl. Genet. 93:679-684.

Chiang, M. S., and M. Hudson. 1973. Inheritance of resistance to the European corn borer in grain corn. Can. J. Plant Sci. 53:779-782.

Cook, J. P., D. M. Wichman, J. M. Martin, P. L. Bruckner, and L. E.Talbert. 2004. Identification of microsatellite markers associated with a stem solidness locus in wheat. Crop Sci. 44:1397-1402.

Cox, T. S., and J. H. Hatchett. 1986. Genetic model for wheat/Hessian fly (Diptera: Cecidomyiidae) interaction: Strategies for deployment of resistance genes in wheat cultivars. Environ. Entomol. 15:24-31.

Cox, T. S., W. W. Bockus, B. S. Gill, R. G. Sears, T. L. Harvey, S. Leath, and G. L. Brown-Guedira. 1999. Registration of KS96WGRC40 hard red winter wheat germplasm resistant to wheat curl mite, stagonospora leaf blotch, and septoria leaf blotch. Crop Sci. 39:597.

Cuthbert, F. P., Jr., and A. Jones. 1972. Resistance in sweet potatoes to Coleoptera increased by recurrent selection. J. Econ. Entomol. 65:1655-1658.

DaCosta, C. P., and C. M. Jones. 1971. Cucumber beetle resistance and mite susceptibility controlled by the bitter gene in *Cucumis sativus* L. Science. 172:1145-1146.

Dahms, R. G. 1972. Techniques in the evaluation and development of host-plant resistance. J. Environ. Qual. 1:254-258.

Daubeny, H. A. 1966. Inheritance of immunity in the red raspberry to the North American strain of the aphid, *Amphorophora rubi* (Kalt.). Proc. Am. Soc. Hort. Sci. 88:346-351.

Davis, G. L., M. D. McMullen, C. Baysdorfer, T. Musket, D. Grant, M. Staebell, G. Xu, M. Polacco, L. Koster, S. Melia-Hancock, K. Houchins, S. Chao, and E. H. Coe, Jr. 1999. A maize map standard with sequenced core markers, grass genome reference points, and 932-ESTs in a 1736-locus map. Genetics. 152:1137-1172.

de Ponti, O. M. B. 1979. Resistance in *Cucumis sativus* L. to *Tetranychus urticae* Koch. 5. Raising the resistance level by the exploitation of transgression. Euphytica. 28:569-577.

Dubcovsky, J., A. J. Lukaszewski, M. Echaide, E. F. Antonelli, and D. R. Porter. 1998. Molecular characterization of two *Triticum speltoides* interstitial translocations carrying leaf rust and greenbug resistance genes. Crop Sci. 38: 1655-1660.

Duffey, S. S., and M. J. Stout. 1996. Antinutritive and toxic components of plant defense against insects. Arch. Insect Biochem. 32:3-37.

du Toit, F. 1987. Resistance in wheat (*Triticum aestivum*) to *Diuraphis noxia* (Hemiptera: Aphididae). Cer. Res. Comm. 15:175-179.

du Toit, F. 1988. Another source of Russian wheat aphid (*Diuraphis noxia*) resistance in *Triticum aestivum*. Cer. Res. Comm. 16:105-106.

du Toit, F. 1989. Inheritance of resistance in two *Triticum aestivum* lines to Russian wheat aphid (Homoptera: Aphididae). J. Econ. Entomol. 82:1251-12153.

Dweikat, I., H. Ohm, S. MacKenzie, F. Patterson, S. Cambron, and R. Ratcliffe. 1994. Association of a DNA marker with Hessian fly resistance gene *H9* in wheat. Theor. Appl. Genet. 89:964-968.

Dweikat, I., H. Ohm, F. Patterson, and S. Cambron. 1997. Identification of RAPD markers for 11 Hessian fly resistance genes in wheat. Theor. Appl. Genet. 94:419-423.

Eenink, A. H., F. L. Dieleman, and R. Groenwold. 1982b. Resistance of lettuce (*Lactuca*) to the leaf aphid *Nasonoria ribis nigri*. 2. Inheritance of the resistance. Euphytica. 31:301-304.

Eenink, A. H., R. Groenwald, and F. L. Dieleman. 1982a. Resistance of lettuce (*Lactuca*) to the leaf aphid, *Nasonovia ribis nigri* 1. Transfer of resistance from *L. virosa* to *L. sativa* by interspecific crosses and selection of resistant breeding lines. Euphytica. 31:291-300.

Elden, T. C., and J. H. Elgin, Jr. 1987. Recurrent seedling and individual-plant selection for potato leafhopper (Homoptera: Cicadellidae) resistance in alfalfa. J. Econ. Entomol. 80:690-695.

Elden, T. C., J. H. Elgin, Jr., and J. F. Soper. 1986. Inheritance of pubescence in selected clones from two alfalfa populations and relationship to potato leafhopper resistance. Crop Sci. 26:1143-1146.

Fatunla, T., and K. Badaru. 1983. Inheritance of resistance to cowpea weevil (*Callosobruchus maculatus* Fabr.). J. Agric. Sci. Camb. 101:423-426.

Fery, R. L., and F. P. Cuthbert, Jr. 1975. Inheritance of pod resistance to cowpea curculio infestation in southern peas. J. Hered. 66:43-44.

Fujita, D., A. Yoshimura, and H. Yashi. 2003. Detrection of QTLs associated with antibiosis to green rice leafhopper, *Nephotettix cincticeps* Uhler, in four Indica rice varieties. Rice Genet. Newslttr. 19:38-39.

Fukuta, Y., K. Tamura, M. Hirae, and S. Oya. 1998. Genetic analysis of resistance to green rice leafhopper (*Nephotettix cincticeps* Uhler) in rice parental line, Norin-PL6, using RFLP markers. Breed. Sci. 48: 243-249.

Gallun, R. L., and F. L. Patterson. 1977. Monosomic analysis of wheat for resistance to Hessian fly. J. Hered. 68:223-226.

Gatehouse, A. M. R., D. Boulter, and V. A. Hilder. 1994. Potential of plant-derived genes in the genetic manipulation of crops for insect resistance. Plant Genetic Manipulation for Crop Protection. 7:155-181.

Gibson, P. T., and R. K. Maiti. 1983. Trichomes in segregating generations of sorghum matings. I. Inheritance of presence and density. Crop Sci. 23:73-75.

Gibson, R. W. 1979. The geographical distribution, inheritance and pest-resisting properties of sticky-tipped foliar hairs on potato species. Potato Res. 22:223-237.

Gill, B. S., J. H. Hatchett, and W. J. Raupp. 1987. Chromosomal mapping of Hessian fly-resistance gene *H13* in the D genome of wheat. J. Hered. 78:97-100.

Glover, D. V., and E. H. Stanford. 1966. Tetrasomic inheritance of resistance in alfalfa to the pea aphid. Crop Sci. 6:161-165.

Glover, E., and B. Melton. 1966. Inheritance patterns of spotted alfalfa aphid resistance in *Zea* plants. N. M. Agric. Exp. Sta. Res. Rep. 127:4.

Gorz, H. J., G. R. Manglitz, and F. A. Haskins. 1979. Selection for yellow clover aphid and pea aphid resistance in red clover. Crop Sci. 19:257-260.

Gould, F. 1986a. Simulation models for predicting durability of insect-resistant germplasm: A deterministic diploid, two-locus model. Environ. Entomol. 15:1-10.

Gould, F. 1986b. Simulation models for predicting durability of insect-resistant germ plasm: Hessian fly (Diptera: Cecidomyiidae)- resistant winter wheat. Environ. Entomol. 15:11-23.

Graham, J. H., R. R. Hill, Jr., D. K. Barnes, and C. H. Hanson. 1965. Effect of three cycles of selection for resistance to common leaf spot in alfalfa. Crop Sci. 5:171-173.

Groh, S, D. Gonzalez-deLeon, M. M. Khairallah, C. Jiang, D. Bergvinson, M. Bohn, D. A. Hoisington, and A. E. Melchinger. 1998. QTL mapping in tropical maize: III. Genomic regions for resistance to *Diatraea* spp and associated traits in two RIL populations. Crop Sci. 38:1062-1072.

Guo, B. Z., Z. J. Zhang, R. G. Li, N. W. Widstrom, M. E. Snook, R. E. Lynch, and D. Plaisted. 2001. Restriction fragment length polymorphism markers associated with silk maysin, antibiosis to corn earworm (Lepidoptera: Noctuidae) larvae in a dent and sweet corn cross. J. Econ. Entomol. 94:564-571.

Gupta, A. K., and K. K. Shukla. 1986. Sources and inheritance of resistance to whitebacked planthopper, *Sogatella furcifera* in rice. In. Rice Genetics. Proc. Intl. Rice Genetics Symp. May 27-31, 1985. Island Publ. Co., Manila. pp. 529-535.

Haldane, J. B. S. 1919. The combination of linkage values, and the calculation of distances between the loci of linked factors. J. Genet. 8:299-309.

Hanson, C. H., J. H. Busbice, R. R. Hill, Jr., O. J. Hunt, and A. J. Oakes. 1972. Directed mass selection for developing multiple pest resistance and conserving germplasm in alfalfa. J. Environ. Qual. 1:106-111.

Harlan, S. C. 1916. The inheritance of immunity to leaf blister mite (*Eriophyes gossypii* Banks) in cotton. West Indian Bull. 17:162-166.

Harvey, T. L., K. D. Kofoid, T. J. Martin, and P. E. Sloderbeck 1991. A new greenbug virulent to E-biotype resistant sorghum. Crop Sci. 31:1689-1691.

Harvey, T. L., G. E. Wilde, and K. D. Kofoid. 1997. Designation of a new greenbug biotype, biotype K, injurious to resistant sorghum. Crop Sci. 37:989-991.

Harvey, T. L., T. J. Martin, and R. W. Livers. 1980. Resistance to biotype C greenbug in synthetic hexaploid wheats derived from *Triticum tauschii*. J. Econ. Entomol. 73:387-389.

Harvey, T.L., and T. J. Martin. 1990. Resistance to Russian wheat aphid, *Diuraphis noxia*, in wheat (*Triticum aestivum*). Cereal Res. Commun. 18:127-129.

Hays, D. B., D. R. Porter, J. A. Webster, and B. F. Carver. 1999. Feeding behavior of biotypes E and H greenbug (Homoptera: Aphididae) on previously infested near-isolines of barely. J. Econ. Entomol. 92:1223-1229.

Heinrichs, E. A. 1986. Perspectives and directions for the continued development of insect-resistant rice varieties. Agric., Ecosyst. Environ. 18:9-36.

Hernandez, J. E., and G. S. Khush. 1981. Genetics of resistance to whitebacked planthopper in some rice (*Oryza sativa* L.) varieties. Oryza. 18:44-50.

Hernández, P., G. Dorado, P. Prieto, M. J. Giménez, M. C. Ramírez, D. A. Laurie, J. W. Snape, and A. Martín. 2001. A core genetic map of *Hordeum chilense* and comparisons with maps of barley (*Hordeum vulgare*) and wheat (*Triticum aestivum*). Theor. Appl. Genet. 102: 1259-1264.

Hernández, P., A. Martin, and G. Dorado. 1999. Development of SCARs by direct sequencing of RAPD products: a practical tool for the introgression and marker-assisted selection of wheat. Mol. Breed. 5:245-253.

Hirabayashi, H., and T. Ogawa. 1995. RFLP mapping of *Bph-1* (Brown planthopper resistance gene) in rice. Breeding Sci. 45:369-371.

Hirabayashi, H., T. Ogawa, D. S. Brar, E. R. Angeles, and G. S. Khush. 1999. A gene for resistance to brown plant hopper introgrssed from *Oryza officinalis* was mapped between RFLP markers *G271* and *R93* on chromosome 4. Breed. Res. (Suppl. 1): 48.

Hittalmani, S., A. Parco, T. W. Mew, R. S. Zeigler, and N. Huang. 2000. Fine mapping and DNA marker-assisted pyramiding of the three major genes for blast resistance in rice. Theor. Appl. Genet. 100:1121-1128.

Hollenhorst, M. M., and L. R. Joppa. 1983. Chromosomal location of genes for resistance to greenbug in 'Largo' and 'Amigo' wheats. Crop Sci. 23:91-93.

Huang, N., A. Parco, T. Mew, G. Magpantay, S. Mccouch, E. Guiderdoni, J. C. Xu, P. Subudhi, E. R. Angeles, and G. S. Khush. 1997. RFLP mapping of isozymes, RAPD and QTLs for grain shape, brown planthopper resistance in a doubled haploid rice population. Mol. Breeding. 3:105-113.

Huang, Z., G. He, L. Shu, X. Li, and Q. Zhang. 2001. Identification and mapping of two brown planthopper resistance genes in rice. Theor. Appl. Genet.102: 929-934.

Ikeda, R., and C. Kaneda. 1981. Genetic analysis of resistance to brown planthopper, *Nilaparvata lugens* Stal, in rice. Japan J. Plt. Breed. 31:279-285.

Ishii, T., D. S. Brar, D. S. Multani, and G. S. Khush. 1994. Molecular tagging of genes for brown planthopper resistance and earliness introgressed from *Oryza australiensis* into cultivated rice, *O. sativa*. Genome. 37:217-221.

Jampatong, C., M. D. McMullen, D. B. Barry, L. L. Darrah, P. F. Byrne, and H. Kross. 2002. Quantitative tratit loci for first- and second-generation European corn borer resistance derived from the maize inbred Mo47. Crop Sci. 41:584-593.

Jena, K. K., I. C. Pasalu, Y. K. Rao, Y. Varalaxmi, K. Krishnaiah, G. S. Khush, and G. Kochert. 2003. Molecular tagging of a gene for resistance to brown planthopper in rice (*Oryza sativa* L.). Euphytica. 129: 81-88.

Jennings, C. W., W. A. Russell, and W. D. Guthrie. 1974. Genetics of resistance in maize to first- and second-brood European corn borer. Crop Sci. 14:394-398.

Johnson, A. L., and B. H. Beard. 1977. Sunflower moth damage and inheritance of the phytomelanin layer in sunflower achenes. Crop Sci. 17:369-372.

Jones, A., and F. P. Cuthbert, Jr. 1973. Associated effects of mass selection for soil-insect resistance in sweet potato. J. Am. Soc. Hort. Sci. 98:480-482.

Jones, A., P. D. Dukes, and F. P. Cuthbert, Jr. 1976. Mass selection in sweet potato: breeding for resistance to insects and diseases and for horticultural characteristics. J. Am. Soc. Hort. Sci. 101:701-704.

Jones, N., H. Ougham, and H. Thomas. 1997. Markers and mapping: we are all geneticists now. New Phytologist. 137: 165-177.

Jones, A. T., W. J. McGavin, and A. N. E Birch. 2000. Effectiveness of resistance genes to the large raspberry aphid, *Amphorophora idaei* Borner, in different raspberry (*Rubus idaeus* L.) genotypes and under different environmental conditions. Ann. Appl. Biol. 136:107-113.

Joppa, L. R., R. G. Timian, and N. D. Williams. 1980. Inheritance of resistance to greenbug toxicity in an amphiploid of *Triticum turgidum/T. tauschi.* Crop Sci. 20:343-344.

Joppa, L. R. 1987. Aneuploid analysis in tetraploid wheat, In: G. E. Heyne (Ed.), Wheat and Wheat Improvement, Agron. Monograph 13, ASA, CSSA, & SSSA, Madison, 255.

Kabir, M. A., and G. S. Khush. 1988. Genetic analysis of resistance to brown planthopper in rice (*Oryza sativa* L.). Plant Breed. 100:54-598.

Katiyar, S. K., Y. Tan, B. Huang, G. Chandel, Y. Xu, Y. Zhang, Z. Xie, and J. Bennett. 2001. Molecular mapping of gene *Gm-6(t)* which confers resistance against four biotypes of Asian rice gall midge in China. Theor. Appl. Genet. 103:953-961.

Katsar, C. S., A. H. Paterson, G. L. Teetes, and G. C. Peterson. 2002. Molecular analysis of sorghum resistance to the greenbug (Homoptera: Aphididae). J. Econ. Entomol. 95:448-457.

Kawaguchi, M., K. Murata, T. Ishii, S. Takumi, N. Mori, and C. Nakamura. 2001. Assignment of a brown planthopper (*Nilaparvata lugens* Stal) resistance gene *bph4* to the rice chromosome 6. Breed. Sci. 51:13-18.

Khairallah, M. M., M. Bohn, C. Jiang, J. A. Deutsch, D. C. Jewell, J. A. Mihm, A. E. Melchinger, D. Gonzalez-deLeon, and D. A. Hoisington. 1998. Molecular mapping of QTL for southwestern corn borer resistance, plant height and flowering in tropical maize. Plant Breed. 117:309-318.

Khush, G. S. 1977. Disease and insect resistance in rice. Adv. Agron. 29:265-341.

Khush, G. S. 1978. Biology techniques and procedures employed at IRRI for developing rice germplasm with multiple resistance to diseases and insects. Jap. Trop. Agric. Res. Ctr. Res. Ser. #11. 8 pp.

Khush, G. S. 1979. Genetics and breeding for resistance to the brown planthopper. In Proc. Symp. Brown Plant Hopper. Threat to Rice Production in Asia. International Rice Research Institute, Los Banos, Philippines. pp. 321-332.

Khush, G. S. 1980. Breeding rice for multiple disease and insect resistance. In: Rice Improvement in China and Other Asian Countries. International Rice Research Institute, Los Banos, Philippines. pp. 219-238.

Khush, G. S. 1984. Breeding rice for resistance to insects. Prot. Ecol. 7:147-165.

Khush, G. S., and D. S. Brar. 1991. Genetics of resistance to insects in crop plants. Adv. Agron. 45:223-274.

Kilen, T. C., J. H. Hatchett, and E. E. Hartwig. 1977. Evaluation of early generation soybeans for resistance to soybean looper. Crop Sci. 17:397-398.

Klenke, J. R., W. A. Russell, and W. D. Guthrie. 1986. Distributions for European corn borer (Lepidoptera: Pyralidae) ratings of S_1 lines from 'BS9' corn. J. Econ. Entomol. 79:1076-1081.

Kobayashi, A., C. Kaneda, R. Ikeda, and H. Ikehashi. 1980. Inheritance of resistance to green rice leafhopper, *Nephotettix cincticeps*, in rice. Jpn. J. Breed. 30, Suppl. 1: 56-57.

Kofoid, K. D., T. L. Harvey, and P. E. Sloderbeck. 1991. A new greenbug, biotype I damaging sorghum. In. Proc., 46th Annual Corn and Sorghum Research Conference of the American Seed Trade Association, Washington, DC. pp. 130-140.

Kornegay, J. L., and S. R. Temple. 1986. Inheritance and combining ability of leafroller defense mechanisms in common bean. Crop. Sci. 26:1153-1158.

Kornegay, J. L., and C. Cardona 1991. Inheritance of resistance to *Acanthoscelides obtectus* in a wild common bean accession crossed to commercial bean cultivars. Euphytica. 52:103-111.

Kornegay, J., C. Cardona, and C. E. Posso. 1993. Inheritance of resistance to Mexican bean weevil in common bean, determined by bioassay and biochemical tests. Crop Sci. 33:589-594.

Kosambi, D. D. 1944. The estimation of map distances from recombination values. Ann. Eugen. 12:172-175

Korzun, V., S. Malyshev, A. V. Voylokov, A. and Börner. 2001. A genetic map of rye (*Secale cereale* L.) combining RFLP, isozyme, protein, microsatellite and gene loci. Theor. Appl. Genet. 102:709-717.

Krakowsky, M. D., M. J. Brinkman, W. L. Woodman-Clikeman, and M. Lee. 2002. Genetic components of resistance to stalk tunneling by the European corn borer in maize. Crop Sci. 42: 1309-1315.

Kretshmer, J. M., K. J. Chalmers, S. Manning, A. Karakousis, A. R. Barr, M. R. Islam, S. J. Logue, Y. W. Choe, S. J. Barker, R. C. M. Lance, and P. Langridge. 1997. RFLP mapping of the *Ha2* cereal cyst nematode resistance gene in barley. Theor. Appl. Genet. 94:1060-1064.

Kumar, A., M. N. Shrivasta and R. K. Sahu. 1998. Geentic analysis for gall midge resistance - a reconsideration. Rice Genet. Newslttr. 15:142-143.

Kumar, A., M. N. Shrivasta and B. C. Shukla. 2000a. Genetic analysis of gall midge (*Orseolia oryzae* Wood Mason) biotype 1 resistance in the rice cultivar RP 2333-156-8. Oryza. 37:79-80.

Kumar, A., S. Bhandarkar, D. J. Pophlay, and M. N. Shrivasta. 2000b. A new gene for gall midge resistance in rice accession Jhitpiti. Rice Genet. Newslttr. 17:83-84.

Kurata, N., Y. Nagamura, K. Yamamoto, Y. Harushima, N. Sue, J. Wu, B. A. Antonio, A. Shomura, T. Shimizu, Y. Kuboki, T. Toyama, M. Miyamoto, T. Kirihara, K. Hayasaka, A. Miyao, L. Monna, H. S. Zhong, Y. Tamura, Z. X. Wang, T. Momma, M. Yano, T. Sasaki, and Y. Minobe. 1994. A 300 kilobase interval map of rice including 883 expressed sequences. Nature Genet. 8:365-372.

Lakshminarayana, A. and G. S. Khush. 1977. New genes for resistance to the brown planthopper in rice. Crop Sci. 17:96-100.

Lamb. E. M., D. W. Davis, and D. A. Andow. 1994. Mid-parent heterosis and combining ability of European corn borer resistance in maize. Euphytica. 72:65-72.

Lander, E. S., P. Green, J. Abrahamson, A. Barlow, M. J. Daly, S. E. Lincoln, and L. Newburg. 1987. MAPMAKER: an interactive computer package for constructing primary genetic maps of experimental and natural populations. Genomics. 1:174-181.

Lee, E. A., P. F. Byrne, M. D. McMullen, M. E. Snook, B. R. Wiseman, N. W. Widstrom, and E. H. Coe. 1998. Genetic mechanisms underlying apimaysin and maysin synthesis and corn earworm antibiosis in maize (*Zea mays* L.) Genetics. 149:1997-2006.

Lincoln, S. E., M. J. Daly, and E. S. Lander. 1993. Mapping genes controlling quantitative traits using MAPMAKER/ QTL version 1.1: a tutorial and reference manual. 2nd ed. Cambridge, MA: Whitehead Institute for Biometrical Research.

Liu, X. 2001. Molecular Mapping of Wheat Genes Expressing Resistance to the Russian Wheat Aphid, *Diuraphis noxia* (Mordvilko) (Homoptera: Aphididae) Ph.D. Dissertation. Kansas State University. 144 Pages.

Liu, X. M., C. M. Smith, and B. S. Gill. 2002. Mapping of microsatellite markers linked to the *Dn4* and *Dn6* genes expressing Russian wheat aphid resistance in wheat. Theor. Appl. Genet. 104:1042-1048.

Liu, X. M., C. M. Smith and B. S. Gill. 2005. Allelic relationships among Russian wheat aphid resistance genes. *Crop Sci.* 45: 2273-2280.

Liu, X., C. M. Smith, B. S. Gill, and V. Tolmay. 2001. Microsatellite markers linked to six Russian wheat aphid resistance genes in wheat. Theor. Appl. Genet. 102:504-510.

Livers, R. W., and T. L. Harvey. 1969. Greenbug resistance in rye. J. Econ. Entomol. 62:1368-1370.

Luedders, V. D., and W. A. Dickerson. 1977. Resistance of selected soybean genotypes and segregating populations to cabbage looper feeding. Crop Sci. 17:395-396.

Lyman, J. M., and C. Cardona. 1982. Resistance in lima beans to a leafhopper, *Empoasca krameri*. J. Econ. Entomol. 75:281-286.

Ma, Z.Q., B. S. Gill, M. E. Sorrells, and S. D. Tanksley. 1993. RFLP markers linked to 2 Hessian fly-resistance genes in wheat (*Triticum aestivum* L. from *Triticum tauschii* (Coss) Schmal. Theor. Appl. Genet. 85:750-754.

Ma, Z.-Q., A. Saidi, J. S. Quick, and N. L. V. Lapitan. 1998. Genetic mapping of Russian wheat aphid resistance genes *Dn2* and *Dn4* in wheat. Genome. 41:303-306.

Maas III, F. B., F. L. Patterson, J. E. Foster, and J. H. Hatchett. 1987. Expression and inheritance of resistance of 'Marquillo' wheat to Hessian fly biotype D. Crop Sci. 27:49-52.

Malik, R., C. M. Smith, T. L. Harvey, G. L. Brown-Guedira. 2003. Genetic mapping of wheat curl mite resistance genes *Cmc3* and *Cmc4* in common wheat. Crop Sci. 43:644-650.

Manglitz, G. R., and H. J. Gorz. 1968. Inheritance of resistance in sweetclover to the sweetclover aphid. J. Econ. Entomol. 61:90-94.

Marais, G. F., and F. A. duToit. 1993. A monosomic analysis of Russian wheat aphid resistance in the common wheat PI294994. Plant Breed. 111:246-248.

Marais, G. F., M. Horn, and F. du Toit F. 1994. Intergeneric transfer (rye to wheat) of a gene(s) for Russian wheat aphid resistance. Plant Breed. 113:265-271.

Marais, G. F., W. G. Wessels, and M. Horn. 1998. Association of a stem rust resistance gene (Sr45) and two Russian wheat aphid resistance genes (*Dn5* and *Dn7*) with mapped structural loci in common wheat. S. Afr. J. Plant Soil 15:67-71.

Martin-Sanchez, J. A., M. Gomez-Colmenarejo, J. Del Moral, E. Sin, M. J. Montes, C. Gonzalez-Belinchon, I. Lopez-Brana, and A. Delibes. 2003. A new Hessian fly resistance gene (H30) transferred from the wild grass *Aegilops triuncialis* to hexaploid wheat. Theor. Appl. Genet. 106:1248-55.

McMullen, M. D., and K. D. Simcox. 1995. Genomic organization of disease and insect resistance genes in maize. Mol. Plant Microbe Interaction. 8:811-815.

Mehlenbacher, S. A., R. L. Plaisted, and W. M. Tingey. 1983. Inheritance of glandular trichomes in crosses with *Solanum berthaultii*. Am. Pot. J. 60:699-708.

Mehlenbacher, S. A., R. L. Plaisted, and W. M. Tingey. 1984. Heritability of trichome density and droplet size in interspecific potato hybrids and relationship to aphid resistance. Crop. Sci. 24:320-322.

Miller, C. A., A. Altinkut, and N. L. V. Lapitan. 2001. A microsatellite marker for tagging *Dn2*, a wheat gene conferring resistance to the Russian wheat aphid. Crop Sci. 41:1584-1589.

Mohan, M., S. Nair, J. S. Bentur, U. P. Rao, and J. Bennett. 1994. RFLP and RAPD mapping of the rice *Gm2* gene confers resistance to biotype 1 of gall midge (*Orseolia oryzae*). Theor. Appl. Genet. 87:782-788.

Moharramipour, S., H. Tsumki, K. Sato, and H. Yoshida. 1997. Mapping resistance to cereal aphids in barley. Theor. Appl. Genet. 94:592-596.

Monet, R., and G. Massonie. 1994. Genetic determination of resistance to the green aphid (*Myzus persicae*) in the peach. Agronomie. 2:177-182.

Moreira, L. A., C. Mollema, and S. van Heusden. 1999. Search for molecular markers linked to *Liriomyza trifolii* resistance in tomato. Euphytica. 109:149-156.

Mornhinweg, D. W., D. R. Porter, and J. A. Webster. 1995. Inheritance of Russian wheat aphid resistance in spring barley. Crop Sci. 35:1368-1371.

Mornhinweg, D. W., D. R. Porter, and J. A. Webster. 2002. Inheritance of Russian wheat aphid resistance in spring barley germplasm line STARS-9577B. Crop Sci. 42: 1891-1893.

Mudge, J., P. B. Cregan, J. P. Kenworthy, W. J. Kenworthy, J. H. Orf, N. D. Young. 1997. Two microsatellite markers that flank the major soybean cyst nematode resistance locus. Crop Sci. 37:1611-1615.

Mullis, K. 1990. The unusual origin of the polymerase chain reaction. Sci. Amer. April 56-65.

Murai, H., Z. Hashimoto, P. N. Sharma, T. Shimizu, K. Murata, S. Takumi, N. Mori, S. Kawasaki, S., and C. Nakamura. 2001. Construction of a high resolution linkage map of a rice brown planthopper (*Nilaparvata lugens* Stal) resistance gene *bph2*. Theor. Appl. Genet. 103:526-532.

Murata, K., C. Nakamura, M. Fujiwara, C. Mori, and C. Kaneda 1997. Tagging and maping of brown planthopper resistance genes in rice. In: J-C. Su (Ed.), Proc 5th Int. Symp. on Rice Molecular Biology. Yi-Hsien Pub, Taipei, Taiwan, pp. 217-231.

Murata, K., M. Fujiwara, C. Kaneda, S. Takumi, C. Mori, and C. Nakamura. 1998. RFLP mapping of a brown planthopper (*Nilaparvata lugens* Stal) resistance gene bph2 of *indica* rice introgressed into a *japonica* breeding line 'Noein-PL4'. Genes Genet. Systm. 73:359-364.

Murata, K., M. Fujiwara, H. Murai, S. Takumi, C. Mori, and C. Nakamura. 2000. A dominant brown planthopper resistance gene, *Bph9*, locates on the long arm of rice chromosome 12. Rice Genet. Newsltt. 17:84-86.

Myers, G. O., C. A. Fatokun, and N. D. Young. 1996. RFLP mapping of an aphid resistance gene in cowpea (*Vigna unguiculata* L. Walp). Euphytica. 91: 181-187.

Nagaraj, N., J. C. Reese, M. R. Tuinstra, C. M. Smith, P. St. Amand, M. B. Kirkham, K. D. Kofoid, L. R. Campbell, and G. E. Wilde. 2005. Molecular mapping of sorghum genes expressing tolerance to damage by the greenbug (Homoptera: Aphididae). J. Econ. Entomol. 98:accepted.

Nair, S., J. S. Bentur, U. P. Rao, and M. Mohan. 1995. DNA markers tightly linked to a gall midge resistance gene (*Gm2*) are potentially useful for marker-aided selection in rice breeding. Theor. Appl. Genet. 91:68-73.

Nair, S., A. Kumar, M.N. Srivastava, M. Mohan. 1996. PCR-based DNA markers linked to a gall midge resistance gene, *Gm4t*, has potential for marker-aided selection in rice. Theor. Appl. Genet. 92:660-665.

Nair, R. V., T. M. Masajo, and G. S. Khush. 1982. Genetic analysis of resistance to whitebacked planthopper in twenty-one varieties of rice, *Oryza sativa* L. Theor. Appl. Genet. 61:19-22.

Narvel, J. A., D. R. Walker, B. G. Rector, J. N. All, W. A. Parrott, and R. Boerma. 2001. A retrospective DNA marker assessment of the development of insect resistant soybean. Crop Sci. 41:1931-1939.

Nemamoto, H., R. Ikeda, and C. Kaneda 1989. New genes for resistance to brown planthopper, *Nilaparvata lugens* Stal., in rice. Jpn. J. Rice Breed. 39:23-28.

Nemoto, H., and M. Yokoo. 1994. Experimental selection of a brown plathopper population on mixtures of resistant rice lines. Breeding Science. 44:133-1336.

Nieto-Lopez, R. M., and T. K. Blake. 1994. Russian wheat aphid resistance in barley: Inheritance and linked molecular markers. Crop Sci. 34:655-659.

Nkongolo, K.K., Quick, J.S., Limin, A.E. and Fowler, D.B. 1991a. Sources and inheritance of resistance to Russian wheat aphid in *Triticum* species amphiploids and *Triticum tauschii*. Can. J. Plant Sci. 71:703-708.

Nkongolo, K.K., J. S. Quick, W. L. Meyers, and F. B. Peairs. 1989. Russian wheat aphid resistance of wheat, rye, and triticale in greenhouse tests. Cer. Res. Comm. 17:227-232.

Nkongolo, K.K., Quick, J.S., Peairs, F.B. and Meyer W.L. 1991b. Inheritance of resistance of PI 373129 wheat to the Russian wheat aphid. Crop Sci. 31:905-906.

Ohm, H. W., H. C. Sharma, F. L. Patterson, R. H. Ratcliffe, and M. Obani. 1995. Linkage relationships among genes on wheat chromosome 5A that condition resistance to Hessian fly. Crop Sci. 35:1603-1607.

Oka, I. N. 1983. The potential for the integration of plant resistance, agronomic, biological, physical/mechanical techniques, and pesticides for pest control in farming systems. In L. W. Shemitt (Ed.), Chemistry and World Supplies: The New Frontiers, CHEMRAWN II. Pergamon Press, Oxford, pp. 173-184.

Onukogu, F. A., W. D. Guthrie, W. A. Russell, G. L. Reed, and J. C. Robbins. 1978. Location of genes that condition resistance in maize to sheath collar feeding by second-generation European corn borers. J. Econ. Entomol. 71:1-4.

Ortiz, R. M. Iwanaga, K.V. Raman, and M. Palacios. 1990. Breeding for resistance to potato tuber moth, *Phthorimaea operculella* (Zeller), in diploid potatoes. Euphytica. 50:119-125.

Paterson, A. H., S. D. Tanksley, and M. E. Sorrells. 1991. DNA markers in plant improvement. Adv. Agron. 46:39-90.

Park, S. J., P. R. Timmins, D. T. Quiring, and P. Y. Jul. 1994. Inheritance of leaf area and hooked trichome density of the first trifoliolate leaf in common bean (*Phaseolus vulgaris* L). Can. J. Plant Sci. 74:235-240.

Pathak, R. S. 1990. Genetics of sorhgum, maize, rice and sugar-cane resistance to the cereal stem borer, *Chilo* spp. Insect Sci. Applic. 11:689-699.

Pathak, R. S. 1991. Genetic expression of the spotted stem borer, *Chilo partellus* (Swinhoe) resistance in three maize crosses. Insect Sci. Applic. 12:147-151.

Pathak, M. D., and Z. R. Khan. 1994. Insect Pests of Rice. International Rice Research Institute, The Philippines. 33 pp.

Pathak, R. S., and J. C. Olela. 1983. Genetics of host plant resistance in food crops with special reference to sorghum stem-borers. Insect Sci. Applic. 4:127-134.

Peterson, G. C., J. W. Johnson, G. L. Teetes, and D. T. Rosenow. 1984. Registration of Tx2783 greenbug resistant sorghum germplasm line. Crop Sci. 24: 390.

Porter, K. B., G. L. Peterson, and O. Vise. 1982. A new greenbug biotype. Crop Sci. 22: 847-850.

Porter, D. R., J. D. Burd, K. A. Shufran, J. A. Webster, and G. L. Teetes. 1997. Gereenbug (Homoptera:Aphididae) biotypes: Selected by resistant cultivars or preadapted opportunists? J. Econ. Entomol. 90:1055-1065.

Porter, D. R., J. A. Webster, and B. Friebe. 1994. Inheritance of greenbug biotype G resistance in wheat. Crop Sci. 34:625-628.

Porter, D. R., J. D. Burd, K. A. Sufran, and J. A. Webster. 2000. Efficacy of pyramiding greenbug (Homoptera: Aphididae) resistance genes in wheat. J. Econ. Entomol. 93:1315-22.

Powell, W., M. Morgante, C. Andre, M. Hanafey, J. Vogel, S. Tingey, and A. Rafalski. 1996. The comparison of RFLP, RAPD, AFLP and SSR (microsatellite) markers for germplasm analysis. Mol. Breeding. 2:225-238.

Puterka, G. J., and D. C. Peters. 1995. Genetics of greenbug (Homoptera: Aphididae) virulence to resistance in sorghum. J. Econ. Entomol. 88:421-429.

Rajyashri, K. R., S. Nair, N. Ohmido, K. Fukui, N. Kurata, T. Sasaki and M. Mohan. 1998. Isolation and FISH mapping of yeast artificial chromosomes (YACs) encompassing an allele of the *Gm2* gene for gall midge resistance in rice. Theor. Appl. Genet. 7:507-514.

Rana, B. S., M. G. Jotwani, and N. G. P. Rao. 1981. Inheritance of host plant resistance to the sorghum shootfly. Insect. Sci. Appl. 2:105-110.

Rapusas, H. R., and E. A. Heinrichs. 1985. Whitebacked planthopper growth and development on rices with monogenic or digenic resistance. Rice Genet. Newslttr 10(5):9-10.

Resende, J. T. V., W. R. Maluf, M. G. Cardoso, D. L. Nelson, and M. V. Faria. 2002. Inheritance of acylsugar contents in tomatoes derived from an interspecific cross with the wild tomato *Lycopersicon pennellii* and their effect on spider mite repellence. Genet. Mol. Res. 1: 106-116.

Ratcliffe, R. H., and J. H. Hatchett. 1997. Biology and genetics of the Hessian fly and resistance in wheat. In: K. Bobdari (Ed.), New Developments in Entomology, Research Signpost, Scientific Information Guild, Trivandrum.

Rector, B. G., J. N. All, W. A. Parrott, and H. R. Boerma. 1998. Identification of molecular markers linked to quantitative trait loci for soybean resistance to corn earworm. Theor. Appl. Genet. 96:786-790.

Rector, B. G., J. N. All, W. A. Parrott, and H. R. Boerma. 1999. Quantitative trait loci for antixenosis resistance to corn earworm in soybean. Crop Sci. 39:531-538.

Rector, B. G., J. N. All, W. A. Parrott, and H. R. Boerma. 2000. Quantitative trait loci for antibiosis resistance to corn earworm in soybean. Crop Sci. 40:233-238.

Redden, R. J., P. Dobie, and A. Gatehouse. 1983. The inheritance of seed resistance to Callosobruchus maculatus F. in cowpea (Vigna unguiculata (L.) Walp.) I. Analyses of parental, F1, F2 and backcross seed generation. Aust. J. Agric. Res. 34:681-695.

Redden, R. J., S. R. Singh, and M. J. Luckfahr. 1984. Breeding for cowpea resistance to bruchids at IITA. Prot. Ecol. 7:291-303.

Renganayaki, K., A. K. Fritz, S. Sadasivam, S. Pammi, S. E. Harrington, S. R. McCouch, S. M. Kumar, and A. S. Reddy. 2002. Mapping and progress toward map-based cloning of brown planthopper biotype-4 resistance gene introgressed Oryza officinalis into cultivated rice, O. sativa. Crop Sci. 42:2112-2117.

Rezaul Kamin, A. N. M., and M. D. Pathak. 1982. New genes for resistance to green leafhopper, Nephotettix virescens (Distant) in rice, Oryza sativa L. Crop Prot. 1:483-490.

Roche, P., F. H. Alston, C. Maliepaard, K.M. Evans, R. Vrielink, F. Dunemann, T. Markussen, S. Tartarini, L. M. Brown, C. Ryder, and G. J. King. 1997. RFLP and RAPD markers linked to the rosy leaf curling aphid resistance gene (Sd1) in apple. Theor. Appl. Genet. 94: 528-533.

Romero Andreas, J., B. S. Yandell, F. A. Bliss. 1986. Bean arcelin. 1. Inheritance of a novel seed protein of Phaseolus vulgaris L. and its effect on seed composition. Theoret. Appl. Genet. 72:123-128.

Porter, D. R., J. D. Burd, K. A. Shufran, and J. A. Webster. 2000. Efficacy of pyramiding greenbug (Homoptera: Aphididae) resistance genes in wheat. J. Econ. Entomol. 93:1315-1318.

Rossetto, C. J., and T. Igue. 1983. Heranca de resistencia variedade de sorgo AF28 a Contarinia sorghicola Coquillett. Bragantia. 42:211-219.

Russell, W. A., G. D. Lawrence, and W. D. Guthrie. 1979. Effects of recurrent selection for European corn borer resistance on other agronomic characters in synthetic cultivars of maize. Maydica. 24:33-47.

Sahu, V. N., and R. K. Sahu. 1989. Inheritance and linkage relationships of gall midge resistance with purple leaf, apiculus and scent in rice. Oryza. 26:79-83.

Saidi, A., and J. S. Quick. 1996. Inheritance and allelic relationships among Russian wheat aphid resistance genes in winter wheat. Crop Sci. 36:256-258.

Saka, N., T. Toyama, T. Tuji, H. Nakamae and T. Izawa. 1997. Fine mapping of green ricehopper resistant gene Grh-3 (t) and screening of Grh-3 (t) among green ricehopper resistant and green leafhopper resistant cultivars in rice. Breed. Sci. 47 (suppl.1): 55 (in Japanese).

Sambrook, J., P. MacCallum, and D. Russell. 2001. Molecular Cloning: A Laboratory Manual. Cold Spring Harbor Laboratory Press, Woodbury, NY USA 2,344 pp.

Sams, D. W., F. I. Lauer, and E. B. Radcliffe. 1976. Breeding behavior of resistance to green peach aphid in tuber-bearing *Solanum* germplasm. Am. Potato J. 53:23-29.

Sanford, L. L. 1997. Tuber yields and specific gravities in *Solanum tuberosum* populations recurrently selected for resistance to potato leafhopper. Amer. Potato J. 74:65-73.

Sanford, L. L., K. L. Deahl, S. L. Sinden, and T. L. Ladd, Jr. 1992. Glykoalkaloid contents in tubers from *Solanum tuberosum* populations selected for potato leafhopper resistance. Am. Potato J. 69:693-703.

Sanford, L. L., and T. L. Ladd, Jr. 1983. Selection for resistance to potato leafhopper in potatoes. III. Comparison of two selection procedures. Am. Potato J. 60:653-659.

Sanford, L. L., and T. L. Ladd, Jr. 1987. Genetic transmission of potato leafhopper resistance from recurrent selection populations in potato *Solanum tuberosum* L. Gp. *tuberosum*. Am. Pot. J. 64:655-662.

Sardesai, N., A. Kumar, K. R. Rajyashri, S. Nair, and M. Mohan. 2001. Identification and mapping of an AFLP marker linked to *Gm7*, a gall midge resistance gene and its conversion to a SCAR marker for its utility in marker aided selection in rice. Theor. Appl. Genet. 105:691-698.

Sastry, M. V. S., and P. S. Prakasa Rao. 1973. Inheritance of resistance to rice gall midge *Pachydiplosis oryzae* Wood-Mason. Current Sci. 42:652-653.

Satyanarayanaiah, K., and M. V. Reddi. 1972. Inheritance of resistance to insect gall midge (*Pachydiplosis oryzae*, Wood Mason) in rice. Andhra Agric. J. 19:1-8.

Schlegel, R., and R. Kynast. 1987. Confirmation of 1A/1R wheat-rye chromosome translocation in the wheat variety 'Amigo'. Plant Breed. 98:57-60.

Schroeder-Teeter, S., R. S. Zemetra, D. J. Schotzko, C. M. Smith, and M. Rafi. 1994. Monosomic analysis of Russian wheat aphid (*Diuraphis noxia*) resistance in *Triticum aestivum* line PI137739. Euphytica 74:117-120.

Scott, D. R. 1977. Selection for *Lygus* bug resistance in carrot. HortSci. 12:452.

Scott, G. E., and W. D. Guthrie. 1967. Reactions of permutations of maize double crosses to leaf feeding of European corn borers. Crop Sci. 7:233-235.

Scott, G. E., F. F. Dicke, and G. R. Pesho. 1966. Location of genes conditioning resistance in corn to leaf feeding of the European corn borer. Crop Sci. 4: 444-446.

Sebesta, E. E., and E. A. Wood, Jr. 1978. Transfer of greenbug resistance from rye to wheat with x-rays. Agron. Absts. 61-62.

Selvi, A., P. Shanmugasundaram, S. Mohan Kumar, and J. A. J. Raja. 2002. Molecular markers for yellow stem borer *Scirpophaga incertulas* (Walker) resistance in rice. Euphytica. 124:371-377.

Sharma, G. C., and C. V. Hall. 1971. Cucurbitacin B and total sugar inheritance in *Cucurbita pepo* L. related to spotted cucumber beetle feeding. J. Am. Soc. Hort. Sci. 96:750-754.

Sharma, H. C., M. V. Satyanarayana, S. D. Singh, and J. W. Stenhouse. 2000. Inheritance of resistance to head bugs and its interaction with grain molds in *Sorghum bicolor.* Euphytica. 112:167-173.

Sharma, P. N., Y. Ketipearachchi, K. Murata, A Torii, S. Takumi, N. Mori, and C. Nakamura. 2003. RFLP/AFLP mapping of a brown planthopper (*Nilaparvata lugens* Stål) resistance gene *Bph1* in rice. Euphytica. 129:109-117.

Sharopova, N., M. D. McMullen, L. Schultz, S. Schroeder, H. Sanchez-Villeda, J. Gardiner, D. Bergstrom, K. Houchins, S. Melia-Hancock, T. Musket, N. Duru, M. Polacco, K. Edwards, T. Ruff, J. C. Register, C. Brouwer, R. Thompson, R. Velasco, E. Chin, M. Lee, W. Woodman-Clikeman, M. J. Long, E. Liscum, K. Cone, G. Davis, and E. H. Coe, Jr. 2002. Development and mapping of SSR markers for maize. Plant Mol. Biol. 48:463-481.

Shrivastava, M. N., A. Kumar, S. Bhandarkar, B. C. Shukla, and K. C. Agrawal. 2003. A new gene for resistance in rice to Asian rice gall midge (*Orseolia oryzae* Wood Mason) biotype 1 population at Raipur, India. Euphytica. 130:143-145.

Shon, C. C., M. Lee, A. E. Melchinger, W. D. Guthrie, and W. L. Woodman. 1993. Mapping and cahracterization of quantitiaive trait loci affecting resistance against second generation European corn borer in maize with the aid of RFLPs. Heredity. 70:648-659.

Sidhu, G. S., G. S. Khush, and F. G. Medrano. 1979. A dominant gene in rice for resistance to whitebacked planthopper and its relationship to other plant characteristics. Euphytica. 28:227-232.

Sifuentes, J. A., and R. H. Painter. 1964. Inheritance of resistance to western corn rootworm adults in field corn. J. Econ. Entomol. 57:475-477.

Singh, D. P. 1986. Breeding for Resistance to Diseases and Insect Pests. Springer. New York. 222 pp.

Sisson, V. A., P. A. Miller, W. V. Campbell, and J. W. Van Duyn. 1976. Evidence of inheritance of resistance to the Mexican bean beetle in soybeans. Crop Sci. 16:835-837.

Siwi, B. H., and G. S. Khush. 1977. New genes for resistance to the green leafhopper in rice. Crop Sci. 17:17-20.

Smith, C. M., and C. A. Brim. 1979. Field and laboratory evaluations of soybean lines for resistance to corn earworm leaf feeding. J. Econ. Entomol. 72:78-80.

Somers, D. J., P. Isaac, and K. Edwards. 2004. A high-density microsatellite consensus map for bread wheat (*Triticum aestivum* L.) . Theor. Appl. Genet. 109:1105-1114.

Song, Q.J., L. F. Marek, R. C. Shoemaker, K. G. Lark, V. C. Concibido, X. Delannay, J. E. Specht, and P. B. Cregan. 2004. A new integrated genetic linkage map of the soybean. Theor. Appl. Genet. 109:122-128.

Soper, J. F., M. S. McIntosh, and T. C. Elden. 1984. Diallel analysis of potato leafhopper resistance among selected alfalfa clones. Crop Sci. 24:667-670.

Srivastava, M. N., A. Kumar, S. K. Shrivastava, and R. K. Sahu. 1993. A new gene for resistance to rice gall midge in rice variety Abhaya. Rice Genet. Newsl. 10:79-80.

Staub, J. E., F. C. Serquen, and M. Gupta. 1996. Genetic markers, map construction, and their application in plant breeding. Hortscience. 31:729-741.

Stebbins, N. B., F. L. Patterson, and R. L. Gallun. 1983. Inheritance of resistance of PI94587 wheat to biotypes B and D of Hessian fly. Crop Sci. 23:251-253.

Stebbins, N. B., F. L. Patterson, and R. L. Gallun. 1982. Interrelationships among wheat genes H3, H6, H9, and Hl0 for Hessian fly resistance. Crop Sci. 22:1029-1032.

Tamura, K., Y. Fukuta, M. Hirae, S. Oya, I. Ashikawa, and T. Yagi. 1999. Mapping of the *Grh1* locus for green rice leafhopper resistance in rice using RFLP markers. Breed. Sci. 49:11-14.

Tao, Y. Z., A. Hardy, J. Drenth, R. G. Henzell, B. A. Franzmann, D. R. Jordan, D. G. Butler, and C. L. McIntyre. 2003. Identifications of two different mechanisms for sorghum midge resistance through QTL mapping. Theor. Appl. Genet. 107: 116-122.

Teetes, G. L., R. M. Anderson, and G. C. Peterson. 1994. Exploitation of sorghum midge nonpreference resistance in sorghum midge (Diptera: Cecidomyiidae) using mixed plantings of resistant and susceptible sorghum hybrids. *J. Econ. Entomol.* 87:826-831.

Terry, L. I., K. Chase, J. Orf, T. Jarvik, L. Mansur, and K.G. Lark. 1999. Insect resistance in recombinant inbred soybean lines derived from non-resistant parents. Entomol. Exp. Appl. 91:465-476.

Theurer, J. C., C. C. Blickenstaff, G. G. Mahrt, and D. L. Doney. 1982. Breeding for resistance to the sugarbeet root maggot. Crop Sci. 22:641-645.

Thomas, J. B., and R. L. Conner. 1986. Resistance to colonization by the wheat curl mite in *Aegilops squarrosa* and its inheritance after transfer to common wheat. Crop Sci. 26:527–530.

Toenniessen, G. H., J. C. O'Toole, and J. DeVries. 2003. Advances in plant biotechnology and its adoption in developing countries. Curr. Opin. Plant Biol. 8:191-198.

Tomaka, N., C. Lairungreang, P. Nakeeraks, Y. Egawa, and C. Thavarasook. 1992. Development of bruchid-resistant mungbean line using wild mungbean germplasm in Thailand. Plant Breed. 109:60-66.

Tomar, J. B., and S. C. Prasad. 1992. Genetic analysis of resistance to gall midge (*Orseolia oryzae* Wood Mason) in rice. Plant Breed. 109:159-167.

Tuinstra, M. R., G. E. Wilde, and T. Krieghauser. 2001. Geentic analysis of biotype I resistance in sorghum. Euphytica. 121:87-91.

Tyler, J. M., J. A. Webster, and O. G. Merkle. 1987. Designations for genes in wheat germplasm conferring greenbug resistance. Crop. Sci. 27:526-527

Venter, E., and A. M. Botha. 2000. Development of markers linked to *Diuraphis noxia* resistance in wheat using a novel PCR-RFLP approach. Theor. Appl. Genet. 100:965-970.

Vos, P., R. Hogers, M. Bleeker, M. Rijans, T. Van de Lee, M. Hornes, A. Frijters, J. Pot, M. Kuiper, and M. Zabeau. 1995. AFLP: A new technique for DNA fingerprinting. Nucleic Acids Res. 23:4407-4414.

Wang, C., H. Yasui, A. Yoshimura, C. Su, H. Zhai, and J. Wan. 2003. Green rice leafhopper resistance gene transferred through backcrossing and CAPs marker assisted selection. Chinese Agric. Sci. 2:13-18.

Wang, C., H. Yasui, A. Yoshimura, H. Zhai, and J. Wan. 2004. Inheritance and QTL mapping of antibiosis to green leafhopper in rice. Crop Sci. 44:389-393.

Warnock, D. F., D. W. Davis, and G. R. Gingera. 1998. Inheritance of ear resistance to European corn borer in 'Apache' sweet corn. Crop Sci. 38:1451-1457.

Weibel, D. E., K. J. Starks, E. A. Wood, Jr., and R. D. Morrison. 1972. Sorghum cultivars and progenies rated for resistance to greenbugs. Crop Sci. 12:334-336.

Weiss, M. J., N. R. Riveland, L. L. Reitz, and T. R. Olson. 1990. Influence of resistant and susceptible cultivar blends of hard red spring wheat on wheat stem sawfly (Hymenoptera: Cephidae) damage and wheat quality parameters. J. Econ. Entomol. 83:255-259.

Weng, Y., and M. D. Lazar. 2002. Amplified fragment length polymorphism- and simple sequence repeat-based molecular tagging and mapping of greenbug resistance gene *Gb3* in wheat. Plant Breeding. 121:218-223

Whelan, E. D. P., and G. E. Hart. 1988. A spontaneous translocation that transfers wheat curl mite resistance from decaploid *Agropyron elongatum* to common wheat. Genome. 30:289-292

Whelan, E. D. P., and J. B. Thomas. 1989. Chromosomal location in common wheat of a gene (*Cmc1*) from *Aegilops squarrosa* that conditions resistance to colonization by the wheat curl mite. Genome. 32:1033-1036.

Widstrom, N. W. 1972. Reciprocal differences and combining ability for corn earworm injury among maize single crosses. Crop Sci. 12:245-247.

Widstrom, N. W. 1989. Breeding methodology to increase resistance in maize to corn earworm, fall armyworm, and maize weevil. In. Toward Insect Resistant Corn for the Third World: Proceedings of the International Symposium on Methodologies for Developing Host Plant Resistance to Corn Insects, Mexico CIMMYT, Mexico. pp. 209-219.

Widstrom, N. W., and J. J. Hamm. 1969. Combining abilities and relative dominance among maize inbreds for resistance to earworm injury. Crop Sci. 9:216-219.

Widstrom, N. W., and W. W. McMillian. 1973. Genetic factors conditioning resistance to earworm in maize. Crop Sci. 13:459-461.

Widstrom, N. W., B. R. Wiseman, and W. W. McMillian. 1972. Resistance among some maize inbreds and single crosses to fall armyworm injury. Crop Sci. 12:290-292.

Widstrom, N. W., B. R. Wiseman, and W. W. McMillian. 1982. Responses to index selection in maize for resistance to ear damage by the corn earworm. Crop Sci. 22:843-846.

Widstrom, N. W., B. R. Wiseman, and W. W. McMillian. 1984. Patterns of resistance in sorghum to the sorghum midge. Crop Sci. 24:791-793.

Widstrom, N. W., K. Bondari, and W. W. McMillian. 1992a. Hybrid performance among maize populations selected for resistance to insects. Crop Sci. 32:85-89.

Widstrom N. W., W. P. Williams, B. R. Wiseman, and F. M. Davis. 1992b. Recurrent selection for resistance to leaf feeding by fall armyworm on maize. Crop Sci. 32:1171-1174.

Willcox, M. C., M. M. Khairallah, D. Bergvinson, J. Crossa, J. A. Deutsch, G. O. Edmeades, D. Gonzalez-de-Leon, C. Jiang, D. C. Jewell, J. A. Mihm, W. P. Williams, and D. Hoisington. 2002. Selection for resistance to southwestern corn borer using marker-assisted selection and conventional backcrossing. Crop Sci. 42:1516-1528.

Williams, W. P., P. F. Buckley, and F. M. Davis. 1995. Combining ability in maize for fall armyworm and southwestern corn borer resistance based on a laboratory bioassay for larval growth. Theor. Appl. Genet. 90:275-278.

Williams, W. P., F. M. Davis, and P. F. Buckley. 1998. Resistance to southwestern corn borer in corn after anthesis. Crop Sci. 38:1514-1517.

Wilson, F. D., and B. W. George 1983. A genetic and breeding study of pink bollworm resistance and agronomic properties in cotton. Crop Sci. 23:1-4.

Wilson, F. D., and B. W. George. 1979. Combining ability in cotton for resistance to pink bollworm. Crop Sci. 19:834-836.

Wilson, F. D., and J. A. Lee. 1971. Genetic relationships between tobacco budworm feeding response and gland number in cotton seedlings. Crop Sci. 11:419-421.

Wilson, F. D., and J. N. Smith. 1977. Variable expressivity and gene action of gland-determining alleles in *Gossypium hirsutum* L. Crop Sci. 17:539-43.

Wiseman, B. F., and K. Bondari. 1992. Genetics of antibiotic resistance in corn silks to the corn earworm (Lepidoptera: Noctuidae). J. Econ. Entomol. 85:293-298.

Wiseman, B. R., and K. Bondari. 1995. Inheritance of resistance in maize silks to the corn earworm. Entomol. Exp. Appl. 77:315-321.

Wright, R. J., M. B. Dimock, W. M. Tingey, and R. L. Plaisted. 1985. Colorado potato beetle (Coleoptera: Chrysomelidae): Expression of resistance in *Solanum berthaultii* and interspecific potato hybrids. J. Econ. Entomol. 78:576-582.

Wu, C. F., and G. S. Khush. 1985. A new dominant gene for resistance to whitebacked planthopper in rice. Crop Sci. 25: 505-509.

Xu, X. F., H. W. Mei, L. J. Luo, X. M. Cheng, and Z. K. Li. 2002. RFLP-facilitated investigation of the quantitative resistance of rice to brown planthopper (*Nilaparvata lugens*). Theor. Appl. Genet. 104:248-253.

Yang, D., A. Parco, S. Nandi, P. Subudh, Y. Zhu, G. Wang, and H. Huang N. 1997. Construction of a bacterial artificial chromosome (BAC) library and identification of overlapping BAC clones with chromosome 4-specific RFLP markers in rice. Theor. Appl. Gen. 95:1147-1154.

Yazawa, S., H. Yasui, A. Yoshimura, and N. Iwata, 1998. RFLP mapping of genes for resistance to green rice leafhopper (*Nephotettix cincticeps* Uhler) in rice cultivar DV85 using near isogenic lines. Sci. Bull. Fac. Agr. Kyushu Univ. 52: 169-175.

Yencho, G. C., M. W. Bonierbale, W. M. Tingey, R. L. Plaisted, and S. D. Tanksley. 1996. Molecular markers locate genes for resistance to the Colorado potato beetle, *Leptinotarsa decemlineata*, in hybrid *Solanum tuberosum x S. berthaultii* potato progenies. Entomol. Exp. Appl. 81:141-154.

Yencho, G. C., M. B. Cohen, and P. F. Byrne. 2000. Applications of tagging and mapping insect resistance loci in plants. Ann. Rev. Entomol. 45:393-422.

Young, N. D., L. Kumar, D. Menancio-Hautea, D. Danesh, N. S. Talekar, S. Shanmugasundarum, and D.-H. Kim. 1992. RFLP mapping of a major bruchid resistance gene in mungbean (*Vigna radiata*, L. Wilczek). Theor. Appl. Genet. 84:839-844.

Zhang Y., J. S. Quick, and S. Liu. 1998. Genetic variation in PI 294994 wheat for resistance to Russian wheat aphid. Crop Sci. 38:527-530.

Zhu, L. C., C. M. Smith, A. Fritz, E. V. Boyko, and M. B. Flinn. 2004. Genetic analysis and molecular mapping of a wheat gene conferring tolerance to the greenbug (*Schizaphis graminum* Rondani). Theor. Appl. Genet. 109:289-293.

CHAPTER 9

CONSTITUTIVE AND INDUCED RESISTANCE GENES

1. CONSTITUTIVE ARTHROPOD RESISTANCE

1.1. Constitutive Resistance Genes, Defense Response Genes, and Resistance Gene Analogs

Two classes of plant genes contribute to abiotic or biotic resistance reactions. Resistance (R) genes are involved in the recognition of plant stress. Defense response (DR) genes are involved in plant defense responses following the recognition of a stress. Resistance gene analogs (RGAs), conserved amino acid motifs derived from sequence comparisons of predominant classes of R genes, have been used to identify arthropod resistance genes. The structure, expression and function of many R and DR genes involved in plant disease and nematode resistance have been extensively studied (Bryngelsson et al. 1994, Chittoor et al. 1997, Feuillet and Keller 1999, Lagudah et al. 1997, Mingeot and Jaquemin 1997, Seah et al. 1998).

1.2. Cloning and Sequencing Plant Resistance Genes

There are five generally recognized classes of resistance gene products (Martin et al. 2003) (Table 9.1). Genes in the *Pto* (tomato *Pto* gene conferring resistance to *Pseudomonus syringae* pv. *tomato* bacterial speck) resistance gene class (Class I) encode serine/threonine protein kinases with a short sequence of amino acids for membrane association (myristylation motif) and no transmembrane domain for attachment to membranes. The second class of resistance genes encodes three different amino acid regions consisting of: leucine rich repeats (LRR), nucleotide binding sites (NBS), and a leucine zipper (LZ) or coiled coil (CC) region. Class II genes contain no transmembrane domain and are commonly referred to as NBS-LRR genes. A third class of resistance genes are similar to NBS-LRR genes except that instead of the CC region, these genes have a TIR region that is similar to the Toll and Interleukin 1 receptor proteins involved in the innate immunity of animals. Class III resistance genes also have no transmembrane domain. The lack of transmembrane domains in classes I, II or III seems to indicate that these resistance gene proteins reside in the plant cytoplasm and interact there with arthropod or disease avirulence gene products.

Table 9.1. Classes of plant resistance gene proteins. Reprinted with permission, from the Annual Review of Plant Biology, Volume 54, copyright 2003 by Annual reviews www.annualreviews.org

Gene Class (Name)	Amino Acid Regions(s)	Resistant to
I (*Pto*)	Serine /threonine protein kinase	Pathogenic bacteria
II (NBS-LRR)	Leucine rich repeats (LRR), nucleotide binding sites (NBS), & leucine zipper (LZ) or coiled coil (CC) region	Pathogenic bacteria, fungi, nematodes & insects
III (TIR LZ)	LRR, NBS, & LZ or Toll & Interleukin 1 (TIR) receptor proteins	Pathogenic fungi
IV (*Cf*)	Extracellular LRR & transmembrane domain	Pathogenic fungi
V (*Xa21*)	Extracellular LRR, transmembrane domain & serine/threonine kinase	Pathogenic bacteria

Class IV resistance genes are characterized by *Cf* (*Cladosporium fulvum* tomato fungus) proteins consisting of an extracellular LRR component and a transmembrane domain. Class V resistance genes are typlified by the *Xanthonomus oryzae* bacteria resistance gene of rice (*Xa21*). Class V genes have an extracellular LRR a transmembrane domain and an intraclular serine/threonine kinase domain. Several other resistance gene proteins are of unknown function and do not conform to any of the five established classes of resistance gene products. The *Hm1* gene controlling resistance to the *Cochliobolus carbonum* fungal toxin in maize, *Zea mays* L., is in this group. For additional, in depth information about resistance gene classes, readers are referred to reviews of Ellis et al. (2000) and Pan et al. (2000).

Two arthropod resistance genes have been identified. The *Mi-1.2* gene (Figure 9.1) from wild tomato, *Lycopersicon peruvianum* (L.) P. Mill., confers resistance to the potato aphid, *Macrosiphum euphorbiae* (Thomas), (Kaloshian et al. 1995, 1997) and to three species of the root knot nematode, *Meloidogyne* spp. (Roberts and Thomason 1986, Rossi et al. 1998, Vos et al. 1998). The *Vat* (virus aphid transmission) gene from melon, *Cucumis melo* L., encodes a cytoplasmic protein and expresses resistance to the melon aphid, *Aphis gossypii* Glover, (Klingler et al. 1998) and to transmission of some non-persistent viruses vectored by *A. gossypii* (Pitrat and Lecoq 1980). Both *Mi-1.2* and *Vat* are members of the NBS-LRR Class II family of disease and nematode resistance genes (Milligan et al. 1998, Brotman et al. 2002, Dogimont et al. 2003). The LRR region of *Mi-1.2* functions to signal localized cell death and programmed cell death (Hwang et al. 2000, Wang et al. 2001). There are indications that a *Dn* gene from wheat controlling resistance to the Russian wheat aphid, *Diurpahis noxia* (Mordvilko), is very similar in function to Class I genes (Boyko and Smith 2004).

Similarities in the sequence and function of other pest resistance genes are also apparent. Resistance gene analog sequences from barley, *Hordeum vulgare* L., map near loci in involved in resistance to the corn leaf aphid, *Rhopalosiphum maidis* (Fitch). These same sequences are similar to the NBS-LRR *Cre3* gene in wheat, *Triticum aestivum* L., for resistance to the cereal cyst nematode, *Heterodera avenae* (Lagudah et al. 1997).

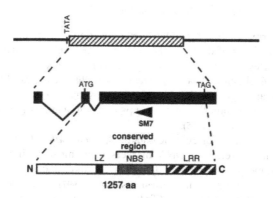

Figure 9.1. Mi-1.2, a gene from Lycopersicon peruvianum conferring resistance to Macrosiphum euphorbiae and to three Meloidogyne species. From Rossi, M., F. L. Goggin, S. B. Milligan, I. Klaoshian, D. E. Ullman, and V. M. Williamson. 1998. The nematode resistance gene Mi of tomato confers resistance against the potato aphid. Proc. Natl. Acad. Sci. USA. 95:9750-975, Copyright 1998, National Academy of Sciences, U.S.A.

1.3. Functional Genomics

With the sequencing of the genomes of *Arabidopsis thaliana* and rice, *Oryza sativa* (L.), many opportunities now exist for in-depth studies of the molecular bases of plant resistance to arthropods. Efforts are underway to sequence much larger crop genomes such as barley, maize and wheat. In the meantime, plant resistance researchers have begun to "mine" data relating to *Arabidopsis* and rice defense gene sequence, function, and expression, in order to provide new information about the biochemical and physiological pathways involved in the resistance of plants to arthropods.

Resistance gene analogs have been isolated in *Arabidopsis*, barley, lettuce, *Lactuca sativa* L., maize, rice, soybean, *Glycine max* (L.) Merr., and wheat (Graham et al. 2000, Leister et al. 1999, Mago et al. 1999, Seah et al. 1998, Shen et al. 1998, Speulman et al. 1998, Tada 1999). Many cereal crop RGAs map to orthologous positions in different cereal species. The fact that *Mi-1.2* is active against two organisms as distantly related as aphids and nematodes, supports the hypothesis that RGAs can also be used to clone or design genes for arthropod resistance. Map positions of RGAs in the Triticeae indicate that these genes occur in clusters (similar to the arthropod resistance gene clusters described in Chapter 8) and are more closely linked physically than genes in other regions with similar genetic distances (Boyko et al. 2002, Feuillet and Keller 1999). This arrangement fits the general relationship between physical and genetic distances on chromosomes. Genetic distances do not necessarily reflect physical distances, and in some different chromosome regions, the genetic distance and physical distances are very different. For this reason, knowledge of the chromosome locations and genome organizations of RGAs in crops are of great value in the analyses of potential (candidate) resistance genes.

1.4. Resistance-Related Plant Gene Homology

Additional information on resistance gene location, order, and function can be received from analyses of data on related crop species. Comparative mapping of the genomes of barley, maize, rice, oat, rye, *Secale cereale* L., sorghum, *Sorghum bicolor* (L.) Moench, and wheat has revealed that chromosomes of these grasses are partially collinear in gene order. Many chromosome regions of these crops have conserved linkages among the same groups of DNA markers (Boyko et al. 1999, 2002, Hulbert et al. 1990, Paterson et al. 1995).

The chromosomes of barley, wheat and oat, *Avena sativa* (L.), are homeologous, and contain a linear series of DNA markers mapping to a similar series of loci. Their chromosome pairs are derived from a common ancestor, do not recombine during meiosis and are syntenic (Linde-Laursen et al. 1997). The extensive homeology between barley, oat, rye and wheat (Triticeae) chromosomes allow the same sets of molecular probes to be used for tagging genes for important traits among these crop plants (Devos et al. 1992, Hohmann et al. 1995, Namuth et al. 1994). Additional orthologous relationships have been shown to exist between Triticeae, rice, and maize chromosomes (Ahn et al. 1993, Van Deynze et al. 1995a,b).

Resistance gene location in barley, sorghum, rice and wheat demonstrate the synteny (similar conserved gene composition and order) among loci of these crops linked to genes expressing resistance to several species of pest arthropods (Figure 9.2). The location of aphid resistance genes on Triticeae chromosome 7 is syntenic to QTLs for resistance in rice to the brown planthopper, *Nilaparvata lugens* Stal, located on rice chromosome 6 (see chapter 8) (Alam and Cohen 1998, Boyko et al. 2004, Liu et al. 2001, 2002, Moharramipour et al. 1997, Nieto-Lopez and Blake 1994, Seah et al. 1998, Weng and Lazar 2002, Xhu et al. 2004). A QTL for resistance to *R. maidis* (*Rsg1a*), as well as several disease and nematode RGAs have also been mapped to the short arm of barley chromosome 7 (Moharramipour et al. 1997, Seah et al. 1998). QTLs for resistance to the European corn borer, *Ostrinia nublialis* Hübner, the southwestern corn borer, *Diatrea grandiosella* (Dyar), and the sugacane borer, *Diatrea saccharalis* (F.), occur on maize chromosomes 5 and 9 (see chapter 8), which are homeologous to rice chromosome 6 and the short arm of wheat chromosome 7.

The exploitation of such conserved gene order to identify pest resistance loci of interest is stimulating efforts to clone arthropod resistance genes in cereals and other crops as functional genomics becomes utilized in agricultural research. For example, specific rice bacterial artificial chromosome (BAC) contiguous segments can be subjected to *in silico* analyses to identify sequences in syntenous areas in barley, rice and wheat chromosomes where resistance genes have been mapped.

As pointed out by Bennetzen and Freeling (1993) for Graminae genomes in general, exploiting this conserved gene order in the Triticeae to identify loci of interest may enhance efforts to clone arthropod resistance gene families in cereals.

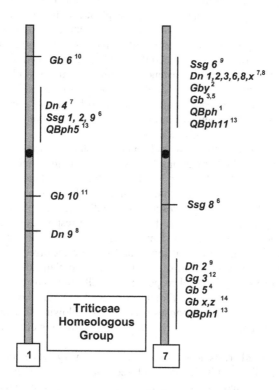

Figure 9.2. Heteroptera resistance gene loci in barley, rice, sorghum and wheat on Triticeae homoeologous chromosome groups 1 and 7 (relative loci positions for illustration only, not ordered). Gb - Schizaphis graminum resistance genes in wheat; Dn - Diuraphis noxia resistance genes in wheat; Ssg - S. graminum resistance QTLs in sorghum; QBph - Main effect Nilapartava lugens resistance QTLs in rice; Grh - Nephotettix cincticeps resistance genes in rice. (1 - Alam and Cohen 1998, 2 - Boyko et al. 2004, 3 - Castro et al. 2001, 4 - Dubcovsky et al. 1998, 5 - Moharramipour et al. 1997, 6 - Katsar et al. 2002, 7 - Liu et al. 2002, 8 - 2001, 9 - Miller et al. 2001, 10 - Smith et al. unpubl., 11 - Tamura et al. 1999, 12 - Weng and Lazar 2002, 13 - Xu et al. 2002, 14 - Zhu et al. 2004a. (From Smith 2004. Reprinted with permission from CABI Publishing, CAB International)

2. INDUCED ARTHROPOD RESISTANCE

Plants use both constitutive and induced defenses to protect themselves from arthropod attack. Constitutive plant resistance genes are often identified in breeding programs. The expression of plant resistance to arthropods is also affected by previous exposure to various stimuli. Prior wounding by arthropod or mechanical means induces increased levels of resistance of many crop plants to arthropod damage.   Kogan and Paxton (1983) define induced plant resistance as the "quantitative or qualitative enhancement of a plant's defense mechanism against pests in response to extrinsic physical or chemical stimuli".   Reviews by Haukioja (1991) and Karban (1992) followed that detailed many specific examples of induced resistance. Kloepper et al. (1992) defined the corollary for induced plant resistance to disease as "the process of active resistance dependent on the host plant's physical or chemical barriers, activated by biotic or abiotic (inducing) agents."   Wound-induced responses in plants to mechanical damage may also be part of a general defensive reaction, since some allelochemicals are similar to those developed during pathogen infection (Edwards and Wratten 1983, Rhoades 1979).

The duration of induction of arthropod resistance varies greatly between different plant-arthropod combinations, and is affected by the same general variables described and discussed in Chapter 7 for constitutive resistance.   Baldwin (1989) proposed that plant induction be studied on three different time scales.   These include pre-formed induced plant responses, rapidly induced plant responses, and delayed induced plant responses.   Pre-formed induced responses occur immediately after damage and are restricted to damaged tissues.   Rapidly induced responses occur within hours or days of plant injury and may be localized or systemic.   The hypersensitive necrotic reactions of several types of crop plants discussed in Chapter 3 are examples of rapidly induced responses. Rapidly induced responses can occur in as little as 1 to 2 hrs and may remain in effect for a few days to several days (Table 9.2). Delayed-induced responses occur in the following season's foliage, and delayed-induced responses in some tree species may remain in effect for as long as 3 years (Table 9.2).

Arthropod-induced resistance occurs across a very wide range of plants, including cereal, fiber, fruit, legume, oil seed and vegetable crops (Table 9.2). Feeding of the twospotted spider mite, *Tetranychus urticae* Koch, on soybean plants induces resistance that effectively limits further increase of mite populations (Brown et al. 1991, Hildebrand et al. 1986).

Wheeler and Slansky (1991) compared the effects of constitutive soybean resistance to an occasional soybean pest, the fall armyworm, *Spodoptera frugiperda* (J. E. Smith) to a major pest, the velvetbean caterpillar, *Anticarsia gemmatalis* Hubner, and the resistance induced to each by feeding of either *T. urticae* or *A. gemmatalis*.   Both Lepidopterous pests are much more adversely affected by constitutive resistance than by either type of induced resistance.   Underwood et al. (2000) evaluated the responses of soybean cultivars displaying differential constitutive and induced resistance to the Mexican bean beetle, *Epilachna varivestis* Mulsant, and also found no relationship between induced and constitutive resistance. As noted by Wheeler and Slansky (1991), constitutive resistance and induced resistance are physiologically distinct processes in soybean.   Similar results have been observed in cotton, *Gossypium hirsutum* L. (Brody and Karban 1992), grape, *Vitus* spp. (English-Loeb et al. 1998), and lettuce (Huang et al. 2003).

Differences in the effects of mechanical damage and arthropod damage to plants are not consistent across plant species.   Anderson and Alborn (1999) demonstrated that plants of cotton fed on by larvae of the armyworm, *Spodoptera littoralis* Boisd., were less preferred by moths for oviposition than control plants. Oviposition was not induced in plants receiving mechanical damage. However, Wackers and Wunderlin (1999) found that in plants of *Gossypium herbaceum* (L.), both *S. littoralis* feeding and mechanical damage induce secretion of cotton extrafloral nectary glands as a general plant defense response. Similar results have been observed with *T. urticae*-induced resistance in cotton promoted by the mite and by mechanical damage (Harrison and Karban 1986, Karban and Carey 1984, Karban 1985).

Mechanical abrasion of soybean foliage also raises the level of soybean resistance to the soybean looper, *Pseudoplusia includens* (Walker), in foliage of a *P. includens*-resistant cultivar (Smith 1985). However, Lin et al. (1990) noted that mechanical injury alone to soybean foliage elicited less induced response to *E. varivestis* than that induced by *P. includens* feeding. Srinivas et al. (2001a) reported similar results from experiments with soybean foliage damaged by the bean leaf beetle, *Cerotoma trifurcata* (Forster).

Comparatively little is known about plant cross-resistance to one arthropod species induced as a result of feeding damage by a different arthropod species. When Willamette mites, *Eotetranychus willamettei* (McGregor), are released on young shoots of grape, they induce a systemic grape resistance to Pacific spider mites, *Tetranychus pacificus* McGregor (Hougen-Eitzman and Karban 1995). Srinivas et al. (2001b), found that *P. includens* larval feeding on soybean foliage induces a higher level of soybean cross-resistance to *C. trifurcata* than *C. trifurcata* feeding alone and that conversely, *C. trifurcata* feeding induces soybean cross resistance to defoliation by *P. includens*. Feeding by the corn earworm, *Helicoverpa zea* (Boddie), on foliage of tomato, *Lycopersicon esculentum* Mill., induces resistance to feeding by larvae of the beet armyworm, *Spodoptera exigua* Hübner. The strength of the induced resistance depends on the induction site and the post-induction period before assay (Stout and Duffey 1996).

Many species of trees also become resistant to arthropod damage following induction by arthropod defoliation or mechanical damage (Table 9.2). Among the deciduous species, an early report demonstrated that damage to the outer bark of apple, *Malus domestica* Borkh., trees induced resistance to *T. urticae* (Ferree and Hall 1981).

Long-term field experiments in Finland have demonstrated how foliage of European white birch, *Betula pubescens* Ehrh., infested by the geometrid moth, *Epirrita autumnata* (Bkh.), displays a delayed induced resistance to moth larvae and pupae in following years (Kaitaniemi et al. 1999, Ruohomäki et al. 1992). Laboratory studies have previously demonstrated how foliage of paper birch trees, *Betula papyrifera* Marsh., previously defoliated by larval feeding of the spear-marked black moth, *Rheumaptera hastata* (L.), is more resistant to larval feeding damage (Werner 1979). Similar results were obtained by Wratten et al. (1984), who demonstrated that foliage of European white birch and silver birch, *Betula pendula* Roth, fed on by *S. littoralis* larvae is more resistant to *S. littoralis* larval feeding than uninfested foliage.

Table 9.2. Wound-induced resistance in plants to arthropods

Plant genus	I/M[a]	Time (days) E[b]	D[c]	Arthropod affected	References
Acer	M	3	?	Lepidoptera	Baldwin & Schultz 1983
Alnus	I	27	44	M. californica pluvidae	Rhoades 1983
Beta	I	24	42	P. betae	Rottger & Klinghauf 1976
Betula	I	360	720	L. dispar	Wallner & Walton 1979
	M	2	720	O. autumnata	Werner 1979
	I	360	920	R. hastate	Wratten et al. 1984
	I	4hr	60	S. littoralis	
Glycine	M	1	3	P. includens	Reynolds & Smith 1985
	I	7	14	E. varivestis	Lin et al. 1990
		1	3	S. festinus	Felton et al. 1994a,b
	I	1	14	C. trifurcata	Srinivas et al. 2001a
	I	14	30	T. urticae	Brown et al. 1991
Gossypium	I	1	4	H. zea	Bi et al. 1997
	I, M	2	14	T. urticae	Harrison & Karban 1986
	I, M	?	?	S. littoralis	Wackers & Wunderlin 1999
Lycoperiscon	M	2	7	S. littoralis	Edwards & Wratten 1983
Malus	M	35	?	T. urticae	Ferree & Hall 1981
	I	2	4	M. persicae	Sauge et al. 2002
	I	2 hr	2	M. persicae	Kfoury & Massonie 1995
Picea	I	1-2	30	P. strobe	Alfaro 1995
Pinus	M	7	200	D. frontalis	Nebeker & Hodges 1983
	I	?	360	N. sertifer	Thielges 1968
	I	360	360	P. flammea	Leather et al.1987
Populus	M	1	?	Lepidoptera	Baldwin & Schultz 1983
Quercus	M	60	360	L. dispar	Schultz & Baldwin 1982
	I	40	120	Phyllonorycter	West 1985
Salix	I, M	7	90	P. versicolora	Raupp & Denno 1984
	I	35	?	H. cunea	Rhoades 1983
	I	11	15	M. pluviale	
	I	180	360	M. lapponica	Zvereva et al. 1997

[a] I - arthopod induced; M - mechanically induced, [b] Elicitation, [c] Duration

Induced resistance factors in different clones of *Populus* spp. fed on by larvae of the gypsy moth, *Lymantria dispar* L., have pronounced antifeedant effects on larval feeding (Havill and Raffa 1999). However, clonal variation negated consistent trends in overall induced resistance to *L. dispar*. The induced responses also had no relationship to levels of clonal constitutive resistance.

Foliage of red oak, *Quercus rubra* L., and black oak, *Quercus velutina* Lam., with a history of defoliation to *L. dispar* larvae are more resistant to later infestations than undamaged trees (Schultz and Baldwin 1982, Wallner and Walton 1979). Feeding by the Lepidopteran leaf miner, *Phyllonorycter harrisella* (L.) also induces resistance in *Quercus robur* (West 1985).

Damage to the leaves of several species of willow (*Salix*), including *Salix alba* L., *Salix babylonica* L., *Salix borealis* (Fries.), and *Salix sitchensis* Sanson, has been shown to impart resistance to larvae of Coleoptera and Lepidoptera. These include the willow leaf beetle, *Plagiodera versicolora* (Raupp and Denno 1984), the fall webworm, *Hyphantria cunea* (Drury) (Rhoades 1983), and the leaf beetle, *Melasoma lapponica* L. (Zvereva et al. 1997).

Delayed induced resistance also occurs in coniferous trees. Nebeker and Hodges (1983) mechanically scarred the trunks of loblolly pine, *Pinus taeda* L., trees to demonstrate how resistance to the southern pine beetle, *Dendroctonus frontalis* Zimmerman, increases for several months. More recently, induced defense of white spruce, *Picea glauca* (Moench), to the white pine weevil, *Pissodes strobe* (Peck), has been shown to involve a direct tree response to *P. strobe* attack (Alfaro 1995). Damaged trees exhibit a change in cambium production from normal cells to traumatic resin canals that shunt toxins into weevil galleries and kill weevil immatures. In contrast, several other studies have failed to demonstrate an induced response effect in Scots pine trees, *Pinus sylvestris* L., to feeding by the pine shoot beetles, *Tomicus piniperda* (L.), and *Tomicus minor* (Hart.) (Langstrom and Hellqvist 1993, Lieutier et al. 1995), or in Norway spruce trees, *Picea abies* Karst., to the European spruce bark beetle, *Dendroctonus micans* Kug. (Lieutier et al. 1992).

2.1 Allelochemical Bases of Induced Arthropod Resistance

Plants respond to arthropod damage by ultimately producing many of the allelochemicals described and discussed in Chapters 2 and 3. The level of expression of induced resistance is also affected by numerous factors described in Chapter 7. At the site of damaged tissues in many plants, wounding causes the synthesis of defensive phenols, as well as the oxidation of pre-formed endogenous phenols by polyphenol oxidases, leading to the production of quinones that reduce the value of dietary proteins in the arthropod digestive system.

Different types of arthropod-induced feeding damage differentially induce the production of putative plant defensive compounds. In tomato plants, these include polyphenol oxidases, peroxidases, lipoxygenases and proteinase inhibitors (Stout et al. 1999). The defensive allelochemistry of tomato plants attacked by the carmine spider mite, *Tetranychus cinnabarinus* Boisd., has been extensively studied by Kielkiewicz (2002). Mite feeding on susceptible tomato leaves induces the accumulation of phenolic deposits and increases the concentrations of chlorogenic acid, rutin, polyphenol oxidase, total peroxidase and ascorbate oxidase in damaged leaves.

Similar results have been observed in experiments with soybean plants exposed to feeding by *C. trifurcata*, *H. zea*, and the three cornered alfalfa hopper, *Spissistilus festinus* (Say), (Felton et al. 1994a,b, Bi and Felton 1995). Plants fed on by these arthropods exhibit increased levels of lipoxygenase, perioxidase, ascorbate oxidase, and polyphenol oxidase. *Helicoverpa zea* larvae placed on the induced plants exhibit reduced growth rates. Results of Bi et al. (1997) indicate that *H. zea* feeding damage increases concentrations of phenolics as well as peroxidase, oxidase, and lipoxygenase activity levels in *H. zea*-damaged cotton plants.

Hypersensitive responses in leaves of woody nightshade, *Solanum dulcamara* L., cultivars resistant to the gall mite, *Aceria cladophthirus* (Nalepa) (Westphal et al. 1981) following feeding by *A. cladophthirus* and the rust mite, *Thamnacus solani* Boczek and Michalska, lead to polyphenolic accumulations and greatly increased leaf peroxidase activity (Bronner et al. 1991), resulting in reduced *A. cladophthirus* survival (Westphal et al. 1991). Interestingly, induced resistance to *A. cladophthirus* does not protect woody nightshade plants against attack by *T. urticae* (Westphal et al. 1992). Resistance in raspberry, *Rubus phoenicolasius* Maxim., to the raspberry cane midge, *Resseliella theobaldi* (Barnes), is due to a hypersensitive reaction to midge feeding, that causes the formation of a wound periderm consisting of suberised and lignified cells (McNicol et al. 1983). However, in this interaction, there is no evidence of phenol production in damaged tissues.

Several studies have also demonstrated the involvement of phenols in arthropod damaged tissues of cereal crops. Amudhan et al. (1999) demonstrated an increase in total phenol concentration in the stems of at least some rice cultivars resistant to the Asian rice gall midge, *Orseolia oryzae* (Wood-Mason), after midge infestation. Feeding by the grain aphid, *Sitobion avenae* (F.), on aphid resistant wheat cultivars causes the production of increased levels of phenylalanine lyase (PAL) and tyrosine ammonia lyase (TAL); key enzymes involved in phenol synthesis (Ciepiela 1989). Aphids fed on plants treated with caffeic, ferulic and sinapic acids exhibit greatly reduced ingestion of phloem sap and salivation (Leszczynski et al. 1995). Susceptible and resistant wheat, barley and oat cultivars exhibit different patterns of peroxide activation in response to feeding by *S. graminum*, *D. noxia*, and the bird-cherry oat aphid, *Rhopalosiphum padi* (L.) (Argandona et al. 2001, Forslund et al. 2000, Ni et al. 2001). Interestingly, polyphenol oxidases in aphid saliva react with several cereal plant phenols, which may allow aphids to modify the phenol composition of susceptible plants (Urbanska et al. 1998). Resistance to *R. padi* in wheat is associated with phenol content (Leszczynski et al. 1985) and has both constitutive and induced components. *R. padi* feeding on resistant cultivars induces significantly greater amounts of several cell wall-bound phenolic acids, including salicylic, syringic, sinapic, and vanillic acid (Havlickova et al. 1996, 1998).

Defense responses in wheat, barley and oats induced by *D. noxia*, *R. padi*, or *S. graminum* include increased amounts of total protein, *PR* (pathogenesis-related) proteins, intercellular chitinases, β-1,3-glucosinases, peroxidases, and lipoxygenases (Argandona et al. 2001, Botha et al. 1998, Chaman et al. 2001, Fidantsef et al. 1999, Moran and Thompson 2001, Porter and Webster 2000, Rafi et al. 1996, van der Westhuizen and Botha 1993, van der Westhuizen and Pretorius 1995, 1996, van der Westhuizen et al. 1998a,b). Peroxidase, a key enzyme in plant cell wall building, has several key functions. These include mediating the oxidation of hydrocinnamyl alcohols into free radical intermediates, phenol oxidation, cross-linking of polysaccharides, and lignification (Chittoor et al. 1997, Ni et al. 2001).

Peroxidases are also involved in direct defenses against arthropods as well as the production of reactive oxygen species (see hydrogen peroxide in 9.2.2 below). Oxidative stress may cause direct oxidative injury to both arthropods and disease-causing organisms (Bowels 1990, Boyko and Smith 2004, Mehdy 1994, Orozco-Cardenas and Ryan 1999). The ß-1,3-glucanases release oligosaccharides from the plant cell wall that also trigger defense reactions in plants, as well as hydrolyzing callose tissue and permitting the continuous flow of nutrients in the phloem (Dixon and Lamb 1990). Chitinase has been proposed to hydrolyze chitin in the arthropod gut (Broadway et al. 1998).

Plants also produce proteinase inhibitors that retard arthropod growth (Stotz et al. 1999). Proteinase inhibitors accumulate in barley leaves following infestation by *R. padi* and decrease *R. padi* survival (Casaretto and Corcuera 1998). General and specific elicitors may also be synthesized by an arthropod or may be products of arthropod endosymbiotic bacteria (Boyko and Smith 2004, Walling 2000).

Increased concentrations of phenolic compounds also occur in the foliage of arthropod-resistant cultivars or clones of trees in association with induced hypersensitive responses to arthropod feeding. Several examples exist involving arthropod-resistant *Acer, Betula, Picea, Pinus* and *Populus* (Baldwin and Schultz 1983, Niemela et al. 1979, Rohfritsch 1981, Tjia and Houston 1975, Wratten et al. 1984).   Phenol concentrations increase in the foliage of lodgepole pine, *Pinus contorta* Douglas. ex Loud., trees attacked by three different arthropods.   The European pine sawfly, *Neodiprion sertifer* (Geoff.), the pine beauty moth, *Panolis flammea*, and the wasp, *Sirex noctilio* (F.) each induce production of phenols that accumulate to higher concentrations in pine trees than those in non-attacked trees (Thielges 1968, Shain and Hillis 1972, Leather et al. 1987).   Concentrations of both flavonoids and hydroxycinnamic acids increase in foliage of Scots pine attacked by *T. piniperda* (Lieutier et al. 1991). Total phenolic concentrations in foliage of *Betula* increase in trees under attack by *E. autumnata* (Kaitaniemi et al. 1998).   Levels of polyphenol oxidase increase in foliage of hybrid poplar, *(Populus trichocarpa* x *P. deltoides)*, following feeding damage by the forest tent caterpillar, *Malacosoma disstria* Hubner (Constabel et al. 2000).   A specific unique phenolic acid, 3-0-*trans-p*-coumaroyltormentic acid, accumulates in leaves of the pear tree, *Pyrus communis* L., attacked by the pear psylla, *Psylla pyricola* Foerster (Scutareanu et al. 1999).

In cruciferous plants, mechanical wounding and feeding damage by the flea beetle, *Phyllotreta cruciferae* (Goeze), reduces subsequent *P. cruciferae* damage (Bodnaryk 1992, Palaniswamy and Lamb 1993). However, resistance is not correlated to increased concentrations of 3-indomethyl glucosinolate (a known crucifer defensive compound) in seedlings of oilseed rape, *Brassica para* L. (Bodnaryk 1992, Koritsas et al. 1991). Agrawal (1998, 1999) induced increased arthropod resistance in wild radish plants by exposing leaves to defoliation by larvae of the cabbage butterfly, *Pieris rapae* (L.).   In contrast to flea beetle feeding, *P. rapae* larval feeding results in increased overall glucosinolate concentrations and decreased herbivory by Coleopteran, Homopteran, and Orthopteran arthropod pests.

Terpenoid plant defensive allelochemicals are also produced in response to arthropod herbivory. *S. exigua* larval feeding is reduced on leaves of cotton plants previously sustaining *S. exigua* defoliation (McAuslane et al. 1997). The degree of reduction increases with the duration of damage. A corresponding increase in the number of terpene aldehyde-containing pigment glands occurs on leaves of damaged cotton plants, compared to undamaged leaves. At 7 days post-induction, greater amounts of the constitutive defensive compounds gossypol and hemigossypolone (Chapter 3) are produced by damaged leaves than by undamaged leaves (McAuslane et al. 1997). McCall et al. (1994) observed a similar trend in the production of cotton leaf volatiles after *H. zea* larval feeding. Volatile compounds in older damaged leaves are unique to *H. zea*-resistant cotton cultivars.

Experiments comparing cotton isolines with and without glands producing gossypol and other terpene aldehydes (see Chapter 3) have shown that *S. exigua* larval feeding is greatly reduced on glanded damaged plants compared to glanded undamaged plants, and only moderately reduced on glandless damaged tissue versus glandless undamaged tissue (McAuslane and Alborn 1998). Reductions are presumably due to significantly increased concentrations of mono-and sesquiterpenes in glanded plants.

The total concentration of volatile monoterpenes in both lodgepole pine and Scots pine also increases following herbivory by key pest arthropods (Watt et al. 1991, Raffa and Smelley 1995). A similar trend is evident in potato plants subjected to damage by the Colorado potato beetle, *Leptinotarsa decimlineata* Say, and the potato leafhopper, *Empoasca fabae* Harris. Plants fed on by *L. decimlineata* produce greater amounts of glycoalkaloids but those fed on by *E. fabae* do not (Hlywka et al. 1994) (Fig. 9.3).

2.2 Elicitors of Arthropod-Induced Resistance

Plant reactions to both arthropod and disease attack may include hypersensitive cell death, as in the case of *Mi-1.2*, activation of defense response (DR) genes, and the redirection of normal cell maintenance genes toward plant defense. In DR gene activation, plants produce elicitors that activate defense gene expression as well as the synthesis of volatile and non-volatile allelochemicals. The similarities of the elicitors produced by plants in response to attacks by different arthropods may be the result of common arthropod salivary enzymes, although several elicitors regulate very species-specific arthropod responses (see below, Walling 2000, van de Ven et al. 2000). In addition, chewing arthropods cause extensive plant tissue damage that elicits different plant responses than those induced in response to feeding by piercing/sucking arthropods. The latter cause comparatively less tissue damage (Fidantsef et al. 1999, Stout et al. 1999, Walling 2000). As discussed in Section 9.2, plant responses to mechanical damage alone also differ from those involved in response to arthropod damage (Botha et al. 1998, Forslund et al. 2000, Halitschke et al. 2001, Winz and Baldwin 2001).

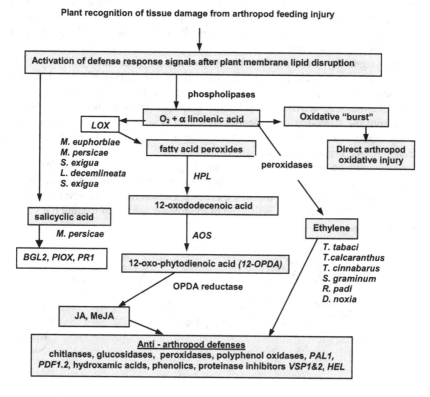

Figure 9.3. A general model of known induced plant resistance elicitors, genes and gene products produced in plant defensive responses to arthropod attack. Substrates shown in shaded grey boxes. Capitalized italics indicate defense response gene(s) involved in arthropod resistance, arthropod names indicate species inducing expression of associated enzyme or allelochemical. (see Section 2.2 for detailed explanation)

Plant tissues that die during the arthropod feeding process stimulate the degradation of linolenic acid, which triggers the production of a "burst" of reactive oxygen species (ROS) of molecules (Figure 9.3). The involvement of ROS in disease resistance is well known, and these compounds may also have direct adverse affects on arthropod midgut tissues. Linolenic acid degradation also stimulates several different signal pathways to produce the defensive proteinase inhibitors, phenolics, and enzymes described in Section 9.2. In plant pathogen resistance, active defense responses also often involve a hypersensitive reaction pattern of localized cell death in plant tissues at the site of infection. Localized cell death blocks pathogen growth through the effects of ROS. In at least one instance, a hypersensitive reaction between an avirulent arthropod biotype and a resistant host plant has been identified. Shukle et al. (1992) demonstrated that larval feeding of the virulent Hessian fly, *Mayetiola destructor* biotype L (see Chapter 11) causes a general increase in cuticular-memebrane permeability in the lower leaf sheath cells of susceptible wheat plants. On resistant plants however, feeding by avirulent *M. destructor* biotype GP larvae causes only localized responses and the production of low molecular weight proteins.

Plant signaling pathways driven by jasmonates (jasmonic acid and methyl jasmonate), salicylic acid and ethylene control the production of plant defenses to arthropod attack. In tobacco and tomato plants, both arthropod damage and mechanical damage activate plant defense response genes mediated by a systemic signal from the specialized polypeptide systemin (Figure 9.4) (Bergey et al. 1996, Ryan and Pearce 2003). Systemin activates the release of linolenic acid and eventually the production of a jasmonic acid signal.

Many induced responses in plants resulting from arthropod attack involve the jasmonate pathway (Kessler and Baldwin 2001, Turner et al. 2002). Application of jasmonic acid to young tomato plants induces increased concentrations of several defensive compounds and increases the resistance of the treated foliage to feeding by *S. exigua* (Thaler et al. 1996).

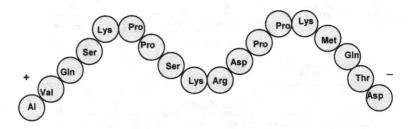

Figure 9.4. The amino acid sequence of the polypeptide signal systemin. From Bergey, D. R., G. A. Howe, and C. A. Ryan. 1996. Polypeptide signaling for plant defensive genes exhibits analogies to defense signaling in animals. Proc. Natl. Acad. Sci. USA. 93:12053-58, Copyright 1996, National Academy of Sciences, U.S.A.

Methyl jasmonic acid-treated wheat plants produce increased amounts of the defensive hydroxamic acids (Chapter 3) and reduce phloem ingestion by *R. padi* relative to that on control plants (Slesak et al. 2001). In barley and maize, methyl jasmonic acid-induced accumulations of ferulic acid and phenolic polymers lead to cell wall strengthening and increased arthropod resistance (Bergvinson et al. 1994, Lee et al. 1997).

Jasmonic acid-regulated pathways likely protect *S. graminum*-susceptible sorghum plants against *S. graminum* feeding (Zhu-Salzman et al. 2004). However, as with *D. noxia* feeding on wheat (Boyko and Smith 2004), certain transcripts exclusively activated by *S. graminum* feeding on sorghum are unique and independent of pathways regulated by jasmonic acid and salicylic acid. For example, both the hydroxymethyltransferase and cytochrome P450 monooxygenase genes are differentially expressed in aphid-infested sorghum and wheat plants.

In field experiments, jasmonic acid applications to tomato plants decreased the population of a pest complex consisting of Coleoptera, Homptera, Lepidoptera, and

Thysanoptera (Thaler 1999, Thaler et al. 2001). However, the utility of jasmonic acid as an integrated pest management tactic remains to be determined.

Lipoxygenases function in cell membrane lipid degradation and also contribute to the production of plant defense response signaling molecules such as jasmonic acid. Transcripts encoding the lipoxygenase (*LOX*) genes are strongly induced by feeding of the *M. euphorbiae* on tomato (Fidantsef et al. 1999) and *M. persicae* feeding on *Arabidopsis* (Moran and Thompson 2001). *LOX H3* genes in potato regulate resistance to *S. exigua* and *L. decemlineata*, but the resistance is not regulated by jasmonic acid (Rojo et al. 1999). The related enzyme hydroperoxide lyase plays a role in the resistance of potato to *M. persicae* (Vancanneyt et al. 2001). Resistant cotton plants damaged by *S. exigua* produce more lipoxygenase and hydroperoxide lyase than susceptible plants (Loughrin et al. 1995). In some plants, wounding also induces increased ethylene production that blocks the jasmonic acid signal and allene oxide synthese production. Nevertheless, increased allene oxide synthese levels induced by *M. sexta* feeding may promote jasmonic acid production (Ziegler et al. 2001).

Salicylic acid promotes the development of systemic acquired resistance, a broad-range resistance against pathogens and some aphid species. Release of methyl salicylate, a strong aphid repellent (Hardie et al. 1994), is triggered by *S. graminum* feeding in sorghum (Zhu-Salzman et al. 2004). Experiments conducted by Moran and Thompson (2001) demonstrate that *M. persicae* feeding on *Arabidopsis* induces major increases in the expression of the *PR1* (pathogenesis resistance) and *BGL* (β-glucosidase) genes, both of which are associated with the salicylic acid defense signaling pathway.

Relatively little research has been conducted on the involvement of ethylene in the induced defense response of plants to arthropods but there is solid evidence that ethylene emissions increase as a result of arthropod herbivory. Feeding by onion thrips, *Thrips tabaci* Lindeman, on foliage of onion, *Allium cepa* L., and feeding by the introduced basswood thrips, *Thrips calcaratus* Uzel, on the foliage of basswood, *Tilia americana* L., cause significant increases in foliar ethylene production (Kendall and Bjostad 1990, Rieske and Raffa 1995). Kielkiewicz (2002) demonstrated that feeding by *T. cinnabarinus* on tomato leaves also causes significant increases in ethylene. In two different studies, ethylene production has shown been shown to increase significantly more in aphid-resistant barley cultivars than in susceptible cultivars. Argandona et al. (2001) observed this reaction in barley fed on by both *S. graminum* and *R. padi*. Miller et al. (1994) noted the same interaction in barley fed on by *D. noxia*.

Jasmonic acid and ethylene often act synergistically, inducing distinct plant defense responses (Alonso et al. 1999, Bostock 1999, Pieterse and van Loon 1999, Stotz et al. 2000, Walling 2000). For example, both jasmonic acid and ethylene induce a metallopeptidase-like protein (*SLW1*) in squash, *Cucurbita moschata* Duchesne, for resistance to feeding by the silverleaf whitefly, *Bemisia argentifolii* Bellows and Perring. An additional gene (*SLW3*) is regulated by an unknown elicitor (van de Ven et al. 2000). Jasmonic acid /abscissic acid signal synergism has also been reported in plant defensive responses to disease infection (Chao et al. 1999). More importantly, plant defense responses to arthropod and disease induced by jasmonic acid and ethylene may be antagonized by those induced by salicylic acid (Reymond and Farmer 1998, Stotz et al. 2002, Thaler et al. 1999). Nevertheless, plant signals elicited by both jasmonic acid and salicylic acid are perceived by *H. zea* larvae and activate specific *H. zea* P450 monooxygenase genes used to detoxify allelochemicals ultimately produced by infested plants (Li et al. 2002).

Genes encoding specific defense compounds have been identified in the transcriptomes of arthropod-challenged plants of *Arabidopsis, Cucurbita*, and *Nicotiana.* These include *SLW1*, the vegetative storage proteins *VSP1* and *VSP2*, *BGL1*, the defensin peptide *PDF1.2*, the pathogen-inducible α-dioxygenase gene *PIOX*, phenylalanine ammonia lyase (*PAL1,)* a Thr deaminase (*TD*) gene involved in synthesis of defensive allelochemicals, and a monosaccharide symporter (*STP4)* (Bell et al. 1995, Hermsmeier et al. 2001, Moran and Thompson 2001, Stotz et al. 2002, van de Ven et al. 2000).

Whitman and Eller (1990) conducted some of the first research on the production of green leaf volatiles (see Chapter 2) produced in plants damaged by pest arthropods and the attraction of pest natural enemies to these plants. In their experiments, females of the braconid wasp, *Microplitis croceipes* (Cresson), and the ichneumonid wasp, *Netelia heroica* Townes, were more strongly attracted to plants of cowpea, *Vigna unguiculata* (L.) Walp., damaged by *H. zea* larval feeding than undamaged plants. Female *Cotesia rubecula* Marshall, are also attracted to plants of Brussels sprouts, *Brassica oleracea* var. *gemmifera* Zenker., damaged by larval feeding of *P. rapae* (Geervliet et al. 1994).

In a series of similar experiments, Turlings et al. (1991a,b) demonstrated a similar phenomenon in maize plants damaged by *S. exigua* larval feeding and the attraction of female *Cotesia marginiventris* (Cresson) parasitic wasps to damaged plants. An eleven compound blend of volatiles occurs in significant amounts in maize leaves fed on by *S. exigua* larvae (Turlings et al. 1993). In *S. exigua*-damaged cotton leaves, volatiles are released very rapidly via *de novo* syntheses, instead of converting previously synthesized precursors from a storage site (Pare and Tumlinson 1997).

Dicke and Dijkman (1993) also observed the attraction of the predatory mite, *Phytoseiulus persimilis* Athias-Henriot to volatiles produced in foliage of lima bean, *Phaseolus lunatus* L., in response to feeding by *T. urticae*. When jasmonic acid is applied exogenously to lima bean petioles, a volatile blend occurs that is similar, but not identical, to that produced by *T. urticae*-infested plants (Dicke et al. 1999). Although predatory mites are attracted to volatiles from jasmonic acid-treated plants, they are more strongly attracted to volatiles from *T. urticae*-infested plants. Similar responses occur in natural enemies of the tea aphid, *Toxoptera aurantii* (Boyer de Fonscolombe), to leaf volatiles of tea foliage. Aphid natural enemies are much more strongly attracted to volatiles from aphid-infested leaves than to mechanically-damaged leaves (Han and Chen 2002).

Volatiles in frass produced by arthropods feeding on plants can also attract natural enemies, and these differences can be correlated to feeding on resistant and susceptible plants. The braconid parasite *Microplitis demolitor* (Wilkinson) is attracted to frass produced by *P. includens* larvae fed soybean and lima bean foliage (Ramachandran et al. 1991). The volatile guaiacol, one of the three major components in frass, is unique to *P. includens* and highly attractive to *M. demolitor*. The quantity of guaiacol present in frass from larvae fed the resistant soybean genotype PI227687 is 5 times greater than the guaiacol content of frass collected from larvae fed foliage of a susceptible soybean cultivar. The increased production of guaiacol after feeding on PI227687 foliage may explain the additive effects of PI227687-based soybean resistance to *P. includens* and parasitism by *M. demolitor* observed by Yanes and Boethel (1983) (see Chapter 12).

Feeding damage from the pest arthropods described above is often insufficient to produce maximal plant defense responses. Herbivore-specific elicitors that induce plant defense responses, and at the same time attract pest natural enemies, have been isolated from oral secretions of lepidopterous larvae. β-glucosidase, a lytic enzyme elicitor contained in larval salivary secretions of the imported cabbageworm, *Pieris brassicae* L., attracts natural enemies (Mattiacci et al. 1995). The fatty acid conjugate (FAC) N-(17-hydroxylinolenyl)-L-glutamine (volicitin) is an elictor isolated from the larval regurgitant of *S. exigua* that induces maize plants to emit the volatile compounds described above (Alborn et al. 1997, Frey et al. 2000, Turlings et al. 1993). Volicitin is biosynthetically related to jasmonic acid (Boland et al. 1998). The application of FACs from tobacco hornworm, *Manduca sexta* (L.), larvae to *Arabidopsis* leaves wounded during larval feeding also elicits the systemic release of volatiles, a burst of jasmonic acid production and the accumulation of plant defense related compounds (Halitschke et al. 2000, 2001).

Schmeltz et al. (2003) demonstrated that *S. exigua* feeding stimulates maize jasmonic acid production, jasmonic acid regulation of maize defense allelochemicals and the associated attraction of *S. exigua* natural enemies. *Arabidopsis* plants treated with jasmonic acid also attract significantly more *C. rubecula* than plants treated with salicylic acid (van Poecke et al. 2002). Intra-plant communication involving synergism between plant green leaf volatiles and jasmonic acid signals have also been demonstrated in maize by Engelberth et al. (2004). Maize plants previously exposed to green leaf volatiles from neighboring damaged plants produce significantly more jasmonic acid and volatile sesquiterpenes when induced with *S. exigua* regurgitant than plants not exposed to green leaf volatiles. Additional discussions of the interactions between different plant defense elicitors are available in the excellent reviews of Heil and Bostock (2002) and Kessler and Baldwin (2001).

2.3. Identification of Unique Expressed Genes in Arthropod-Induced Resistant Plants

In damaged plant tissues, messenger RNA (mRNA) sequences are translated to proteins. Unique mRNA gene transcripts expressed in resistant plants can be isolated and RNA molecules copied back into their complementary(c) DNA using reverse transcriptases. cDNA populations from infested and uninfested plants can be hybridized and sequences common to both populations are removed. The un-hybridized sequences unique to the resistant plant become a "subtracted" library of resistance-related cDNAs that is sequenced to begin to determine the putative functions of the subtracted cDNAs (Diatchenko et al. 1996).

Ren et al. (2004) used suppressive subtractive hybridization (SSH) to identify rice genes differentially regulated by *N. lugens* feeding in *N. lugens*-resistant plants. The identified sequences were mapped to known chromosome locations using RFLPs and *in silico* searches of the rice genome database. The expressed gene sequences show some clustering, and some are similar to some of the *N. lugens* resistance genes described in Chapter 8. Boyko and Smith (2004) used the SSH approach to develop baseline information about genes differentially expressed in aphid-resistant wheat infested by *D. noxia,* and identified more than 40 groups of sequences with various plant resistance and plant defense response functions. Hermsmeier et al. (2001) used a related technique, mRNA differential display (Ling and Pardee 1992), to identify unique mRNA transcripts in coyote tobacco, *Nicotiana attenuata* Torrey ex S. Watson, produced after feeding by *M. sexta* larvae. These findings suggest that the identification of differentially induced ESTs (expressed sequence tags) may be a very beneficial way to identify candidate plant defense response genes, which, in turn, may be used to create arthropod resistant plant cultivars.

Unique cDNAs may also be used to probe microarrays of oligonucleotides of known function to determine resistance gene function based on mRNA expression levels. Reymond et al. (2000) constructed a small-scale microarray to identify genes implicated in *Arabidopsis* defense against feeding by *P. rapae* larvae. Arimura et al. (2000) probed a larger array of more than 2,000 known genes in lima bean foliage with expressed cDNAs from leaves infested with *T. urticae* or exposed to volatiles produced by *T. urticae*-infested leaves to detect defense-related genes activated by mite feeding. Expressed genes related to herbivory include among others, those in metabolic pathways related to ethylene and flavonoid biosynthesis, chaperones, secondary wound signals and membrane transporters.

Commercial oligonucleotide microarrays contain several thousand expressed sequences and allow rapid screening of plant resistance-related cDNAs. Arrays exist for *Arabidopsis* (~24,000 transcripts), *Glycine* (~37,500 *Glycine max* transcripts, 15,800 *Phytophthora sojae* transcripts, 7,500 *Heterodera glycines* transcripts), barley (~25,000 transcripts), tomato (a partial array of ~20,000 transcripts), and wheat (~60,000 transcripts). Although these arrays contain few specific genes derived from arthropod induction, their genome wide representations of plant genes will provide essential information about those genes involved in defense responses to arthropod attacks.

3. CONCLUSIONS

Molecular genetic and genomic technologies have opened completely new avenues of research in plant resistance to arthropods during the past 10 years. Knowledge about the location(s) of existing and new arthropod resistance genes in several major crop plants species is rapidly evolving. The sequencing of the genomes of crop (rice) and non-crop (*Arabidopsis*) plants has begun to provide the first real insights into plant arthropod resistance gene structure, function and location. As additional plant genomes are sequenced, existing and new information about resistance gene synteny will be used to make foresighted decisions in crop plant breeding. The development of future arthropod-resistant crop cultivars should rely on knowledge about the sequences of resistance genes from different resistance sources. In this way, cultivars with resistance genes of diverse sequence and function can be released and deployed to sustain resistance and help delay the development of avirulent, resistance-breaking arthropod biotypes (see Chapter 11).

The existence of RGAs in many crop plants suggests that current and future plant resistance researchers should increasingly utilize these genetic resources to provide *in silico* information about the location and function of candidate resistance genes. As a more complete knowledge of crop plant genomes develops, genomic microarrays will provide valuable information about the identity of resistance genes and the gene products mediating their function.

Elicitor-induced responses do play a role in induced plant resistance to arthropods, as discussed and described in this chapter. However, elicitor-induced responses such as those involving jasmonic acid may also lower plant fitness and reduce seed yields, suggesting that plants bred for induced arthropod resistance responses may be counterproductive to efficient crop production (Baldwin 1998, Baldwin et al. 1997). There are many gaps in the level and extent of knowledge about elicitor-induced plant resistance to arthropods. Additional research at both the molecular and plant level is critical to better understand how different species of plants integrate elicitor signals generated as part of defense responses against both arthropods and diseases. The wide variety of specific gene products in both resistant and susceptible plants attacked by arthropods indicates that there are few general plant elicitors of arthropod resistance across the plant kingdom. This is not surprising, given the tremendous variation in the differing degrees of arthropod-host specificity (polyphagy versus monophagy) and the differing degrees of this specialization that have occurred between different arthropod orders.

Our knowledge of how plants recognize the signals generated by arthropod feeding and the elicitors these signals produce is rapidly increasing. Flexible, evolving models of these processes will be necessary, in order to ensure a better understanding of the metabolism of plants induced to defend themselves against arthropod attack. This understanding will also be necessary, in order to use induced crop plant resistance in arthropod pest management programs.

REFERENCES CITED

Agrawal, A. A. 1998. Induced responses to herbivory and increased plant performance. Science. 279:1201–1202.

Agrawal, A. A. 1999. Induced seasonal responses to herbivory in wild radish: Effects on several herbivores and plant fitness. Ecology. 80:1713–1723.

Ahn, S. N., J. A. Anderson, M. E. Sorrells, and S. D. Tanksley. 1993. Homoeologous relationships of rice, wheat and maize chromosomes. Mol. Gen. Genet. 241:483–490.

Alam, S. N., and M. B. Cohen. 1998. Detection and analysis of QTLs for resistance to the brown planthopper, *Nilaparvata lugens*, in a doubled-haploid rice population. Theor. Appl. Genet. 97:1370–1379.

Alborn, T., T. C. J. Turlings, T. H. Jones, G. Stenhagen, J. H. Loughrin, and J. H. Tumlinson. 1997. An elicitor of plant volatiles from beet armyworm oral secretion. Science. 276:945–949.

Alfaro, R. I. 1995. An induced defense reaction in white spruce to attack by the white pine weevil, *Pissodes strobi*. Can. J. Forest Res. 25:1725–1730.

Alonso, J.M., T. Hirayama, G. Roman, S. Nourizadeh, and J. R. Ecker. 1999. EIN2, a bifunctional transducer of ethylene and stress responses in *Arabidopsis*. Science. 284:2148–2152.

Amudhan, S., U. P. Rao, and J. S. Bentur. 1999. Total phenol profile in some rice varieties in relation to infestation by Asian rice gall midge *Orseolia oryzae* (Wood-Mason). Current Sci. 76:1577–1580.

Anderson, P., and H. Alborn. 1999. Effects on oviposition behaviour and larval development of *Spodoptera littoralis* by herbivore-induced changes in cotton plants. Entomol. Exp. Appl. 92:45–51.

Argandona, V.H., M. Chaman, L. Cardemil, O. Munoz, G. E. Zuniga, and L. J. Corcuera. 2001. Ethylene production and peroxidase activity in aphid- infested barley. J. Chem. Ecol. 27: 53–68.

Arimura, G.-I., K. Tashiro, S. Kuhara, T. Nishioka, R. Ozawa, and J. Takabayashi. 2000. Gene responses in bean leaves induced by herbivory and herbivore-induced volatiles. Biochem. Biophys. Res. Commun. 277:305–310.

Baldwin, I. T. 1989. The mechanism of damage-induced alkaloids in wild tobacco. J. Chem Ecol. 15:1661–1669.

Baldwin, I. T. 1998. Jasmonate-induced responses are costly but benefit plants under attack in native populations. Proc. Natl. Acad. Sci. USA. 95:8113–8118.

Baldwin, I. T., Z.-P. Zhang, N. Diab, T. E. Ohnmeiss, E. S. McCloud, G. Y. Lynds, and E. A. Schmelz. 1997. Quantification, correlations and manipulation of would-induced changes in jasmonic acid and nicotine in *Nicotiana sylvestris*. Planta. 210:397–404.

Baldwin, I. T., and J. C. Schultz. 1983. Rapid changes in tree leaf chemistry induced by damage: Evidence for communication between plants. Science. 221:277–279.

Bell, E., R. A. Creelman, and J. E. Mullet. 1995. A chloroplast lipoxygenase is required for wound-induced jasmonic acid accumulation in *Arabidopsis*. Proc. Natl. Acad. Sci. USA. 92:8675–8679.

Bennetzen, J. L., and M. Freeling. 1993. Grasses as a single genetic system: Genome composition, collinearity, and compatibility. Trends Genet. 9:259–261.

Bergvinson, D. J., J. T. Arnason, and L.N. Pietrzak. 1994. Localization and quantification of cell wall phenolics in European corn borer resistant and susceptible maize inbreds. Can. J. Bot. 72:243–1249.

Bergey, D. R., G. A. Howe, and C. A. Ryan. 1996. Polypeptide signaling for plant defensive genes exhibits analogies to defense signaling in animals. Proc. Natl. Acad. Sci. USA. 93:12053–12058.

Bi, J. L., and G. W. Felton. 1995. Foliar oxidative stress and insect herbivory: Primary compounds, secondary metabolites, and reactive oxygen species as components of induced resistance. J. Chem. Ecol. 21:1511–1530.

Bi, J. L., J. B. Murphy, and G. W. Felton. 1997. Antinutritive and oxidative components as mechanisms of induced resistance in cotton to *Helicoverpa zea*. J. Chem. Ecol. 23:97–117.

Bodnaryk, R. P. 1992. Effects of wounding on glucosinolates in the cotyledons of oilseed rape and mustard. Phytochem. 31:2671–2677.

Boland, W., J. Hopke, and J. Piel. 1998. Biosynthesis of jasomates. In: P. Schreier, M. Herderich, H.-U. Humpf and W. Schwab (Eds.), Natural Product Analysis; Chromotography, Spectroscopy, Biological Testing. Friedr. Vieweg, Braunschweig/Wiesbaden. pp. 255–269.

Bostock, R.M. 1999. Signal conflicts and synergies in induced resistance to multiple attackers. Physiol. Mol. Plant Path. 55:99–109.

Botha, A.M., M.A.C. Nagel, A. J. Van der Westhuizen, and F. C. Botha. 1998. Chitinase isoenzymes in near-isogenic wheat lines challenged with Russian wheat aphid, exogenous ethylene and mechanical wounding. Bot. Bull. Acad. Sin. 39:99–106.

Bowels, D. J. 1990. Defense-related proteins in higher plants. Ann. Rev. Biochem. 58:837–907.

Boyko, E. V., and C. M. Smith 2004. Expression of *Pto* and *Pti*-like genes is involved in wheat resistance response to aphid attack. In: Plant & Animal Genome XII. Final Abstracts Guide. Workshop Abstracts. January 10–14, 2004, San Diego, CA, W200.

Boyko, E. V., K. S. Gill, L. Mickelson-Young, S. Nasuda, W. J. Raupp, J. N. Ziegler, S. Singh, D. S. Hassawi, A. K. Fritz, D. Namuth, N. L. V. Lapitan, and B. S. Gill. 1999. A high-density genetic linkage map of *Aegilops tauschii*, the DS-genome progenitor of bread wheat. Theor. Appl. Genet. 99:16–26.

Boyko, E. V., R. Kalendar, V. Korzun, A. Korol, A., Schulman, and B. S. Gill. 2002. A high density genetic map of *Aegilops tauschii* includes genes, retro-transposons, and microsatellites which provide unique insight into cereal chromosome structure and function. Plant Mol. Biol. 48:767–790.

Boyko, E., S. Starkey, and C. M. Smith. 2004. Molecular genetic mapping of *Gby*, a new greenbug resistance gene in bread wheat. Theor. Appl. Genet. 109:1230–1236.

Broadway, R. M., C. Gongore, W. C. Kain, J. P. Sanderson, J. A. Monroy, K. C. Bennett, J. B. Warner, and M. P. Hoffman. 1998. Novel chitinolytic enzymes with biological activity against insects. J. Chem. Ecol. 24:985–998.

Brody, A. K. and R. Karban. 1992. Lack of a tradeoff between constitutive and induced defenses among varieties of cotton. Oikos. 65:301–306.

Bronner, R., E. Westphal, and F. Dreger. 1991. Enhanced peroxidase activity associated with the hypersensitive response of *Solanum dulcamara* to the gall mite *Aceri cladophthirus* (Acari, Eriophyoidea). Can. J. Bot. 69:2192–2196.

Brotman, Y., L. Silberstein, I. Kovalski, C. Perin, C. Dogimont, M. Pitrat, J. Klingler, G. A. Thompson, and R. Perl-Treves. 2002. Resistance gene homologues in melon are linked to genetic loci conferring disease and pest resistance. Theor. Appl. Genet. 104: 1055–1063.

Brown, G. C., F. Nurdin, J. G. Rodriguez, and D. F. Hildebrand. 1991. Inducible resistance of soybean (var 'Williams') to two spotted spider mite (*Tetranychus urticae* Koch). J. Kanas Entomol. Soc. 64:388–393.

Bryngelsson, T., J. Sommer-Knudsen, P.L. Gregersen, D. B. Collinge, B. Ek, and H. Thordal-Christensen. 1994. Purification, characterization, and molecular cloning of basic PR-1-type pathogenesis-related proteins from barley. Molecular Plant-Microbe Interactions 7:267–275.

Casaretto, J. A., and L. J. Corcuera. 1998. Proteinase inhibitor accumulation in aphid-infested barley leaves. Phytochem. 49:2279–2286.

Castro, A. M., S. Ramos, A. Vasicek, A. Worland, D. Gimenez, A. A. Clua, and E. Suarez. 2001. Identification of wheat chromosomes involved with different types of resistance to greenbug (*Schizaphis graminum* Rond.) and the Russian wheat aphid (*Diuraphis noxia* Mordvilko). Euphytica. 118:131–137.

Chaman, M. E., L. J. Corcuera, G. E. Zuniga, L. Cardemil, and V. H. Argandona. 2001. Induction of soluble and cell wall peroxidases by aphid infestation in barley. J. Agric. Food Chem. 49:2249–53.

Chao, W. S., Y-Q., Gu, V. Pautot, E. A. Bray, and L. L. Walling. 1999. Leucine aminopeptidase RNAs, proteins, and activities increase in response to water deficit, salinity, and the wound signals systemin, methyl jasmonate, and abscisic acid. Plant Physiol. 129:979–992.

Chittoor, J. M., J. E. Leach and F. F. White. 1997. Differential induction of a peroxidase gene family during infection of rice by *Xanthomonus oryzae pv. oryzae*. Mol. Plant-Microbe Interact. 10:861–871.

Ciepiela, A. 1989. Biochemical basis of winter wheat resistance to the grain aphid, *Sitobion avenae*. Entomol. Exp. Appl. 51:269–275.

Constabel, C. P., L. Yip, J. J. Patton, and M. E. Chistopher. 2000. Polyphenol oxidase from hybrid poplar. Cloning and expression in response to wounding and herbivory. Plant Physiol. 124:285–295.

Devos, K. M., M. Atkinson, C. N. Chinoy, C. Liu, and M. D. Gale. 1992. RFLP-based genetic map of the homoeologons group 3 chromosomes of wheat and rye. Theor. Appl. Genet. 83:931–939.

Diatchenko, L., Y.-F. C. Lau, A. P. Campbell, A. Chenchik, F. Moqadam, B. Huang, S. Lukyanov, K. Lukyanov, N. Gurskaya, E. D. Sverdlov, and P. D. Siebert. 1996. Suppression subtractive hybridization: a method for generating differentially regulated or tissue-specific cDNA probes and libraries. Proc. Natl. Acad. Sci. USA. 93:6025–6030.

Dicke, M., and H. Dijkman. 1993. Herbivory induces systemic production of plant volatiles that attract predators of the herbivore: Extraction of endogenous elicitor. J. Chem. Ecol. 19:581–599.

Dicke, M., R. Gols, D. Ludeking, and M. A. Posthumus. 1999. Jasmonic acid and herbivory differentially induce carnivore-attracting plant volatiles in lima bean plants J. Chem. Ecol. 25:1907–1922

Dixon, R.A., and C. L. Lamb. 1990. Molecular communication in interactions between plants and microbial pathogens. Ann. Rev. Plant Physiol. Plant. Mol. Biol. 41:339–367.

Dogimont, C., A. Bendahmane, J. Pauquet, E. Burget, S. Desloire, L. Hagen, M. Caboche, and M. Pitrat. 2003. Map-based cloning of the *Vat* melon gene that confers resistance to both aphid colonization and virus transmission. Proc. 11th International Congress on Molecular Plant-Microbe Interactions, July 18–26, 2003, St. Petersburg, Russia.

Dubcovsky, J., A.J. Lukaszewski, M. Echaide, E.F. Antonelli, and D.R. Porter, 1998. Molecular characterization of two *Triticum speltoides* interstitial translocations carrying leaf rust and greenbug resistance genes. Crop Sci. 38:1655–1660.

Edwards, P. J., and S. D. Wratten. 1983. Wound induced defences in plants and their consequences for patterns of insect grazing. Oecologia. 59:88–93.

Ellis, J., P. N. Dodds, and T. Pryor. 2000, Structure, function and evolution of plant disease resistance genes. Curr. Opinion Plant Biol. 3:278–284.

Engelberth, J., H. T. Alborn, E. A. Schmelz, and J. H. Tumlinson. 2004. Airborne signals prime plants against insect herbivore attack. Proc. Natl. Acad. Sci. 101:1781–1785.

English-Loeb, G., R. Karban, and M. A. Walker. 1998. Genotypic variation in constitutive and induced resistance in grapes against spider mite (Acari: Tetranychidae) herbivores. Environ. Entomol. 27:297–304.

Felton, G. W., J. L. Bi, C. B. Summers, A. J. Mueller, and S. S Duffey. 1994a. Potential role of lipoxygenases in defense against insect herbivory. J. Chem. Ecol. 20:651–666.

Felton, G. W., C. B. Summers, and A. J. Mueller. 1994b. Oxidative responses in soybean foliage to herbivory by bean leaf beetle and three-cornered alfalfa hopper. J. Chem. Ecol. 20:639–650.

Ferree, D. C., and F. R. Hall. 1981. Influence of physical stress on photosynthesis and transpiration of apple leaves. J. Amer. Soc. Hort. Sci. 106:348–351.

Feuillet, C., and B. Keller. 1999. High genome density is conserved at syntenic loci of small and large grass genomes. Proc. Natl. Acad. Sci. USA. 96: 8265–8270.

Fidantsef, A.L., M. J. Stout, J. S. Thaler, S. S. Duffey, and R. M. Bostock. 1999. Signal interactions in pathogen and insect attack: expression of lipoxygenase, proteinase inhibitor II, and pathogenesis-related protein P4 in the tomato, *Lycopersicon esculentum*. Physiol. Mol. Plant. Pathol. 54: 97–114.

Forslund, K., J. Perrersson, T. Bryngelsson, and L. Jonsson. 2000. Aphid infestation induces PR-proteins differentially in barley susceptible or resistant to the bird cherry-oat aphid. Physiol. Plant. 110:496–502.

Frey, M., C. Stettner, P. W. Pare, E. A. Schmelz, J. H. Tumlinson, and A. Gierl. 2000. An herbivore elicitor activates the gene for indole emission in maize. Proc. Natl. Acad. Sci. 97:14801–14806.

Geervliet, J. B. F., L. E. M. Vet, and M. Dicke. 1994. Volatiles from damaged plants as major cues in long-range host-searching by the specialist parasitoid *Cotesia rubecula*. Entomol. Exp. Appl. 73:289–297.

Graham, M. A., L. F. Marek, D. Lohnes, P. Cregan, and R. Shoemaker. 2000. Expression and genome organization of resistance gene analogs in soybean. Genome. 43:86–90.

Halitschke, R., A. Kessler, J. Kahl, A. Lorenz, and I. T. Baldwin. 2000. Ecophysiological comparison of direct and indirect defenses in *Nicotiana attenuata*. Oecologia. 124: 408–417.

Haltischke, R., U. Schittko, G. Pohnert, W. Boland, and I. T. Baldwin. 2001. Molecular interactions between the specialist herbivore *Manduca sexta* (Lepidoptera, Sphingidae) and its natural host *Nicotiana attenuata*. III. Fatty amino-acid conjugates in herbivore oral secretions are necessary and sufficient for herbivore-specific plant responses. Plant Physiol. 125:711–717.

Han, B. Y., and Z. M. Chen. 2002. Composition of the volatiles from intact and mechanically pierced tea aphid-tea shoot complexes and their attraction to natural enemies of the tea aphid. J. Chem. Ecol. 50:2571–2575.

Hardie, J. R., R. Issacs, J. A. Pickett, L. J. Wadhams, and C. M. Woodcock. 1994. Methyl salicylate an (-) - (1R,5S) - myrtenal are plant-derived repellents from the black bean aphid, *Aphis fabae* Scop. (Homoptera : Aphididae). J. Chem. Ecol. 20:2847–2855.

Harrison, S., and R. Karban. 1986. Behavioral response of spider mites (*Tetranychus urticae*) to induced resistance of cotton plants. Ecol. Entomol. 11:181–188.

Haukioja, E. 1991. Induction of defenses in trees. Ann. Rev. Entomol. 36:25–42.

Havill, N. P., and K. F. Raffa. 1999. Effects of elicitation treatment and genotypic variation on induced resistance in *Populus*: impacts on gypsy moth (Lepidoptera : Lymantriidae) development and feeding behavior. Oecologia. 120:295–303.

Havlickova, H., M. Cvikrova, and J. Eder. 1996. Phenolic acids in wheat cultivars in relation to plant suitability for and response to cereal aphids. Z. Pflanzenkr. Pflanzensch. 103:535–542.

Havlickova, H., M. Cvikrova, J. Eder, and M. Hrubcova. 1998. Alterations on the levels of phenolics and peroxidases activities induced by *Rhopalosiphum padi* (L.) in two winter wheat cultivars. Z. Pflanzenkr. Pflanzensch. 105:140–148.

Heil, M., and R. M. Bostock. 2002. Induced systemic resistance (ISR) against pathogens in the context of induced plant defenses. Ann. Bot. 89:503–512.

Hermsmeier, D., U. Schittko, and I. T. Baldwin. 2001. Molecular interactions between the specialist herbivore *Manduca sexta* (Lepidoptera, Sphingidae) and its natural host *Nicotiana attenuata*. I. Large-scale changes in the accumulation of growth-and defense-related plant mRNAs. Plant Physiol. 125:683–700.

Hildebrand, D. F, J. G. Rodriquez, G. C. Brown, K. T. Lu, and C. S. Volden. 1986. Peroxidative responses of leaves in two soybean genotypes injured by twospotted spider mites (Acari: Tetranychidae). J. Econ. Entomol. 79:1459–1465.

Hlywka, J. J., G. R. Stephenson, M. K. Sears, and R. Y. Yada. 1994. Effects of insect damage on glycoalkoloid content in potatoes (*Solanum tuberosum*). J. Agric. Food Chem. 42:2545–2550.

Hohmann, U., A. Graner, T. R. Endo, B. S. Gill, and R. G. Herrmann. 1995. Comparison of wheat physical maps with barley linkage maps for group 7 chromosomes. Theor. Appl. Genet. 91:618–626.

Hougen-Eitzman, D., and R. Karban. 1995. Mechanisms of interspecific competition that results in successful control of Pacific mites following inoculations of Willamette mites on grapevines. Oecologia. 103:157–161.

Huang, J., H. J. McAuslane, and G. S. Nuessly. 2003. Resistance in lettuce to *Diabrotica balteata* (Coleoptera: Chrysomelidae): The roles of latex and inducible defense. Environ. Entomol. 32:9–16.

Hulbert, S. H., T. E. Richter, J. D. Axtell, and J. L. Bennetzen. 1990. Genetic mapping and characterization of sorghum and related crops by means of maize DNA probes. Proc. Natl. Acad. Sci. USA. 87:4251–4255.

Hwang, C. F., A. V. Bhakta, G. M. Truesdell, W. M. Pudlo, and V. M. Williamson. 2000. Evidence for a role of the N terminus and leucine-rich repeat region of the *Mi* gene product in regulation of localized cell death. Plant Cell. 12:1319–1329.

Kaitaniemi, P., K. Ruohomäki, V. Ossipov, E. Haukioja, and K. Pihlaja. 1998. Delayed induced changes in the biochemical composition of host plant leaves during an insect outbreak. Oecologia. 116:182–190.

Kaitaniemi, P., K. Ruohomäki, T. Tammaru, and E. Haukioja. 1999. Induced resistance of host tree foliage during and after a natural insect outbreak. J. Animal Ecol. 68:382–389.

Kaloshian, I., M. G. Kinser, D. E. Ullman, and V. M. Willamson. 1997. The impact of *Meu1*-mediated resistance in tomato on longevity, fecundity, and behavior of the potato aphid, *Macrosiphum euphorbiae.* Entomol. Exp. Appl. 83:181–187.

Kaloshian, I., W. H. Lange, and V. M. Willamson. 1995. An aphid-resistance locus is tightly linked to the nematode-resistance gene, *Mi,* in tomato. Proc. Natl. Acad. Sci. USA. 92:622–625.

Karban, R. 1985. Resistance against spider mites in cotton induced by mechanical abrasion. Entomol. Exp. Appl. 37:137–141.

Karban, R. 1992. Inducible resistance in agricultural systems. In: D. W. Tallamy and M. J. Raupp (Eds.), Phytochemical Induction by Herbivores, Wiley, New York. pp. 403–419.

Karban, R., and J. R. Carey. 1984. Induced resistance of cotton seedlings to mites. Science. 225:53–54.

Katsar, C. S., A. H. Paterson, G. L. Teetes, and G. C. Peterson. 2002. Molecular analysis of sorghum resistance to the greenbug (Homoptera: Aphididae). J. Econ. Entomol. 95:448–457.

Kendall, D. M., and L. B. Bjostad. 1990. Herbivory by *Thrips tabaci* induces greater ethylene production in intact onions than mechanical damage alone. J. Chem. Ecol. 16:981–991.

Kessler, A., and I. T. Baldwin. 2001. Defensive function of herbivore-induced plant volatile emissions in nature. Science. 291:2141–2144.

Kfoury, L., and G. Masonie. 1995. Characteristics of the resistance of the peach cultivar Rubira to *Myzus persicae* Sulzer. Agronomie. 15:277–284.

Kielkiewicz, M. 2002. Influence of carmine spider mite *Tetranychus cinnabarinus* Boisd. (Acarida: Tetranychidae) feeding on ethylene production and the activity of oxidative enzymes in damaged tomato plants. In: F. Bernini, R. Nannelli, G. Nuzzaci, and E. de Lillo (Eds.), Acarid Phylogeny and Evolution. Adaptations in mites and ticks. Kluwer Academic Publishers, pp. 389–392.

Klingler, J., G. Powell, G. A. Thompson, and R. Isaacs. 1998. Phloem specific aphid resistance in *Cucumis melo* line AR 5: effects on feeding behaviour and performance of *Aphis gossypii.* Entomol. Exp. Appl. 86:79–88.

Kloepper, J. W., S. Tuzun, and J. A. Kuc. 1992. Proposed definitions related to induced disease resistance. Biocontrol Sci. Technol. 2:349–351.

Kogan, M., and J. Paxton. 1983. Natural inducers of plant resistance to insects. In: P. A. Hedin (Ed.), Plant Resistance to Insects. Am. Chem. Soc. Symp. Series 208, American Chemical Society, Washington, DC. pp. 153–171.

Koritsas, V. M., J. A. Lewis, and G. R. Fenwick. 1991. Glucosinloate responses of oilseed rape, mustard and kale to mechanical wounding and infestation by cabbage stem flea beetle (*Psylliodes chrysocephala*). Ann. Appl. Biol. 118:209–221.

Lagudah, E. S., O. Moullet, and R. Appels. 1997. Map-based cloning of a gene sequence encoding a nucleotide binding domain and a leucine-rich region at the *Cre3* nematode resistance locus of wheat. Genome. 40:659–665.

Langstrom, B., and C. Hellqvist. 1993. Induced and spontaneous attacks by pine shoot beetles on young Scots pine trees: tree mortality and beetle performance. J. Appl. Entomol. 115:25–36.

Leather, S. R., A. D. Watt, and G. I. Forrest. 1987. Insect-induced chemical changes in young lodgepole poine (*Pinus contorta*): the effect of previous defoliation on oviposition, growth and survival of the pine beauty moth, *Panolis flammea*. Ecol. Entomol. 12:275–281.

Lee, J. E., T. Vogt, B. Hause, and M. Lebler. 1997. Methyl jasmonate induces an *O*-methyltransferase in barley. Plant Cell Physiol. 38:851–862.

Leister, D., J. Kurth, D. A. Laurie, M. Yano, T. Sasaki, A. Graner, and P. Schulze-Lefert. 1999. Rflp- and physical mapping of resistance gene homologues in rice (*O. sativa*) and barley (*H. vulgare*). Theor. Appl. Genet. 98: 509–520.

Leszczynski, B. 1985. Changes in phenols content and metabolism in leaves of susceptible and resistant winter wheat cultivars infested by *Rhopalosiphum padi* (L.) (Hom., Aphididae). Z. Angew. Entomol. 100: 343–348.

Leszczynski, B., W. F. Tjallingii, A. F. G. Dixon, and R. Swiderski. 1995. Effect of methoxyphenols on grain aphid feeding behaviour. Entomol. Exp. Appl. 76:157–162.

Li, X., M. A. Schuler, and M. R. Berenbaum. 2002. Jasmonate and salicylate induce expression of herbivore cytochrome P450 genes. Nature. 419:712–715.

Lieutier, F., J. Garcia, A. Yart, and P. Romary. 1995. Wound reactions of Scots pine (*Pinus sylvestris* L.) to attacks by *Tomicus piniperda* L. and *Ips sexdentatus* Boern (Gal, Scolytidae). J. Appl. Entomol. 119:591–600.

Lieutier, F., G. Vouland, M. Pettinetti, J. Garcia, P. Romary, and A. Yart. 1992. Defence reactions of Norway spruce (*Picea abies* Karst) to artificial insertion of *Dendroctonus micans* Kug (Col. Scolytidae). J. Appl. Entomol. 114:174–186.

Lieutier, F., A. Yart, C. Jay-Allemand, and L. Delmorme. 1991. Preliminary investigations on phenolics as a response of Scots pine phloem to attack by bark beetles and associated fungi. European J. For. Pathol. 21:354–364.

Lin, H. M. Kogan, and D. Fischer. 1990. Induced resistance in soybean to the Mexican bean beetle (Coleoptera: Coccinellidae): Comparisons of inducing factors. Environ. Entomol. 19:1852–1857.

Linde-Laursen, I., J. S. Heslop-Harrison, K. W. Shepherd, and S. Taketa. 1997. The barley genome and its relationship with the wheat genomes. A survey with an internationally agreed recommendation for barley chromosome nomenclature. Hereditas. 126:1–16.

Liang, P. and A. B. Pardee. 1992. Differential display of eukaryotic mRNA by means of the polymerase chain reaction. Science. 257:967–971.

Liu, X. M., C. M. Smith, and B. S. Gill. 2002. Identification of microsatellite markers linked to Russian wheat aphid resistance genes *Dn4* and *Dn6*. Theor. Appl. Genet. 104:1042–1048.

Liu, X.M., C. M. Smith, B. S. Gill, and V. Tolmay. 2001. Microsatellite markers linked to six Russian wheat aphid resistance genes in wheat. Theor. Appl. Genet. 102:504–510.

Loughrin, J. H., A. Manukian, R. A. Heath, and J. H. Tumlinson. 1995. Volatiles emitted by different cotton varieties damaged by feeding beet armyworm larvae. J. Chem. Ecol. 21:1217–1227.

Mago, R., S. Nair, and M. Mohan. 1999. Resistance gene analogues from rice: cloning, sequencing and mapping. Theor. Appl. Genet. 99:50–57.

Martin, G. B., A. J. Bogdanove, and G. Sessa. 2003. Understanding the functions of plant disease resistance proteins. Ann. Rev. Plant Biol. 54: 23–61. ·

Mattiacci, L., M. Dicke, and M. A. Posthumas. 1995. Beta-glucosidase – an elicitor of herbivore-induced plant odor that attracts host-searching parasitic wasps. Proc. Natl. Acad. Sci. USA. 92: 2036–2040.

McAuslane, H. J., and H. T. Alborn. 1998. Systemic induction of allelochemicals in glanded and glandless isogenic cotton by *Spodoptera exigua* feeding. J. Chem. Ecol. 24:399–416.

McAuslane, H. J., H. T. Alborn, and J. P. Toth. 1977. Systemic induction of terpenoid aldehydes in cotton pigment glands by feeding of larval *Spodoptera exigua*. J. Chem. Ecol. 23:2861–2879.

McCall, P. J., T. C. J. Turlings, J. Loughrin, A. D. Proveaux, and J. H. Tumlinson. 1994. Herbivore-induced volatile emission from cotton (*Gossypium hirsutum* L.) seedlings. J. Chem. Ecol. 20:3039–3050.

McNicol, R. J., B. Williamson, P. L. Jennings, and J. A. T. Woodford. 1983. Resistance to raspberry cane midge (*Resseliella theobaldi*) and its association with wound periderm in *Rubus crataegifolius* and its red raspberry derivatives. Ann. Appl. Biol. 103:489–495.

Mehdy, M. C. 1994. Active oxygen species in plant defense against pathogens. Plant Physiol. 105:467–472.

Miller, H. L., P. A. Neese, D. L Ketring, and J. W. Dillwith. 1994. Involvement of ethylene in aphid infestation of barley. J. Plant Growth Reg. 13:167–171.

Miller, C. A., A. Altinkut, and N. L. V. Lapitan. 2001. A microsatellite marker for tagging *Dn2*, a wheat gene conferring resistance to the Russian wheat aphid. Crop Sci. 41:1584–1589.

Milligan, S., J. Bodeau, J. Yaghoobi, I. Kaloshian, P. Zabel, and V. Williamson. 1998. The root-knot nematode resistance gene *Mi* from tomato is a member of the leucine zipper, nucleotide binding, leucine-rich repeat family of plant genes. Plant Cell. 10:1307–1319.

Mingeot, D., and J. M. Jacquemin. 1997. A wheat cDNA coding for a thaumatin-like protein reveals a high level of RFLP in wheat. Theor. Appl. Genet. 95:822–827.

Moharramipour, S., H. Tsumki, K. Sato, and H. Yoshida. 1997. Mapping resistance to cereal aphids in barley. Theor. Appl. Genet. 94:592–596.

Moran, P. J., and G. A. Thompson. 2001. Molecular responses to aphid feeding in Arabidopsis in relation to plant defense pathways. Plant Physiol. 125:1074–1085.

Namuth, D. M., N.L.V. Lapitan, K.S. Gill, and B.S. Gill. 1994. Comparative mapping of *Hordeum vulgare* and *Triticum tauschii*. Theor. Appl. Genet. 89:865–872.

Nebeker, T. E. and J. D. Hodges. 1983. Influence of forestry practices on host-susceptibility to bark beetles. Z. Angew. Entomol. 96:194–208.

Ni, X., S. S. Quisenberry, T. Heng-Moss, J. Markwell, G. Sarath, R. Klucas, and F. Baxendale. 2001. Oxidative responses of resistant and susceptible cereal leaves to symptomatic and nonsymptomatic cereal aphid (Hemiptera: Aphididae) feeding. J. Econ. Entomol. 94:743–751.

Niemela, P., E. M. Aro, and E. Haukioja. 1979. Birch leaves as a resource for herbivores. Damaged-induced increase in leaf phenols with trypsin-inhibiting effects. Rep. Kevo Subarctic Res. Stat. 15: 37–40.

Nieto-Lopez, R. M., and T. K. Blake. 1994. Russian wheat aphid resistance in barley: Inheritance and linked molecular markers. Crop Sci. 34:655–659.

Orozco-Cardenas, M. and C. A. Ryan. 1999. Hydrogen peroxide is generated systemically in plant leaves by wounding and systemin via octadecanoid pathway. Proc. Natl. Acad. Sci. USA. 96:6553–6557.

Palaniswamy, P., and R. J. Lamb. 1993. Wound-induced antixenotic resistance to flea beetles, *Phyllotreta cruciferae* (Goeze) (Coleoptera: Chrysomelidae), in crucifers. Can. Entomol. 125:903–912.

Pan, Q., J. Wendel, and R. Fluhr. 2000. Divergent evolution of plant NBS-LRR resistance gene homologues in dicot and cereal genomes. J. Mol. Evol. 50:203–213.

Pare, P. M. and J. H. Tumlinson. 1997. Induced synthesis of plant volatiles. Nature. 385:30–31.

Paterson, A., Y.-R., Lin, , Z. Li, K. F. Scherta, J. F. Doebley, S. R. M. Pinson, S.-C. Liu, J. W. Stansel, and J. E. Irvine. 1995. Convergent domestication of cereal crops by independent mutations at corresponding genetic loci. Science. 269:1714–1718.

Pieterse, C. M., and L. C. van Loon. 1999. Salicylic acid-independent plant defense pathways. Trends Plant Sci. 4:52–58.

Pitrat, M., and H. Lecoq. 1980. Inheritance of resistance to cucumber mosaic virus transmission by *Aphis gossypii* in *Cucumis melo*. Phytopathol. 70:958–961.

Porter, D. R., and Webster, J. A. 2000. Russian wheat aphid-induced protein alterations in spring wheat. Euphytica. 111:199–203.

Rafi, M. M., R. S. Zemetra, and S. S. Quisenberry. 1996. Interaction between Russian wheat aphid (Homoptera:Aphidadae) and resistant and susceptible genotypes of wheat. J. Econ. Entomol. 89:239–246.

Raffa, K. F., and E. B. Smelley. 1995. Interaction of pre-attack and induced monoterpene oncentrations in host conifer defense against bark beetle-fungus complexes. Oecologia. 102:285–295.

Ramachandran, R., D. M. Norris, J. K. Phillips, and T. W. Phillips. 1991. Volatiles mediating plant-herbivore-natural enemy interactions: Soybean looper frass volatiles, 3-octanone and guaiacol, as kairomones for the parasitoid *Microplitis demolitor*. J. Agric. Food Chem. 39:2310–2317.

Raupp, M. J., and R. F. Denno. 1984. The suitability of damaged willow leaves as food for the leaf beetle, *Plagiodera versicolora*. Ecol. Entomol. 9: 443–448.

Ren, X., X. Wang, H. Yuan, Q. Weng, L. Zhu, and G. He. 2004. Mapping quantitative trait loci and expressed sequence tags related to brown planthopper resistance in rice. Plant Breeding. 123:342–348.

Reymond, P., and E. E. Farmer. 1998. Jasmonate and salicylate as global signals for defense gene expression. Curr. Opin. Plant Biol. 1:404–11.

Reymond, P., H. Weber, M. Damond, and E. E. Farmer. 2000. Differential gene expression in response to mechanical wounding and insect feeding in *Arabidopsis*. Plant Cell. 12:707–719.

Reynolds, G. W., and C. M. Smith. 1985. Effects of leaf position, leaf wounding, and plant age of two soybean genotypes on soybean looper (Lepidoptera: Noctuidae) growth. Environ. Entomol. 14: 475–478.

Rhoades, D. F. 1979. Evolution of a plant chemical defense against herbivores. In: G. A. Rozenthal and D. Janzen (Eds.), Herbivores: Their Interaction with Secondary Plant Metabolites. Academic Press, New York. pp. 3–54.

Rhodes, D. F. 1983. Herbivore population dynamics and plant chemistry. In: R. F. Denno and M. S. McClure (Eds.), Variable Plants and Herbivores in Natural and Managed Systems. Academic Press, New York. pp. 155–220.

Rieske, L. K., and K. F. Raffa. 1995. Ethylene emission by a deciduous tree, *Tilia americana*, in response to feeding by introduced basswood thrips, *Thrips calcaranthus*. J. Chem Ecol. 21:187–197.

Roberts, P. A., and I. J. Thomason. 1986. Variability in reproduction of isolates of *Meloidogyne incognita* and *M. javanica* on resistant tomato genotypes. Plant Disease. 70:547–551.

Rohfritsch, O. 1981. A defense mechanism of *Picea excelsa* L. against the gall former *Chormes abietis* L. (Homoptera: Adelgidae). Z. Angew. Entomol. 92:18–26.

Rojo, E., J. Leon, and J. J. Sanchez-Serrano. 1999. Cross-talk between wound signaling pathways determines local versus systemic gene expression in *Arabidopsis thaliana*. Plant J. 20:135–142.

Rossi, M., F. L. Goggin, S. B. Milligan, I. Klaoshian, D. E. Ullman, and V. M. Williamson. 1998. The nematode resistance gene *Mi* of tomato confers resistance against the potato aphid. Proc. Natl. Acad. Sci. USA. 95:9750–9754.

Rottger, V. U., and F. Klinghauf. 1976. Anderung im stoffwechsel von zuckerruben blattern durch befall mit *Pegomya bettae* Curt (Muscidae: Anthomyidae). Z. Angew. Entomol. 82:220–227.

Ruohomäki, K. S. Hanhimäki, E. Haukioja, L. Iso-Iivari, S. Neuvonen, P. Niemelä, and J. Suomela. 1992. Variability in the efficiacy of delayed inducible resistance in mountain birch. Entomol. Exp. Appl. 62:107–115.

Ryan, C. A., and G. Pearce. 2003. Systemins: A functional defined family of peptide signals that regulate defensive genes in Solanaceae species. Proc. Natl. Acad. Sci. USA. 100:14577–14580.

Sauge, M.-H., J.-P. Lacroze, J. - L. Poessël T. Pascal, and J. Kervella. 2002. Induced resistance by *Myzus persicae* in the peach cultivar 'Rubira'. Entomol. Exp. Appl. 102:29–37.

Schmeltz, E. A. H. T. Alborn, E. Banchio, and J. H. Tumlinson. 2003. Quantitaive relationships between induced jasmonic acid levels and volatile emission in *Zea mays* during *Spodoptera exigua* herbivory. Planta. 216:665–673.

Schultz, J. C., and I. T. Baldwin. 1982. Oak leaf quality declines in response to defoliation by gypsy moth larvae. Science. 221: 149–151.

Scutareanu, P., Y. L. Ma, M. Claeys, R. Dommisse, and M. W. Sabelis. 1999. Induction of a p-coumaroyl trihydroxy triterpene acid in *Psylla*-infested and mechanically damaged pear trees. J. Chem. Ecol. 25:2177–2191.

Seah, S., K. Sivasithamparam, A. Karalousis, and E. S. Lagudah. 1998. Cloning and characterization of a family of disease resistance gene analogs from wheat and barley. Theor. Appl. Genet. 97:937–945.

Shain, L., and W. E. Hillis. 1972. Ethylene production in *Pinus radiata* in response to *Sirex amylostereum* attack. Phytopathol. 62:1407–1409.

Shen, K. A., B. C. Meyers, M. N. Islam-Faridi, D. Chin, D. M. Stelly, and R. W. Michelmore. 1998. Resistance gene candidates identified by PCR with degenerate oligonucleotide primers map to clusters of resistance genes in lettuce. Mol. Plant-Microbe Interact. 11:815–823.

Shukle, R. H., P. B. Glover, Jr., and G. Mocelin. 1992. Responses of susceptible and resistant wheat associated with Hessian fly (Diptera : Cecidomyiidae) infestation. Environ. Entomol. 21:845–853.

Slesak, E., M. Slesak, and B. Gabrys. 2001. Effect of methyl jasmonate on hydroxamic acid, protease activity, and bird cherry-oat aphid *Rhoplaosiphum padi* L. probing behavior. J. Chem. Ecol. 12:2529–2543.

Smith, C. M. 1985. Expression, mechanisms, and chemistry of resistance in soybean, *Glycine max* L. (Merr.) to the soybean looper, *Pseudoplusia includens* (Walker). Insect Sci. Appl. 6:243–248.

Smith, C. M. 2004. Plant resistance against pests: Issues and strategies, In: O. Koul, G. S. Dhaliwal and G. Cuperus, (Eds.), Integrated Pest Management: Potential, Constraints And Challenges. CABI Publ., Oxon, UK, pp. 147–167.

Speulman, E., D. Bouchez, E. Holub, and J. L. Beynon. 1998. Disease resistance gene homologs correlate with disease resistance loci of *Arabidopsis thaliana*. Plant J. 14: 467–474.

Srinivas, P., S. D. Danielson, C. M. Smith, and J. D. Foster. 2001a. Induced resistance to bean leaf beetle (Coleoptera: Chrysomelidae) in soybean. J. Entomol. Sci. 36:438–444.

Srinivas, P., S. D. Danielson, C. M. Smith, and J. D. Foster. 2001b. Cross-resistance and resistance longevity as induced by bean leaf beetle, *Cerotoma trifurcata* andsoybean looper, *Pseudoplusia includens* herbivory on soybean. J. Insect Science. 1.5.

Stotz, H. U., T. Koch, A. Biedermann, K. Weniger, W. Boland, and T. Mitchell-Olds. 2002. Evidence for regulation of resistance in *Arabidodpsis* to Egyptian cotton worm by salicylic and jasmonic acid signaling pathways. Planta. 214:648–652.

Stotz, H. U., J. Kroymann, and T. Mitchell-Olds. 1999. Plant-insect interactions. Curr. Opin. Plant Biol. 2:268–272.

Stotz, H. U., B. R. Pittendrigh, J. Kroyman, K. Weniger, J. Fritsche, A. Bauke, and T. Mitchell-Olds. 2000. Induced plant defense responses against chewing insects. Ethylene signaling reduces resistance of *Arabidopsis* against Egyptian cotton worm but not diamondback moth. Plant Physiol. 124:1007–1017.

Stout, M. J, and S. S. Duffey. 1996. Characterization of induced resistance in tomato plants. Entomol. Exp. Appl. 1996 79:273–283.

Stout, M.J., A. L. Fidantsef, S. S. Duffey, and R. M. Bostock. 1999. Signal interactions in pathogen and insect attack: systemic plant-mediated interactions between pathogens and herbivores of the tomato, *Lycopsericon esculentum*. Physiol. Mol. Plant Pathol. 54:115–130.

Tada, T. 1999. PCR-amplified resistance gene analogs link to resistance loci in rice. Breed. Sci. 49:267–273.

Thaler, J. S. 1999. Induced resistance in agricultural crops: Effects of jasmonic acid on herbivory and yield in tomato plants. Environ. Entomol. 328:30–37.

Thaler, J. S., A. L. Fidantsef, S. S. Duffey, and R. M. Bostock. 1999. Trade-offs in plant defense against patogens and herbivores: A field demonstration of chemical elicitors of induced resistance. J. Chem. Ecol. 25:1597–1609.

Thaler, J. S., M. J. Stout, R. Karban, and S. S. Duffey. 1996. Exogenous jasmonates simulate insect wounding in tomato plants (*Lycopersicon esculentum*) in the laboratory and field. J. Chem. Ecol. 22:1767–1781.

Thaler, J. S., M. J. Stout, R. Karban, and S. S. Duffey. 2001. Jasmonate-mediated induced plant resistance affects a community of herbivores. Ecol. Entomol. 26:312–324.

Thielges, B. A. 1968. Altered polyphenol metabolism in the foliage of *Pinus sylvestris* associated with European pine sawfly attack. Can. J. Bot. 46: 724–725.

Tjia, B., and D. B. Houston. 1975. Phenolic constituents of Norway spruce resistant or susceptible to the eastern spruce gall aphid. For. Sci. 211: 180–184.

Turlings, T. C. J., P. J. McCall, H. T. Alborn, and J. H. Tumlinson. 1993. An elicitor in caterpillar oral secretions that induces corn seedlings to emit chemical signals attractive to parasitic wasps. J. Chem. Ecol. 19:411–425.

Turlings, T. C. J., J. H. Tumlinson, F. J. Eller, and W. J. Lewis. 1991a. Larval-damaged plants: source of volatile synomones that guide the parasitoid *Cotesia marginiventris* to the micro-habitat of its host. Entomol. Exp. Appl. 58:75–82.

Turlings, T. C. J., J. H. Tumlinson, R. R Heath, A. T. Proveaux, and R. E. Doolittle. 1991b. Isolation and identification of allelochemicals that attract the larval parasitoid, *Cotesia marginiventris* (Cresson), to the microhabitat of one of its hosts. J. Chem. Ecol. 17:2235–2251.

Turner, J. G., C. Ellis, and A. Devoto. 2002. The jasmonate signal pathway. Plant Cell. 14:153–164.

Underwood, N., W. Morris, K. Gross, and J. R. Lockwood. 2000. Induced resistance to Mexican bean beetles in soybean: variation among genotypes and lack of correlation with constitutive resistance. Oecologia. 122:83–89.

Urbanska, A., T. C. J., Turlings, W. F. Tjallingii, A. F. G. Dixon, and B. Leszczynski. 1998. Phenol oxidizing enzymes in the grain aphid's saliva. Entomol. Exp. Appl. 86:197–203.

van der Westhuizen, A. J., and F. C. Botha. 1993. Effect of the Russian wheat aphid on the composition and synthesis of water soluble proteins in resistant and susceptible wheat. J. Agron. Crop Sci. 170:322–326.

van der Westhuizen, A. J., and Z. Pretorius. 1995. Biochemical and physiological responses of resistant and susceptible wheat to Russian wheat aphid infestation. Cereal Res. Comm. 23:305–313.

van der Westhuizen, A. J., and Z. Pretorius. 1996. Protein composition of wheat apoplastic fluid and resistance to the Russian wheat aphid. Aust. J. Plant Physiol. 23:645–648.

van der Westhuizen, A.J., X.-M. Qian, and A.-M. Botha. 1998a. β-1,3-glucanases in wheat and resistance to the Russian wheat aphid. Physiol. Plant. 103:125–131.

van der Westhuizen, A.J., X.-M. Qian, and A.-M. Botha. 1998b. Differential induction of apoplastic peroxidase and chitinase activities in susceptible and resistant wheat cultivars by Russian wheat aphid infestation. Plant Cell Reports. 18:132–137.

van de Ven, W. T. G., C. S. LeVesque, T. M. Perring, and L. L. Walling. 2000. Local and systemic changes in squash gene expression in response to silverleaf whitefly feeding. Plant Cell. 12:1409–1424.

Van Deynze, A. E., J. C. Nelson, E. S. Yglesias, S. E. Harrington, D. P. Braga, S. R. McCouch, and M. E Sorrells. 1995a. Comparative mapping in grasses. Wheat relationships. Mol. Gen. Genet. 248:744–754.

Van Deynze, A. E., J. C. Nelson, L. S. O'Donoughue, S. N. Ahn, W. Siripoonwiwat, S. E. Harrington, E. S. Yeglesias., D. P Braga, S. R. McCouch, and M. E Sorrells. 1995b. Comparative mapping in grasses. Oat relationships. Mol. Gen. Genet. 249: 349–356.

Vancanneyt, G., C. Sanz, T. Farmaki, M. Paneque, F. Ortego, P. Castañera, and J. Sánchez-Serrano. 2001. Hydroperoxyde lyase depletion in transgenic potato plants leads to an increase in aphid performance. Proc. Natl. Acad. Sci. USA. 98:8139–8144.

van Poecke, R. M. P., and M. Dicke. 2002. Induced parasitoid attraction by *Arabidopsis thaliana*: involvement of the octadecanoid and the salicylic acid pathway. J. Exp. Bot. 53:1793–1799.

Vos, P., G. Simons, T. Jesse, J. Wijbrandi, L. Heinen, R. Hogers, A. Frijters, J. Groenendijk, P. Diergaarde, M. Reijans, J. Fierens-Onstenk, M. de Both, J. Peleman, T. Liharska, J. Hontelez, and M. Zabeau. 1998. The tomato *Mi-1* gene confers resistance to both root-knot nematodes and potato aphids. Nat. Biotechnol. 16:1315–1316.

Wackers, F. L., and R. Wunderlin. 1999. Induction of cotton extrafloral nectar production in response to herbivory does not require a herbivore-specific elicitor. Entomol. Exp. Appl. 91:149–154.

Walling, L. L. 2000. The myriad plant responses to herbivores. J. Plant Growth Regul. 19:195–216.

Wallner, W. E., and G. S. Walton. 1979. Host defoliation: A possible determination of gypsy moth population quality. Ann. Entomol. Soc. Am. 72:62–67.

Wang, Y. H., D. F. Garvin, and L.V. Kochian. 2001. Nitrate-induced genes in tomato roots. Array analysis reveals novel genes that may play a role in nitrogen nutrition. Plant Physiol. 127:345–359.

Watt, A. D., S. R. Leather, and G. I. Forrest. 1991. The effect of previous defoliation of pole-stage lodgepole pine on plant chemistry, and on the growth and survival of pine beauty moth (*Panolis flammea*) larvae. Oecologia. 86:31–35.

Weng, Y., and M. D. Lazar. 2002. Amplified fragment length polymorphism- and simple sequence repeat-based molecular tagging and mapping of greenbug resistance gene *Gb3* in wheat. Plant Breed. 121:218–223.

Werner, R. A. 1979. Influence of host foliage on development, survival, fecundity, and oviposition of the spear-marked black moth, *Rheumaptera hastata* (Lepidoptera: Geometridae). Can. Entomol. 111:317–322.

West, C. 1985. Factors underlying the late seasonal appearance of the lepidopterous leaf-mining guild on oak. Ecol. Entomol. 10: 111–120.

Westphal, E., R. Bronner, and M. LeRet. 1981. Changes in leaves of susceptible and resistant *Solanum dulcamara* infested by the gall mite *Eriophyes cladophthirus* (Acarina, Eriophyoidea). Can. J. Bot. 59:875–882.

Westphal, E., F. Dreger, and R. Bronner. 1991. Induced resistance in *Solanum dulcamara* triggered by the gall mite *Aceria cladophthirus* (Acari, Eriophyoidea*)*. Exp. Appl. Acarol. 12:111–118.

Westphal, E., M. J. Perrot-Minnot, S. Kreiter, and J. Gutierrez. 1992. Hypersensitive reaction of *Solanum dulcamara* to the gall mite *Aceria cladophthirus* causes an increased susceptibility to *Tetranychus urticae*. Exp. Appl. Acarol. 15:15–26.

Wheeler, G. S., and F. Slansky. 1991. Effect of constitutive and herbivore - induced extractables from susceptible and resistant soybean foliage on nonpest and pest noctuid caterpillars. J. Econ. Entomol. 84:1068–1079.

Whitman, D. W., and F. J. Eller. 1990. Parasitic wasps orient to green leaf volatiles. Chemoecology. 1:69–75.

Winz, R. A., and I. T. Baldwin. 2001. Molecular interactions between the specialist herbivore *Manduca sexta* (Lepidoptera, Sphingidae) and its natural host *Nicotiana attenuata*. IV. Insect-induced ethylene reduces jasmonate-induced nicotine accumulation by regulating *N*- methyltransferase transcripts. Plant Physiol. 125:2189–2202.

Wratten, S. D., P. J. Edwards, and I. Dunn. 1984. Wound-induced changes in the palatability of *Betula pubescens* and *B. pendula*. Oecologia 61:372–375.

Xu, X. F., H. W. Mei, L. J. Luo, X. M. Cheng, and Z. K. Li. 2002. RFLP-facilitated investigation of the quantitative resistance of rice to brown planthopper (*Nilaparvata lugens*). Theor. Appl. Genet. 104:248–253.

Yanes, J., Jr., and D. J. Boethel. 1983. Effect of a resistant soybean genotype on the development of the soybean looper (Lepidoptera: Noctuidae) and an introduced parasitoid, *Microplitis demolitor* Wilkinson (Hymenoptera: Braconidae). Environ. Entomol. 12:1270–1274.

Zhu, L. C., C. M. Smith, A. Fritz, E. V. Boyko, and M. B. Flinn. 2004. Genetic analysis and molecular mapping of a wheat gene conferring tolerance to the greenbug (*Schizaphis graminum* Rondani). Theor. Appl. Genet. 109:289–293.

Zhu-Salzman, K., R. A. Salzman, J-E. Ahn, and H. Koiwa. 2004. Transcriptional regulation of sorghum defense determinants against a phloem-feeding aphid. Plant Physiol. 134:420–431.

Ziegler, J., M. Keinanen, and I. T. Baldwin. 2001. Herbivore-induced allene oxide synthase transcripts and jasmonic acid in *Nicotiana attenuate*. Phytochem. 58:729–738.

Zvereva, E. L., M. V. Kozlov, P. Niemela, and E. Haukioja. 1997. Delayed induced resistance and increase in leaf fluctuating asymmetry as responses of *Salix borealis* to insect herbivory. Oecologia. 109:368–373.

CHAPTER 10

TRANSGENIC ARTHROPOD RESISTANCE

1. DEFINITIONS AND HISTORY

Plants containing genes artificially inserted into them, instead of by plant aquisition through pollination, were originally created in the 1980s. Bacterial genes were inserted into plants, providing them with resistance to antibiotics, such as kanamycin (Bevan et al. 1983, Fraley et al. 1983, Herrera-Estrella et al. 1983). Murai et al. (1983) were the first to generate a transgenic plant based on intra-genomic transfer, after inserting and expressing a *Phaseolus* gene in the genome of sunflower, *Helianthus annuus* L. Much additional research since then has resulted in the development of transgenic plants with resistance to arthropod pests, herbicides and viruses. In the 1990s, adoption of this technology led to a major change in the development of arthropod resistant cultivars. *Agrobacterium* transformation systems and ballistic projection devices were used to transfer genes from bacteria, plants or arthropods into the genomes of various crop plants to create arthropod resistant plants.

2. TYPES OF TRANSGENES

2.1. Bacillus thuringiensis

Transgenes from the soil bacteria, *Bacillus thuringiensis* (*Bt*), which encode insecticidal crystalline (Cry) or cytolytic (Cyt) proteins, have been expressed in the genomes of several different crop plants. In each case, transformed plants resulting from cell and tissue culture were grown to maturity and produced transgenic seed, which was used to create commercial transgenic crop cultivars (Figure 10.1). Ingested *Bt* crystals are solubilized in the alkaline environment of the insect gut, where Cry or Cyt proteins break down to release a toxin, ∂ - endotoxin, that binds to and creates pores in the midgut cells of susceptible larvae. These pores cause an ion imbalance, paralysis of the digestive system, colloid osmotic lysis of midgut cells, and after a few days, insect death. *Bt* has been used as a conventionally applied insecticide for many years, and *Bt*-based insecticides have been considered non-toxic to mammals, birds, and non-target benefical arthropods. *Bacillus thuringiensis* Berliner, was first commercially formulated in France in 1938 for control of the European corn borer, *Ostrinia nubilalis* Hübner (Peferoen 1997).

Different versions of Cry genes have been identified and each may affect the gut of different orders of insects in different ways. However, five regions of amino acid homology exist in *Bt* Cry toxins that are thought to influence the folding of the toxin molecule. The five conserved regions exhibit few differences, suggesting that

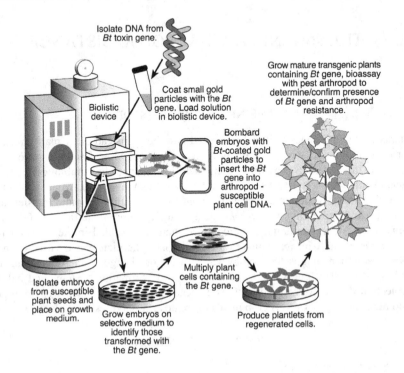

Figure 10.1. Creation of Bt *transgenic arthropod-resistant plants using a DNA micro-particle ballistic projection device.*

many, if not all *Bt* toxins will form similar a structure.

The *Bt* molecule is composed of domains I, II and III (Schnepf et al. 1998, Saraswathy and Kumar 2004). Domain I controls formation of lytic pores in the larval midgut epithelium of targeted arthropods. Domain II likely has a major role in receptor binding and controls the insecticidal specificity of the *Bt* toxin, because of a strong similarity between the three beta-sheet structure of domain II and two lectins (see Section 3.2 below). Domain III, at the C-terminus of the molecule, functions in structural stability of the *Bt* molecule, the determination of insecticidal specificity and binding of the molecule to the brush border membrane of the insect midgut (Van Rie et al. 1990a, deMaagd et al.1996, Schnepf et al. 1998). Over 240 different types of crystal toxins have been identified and are registered in the *Bacillus* Genetic Stock Center http://www.bgsc.org/. A nomenclature for the identity of more than 150 of these toxins was developed by Crickmore et al. (1998).

The *Bt* molecule has been inserted into over 30 different food and fiber crops, and the resulting transgenic crop cultivars have shown to produce insecticidal *Bt* proteins that provide a high level of insect control (Table 10.1) (see review by Huang et al. 1999b). At least 20 *Bt* crops have been approved for field testing by the U. S. Department of Agriculture, and transgenic *Bt* cotton, *Gossypium hirsutum* L., and maize, *Zea mays* L., expressing resistance to several Lepidopteran pests have been produced commercially in the United States since 1996 (Rissler and Mellon 1996).

Cultivars of transgenic *Bt* potato, *Solanum tuberosum* L., were produced in the United States from 1996 until 1998, but production was halted in 1999, when potato processors and producers stopped purchasing *Bt* potatoes, because of un-founded public concerns over the effects of *Bt* on human health.

Table 10.1. Arthropod resistant crop plants expressing Bt *toxic proteins*

Plant	Bt toxin	Arthopods Affected	References
Brassica napus	CryIAc	*Helicoverpa zea*	Ramachandran et al. 1998
		Plutella xylostella	Stewart et al. 1996b
		Trichoplusia ni	
Brassica oleracea	CryIAc,	Lepidoptera	Cao et al. 2002
capitata, cymosa	1Ac +1C	*P. xylostella*	Metz et al. 1995
Cicer arietinum	CryIAc	*Heliothis armigera*	Kar et al. 1997
Diospyros kaki	CryIAc	*Ploida*	Tao et al. 1997
		interpunctella	
		Monema flavescens	
Glycine max	CryIAc	*Anticarsia*	Stewart et al. 1996a
		gemmatalis	Ashfaq et al. 2001
		Heliothis virescens	
		Pseudoplusia	
		includens	
Gossypium	CryIAc	*H. virescens*	Perlak et al. 1990
hirsutum		*Helicoverpa zea*	Jenkins et al. 1997
			Benedict et al.1996
			Flint et al. 1995
	CryIIA	*H. virescens, H. zea*	Jenkins et al. 1997
	CryIAb	*Actebia fennica*	Wilson et al. 1992
		Bucculatrix	Benedict et al. 1996
		thurberiella	
		Pectinophora	
		gossypiella	
	CryIAa	*H. virescens, H. zea*	Halcomb et al. 1996

Table 10.1. Continued

Juglans nigra	CyIA, Ac	*Cydia pomonella* *P. interpunctella*	Dandekar et al. 1994
Larix decidua	CyIA	*Lymantria dispar*	Shin et al. 1994
Lycopersicon	CryIAc	*A. fennica*	Noteborn et al. 1993
esculentum	CryIAb	*H. virescens*	Van DerSalm et al. 1994
	CryIC	*Manduca sexta*	
	CryIA		
	Bt tenebrionis	*Leptinotarsa decemlineata*	Rhim et al. 1995
Nicotiana tabacum	CryIAa	*M. sexta*	Barton et al. 1987
	CryIAb	*M. sexta*	Barton et al. 1987
	CryAc	*H. virescens*	Carozzi et al. 1992 van Aarssen et al. 1995 Warren et al. 1992
	CryIC	*Spodoptera littoralis*	Mazier et al. 1997
	CryIC-	*H. virescens*	Van Der-Salm et al.
	CryIAb	*A. fennica, M. sexta*	1994
	CryIIIA	*L. decemlineata*	Sutton et al. 1992
Oryza sativa	CryIAc	*Chilo suppressalis*	Ghareyazie et al. 1997
	CryIAb	*Scirophaga*	Lee et al. 1997
	CryIAc	*incertulas*	Marfà et al. 2002
	Cry2A	*Dicladispa*	Nayak et al. 1997
	CryIB	*armigera*	Wu et al. 1997
	Cry1C	*Sitophilus oryzae*	Wünn et al. 1996
	CryIIIA		Ye et al. 2001 Johnson et al. 1996
Petunia	CryIIIB	*H. zea*	Iannacone et al. 1997
	CryIAc	*M.sexta* *Spodoptera exigua* *Trichoplusia ni*	Omer et al. 1992
Populus alba	CryIAa	*L. dispar* *Malacosoma disstria*	Kleiner et al. 1995 Robison et al. 1994
Pinus taeda	CryIAc	*Dendrolimus punctatus* *Crypyothelea formosicola*	Tang & Tian 2003

Table 10.1. Continued

Saccharum officinarum	CryIAb	Diatraea saccharalis	Arencibia et al. 1997
Solanum integrifolium	CryIIIB	L. decemlineata	Iannacone et al. 1997
Solanum melongena	CryIIIA	L. decemlineata	Jelenkovic et al. 1998
	CryIIIB	L. decemlineata	Arpaia et al. 1997 Chen et al. 1995 Iannacone et al. 1997
Solanum tuberosum	CryIAb	Phthorimaea operculella	Duck & Evola 1997 Jansens et al. 1995 Peferoen et al. 1990
	CryIAc	Elasmopalpus lignosellus	Singsit et al. 1997
	CryIIIA	L. decemlineata	Perlak et al. 1993 Feldman & Stone 1997
	CryIAb	L. decemlineata P. operculella	Jansens et al. 1995
	CryIAa	Lepidoptera	Chan et al. 1996
	CryIAc	P. operculella Ostrinia nubilalis	Ebora et al. 1994
Zea mays	CryIAb	O. nubilalis H. zea Diatraea grandiosella Spodoptera frugiperda	Armstrong et al. 1995 Koziel et al. 1993 Lynch et al. 1999 Williams et al. 1997
	CryIB-CrylAb fusion	D. grandiosella D. saccharalis S. frugiperda	Bohorova et al. 2001
	Cry9C	Agrotis ipsilon D. grandiosella O. nubilalis Papaipema nebris	Binning & Rice 200 Jansens et al. 19972
	Cry IH	O. nubilalis S. frugiperda	Duck & Evola 1997

Bt crops are currently produced and marketed in 14 countries in Africa, Asia, Australia, Europe, North America and South America. Globally, the area of transgenic *Bt* cotton and maize has increased from less than 500,000 ha in 1996 to approximately 50 million ha in 2003 (Shelton et al. 2002, James 2003). In the United States, the planting of *Bt* cotton increased from 15% of the total crop planted in 1996 to 40% of the crop in 2003. The planting of *Bt* maize has followed a similar trend, increasing from 3% of the total crop planted in 1996 to 28% of the crop in 2003 (Figure 10.2). Pilcher et al. (2002) surveyed over 7,000 midwestern U. S. maize producers and determined that adoption of *Bt* maize cultivars increased from 10% in 1996 to 40% in 1998.

2.2. Non-Bt Transgenes

Many other proteins exhibiting toxicity to or growth inhibition of arthropods have been identified during the past 30 years. Prominent among these are the serine proteinase inhibitors of arthropod endopeptidases. The Bowman-Birk class of proteinsae inhibitor from soybean, *Glycine max* (L.) Merr. (Bowman 1944, Birk 1985) includes inhibitors from other legumes, such as cowpea, *Vigna unguiculata* (L.) Walp., and winged bean, *Psophocarpus tetragonolobus* (L.). Bowman-Birk inhibitors have two reactive sites for different proteinases, normally trypsin and chymotrypsin. A cowpea trypsin inhitor (CpTi) inhibits the growth of several arthropod pests (Gatehouse and Boulter 1983), and the CpTi gene was used to create some of the first transgenic plants of tobacco, *Nicotiana tabacum* L. (Hilder et al. 1987, Hoffman et al. 1992). Kunitz trypsin inhibitors also occur in several legumes and cereal crops such as barley, *Hordeum vulgare* L. Kunitz inhibitors possess one reactive site for trypsin and a weakly reactive site for chymotrypsin. Soybean trypsin

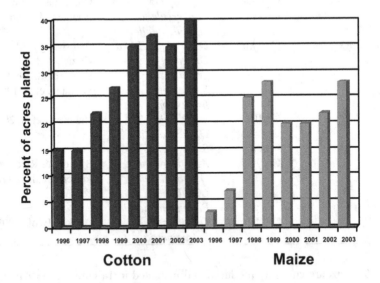

Figure 10.2. Adoption of Bt *cotton and maize in the United States from 1996 to 2004. From Fernandez and McBride (2004)*

inhibitor (SbTi) has antigrowth effects on several arthropod crop pests, including the mealworm, *Tenebrio molitor* (Applebaum et al. 1964), the corn earworm, *Helicoverpa zea* (Boddie), and the beet armyworm, *Spodoptera exigua* (Hübner), (Broadway and Duffey 1986) and the cluster caterpillar, *Spodoptera litura* (F.) (McManus and Burgess 1995).

Trypsin inhibitors have been used to develop numerous transgenic arthropod-resistant crop plants. Transgenic plants of strawberry, *Fragaria* x *ananassa* Duchesne, containing the CpTi gene are resistant to the vine weevil *Otiorhynchus sulcatus* F. (Graham et al. 1997). Transgenic rice plants containing CpTI or trypsin inhibitor genes from soybean and winged bean have been used to transform rice for resistance to the brown planthopper, *Nilaparvata lugens* Stal, and the rice stem borer, *Chilo suppressalis* (Walker), respectively (Xu et al. 1996, Lee et al. 1999, Mochizuki et al. 1999) (Table 10.2). A trypsin inhibitor gene from barley has been used to transform wheat, *Triticum aestivum* L., for resistance to the Angoumois grain moth, *Sitotroga cerealella* (Olivier) (Altpeter et al. 1999). A trypsin inhibitor gene from sweet potato, *Ipomoea batatas* Lam., has been used to create transgenic broccoli, *Brassica oleracea* vara. *botrytis* L., and tobacco plants expressing resistance to *S. litura* (Yeh et al. 1997, Ding et al. 1998b).

Plant cysteine proteinase inhibitors are small (12-18 kDa) proteins that lack disulphide bonds. These inhibitors occur in a wide range of crop plants and inhibit the growth of several species of Coleoptera. Oryzacystatin (OCI), a cysteine proteinase inhibitor from rice, *Oryza satvia* (L.), inhibits the growth of the red flour beetle, *Tribolium castaneum* (Herbst), the rice weevil, *Sitophilus oryzae* (L.), the southern corn rootworm, *Diabrotica undecimpunctata howardi* Barber, and the western corn rootworm *Diabrotica virgifera virgifera* LeConte (Liang et al. 1991, Chen et al. 1992, Edmonds et al. 1996). OCI has been used to transform plants of potato for resistance to the Colorado potato beetle, *Leptinotarsa decemlineata* (Say) (Lecardonnel et al. 1999) and hybrid poplar, *Populus tremula* x *P. tremuloides*, for resistance to the poplar leaf beetle, *Chrysomela tremulae* F. (Leple et al. 1995) (Table 10.2). The gene encoding a cysteine proteinase inhibitor (*Atcys*) from *Arabidopsis thaliana* has been used to create transgenic white poplar, *Populus alba* L., plants that inhibit the digestive proteinase activity of the chrysomelid beetle *Chrysomela populi* L. (Delledonne et al. 2001).

Plant-derived α-amylase inhibitors form complxes in the arthropod midgut to disrupt the breakdown of starches. As described in Chapter 3, the α-amylase inhibitor αAI-1 from common bean, *Phaseolus vulgaris* L., causes antibiotic effects in larvae of the cowpea weevil, *Callosobruchus maculatus* (F.) and the adzuki bean weevil, *Callosobruchus chinensis* (L.) (Ishimoto and Kitamura 1989). Transgenic adzuki bean plants, *Vigna angularis* (Willd.) Ohwi & Ohashi, containing the αAI-1 gene are toxic to *C. maculatus* and *C. chinensis* (Shade et al. 1994, Ishimoto et al. 1996). Schroeder et al. (1995) transformed plants of pea, *Pisum sativum* L., to express the αAI-1 gene for resistance to the pea weevil, *Bruchus pisorum* (L.) (Table 10.2).

Plant lectins are proteins that bind specifically to carbohydrates on the cell surface (Van Damme et al. 1998), and several lectins are known to cause antibiotic effects in Coleoptera, Homoptera, or Lepidoptera (Powell et al. 1993, Leple et al. 1995,

Table 10.2. Non-Bt transgene proteins toxic to arthropods

Transgene/Source	Target Plant	Target Arthropod	References
α-amylase inhibitors			
αAI-1	Vigna	Callosobruchus	Shade et al. 1994
Phaseolus	angularis	maculatus	Ishimoto et al. 1996
vulgaris		C. chinensis	
	Pisum sativum	Bruchus pisorum	Schroeder et al. 1995
Trypsin inhibitors			
CpTi	Fragaria	Otiorhynchus	Graham et al. 1997
Vigna unguiculata	chiloensis	sulcatus	
	Nicotiana	Helicoverpa zea	Hilder et al. 1987
	tabacum		Hoffman et al. 1992
	Oryza sativa	Chilo suppressalis	Xu et al. 1996
		Sesamia inferens	
WTI-1B	O. sativa	C. suppressalis	Mochizuki et al. 1999
Psophocarpus			
tetragonolobus			
SKTI	O. sativa	Nilaparvata	Lee et al. 1999
Glycine max		lugens	
BTI-CMe	Triticum	Sitotroga	Altpeter et al. 1999
Hordeum vulgare	aestivum	cerealella	
TI	B. oleracea	Spodoptera	Ding et al. 1998b
Ipomoea batatas	botrytis	litura	Yeh et al. 1997
	N. tabacum		
Proteinase inhibitors			
OCI	Solanum	Leptinotarsa	Lecardonnel et al.
O. sativa	tuberosum	decemlineata	1999
	Populus spp.	Chrysomela	Leple et al. 1995
		tremulae	
ATCYS	Populus alba	Chrysomela	Delledonne et al.
A. thaliana		populi	2001
Plant lectins			
GNA	S. tuberosum	Myzus persicae	Birch et al. 1999
Galanthus nivalis			
	T. aestivum	Sitobion avenae	Stoger et al. 1999
	O. sativa	N. lugens	Rao et al. 1998
	Saccharum	Diatraea	Nutt et al. 1999
	officinarum	saccharalis	Setamou et al. 2002
WGA	Brassica	Lipaphis	Kanrar et al. 2002
T. aestivum	juncea	erysimi	
Chitinase inhibitor			
Manduca sexta	N. tabacum	M. sexta	Wang et al. 1996
		H. virescens	Ding et al. 1998a

Powell et al. 1998, Fitches et al. 1997, Gatehouse et al. 1997, Machuka et al. 1999). Genes encoding lectins have been used to successfully transform plants for resistance to Homptera and Lepidoptera (Table 10.2). Birch et al. (1999) demonstrated that transgenic potato plants expressing a gene coding for a mannose-specific lectin from the snowdrop plant, *Galanthus nivalis* L., are resistant to the green peach aphid, *Myzus persicae* Sulzer. The GNA (*Galanthus nivalis* agglutinin) gene has also been used to transform wheat for resistance to the grain aphid, *Sitobion avenae* F., (Stoger et al. 1999); rice for resistance to *N. lugens* (Rao et al. 1998); and sugarcane, a complex hybrid of *Saccharum* species, for resistance to *Diatraea saccharalis* (Fabricius) (Nutt et al. 1999, Setamou et al. 2002).

Kanrar et al. (2002) transformed brown mustard, *Brassica juncea* (L.) Czern. & Cosson., to express a gene encoding a N-acetylglucosoamine-specific lectin from wheat germ agglutinin, a lectin from wheat germ for resistance to the mustard aphid, *Lipaphis erysimi* Kalt. Potato plants have also been transformed to contain an α-amylase inhibitor from wheat in combination with a bean chitinase gene or a GNA gene for resistance to *M. persicae* (Gatehouse et al. 1996). Similarly, Maqbool et al. (2001) produced transgenic rice plants containing GNA, Cry1Ac and Cry2A for resistance to *N. lugens*, the rice leaf folder, *Cnaphalocrocis medinalis* (Guenée), and the yellow stemborer, *Scirpophaga incertulas* Walker.

Chitinases exist in virtually all plants (Hamel et al. 1997) and from arthropds, where they play a natural role in the digestion of arthropod cuticle during the moulting process (Kramer et al. 1985). Chitinase genes from both plants and arthropods have been used in attempts to create transgenic arthropod resistant plants. Chitinase inhibitor genes from the tobacco hornworm, *Manduca sexta* (L.), placed into tobacco plants are resistant to *M. sexta* (Wang et al. 1996) and to the tobacco budworm, *Heliothis virescens* (F.) (Ding et al. 1998a).

Murdock and Shade (2002) reviewed the use of lectins and protease inhibitors as plant defenses against arthropods, and described several reasons for the lack of their present commercial devlopment. Both classes of these compounds are toxic to mammals, necessitating extensive processing of food crop plants containing them to remove the toxins, as well as rigorous food safety testing of food products originating from these plants. In addition, the level of resistance in plants containing lectins and protease inhibitors varies widely among different arthropod species, making it difficult to employ them in crop pest management systems against complexes consisting of multiple species of crop pests. Finally, as pointed out by reviews of protease inhibitors by Michaud (1997) and Jongsma and Bolter (1997), many major arthropod pests have the same ability to become resistant to these compounds as to *Bt* transgenes, and to overcome the effects of non-*Bt* transgenic resistant plants. These hindrances, coupled with the overall utility and success of *Bt* transgenic crops, have resulted in transgenic plants containing non-*Bt* toxic proteins remaining uncommercialized and yet to be implemented in arthropod pest management. For additional reading about the development and use of non-*Bt* transgenic plants, see the reviews of Boulter (1993), Jouanin et al. (1997), Schuler et al. (1998) Watt et al. (1999), Sharma et al. (2000), Oppert (2001) and Lawrence and Koundal (2002).

3. ARTHROPOD RESISTANCE TO *BACILLUS THURINGIENSIS*

Bt crop cultivars are in reality transgenic insecticidal plants, and as with any pesticide, natural or synthetic, the amount of toxin applied to the pest population will determine if and when resistance to the transgene will develop in the pest. Only limited research was conducted on arthropod resistance to *Bt* until the Indianmeal moth, *Plodia interpunctella* (Hübner), was show to be resistant to *Bt* in the laboratory (McGaughey 1985) and the diamondback moth, *Plutella xylostella* (L.), was shown to have developed high levels of resistance to *Bt* in the field (Tabashnik et al. 1990, Shelton et al. 1993).   Since then, high levels of resistance have also been documented in several species of arthropods, including such major crop pests as *H. virescens, L. decemlineata, O. nubilalis, S. exigua,* the Eastern spruce budworm, *Choristoneura fumiferana* Clemens, and the Egyptian cotton leafworm, *Spodoptera littoralis* (Boisduval) (Table 10.3).

Selection for resistance to one *Bt* toxin can also lead to resistance to other toxins, but the pattern of cross resistance varies among different virulent arthropod strains.   Nevertheless, strains of *P. interpunctella, P. xylostella, H. virescens, S. exigua, S. littoralis,* and the cottonwood leaf beetle, *Chrysomela scripta* F., demonstrating virulence to one *Bt* toxin have all been shown to exhibit high levels of cross resistance to different *Bt* toxins (Tabashnik et al. 1993, 1994b, McGaughey and Johnson 1994, Bauer 1995, Moar et al. 1995, Müller-Cohn et al. 1996).

3.1. Inheritance of Bacillus thuringiensis Resistance

As commercialization of transgenic *Bt* crops began in the U. S. in the 1990s, there was great concern among scientists about the lack of knowledge about the inheritance of resistance to *Bt*. Early research of McGaughey (1985), and McGaughey and Beeman (1988) found resistance to be autosomal and inherited as a recessive or partially recessive trait in *P. interpunctella*. Attempts to determine whether a single major gene was involved were inconclusive because heterozygous and homozygous genotypes could not be discriminated using the available bioassay methods. Sims and Stone (1991) determined that resistance to *Bt kurstaki* strain HD-1 in *H. virescens* is autosomal, incompletely dominant, and controlled by several genetic factors.   Gould et al. (1992, 1995) reported that *H. virescence* resistance to CryIAc and CryIAb is partially recessive and controlled by a single locus.

Resistance to *Bt kurstaki* in *P. xylostella* is autosomal, and inherited as a recessive or incompletely recessive trait controlled by one or a few major loci (Tabashnik et al. 1992, Hama et al. 1992, Tang et al. 1997). In *L. decemlineata*, resistance to CryIIIA is linked to a single gene inherited as an incompletely dominant trait (Rahardja and Whalon 1995). In *S. littoralis,* resistance to the CryIC toxin is multifactorial and inherited as a partially recessive trait (Chaufaux et al. 1997). Resistance to *Bt kurstaki* in *O. nubilalis* is autosomal and governed a single gene or a few genes and inherited as an incompletely dominant trait (Huang et al. 1999a). Liu et al (2001) evaluated the inheritance of resistance to the Cry1Ac toxin in larvae of the pink bollworm, *Pectinophora gossypiella* Saunders, and determined that resistance is *Bt*-concentration specific. Resistance is codominant at a low Cry1Ac concentration, partially recessive at an intermediate concentration and completely recessive at a comparatively high concentration.

While the genetic basis of *Bt* resistance may differ among arthropod species and specific *Bt* toxins, results of many studies indicate that resistance is linked to one or a few autosomal genes. With the exception of *O. nubilalis*, in which resistance is inherited as an incompletely dominant trait, resistance is normally inherited as a recessive trait. For a more thorough discussion of the biochemistry and genetics of resitance to *Bt*, see the reviews of Frutos et al. (1999) and Ferré and Van Rie (2002).

3.2 Resistance Mode of Action

As described previously, *Bt* toxins cross the arthropod peritrophic membrane after ingestion and bind to target sites in the midgut epithelial membrane. After binding, pores form in cells of the membrane that leads to colloidal osmotic cell lysis. Different *Bt* toxins bind to different receptors in the brush border membrane vesicles of the arthropod gut epithelia (Sacchi et al. 1986) and to the epithelial microvilli of the Malphigian tubules (Denolf et al. 1993). The lack of binding or the reduced binding of toxins to midgut target site tissues is a mechanism of resistance to *Bt* toxins in several species of Lepidoptera. In *O. nubilalis* larvae, Cry1Ab and Cry1Ac are recognized by the same receptor in the brush border membrane of the larval gut epithelium, but the Cry1B toxin is recognized by a different receptor (Denolf et al. 1993). In *M. sexta* larvae, a somewhat different situation exists, where Cry1Ab, Cry1Ac and Cry1C each bind to brush border membane proteins of distinctly different sizes (Garczynski et al. 1991, Vadlamudi et al. 1993, Sanchis and Ellar 1993).

In other Lepidoptera, reduced toxin binding due to a lack of affinity between the toxin and the binding site has also been shown to be at least one mechanism of resistance to *Bt*. The toxicity of Cry1Aa, Cry1Ab and Cry1Ac to *T. ni* larvae are correlated to their binding affinity to the larval midgut brush border epithelial cells. Cry1Ab and Cry1Ac bind to the same receptor site with high affininty and Cry1Aa binds to a different site (Estada et al. 1994). When individuals are selected for resistance to Cry1Ab over several generations, resistance is specific to Cry1Ab only, and not to Cry1Ac, suggesting that other factors such as midgut pore formation or changes in the midgut pH may also contribute to the development of *T. ni* resistance to Cry1Ac. Resistance in *P. interpunctella* to Cry1Ab is also highly correlated to reduced binding of the toxin to midgut membrane tissues (Van Rie et al. 1990b).

Lee et al. (1997) examined the toxicity of Cry1Aa, Cry1Ac, Cry12A and Cry1C to larvae of the *S. incertulas* and *C. suppressalis* in relation to the affininty of each toxin to bind with larval brush border membrane vesicles. Each toxin is toxic to both borer species, but Cry12A and Cry1C have relatively low binding affininties to larval midgut tissues. Binding site competiton assays indicate that Cry1Aa and Cry1Ac recognize the same binding site and that this site is different than the site recognized by Cry2A or Cry1C.

Table 10.3. Occurence of resistance to Bacillus thuringiensis *in plant feeding arthropods*

Arthropod Order & Species	Cry Toxin/ Bt strain	References
Coleoptera		
Chrysomela scripta	CryIIIA	Bauer 1995
Leptinotarsa decemlineata	CryIIIA	Rahardja & Whalon 1995
		Whalon et al. 1993
Lepidoptera		
Cadra cautella	*Bt kurstaki*	McGaughey & Beeman 1988
Heliothis virescens	CryIAb	Elzen 1997, Gould et al. 1992, 1995,
	CryIAc	Heckel et al. 1997, Sims & Stone
		1991, Stone et al. 1989
Homoeosoma electellum	*Bt kurstaki*	Brewer 1991
Ostrinia nubilalis	*Bt kurstaki*	Huang et al. 1997
Pectinophora gossypiella	*Bt*	Bartlett et al. 1995
Plodia interpunctella	Cry1Ab	Johnson et al. 1990
	Cry1Ac	McGaughey 1985
	Cry1C,	McGaughey & Beeman 1988
	Bt aizawai	McGaughey & Johnson 1992, 1994
	Bt kurstaki	Oppert et al. 1997
Plutella xylostella	CryIAb	Cho & Lee 1994, Feng et al. 1996,
	CryIAc	Ferré et al. 1991, Hama et al. 1992,
	Cry1F	Liu et al. 1995, 1998, Perez &
	Cry1J	Shelton 1997, Song 1991, Shelton et
	Bt aizawai	al. 1993, Tabashnik et al. 1990, 1991,
	Bt kurstaki	1995, 1997, Wright et al. 1995
Spodoptera exigua	*Bt kurstaki*	Moar et al. 1995
	CryIAb	
	Cry1C	
	Cry1H	
	Cry2A	
	Cry1E-C	
Spodoptera littoralis	Cry1C	Salama & Matter 1991
	Cry1D	Müller-Cohn et al. 1996
	Cry1E	
	Bt aizawai	
Trichoplusia ni	CryIAb	Estada & Ferré 1994

Reduced toxin binding site activity is also a mechanism of resistance in *H. virescens* larvae. Cry1Ac resistant populations have been developed that exhibit reduced binding of Cry1Ac to larval brush border membrane vesicles (Gould et al. 1992, Lee et al. 1995, Jurat-Fuentes and Adang 2001). Jurat-Fuentes et al. (2003) examined membrane vesicle binding in *H. virescens* larvae from populations selected for resistance to Cry1Ac and for cross- resistance to Cry2Aa, and determined that altered binding site affininty is also a mechanism of *H. virescens* resistance to Cry1Aa. Resistance to Cry2Aa is likely due to a different mechanism, and related to a change in how the toxin is processed in larval midgut tissues (Jurat-Fuentes et al. 2003).

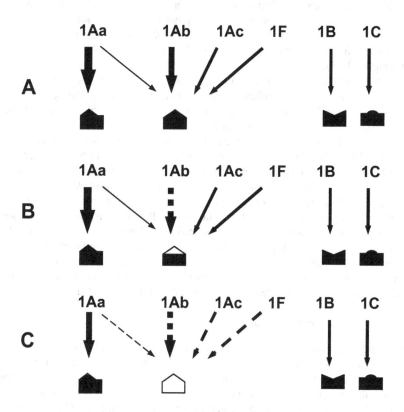

Figure 10.3. A model for the binding of Cry toxins to binding sites in larval midgut epithelial membranes of P. xylostella. (A) susceptible larvae, (B) resistant larvae from the Philippines, (C) resistant larvae from Hawaii and Pennsylvania. Wider arrows indicate greater toxin binding, dashed arrows indicate no binding or extremely reduced binding. (From Ballester 1999, Reprinted with permission of the American Society of Microbiology)

The mechanism of *Bt* toxin resistance has been studied extensively in laboratory and field populations of *P. xylostella*. Here, the reduced binding of either Cry1Ab or Cry1Ac to *P. xylostella* larval midgut membrane receptors is a resistance trait in several different populations (Ferré et al. 1991, Tabashnki et al. 1994a). Wright et al. (1997)

evaluated the binding of CrylAa, CrylAb, CrylAc and CrylC in the brush border membrane vesicles of individuals from *P. xylostella* populations resistant to either *Bt kurstaki* or *Bt aizawai*. Membrane vesicles of larvae from these resistant populations have reduced binding to CrylAb, indicating that reduced midgut tissue binding is only partially responsible for resistance in *P. xylostella* to CrylAb, and that additional *Bt* resistance mechanisms are present in some *P. xylostella* populations.

Ballester et al. (1999) used competitive binding site data for different Cry toxins to larval brush border membrane vesicles from *Bt*-resistant and *Bt*-susceptible strains of *P. xylostella* from Hawaii, the Philippines and Pennsylvania to develop an integrative model of *Bt* toxin binding in *P. xylostella*. The model is based on four sites involved in binding the CrylAa, CrylAb, CrylAc, CrylB, CrylC or CrylF toxins (Figure 10.3).

The CrylA toxins and the CrylF toxin compete for a common binding site. CrylAa binds with reduced affininty to the common binding site shared with CrylAb, CrylAc and CrylF, and binds with high affinnity to a site not shared with other toxins. CrylB and CrylC each bind to different sites.

The common binding site is altered in resistant populations from at least two locations. In a Philippine population, the binding of CrylAb is reduced or eliminated with no detrimental effects on the binding of CrylAa or CrylAc. Genetic studies with the Philippine population indicate that *P. xylostella* resistance to CrylAa is controlled by a different gene than that for resistance to CrylAb (Tabashnik et al. 1997). *P. xylostella* populations from Hawaii and Pennsylvania have even greater reductions in toxin binding, with greatly reduced or no binding of any of the CrylA toxins to the common binding site. A lack of correlation between the differences in toxin binding observed by Ballester et al. (1999) and field observations of resistance by Ferré et al. (1991) and Granero et al. (1996) suggest reduced toxin binding is not the only mechanism of *Bt* resistance in *P. xylostella* (Ballester et al (1999).

3.3 Molecular Bases of Bt Resistance

Determining the molecular genetics of *Bt* resistance was a difficult undertaking because of not only the recessive inheritance of the trait but because of the lack of knowledge about *Bt*. However, research by (Gahan et al. 2001) and (Morin et al. 2003) has provided the first evidence of the genes involved in the expression of arthropod resistance to *Bt* proteins. Two insect proteins are targets of CrylA toxins. Aminopeptidases are involved in arthropod digestion and cadherins are involved in cell-to-cell adhesion. In Lepidoptera, both aminopeptidase and cadherin proteins mediate binding of the CrylA toxin to the midgut epithelium of Lepidopteran larvae and ensuing cell lysis. Gahan et al. (2001) demonstrated how disruption of a cadherin gene by retrotransposon-mediated insertion is linked to high levels of resistance to the CrylAc toxin in *H. virescens*. Morin et al. (2003) identified three mutant alleles of a cadherin gene in field populations of *P. gossypiella*. Each of the three resistance alleles has a deletion that removes several amino acids near the putative toxin-binding region of the cadherin protein. *P. gossypiella* individuals with two of the three resistance alleles in any combination survive the effects of the CrylAc toxin expressed by transgenic cotton plants.

3.4. Bt Transgene Crop Deployment Methods

Commercial transgenic cultivars express *Bt* toxins at high levels in plant leaves, stems or roots for the life of the plant. For this reason, it is not surprising that since the inception of *Bt* transgenic crops, much scientific debate has occurred about how much expression of *Bt* in plants is sufficient to provide a dose of toxin that provides economical pest control without causing the development of resistance. Concerns also exist about the tissues in which toxins are expressed and for how long toxins should be expressed during the life of the transgenic plant.

The high level of *Bt* toxin expression in transgenic plants and the reports described previously of resistance to *Bt* in several important arthropod crop pests stimulated research and discussions about which strategies should be adopted to avoid field resistance to *Bt*. Computer simulation models (Peck et al. 1999) and in some cases, actual field monitoring of arthropod resistance genes led to initial information about the levels of the *Bt* resistance alleles being expressed in arthropods. Gould et al. (1997) conducted one of the first experiments to estimate the frequency of arthropod alleles for *Bt* resistance, and determined a frequency of 1.5×10^{-3} in field populations of *H. virescens* in the southern United States. Andow et al. (1998) used an F_2 screening procedure to estimate the frequency of *Bt* resistance alleles in a wild northern United States (Minnesota) population of *O. nubilalis* to be $<1.3 \times 10^{-2}$ (95% probability). Most recently, Bourguet et al. (2003) used the F_2 screening procedure to estimate the frequency of *Bt* resistance alleles in *O. nubilalis* populations in the maize production regions of France and at several locations in the the northern United States in 1999, 2000 and 2001. The frequency of resistance alleles was $<9.2 \times 10^{-4}$ (95% probability, 80% detection probability) in France and $<4.23 \times 10^{-4}$ (95% probability, 80% detection probability) in the northern United States. Genissel et al. (2003) used the F_2 screening to monitor the initial frequency of alleles conferring resistance to transgenic hybrid poplar, (*Populus tremula x Populus tremuloides*), producing Cry3A *Bt* in a field population of *C. tremulae*. The estimated mean *C. tremulae Bt* resistance-allele frequency from 1999 to 2001 was 3.7×10^{-3} (95% confidence interval, 90% detection probability).

Monitoring the development of *Bt* resistance in arthropods infesting *Bt* plants in the field is essential to the success of any *Bt* transgenic crop plant deployment strategy. However, this is very difficult when *Bt* resistance is inherited as a recessive trait. The data of Gould et al. (1997) and Andow et al. (1998) were of great value in developing computer simulation models used to evaluate different methods of deploying *Bt* in crop plants. DNA-based screening for resistant heterozygotes expressing the cadherin gene to detect the recessive resistance allele is a promising tool that may prove useful in identifying resistance to *Bt* toxins in lepidopterous larvae on *Bt* crops.

3.4.1. Pyramiding Mixtures of Different Bt Toxins

Mixtures composed of different *Bt* toxins in different seeds mixed in a single cultivar, or multiple *Bt* toxins in the same cultivar have also been proposed as a means of *Bt* transgene deployment. However, in order for the development of resistance to be delayed, gene mixtures require a lack of cross-resistance for each toxin. Patterns of cross-resistance to *Bt* toxins differ among arthropod species and these differences may reduce the success of mixtures in delaying resistance. McGaughey and Johnson (1992) and Tabashnik et al. (1991) have shown that *Bt* toxin mixtures do not improve the durability of indidual *Bt* toxins to either *P. interpunctella or P. xylostella*. As was pointed out in Chapter 8, similar comparisons of sequentially-released individual conventional plant resistance genes to pyramids of genes for *M. destructor* and *S. graminum* resistance have shown no net benefit of gene combinations over the deployment of single resistance genes. Because of the possibilities for simultaneous selection for *Bt* cross-resistance, it is likely that little difference exists between sequentially releasing different *Bt* transgenes and pyramiding combinations of genes. Conventional plant arthropod resistance genes in cotton (high terpenoid content), potato (glandular trichomes) and soybean have been successfully combined with *Bt* toxins in transgenic plants (Altman et al. 1996, Sachs et al. 1996, Douches et al. 1998, Walker et al. 2002). This technology has not yet been commercially marketed.

3.4.2. Rotating Deployment of Different Bt Toxins

Delaying *Bt* resistance development by rotating different *Bt* plants with different toxins expressed as independent resistance mechanisms has also been proposed as a resistance avoidance strategy (Roush 1997). This technique is based on the assumption that the frequency of resistance alleles in the pest poulation will decrease when selection pressure ceases or when a different toxin is used. The success of toxin rotation depends on the absence of arthropod cross-resistance to the toxins involved, no linkage between the different resistance genes, and unstable resistance. The utility of *Bt* rotations in resistance management remains in question, however, as studies by McGaughey and Beeman (1988), Tabashnik et al. (1990) and Liu et al. (1996) have determined that *P. interpuctella* and *P. xylostella* maintain resistance to *Bt* for long periods of time, after the removal of, or in the absence of selection for resistance. Roush (1997) indicated that there is little evidence for an advantage in toxin rotation over sequential release of new *Bt* toxins after the first toxin has failed.

3.4.3. High Level Bt Expression Deployed with a Non-Bt Refuge

The high dose-refuge strategy involves deploying very high levels of the *Bt* toxin in plants to kill all homozygous susceptible individuals as well as individuals heterozygous for resistance. The rationale behind the high dose-refuge strategy is that if a transgenic cultivar has a *Bt* toxin level above the lethal dose that kills all but 0.01% of the pest population, there will be sufficiently high mortality of heterozygous individuals to prohit their survival and eventual mating. This deployment strategy assumes that resistance is controlled by one major locus with recessive or at least partially recessive inheritance; that the initial resistance gene frequency is < 0.001; that toxin concentrations are sufficient to kill all heterozygous individuals; and that homozygous resistant individuals mate randomly with homozygous susceptible individuals.

The refuge provides a source of susceptible individuals that mate with any homozygous resistant individuals surviving in the *Bt* maize crop, which results in only heterozygous progeny. Shifting the mortality of larvae heterozygous for resistance from 50% to 95% provides a 10-fold delay in the time required before the development of resistance (Gould 1998). Simulation models developed by Caprio (1998), Gould (1998) and Carrière and Tabashnik (2001) each predict that refuges substantially delay the development of resistance. Rissler and Mellon (1996) recommended a mandatory resistance refuge plan for all *Bt* crops and that producers plant relatively large refuges of non- *Bt* crops close to *Bt* crops. Ostlie et al. (1997) recommended a 20 to 30% non-*Bt* maize refuge for situations where the refuge is not treated with insecticides, or a 40% refuge if it is treated with an insecticide. For the refuge strategy to be effective, insects must emerge from the refuge at the same time as resistant insects and be close enough to mate with resistant insects so that homozygous resistant individuals are not produced.

3.5. Bt Insecticide Resistance Management Programs

Initially, a refuge portion of *Bt* crops was not commercially or sociologically acceptable, but the U. S. Environmental Protection Agency coordinated industry and academic research efforts to develop an integrated management tactic to prevent the occurrence of resistance to *Bt* in key maize insect pests. In 2001, an Insect Resistance Management (IRM) compliance assurance program was created to promote U. S. grower compliance and preserve the effectiveness of *Bt* maize. Manufacturers sponsor an annual survey of *Bt* maize growers, conducted by an independent third party, and growers not in compliance with IRM requirements over two consecutive years are denied additional access to *Bt* maize seed.

Ostlie et al. (1997) proposed different refuge configurations to manage *O. nubilalis* resistance to *Bt* maize. These include strip planting *Bt*- and non-*Bt* maize in the same field, blocks of *Bt*- and non-*Bt* maize in the same field, and planting *Bt* - and non-*Bt* maize in adjacent fields. In U. S. IRM programs, *Bt* crop producers must plant at least a 20% non- *Bt* maize refuge, except in certain cotton growing areas where at least a 50% non- *Bt* maize refuge is required. Refuge planting options include blocks within fields, strips across fields or separate fields, and *Bt* maize fields must be planted within 0.8 km of a refuge. The IRM programs are the first type of government-industry regulated integrated pest management technology, and to date, have been successful. Producer compliance has improved since 1999 (Byrne et al. 2003). The continued success of such producer-regulatory, agency-industry coopertive agreements will depend on a combination of science, communication and sound decision-making about *Bt* crop cultivar selection and refuge composition.

Daly and Wellings (1996) contrasted aspects of conventional and transgenic plant resistance to insects. Conventional plant resistance genes control effects measured in the arthropod (antibiosis or antixenosis) or the plant (tolerance) (Table 10.4) and many biochemical and biophysical mechanisms mediate resistance (see Chapters 2 and 3). Transgenes provide resistance that is exhibited only as antibiosis, due to the effects of a digestive toxin. Both types of resistance are constitutive, but conventional resistance also includes induced components. Both types of resistance genes have high stability, but the major difference between conventional and *Bt*-based transgenes is the resistance management plan for transgenes.

Bt maize with resistance to *D. virgifera*, a more damaging pest than *O. nubilalis*, has also been registered for production by the U. S. Environmental Potection Agency (Knight 2003). A 20% refuge similar to that for *O. nubilalis* has been adopted for initial production purposes, but an advisory panel had recommended a much larger 50% refuge, to sufficiently dilute *Bt* virulence alleles in surviving *D. virgifera* larvae. *O. nubilalis*-resistant *Bt* maize cultivars produce high doses of toxin, but *D. virgifera* resistant maize cultivars cause only 50% larval mortality. This reduced *D. virgifera* *Bt* efficiency and decreased refuge size may lead to a different outcome for *D. virgifera* resistant *Bt* maize than that presently developing for *Bt* cultivars resistant to Lepidopteran larvae. If Lepidoptera *Bt* resistance management programs continue to succeed, they could serve as workable models for the deployment of conventional plant resistance genes expressed as high levels of antibiosis, in order to delay the development of resistant arthropod biotypes.

Table 10.4. Comparisons of conventional and transgenic resistance in crops plants (From Daly and Wellings 1996)

Criteria	Conventional Resistance	Transgenic Resistance
Resistance category	Antibiosis, Antixenosis, Tolerance	Antibiosis
Mechanism(s)	Chemical & Physical	Chemical
Efficacy	Moderate	High
Expression	Constitutive & Induced	Constitutive
Management	Optional	Required
Sociology	Simple	Complex
Stability	High	High
Technology transfer	Moderate	Fast

Reproduced with permission from Frontiers of Population Ecology (RB Floyd, AW Shepard, & P J De Barro), Published by CSIRO PUBLISHING, Melbourne Australia (http://publish.csiro.au/pid/394.htm) Copyright CSIRO 1996.

4. RISKS AND BENEFITS ASSOCIATED WITH PRODUCTION OF TRANSGENIC CROPS

4.1. History and Bases of Societal Concerns

Initial experiments by Losey et al. (1999) indicated that *Bt* maize pollen, applied at abnormally high concentrations to leaves of the milkweed (*Asclepias* spp.) plants, was toxic to larvae of the monarch butterfly, *Danaus plexippus* (Linnaeus), feeding on milkweed. Scientific efforts to respond to Losey's paper were viewed as industry-driven. However scientists from government, industry, universities and environmental groups developed experiments to determine the risks involved in the production of transgenic *Bt* maize (Pew Initiative on Food and Biotechnology 2002). Unfortunately, this initial misunderstanding and mistrust resulted in the destruction of *Bt* crop research faciles and germplasm valued at over $1 million by opponents of transgenic crop plant research (Service 2001).

Since publication of Losey's data, several studies based on more realistic ecological situations have yielded contrary results. Wraight et al. (2000) reported no mortality of the larvae of black swallowtail, *Papilio polyxenes* F., on food plants located at varying distances from field plantings of *Bt* maize, no matter how close the larval food plants were to the pollen-shedding *Bt* maize plants. Subsequent studies by Sears et al. (2001) and Stanley-Horn et al. (2002) concluded that the risk from *Bt* maize to *D. plexippus* is insignificant. One *Bt* maize event grown on only 2 % of the annual U. S. crop hectarage was shown to be toxic to *D. plexippus* larvae and has been eliminated from production. Pimentel and Raven (2000) classified the effects of *Bt* pollen on the food plants of several non-target U. S. butterfly species as relatively insignificant, in comparison to maize pesticide applications and butterfly abiotic mortality factors such as habitat destruction. For example, several hundred million *D. plexippus* adults died in January 2002 during a period of very low temperatures and heavy precipitation in mountain overwintering areas in Mexico.

4.2. Effects on Beneficial Arthropods

Bt toxins are highly selective in killing arthropod pests and have limited effects on beneficial arthropods in agroecosystems. An initial review concluded there were few adverse effects on non-target arthropods such as *Aphis mellifera* L. larvae and coccinellid predators (Anoymous 1995). Results of research by Arpaia (1996) with *A. mellifera* larvae fed a diet containing CryIIIB toxins suggest that transgenic crops producing CryIIIB toxins may represent a suitable environment for pollinators. Additional laboratory studies have also shown no detrimental effects of CryIAb *Bt* maize pollen on the development or survival of the lady beetle, *Coleomegilla maculata* DeGeer, or the insidious flower bug, *Orius insidiosus* (Say) (Pilcher et al. 1997a). Riddick et al. (1998) determined that the biology and predation potential of *C. maculata* is unaffected by consumption of *L. decemlineata* larvae feeding on Cry3A transgenic potato foliage. Al-Deeb et al. (2001) determined that the biology of *O. insidiosus* is unaffected after feeding on *O. nubilalis* larvae that have consumed a diet containing *Bt kurstaki*.

The biology of a collembolan, *Folsomia candida* Willem, and an orbatid mite, *Oppia nitens* Koch, is unaffected after feeding on the leaves of *Bt* cotton (Yu et al. 1997). In addition, Sims and Martin (1997) reported that the CryIAb, CryIAc, CryIIA, and CryIIIA *Bt* toxins have no significant effect on survival or reproduction of *F.*

candia or the collembolan *Xenylla grisea*. Johnson and Gould (1992) found no differences in parasitism of *H. virescens* larvae by *Campoletis sonorensis* (Cameron), or *Cardiochiles nigriceps* Viereck, when *H. virescens* larvae were fed non-*Bt* tobacco plant foliage or foliage of *Bt* tobacco plants containing the Cry1Ab toxin. Finally, large-plot farm field experiments in the southern U. S. state of Tennessee demonstrated the impact of *Bt* cotton on beneficial arthropod populations to be negligible (Van Tol and Lentz 1998).

Nevertheless, *Bt* toxins have been shown to have adverse effects on non-target arthropods at two or three trophic levels. Significant mortality of green lacewing *Chrysoperla carnea* (Stephens), larvae occurs after exposure to or ingestion of the CryIAb toxin in diet or in *O. nubilalis* reared on maize expressing the toxin (Hilbeck et al. 1998a,b, 1999, Dutton et al. 2002, 2003). Similarly, Hafez et al. (1997) demonstrated that the reproductive potential and longevity of *Meteorus laeviventris*, a larval parasite of the black cutworm, *Agrotis ypsilon* (Rott.), are adversely affected when *A. ypsilon* larvae are fed a diet containing the Cry IAb *Bt* toxin. Birch et al. (1998) conducted laboratory studies to show a tritrophic interaction involving transgenic potato plants expressing the snowdrop lectin, *M. persicae,* and the two-spotted ladybird beetle, *Adalia bipunctata* (L.), a *M. persicae* predator. *Adalia bipunctata* fecundity and longevity significantly decrease over 21 days post-infestation but these effects are reversed after switching beetles to *M. persicae* feeding on nontransgenic plants.

4.3. Asessment Risks Associated with Transgenic Crops

The effects of *Bt* transgenes on beneficial and non-target arthropods to date have been shown to be largely negligible. However, the toxicity of the CryIAb toxin to *C. carnea* suggests that additional investigations are necessary to understand the potential toxicity to other non-target organisms in the agroecosystem and adjacent natural ecosystems. There have been few large-scale studies of the effects of *Bt* on non-target organisms across several trophic levels when transgenic crops containing *Bt* and other transgenes are deployed commercially.

Andow and Hilbeck (2004) have proposed an ecologically-based model to assess the risk of *Bt* plants on non-target organisms. The model combines the positive aspects of ecotoxicology models which test chemical efficacy (Forbes and Forbes 1994) and non-indigenous species models that assess risks of technology to introduced biological control agents. In traditional ecotoxicology models, one indicator species is tested for acute toxicity after short-term exposure, using single-chemical, dose-response assays. In non-indigenous species models, several nontarget species potentially at risk are assessed for their ability to survive on hosts exposed to a new host food source, over a much longer period of time.

The combined ecologically-based model tests a smaller number of local, representative nontarget species most likely to be affected by the *Bt* transgene over longer periods of time. Abundant, prevalent species with broad geographic ranges and high temporal overlap with the transgenic crop are strong candidates for assessment in ecologically based models, as are rare, non-target species with significant ecological, economical or symbolic value.

Andow and Hilbeck (2004) recommend that minimum risk assessments of *Bt* plants on non-target orgamisms include six basic criteria. These include: use of identical transgenic products in related field and laboratory studies; use of plant parts containing transgenes at known concentrations; verification of the *Bt* content of test plants throughout the test duration; verification of contact by the non-target arthropod species with the transgene; use of proper non-transgenic controls; and assessment of sufficient numbers of individuals to perform valid statistical analyses. When these criteria were applied to experiments that evaluated the effects of *Bt* toxins on *C. carnea* larvae, Andow and Hilbeck (2004) noted that in experiments that reported no toxicity, larvae were unlikely to have ingested significant amounts of toxin. Experiments that met the minimal criteria described above and reported *C. carnea* larval mortality after exposure to the CryIAb toxin (Hilbeck et al. (1998a,b, 1999, Dutton et al. 2002, 2003) present some of the best evidence to date that assessments of the risk of *Bt* crop plants to ecologically sensitive natural enemies are imperative to the future evaluation and successful use of transgenic plant resistance.

Approvals for the use of transgenic crops in the Netherlands have also followed a similar ecologically-based course, and are based on a "precautionary principle". The principle stipulates that ecological hazards related to the introduction of a transgenic crop require that interim measures be adopted until the risk is assessed and an acceptable risk management plan is developed. A proposed Dutch risk assessment plan (Knols and Dicke 2003) advocates a shift from short-term studies of the effects of *Bt* toxins on target arthropods in one (pest) trophic level, to longer-duration studies of different species in several trophic levels, similar to the Andow and Hilbeck ecologically-based risk assessment model. The proposed Dutch strategy includes components at four levels. Primary (level 1) risk assessment includes identifying the effects of *Bt* toxins on key above ground target and non-target species in multi-trophic interactions and their ecological interactions in the food web. Secondary (level 2) risk assessment includes determining the effects of *Bt* toxins on the interactions between species in below- and above ground trophic levels. Tertiary (level 3) risk assessment involves evaluating community level interactions between *Bt* transgenes with unrelated plants and animals. Finally, the proposed Dutch risk assessment strategy involves modeling (level 4) of the potential ecological effects of *Bt* on key ecosystem species over larger temporal and spatial scales.

Both models demonstrate the importance of in-depth field evaluations of *Bt* transgenic plants before releasing them into the environment. In addition, both models serve as a reminder that transgenic arthropod resistant plants should be developed and field-tested for their broad-spectrum effects on non-target benefical arthropods in the same manner as conventionally bred arthropod-resistant plants are developed.

4.4. Economic and Evironmental Impact of Transgenic Crop Production

Globally, the adoption of *Bt* transgenic crops has increased dramatically. Since 1998, farm trials in India show *Bt* cotton yield increases of approximately 60% over yields of conventional non-*Bt* cultivars (Qaim and Zilberman 2003). Similar results have been reported for planting of *Bt* cotton in South Africa, where *Bt* cotton production shifted from 7% of the total crop hectarage in the 1997-1998 growing season to 90% of the crop in the 2001-2002 growing season on both small and large farms (Kirsten and Gouse 2002). The primary benefits have been increased yields from improved pest Lepidoptera control and related decreased production costs from greatly reduced insecticide usage (Table 10. 5).

One year after the first commercial plantings of *Bt* maize in the Philippines in 2001, the use of *Bt* maize led to a 37% yield increase over non-*Bt* maize. The higher cost of production associated with adopting *Bt* technology was offset by higher net producer incomes. Producers earned an additional ~$US 170/ha and saved ~ $US 3/ha) on insecticide use. Greatest profits were gained when *Bt* maize was planted during the wet season, when stem borer infestations were highest (Yorobe et al. 2004).

Table 10. 5. Adoption and impact of Bt *cotton on farm incomes of small-scale and large-scale South Africa producers (From Kirsten and Gouse 2002)*

	Small-scale producer	Large-scale producer	
Income effect	Dry land (R/ha)	Dry land (R/ha)	Irrigated (R/ha)
Mean yield benefit per ha @ R2.75/kg	498	314	1741
Mean reduced pesticides benefit (chemical costs)	32	114	293
Mean increased *Bt* detriment (seed & technology fee)	(163)	(234)	(570)
Income advantage	367	194	1463

The use of *Bt* transgenic crop cultivars has dramatically reduced the amount of chemical insecticides applied to U. S. cotton and maize. U. S. farmers applied approximately 450,000 kg less insecticide to *Bt* cotton than would have been applied to non-*Bt* cultivars in 1998 (Ferber 1999). In a 10-year study in the southwestern U. S. state of Arizona, Carrière et al. (2003) determined that production of *Bt* cotton significantly suppressed populations of *P. gossypiella* and that the deployment of *Bt* cotton cultivars contributed to reducing the need for insecticide sprays. In 2001, global insecticide applications to cotton were reduced by approximately 50% due to the production of *Bt* cotton (Phipps and Park 2002, James 2004).

The benefits from *Bt* maize, however, have not been as definitive.

Nevertheless, recommended insecticide use against a single chronic pest, *O. nubilalis*, has declined by approximately 30% since the commercialization of *Bt* maize in North America in 1996. *Bt* maize has proven to be a particularly effective means of *O. nubilalis* control. Larvae feed inside maize stalks making them impossible to kill by conventional spray insecticide applications (Rice and Pilcher 1998). The reduced arthropod damage resulting from production of *Bt* maize also serves to reduce the incidence of fungal infection and accumulation of the associated mycotoxins in maize. The incidence of maize fumonisin mycotoxins is greatly reduced in *Bt* corn hybrids (Munkvold et al. 1997, Masoero et al. 1999).

5. SUMMARY AND CONCLUSIONS

Transgenic plants serve an increasingly important role in the development and utilization of crop germplasm for arthropod resistance. The field performance of transgenic cotton and maize hybrids containing *Bt* Cry toxin-based resistance to several major pest species of Coleoptera and Lepidoptera has been impressive and beneficial. These *Bt*-based insecticidal plants are toxic to pests that are difficult to reach with conventional insecticides and have resulted in major reductions in the use of broad-spectrum insecticides in many countries.

Accomplishments such as these represent a major technological advance in cotton and maize arthropod pest management in the United States, China and India, and the adoption of *Bt* technology in these countries is now continuing in many other countries. The results of the introduction and marketing of *Bt* crops in Africa, Central America and South America are examples of the tremendous economic and environmental benefits of transgenic arthropod-resistant crops in developing countries. Major needs remain to be addressed however, in improving food production in the semi-arid tropics, where transgenic crop plant resistance in cereal grains and cool-season legumes, such as sorghum, millet, pigeon pea and chickpea, is yet to occur.

One of the largest impediments to an even greater rate of global adoption of *Bt* arthropod resistant plants relates to the issue of their safety in ecological and agroecological systems, as well as the safety of foods processed from *Bt* plants. There are equally important needs for research and educational programs that inform consumers about the safety of trasnsgenic crop plants, in order to alleviate consumer concerns and fears about the effects of *Bt* crops on human health. For example, production of *Bt* potatoes in the U. S. ceased when processors and producers stopped purchasing *Bt* potatoes, because of un-founded public concerns over the effects of *Bt* on human health.

In spite of the success of *Bt* crop plants, resistance to *Bt* in many different insect species, by selection in the laboratory or field, indicates the reality of resistance developing in the field. The high level of *Bt* expression currently deployed, similar to high doses of conventional insecticide or high-level expression of a conventional resistance gene, leads to the development of insects resistant to *Bt* crop plants. The longevity of *Bt* transgenes is being extended through the use of insect resistance management (IRM) programs centered on non-transgene refuge areas that allow the survival of individual arthropods that are homozygous for susceptibility to the *Bt* toxin.

An improved understanding of the development of resistance will be associated with developing or refining techniques that estimate initial arthropod resistance allele frequencies, in order to understand more about the inheritance of *Bt* resistance, cross-resistance and *Bt* resistance mechanisms. This information will be extremely useful in refining the designs of non-*Bt* resistance refuge structure and arrangement. Therefore, it is essential to the continued success of *Bt* crops to use new data to improve existing *Bt* IRM strategies in order to continue to avoid or delay the development of resistance to *Bt* endotoxins in arthropod pests. It is equally important to develop and implement new scientifically sound IRM strategies that utilize the most contemporary data on *Bt* toxicity to not only target pest arthropods, but to ecologically sensitive non-target arthropods as well. Finally, present *Bt* cultivars are based on the differences in only a few active protein domains of toxin expression. In order to retain the future use of *Bt* crop resistance, additional research is needed to develop transgenic plants containing different types of *Bt* toxic proteins and non-*Bt* toxic proteins in IPM systems. This should include efforts to identify non-*Bt* resistance transgenes with higher, more stable levels of toxicity than lectins and protease inhibitors for example, with no mammalian toxicity.

As with any arthropod resistant crop plant, *Bt* crop cultivars are only one component, albeit a major component, of an integrated crop pest management system. However a *Bt* resistant plant is only as good as the other components of the integrated crop pest management system. The ultimate success or failure of a *Bt* crop cultivar depends on a thorough knowledge of the biology and ecology of the target arthropod, in this case in the presence of the *Bt* plant, as well as the interaction of the *Bt* plant with biological control agents that use the pest arthropod as a host.

Future research on the broad-spectrum effects of *Bt* transgenic plants on such biological control agents as well as other non-target benefical arthropods should follow the primary tenents proposed in the risk assessment models devised by Andow and Hilbeck (2004) and Knols and Dicke (2003). Both models focus on interpretations of data from large-scale studies of the potential toxic effects of *Bt* transgenes expressed at known concentrations on abundant, prevalent non-target organisms in multiple trophic levels in both the agroecosystem and adjacent natural ecosystems. These models are reminders that arthropod resistant plants devised by any means should be evaluated for broad-spectrum effects on non-target benefical organisms before their adoption and use in crop pest management systems.

REFERENCES CITED

Al-Deeb, M. A., G. E. Wilde, and R. A. Higgins. 2001. No effect of *Bacillus thuringiensis* corn and *Bacillus thuringiensis* on the predator *Orius insidiosus* (Say) (Hemiptera: Anthocoridae). Environ. Entomol. 30:625–629.

Altman, D. W., J. H. Benedict, and E. S. Sachs. 1996. Transgenic plants for the development of durable insect resistance. Proc. New York Acad. Sci. 792:106–114.

Altpeter, F., I. Diaz, H. McAuslane, K. Gaddour, P. Carbonero, and I. K. Vasil. 1999. Increased insect resistance in transgenic wheat stably expressing trypsin inhibitor CMe. Mol. Breed. 5:53–63.

Andow, D. A., D. N. Alstad, Y. H. Pang, P. C. Bolin, and W. D. Hutchison. 1998. Using a F_2 screen for resistance alleles to *Bacillus thuringiensis* toxin in European corn borer (Lepidoptera: Crambidae). J. Econ. Entomol. 91:579–584.

Andow, D. and A. Hilbeck. 2004. Science-based risk assessment for nontarget effects of transgenic crops. BioSci. 54:637–649.

Anonymous. 1995. *Bacillus thuringiensis* CryIA(b) d-endotoxin and the genetic material nessary for its production (Plasmid vector pCIB4431) in corn. Pesticide Fact Sheet. U. S. Environmental Protection Agency, Washington, DC. pp. 18.

Applebaum, S. W., B. I. Harpaz, and A. Bondi. 1964. Comparative studies on proteolytic enzymes of *Tenebrio molitor* L. Comp. Biochem. Physiol. 11:85–103.

Arencibia, A., R. Vazquez, D. Prieto, P. Tellez, E. R. Carmona, A. Coego, L. Hernandez, G. De-La-Riva, and G. Selman-Housein. 1997. Transgenic sugarcane plants resistant to stem borer attack. Mol. Breed. 3:247–255.

Armstrong, C. L., G. B. Parker, J. C. Pershing, S. M. Brown, R. P. Sanders, R. D. Duncan, T. Stone, D. A. Dean, D. L. DeBoer, J. Hart, A. R. Howe, F. M. Morrish, M. E. Pajeau, W. L. Petersen, B. J. Reich, S. J. Sate, S. R. Sims, S. Stehling, R. Rodriguez, C. G. Santino, W. Schuler, L. J. Tarochione, and M. E. Fromm. 1995. Field evaluation of European corn borer control in progeny of 173 transgenic corn borer events expressing an insecticidal protein from *Bacillus thuringiensis*. Corp Sci. 35:550–557.

Arpaia, S. 1996. Ecological impact of Bt-transgenic plants: 1. Assessing possible effects of cryIIIB toxin on honey bee (Apis mellifera L.) colonies. J. Genet. Breed. 50:315–319.

Arpaia, S., G. Mennella, V. Onofaro, E. Perri, F. Suuseri, and G. L. Rotino. 1997. Production of transgenic eggplant (*Solanum melongena* L.) resistance to Colorado potato beetle (*Leptinotarsa decemlineata* Say). Theor. Appl. Genet. 95:329–334.

Ashfaq, M., S. Y. Young, and R. W. McNew. 2001. Larval mortality and development of *Pseudoplusia includens* (Lepidoptera: Noctuidae) reared on a transgenic *Bacillus thuringiensis*-cotton cultivar expressing CryIAc insecticidal protein. J. Econ. Entomol. 94:1053–1058.

Ballester, V., F. Granero, B. E. Tabashnik, T. Malvar, and J. Ferré. 1999. Integrative model for binding of *Bacillus thuringiensis* toxins in susceptible and resistant larvae of the diamondback moth (*Plutella xylostella*). Appl. Envir. Microbiol. 65:1413–1419.

Bartlett, A.C. 1995. Resistance of the pink bollworm to Bt transgenic cotton. Proc. 1995 Beltwide Cotton Conference. National Cotton Council. pp.766–768.

Barton, K. A., H. R. Whiteley, N.-S. Yang. 1987. *Bacillus thuringiensis* -endotoxin expressed in transgenic *Nicotiana tabacum* provides resistance to lepidopteran insects. Plant Physiol. 85:1103–1109.

Bauer, L. S. 1995. Resistance: A threat to the insecticidal crystal proteins of *Bacillus thuringiensis*. Florida Entomol. 78:414–443.

Benedict, J. H., E. S. Sachs, D. W. Altman, W. R. Deaton, R. J. Kohel, D. R. Ring, and S. A. Berberich. 1996. Field performance of cottons expressing transgenic CryIA insecticidal proteins for resistance to *Heliothis virescens* and *Helicoverpa zea* (Lepidoptera: Noctuidae). J. Econ. Entomol. 89:230–238.

Bevan, M.W., R.B. Flavell, and M.D. Chilton. 1983. A chimaeric antibiotic resistance gene as a selectable marker for plant cell transformation. Nature. 304:184–187.

Binning, R. R., and M. E. Rice. 2002. Effects of transgenic Bt corn on growth and development of the stalk borer *Papaipema nebris* (Lepidoptera: Noctuidae). J. Econ. Entomol. 95:622–627.

Birch, A. N. E., I. E. Geoghegan, M. E. N. Majerus, J. W. Mc Nicol, C. A. Hackett, A. M. R. Gatehouse, and J. A. Gatehouse. 1999. Tri-trophic interactions involving pest aphids, predatory 2-spot ladybirds and transgenic potatoes expressing snowdrop lectin for aphid resistance. Mol. Breed. 5:75–83.

Birk, Y. 1985. The Bowman-Birk inhibitor. Trypsin and chymotrypsin inhibitor from soybeans. Internat. J. Peptide Protein Res. 25:113–131

Bohorova, N., R. Frutos, M. Royer, P. Estañol, M. Pacheco, Q. Rascón, S. McLean, and D. Hoisington. 2001. Novel synthetic *Bacillus thuringiensis* cry1B gene and the cry1B-cry1Ab translational fusion confer resistance to southwestern corn borer, sugarcane borer and fall armyworm in transgenic tropical maize. Theor. Appl. Genet. 103:817–826.

Boulter, D. 1993. Insect control by copying nature using genetically engineered crops. Phytochem. 34:1453–1466.

Bourguet, D., J. Chaufaux, M. Séguin, C. Buisson, J. L. Hinton, T. J. Stodola, P. Porter, G. Cronholm, L. L. Buschman, and D. A. Andow. 2003. Frequency of alleles conferring resistance to Bt maize in French and US corn belt populations of the European corn borer. Theor. Appl. Genet. 106:1225–1233.

Bowman, D. E. 1944. Fractions derived from soybeas and navy beans which retard tryptic digestion of casein. Proc. Soc. Exp. Phsiol. Med. 57:139–140.

Brewer, G. J. 1991. Resistance to *Bacillus thuringiensis* subsp. *kurstaki* in the sunflower moth (Lepidoptera : Pyralidae). Environ. Entomol. 20:316–322.

Broadway, R. M., and S. S. Duffey. 1986. Plant proteinases inhibitors mechanism of action and effect on the growth and digestive physiology of larval *Heliothi zea* and *Spodoptera exigua.* J. Insect Physiol. 32:827–834.

Byrne, P., S. Ward, J. Harrington, and L. Fuller. 2003. Transgenic plants: An introduction and resource guide. March 2003. http://www.colostate.edu/programs/lifesciences/TransgenicCrops/.

Cao, J., J.-Z. Zhao, J. Tang, A. Shelton, and E. Earle. 2002. Broccoli plants with pyramided cry1Ac and cry1C Bt genes control diamondback moths resistant to Cry1A and Cry1C proteins. Theor. Appl. Genet. 105:258–264.

Caprio, M. A. 1998. Evaluating resistance management strategies for multiple toxins in the presence of external refuges. J. Econ. Entomol. 91:1021–1031.

Carozzi, N. B., G. W. Warren, S. M. Jayne, R. Lotstein, D. A. Rice, S. Evola, and M. G. Koziel. 1992. Expression of a chemeric CaMV 35S *Bacillus thuringiensis* insecticidal protein gene in transgenic tobacco. Plant Mol. Biol. 20:539–548.

Carrière, Y., and B. E. Tabashnik. 2001. Reversing insect adaptation to transgenic insecticidal plants. Proc. R. Soc. London B. 268:1475–148.

Carrière, Y., C. Ellers-Kirk, M. Sisterson, L. Antilla, M. Whitlow, T. J. Dennehy, and B. E. Tabashnik. 2003. Long-term regional suppression of pink bollworm by *Bacillus thuringiensis* cotton. Proc. Natl. Acad. Sci. USA. 100:1519–1523.

Chan, M. T., L. J. Chen, and H. H Chang. 1996. Expression of *Bacillus thuringiensis* (Bt) insecticidal crystal protein gene in transgenic potato. Botan. Bull. Acad. Sin. 37:17–23.

Chaufaux, J., J. Müller-Cohn, C. Buisson, V. Sanchis, D. Lereclus, and N. Pasteur. 1997. Inheritance of resistance to the *Bacillus thuringiensis* CryIC toxin in *Spodoptera littoralis* (Lepidoptera: Noctuidae). J. Econ. Entomol. 90: 873–878.

Chen, M. S., B. Johnson, L. Wen, S. Muthukrishnan, K. J. Kramer, T. D. Morgan, and G. R. Reeck. 1992. Rice cystatin-bacterial expression, purification, cysteine proteinase inhibitory activity and insect growth suppressing activity of a truncated form of the protein. Protein Expression Purification. 3:41–49.

Chen, Q., G. Jelenkovic, C. K. Chin, S. Billings, J. Eberhardt, J. C. Goffreda, and P. Day. 1995. Transfer and transcriptional expression of coleopteran cryIIIB endotoxin gene of *Bacillus thuringiensis* in eggplant. J. Am. Soc. Hort. Sci. 120:921–927.

Cho, Y. S., and S. C. Lee. 1994. Resistance development and cross-resistance of diamondback moth (Lepidoptera: Plutelidae) by single selection of several insecticides. Korean J. Appl. Entomol. 33:242–249.

Crickmore, N., D. R. Zeigler, J. Feitelson, E. Schnepf, J. Van Rie, D. Lerecus, J. Baum, and D. H. Dean. 1998. Revision of the nomenclature for the *Bacillus thuringiensis* pesticidal crystal proteins. *Microbiol. Molecular Biol. Rev.* 62:807–813.

Daly, J. C., and P. W. Wellings. 1996. Ecological constraints to the deployment of arthropod resistant crop plants: A cautionary tale. In: R. B. Floyd, A. W. Shepard, and P. J. De Barro (Eds.), Frontiers of Population Ecology. CSIRO Publishing, Melbourne. pp. 311–323.

Dandekar, A. M., G. H. McGranahan, P. V. Vail, S. L. Uratsu, C. Leslie, and J. S. Tebbets. 1994. Low levels of expression of wild type *Bacillus thuringiensis* var. *kurstaki* cryIA(c) sequences in transgenic walnut somatic embryos. Plant Sci. 96:151–162.

Delledonne, M., G. Allegro, B. Belenghi, A. Balestrazzi, F. Picco, A. Levine, S. Zelasco, P. Calligari, and M. Confalonieri. 2001. Transformation of white poplar (*Populus alba* L.) with a novel *Arabidopsis thaliana* cysteine proteinase inhibitor and analysis of insect pest resistance. Mol. Breed. 7:35 –42.

deMaagd, R.A., M.S.G. Kwa, H. van der Klei, T. Yamamoto, B. Schipper, J. M. Vlak, W. J. Stiekema, and D. Bosch. 1996. Domain III substitution in *Bacillus thuringiensis* delta-endotoxin CryIA(b) results in superior toxicity for *Spodoptera exigua* and altered membrane protein recognition. Appl. Environ. Microbiol. 62:1537–1543.

Denolf, P., S. Jansens, M. Peferoen, D. Degheel, and J. Van Rie. 1993. Two different *Bacillus thuringiensis* δ-endotoxin receptors in the midgut brush border membrane of the European corn borer, *Ostrinia nubilalis* (Hübner)(Lepidoptera: Pyralidae). Appl. Environ. Microbiol. 59:1828–1837.

Ding X., B. Gopalakrishnan, L. Johnson, F. F. White, X. Wang, T. D. Morgan, K. J. Kramer, and S. Muthukrishnan. 1998a. Insect resistance of transgenic tobacco expressing an insect chitinase gene. Transgenic Res. 7:77–84.

Ding, L.-C., C.-Y. Hu, K.-W. Yeh, P.-J. Wang, and K. E. Espelie. 1998b. Development of insect-resistant transgenic cauliflower plants expressing the trypsin inhibitor gene isolated from local sweet potato. Plant Cell Rep. 17:854–860.

Douches, D. S., A. L. Westedt, K. Zarka, and B. Schroeter. 1998. Potato transformation to combine natural and engineered resistance for controlling potato tuber moth. HortScience. 33:1053–1056.

Duck, N., and S. Evola. 1997. Use of transgenes to increase host plant resistance to insects; opportunities and challenges. In: N. Carozzi and M. Koziel (Eds). Advances in Insect Control: The Role of Transgenic Plants. Taylor and Francis, London. pp. 1–18.

Dutton, A., H. Klein, J. Romeis, and F. Bigler. 2002. Uptake of *Bt*-toxin by herbivores feeding on transgenic maize and consequences for the predator *Chrysoperla carnea*. Ecol. Entomol. 27:441–447.

Dutton, A., H. Klein, J. Romeis, and F. Bigler. 2003. Prey-mediated effects of *Bacillus thuringiensis* spray on the predator *Chrysoperla carnea* in maize. Biol. Control. 26:209–215.

Ebora, R .V., M. M. Ebora, and M. B. Sticklen. 1994. Transgenic potato expressing the *Bacillus thuringiensis* CryIA(c) gene effects on the survival and food consumption of *Phthorimea operculella* (Lepidoptera: Gelechiidae) and *Ostrinia nubilalis* (Lepidoptera: Noctuidae). J. Econ. Entomol. 87:1122–1127.

Elzen, G.W. 1997. Changes in resistance to insecticides in tobacco budworm populations in Mississippi 1993–1995. Southwest. Entomol. 22:61–72.

Edmonds, H. S., L. N. Gatehouse, V. A. Hilder, and J. A. Gatehouse. 1996. The inhibitory effects of the cysteine proteinases inhibitor, oryzacystatin, on digestive proteases and on larval survival and development of the southern corn rootworm (*Diabrotica undecimpunctata howardi*). Entomol. Exp. Appl. 78:83–94.

Estada, U., and J. Ferré. 1994. Binding of insecticidal crystal protein of *Bacillus thuringiensis* to the midgut brush border of the cabbage looper, *Trichoplusia ni* (Hübner)(Lepidoptera: Noctuidae), and selection for resistance to one of the crystal proteins. Appl. Environ. Microbiol. 60:3840–3846.

Feldman, J., and T. Stone. 1997. The development of a comprehensive resistance management plan for potatoes expressing the cry3A endotoxin. In: N. Carozzi and M. Koziel (Eds), Advances in Insect Control: The Role of Transgenic Plants. Taylor and Francis, London. pp. 49–61.

Feng, X., H. Chen, Y. Shuai, Q. Xie, and Y. Lu. 1996. A study on the resistance of diamondback moth to *Bacillus thuringiensis* in Guangdong. Acta Entomol. Sinica 39:238–245.

Ferber, D. 1999. Risks and benefits: GM crops in the cross hairs. Science. 286:1662–1666.

Fernandez-Cornejo, J., and W. D. McBride. 2004. Genetically engineered crops for pest management in U. S. agriculture: Farm-level effects. USDA Economic Research Service, Resource Economics Division, Agricultural Economic Report No. 786.

Ferré, J., and J. Van Rie. 2002. Biochemistry and genetics of insect resistance to *Bacillus thuringiensis*. Ann. Rev. Entomol. 47:501–533.

Ferré, J., M. D. Real, J. van Rie, S. Jansens, and M. Peferoen. 1991. Resistance to the *Bacillus thuringiensis* bioinsecticide in a field population of *Plutella xylostella* is due to a putative insect brush border membrane-binding molecules specific to *Bacillus thuringiensis* d-endotoxin by protein blot analysis. Appl. Environ. Microbiol. 57:2816–2820.

Fitches, E., A. M. R. Gatehouse, J. A. Gatehouse. 1997. Effects of snowdrop lectin (GNA) delivered via artificial diet and transgenic plants on the development of tomato moth (*Lacanobia oleracea*) larvae in laboratory and glasshouse trials. J. Insect Physiol. 43:727–739.

Flint, H. M., T. J. Henneberry, F. D. Wilson, E. Holguin, N. Parks, and R. E. Buchler. 1995. The effects of transgenic cotton, *Gossypium hirsutum* L., containing *Bacillus thuringiensis* toxin genes for the control of the pink bollworm *Pectinophora gossypiella* (Saunders) and other arthropods. Southwest. Entomol. 20:281–292.

Fraley, R.T., S.G. Rogers, R.B. Horsch, P.R. Sanders, J.S. Flick, S.P. Adams, M.L. Bittner, L.A. Brand, C.L. Fink, J.S. Fry, G.R. Galluppi, S.B. Goldberg, N.L. Hoffmann, and S.C. Woo. 1983. Expression of bacterial genes in plant cells. Proc. Natl. Acad. Sci. USA. 80:4803–4807.

Gahan, L. J., F. Gould, and D. G. Heckel. 2001. Identification of a gene associated with *Bt* resistance in *Heliothis virescens*. Science. 293:857–860.

Granero, F., V. Ballester, and J. Ferré. 1996. *Bacillus thuringiensis* crystal proteins Cry1Ab and Cry1Fa share a high affininty binding site in *Plutella xylostella* (L.). Biochem. Biophys. Commun. 224:779–784.

Garczynski, S. F., J. W. Crim, and M.J. Adang. 1991. Identification of putative insect brush border membrane binding molecules specific to *Bacillus thuringiensis* δ -endotoxin by protein blot anaysis. Appl. Environ. Microbiol. 57:2816–2820.

Gatehouse, A. M. R., and D. Boulter. 1983. Assessment of the antimetabolic effects of trypsin-inhibitors from cowpea (*Vigna unguiculata)* and other legumes on development of the bruchid beetle *Callosobruchus maculatus.* J. Sci. Food Agric. 34:345–350.

Gatehouse, A. M. R., G. M. Davison, C. A. Newell, A. Merryweather, W. D. O. Hamilton, E. P. J. Burgess, R. J. C. Gilbert, and J. A. Gatehouse. 1997. Transgenic potato plants with enhanced resistance to the tomato moth, *Lacanobia oleracea*: Growth room trials. Mol. Breed. 3:49–63.

Gatehouse, A. M. R., R. E. Down, K. S. Powell, N. Sauvion, Y. Rahbe, C. A. Newell, A. Merryweather, W. D. O. Hamilton, and J. A. Gatehouse. 1996. Transgenic potato plants with enhanced resistance to the peach-potato aphid *Myzus persicae.* Entomol. Exp. Appl. 79:295–307.

Genissel, A., S. Augustin, C. Courtin, G. Pilate, P. Lorme, and D. Bourguet. 2003. Initial frequency of alleles conferring resistance to *Bacillus thuringiensis* poplar in a field population of *Chrysomela tremulae.* Proc. Royal Soc. London Ser. B-Biol. Sci. 270:791–797.

Ghareyazie, B., F. Alinia, C. A. Menguito, L. G. Rubia, J. M. De-Palma, E. A. Liwanag, M. B. Cohen, G. S. Khush, and J. Bennett. 1997. Enhanced resistance to two stem borers in an aromatic rice containing a synthetic cryIA(b) gene. Mol. Breed. 3:401–414.

Gould, F. 1998. Sustainability of transgenic insecticidal cultivars: Integrating pest genetics and ecology. Ann. Rev. Entomol. 43:701–726.

Gould, F., A. Anderson, A. Reynolds, L. Bumgarner, and W. Moar. 1995. Selection and genetic analysis of a *Heliothis virescens* (Lepidoptera: Noctuidae) strain with high levels of resistance to *Bacillus thuringiensis* toxins. J. Econ. Entomol. 88:1545–1559.

Gould, F., A. Anderson, A. Jones, D. Sumerford, D. G. Heckel, J. Lopez, S. Micinski, R. Leonard, and M. Laster. 1997. Initial frequency of alleles for resistance to *Bacillus thuringiensis* toxins in field populations of *Heliothis virescens*. *Proc. Natl. Acad. Sci. USA.* 94:3519–3523.

Gould, F., A. Martinez-Ramirez, A. Anderson, J. Ferré, F. J. Silva, and W. J. Moar. 1992. Broad-spectrum resistance to *Bacillus thuringiensis toxins* in *Heliothis virescens*. Proc. Natl. Acad. Sci. USA. 89:7986–7990.

Graham, J., S. C. Gordon, and R. J. McNicol. 1997. The effect of the CpTi gene in strawberry against attack by vine weevil (*Otiorhynchus sulcatus* F., Coleoptera: Curculionidae). Ann. Appl. Biol. 131:133–139.

Hafez, M. H. S. Samlama, R. Aboul-Ela, F.N. Zaki, and M. Ragaei. 1997. *Bacillus thuringiensis* affecting the larval parasite *Meteorus laeviventris* Wesm. (Hym., Braconidae) associated with *Agrotis ypsilon* (Rott.) (Lep., Noctuidae) larvae. J. Appl. Entomol. 121:535–538.

Halcomb, J.L., J. H. Benedict, B. Cook, and D. R. Ring. 1996. Survival and growth of bollworm and tobacco budworm on nontransgenic and transgenic cotton expressing a cryIA insecticidal protein (Lepidoptera; Noctuidae). Environ. Entomol. 25:250–255.

Hama, H., K. Suzuki, and H. Tanaka. 1992. Inheritance and stability of resistance to *Bacillus thuringiensis* formulations of the diamondback moth, *Plutella xylostella* L. (Lepidoptera: Yponomeutidae). Appl. Entomol. Zool. 27:355–362.

Hamel, F., R. Boivin, C. Tremblay, and G. Bellemare. 1997. Structural and evolutionary relationships among chitinases of flowering plants. J. Mol. Evol. 44:614–624.

Heckel, D. G., L. C. Gahan, F. Gould, and A. Anderson. 1997. Indentification of a linkage group with a major effect on resistance to *Bacillus thuringiensis* CryIAc endotoxin in tobacco budworm (Lepidoptera : Noctuidae). J. Econ. Entomol. 90:75–86.

Herrera-Estrella, L., A. Depicker, M. van Montagu, and J. Schell. 1983. Expression of chimaeric genes transfered into plant cells using a Ti-plasmid-derived vector. Nature. 303:209–213.

Hilbeck, A., M. Baumgartner, P. M. Fried, and F. Bilger. 1998a. Effects of transgenic *Bacillus thuringiensis* corn-fed prey on mortality and development time of immature *Chrysoperla carnea* (Neuroptera: Chrysopidae). Environ. Entomol. 27:480–487.

Hilbeck, A., W. J. Moar, M. Pusztai-Carey, A. Filippini, and F. Bigler. 1998b. Toxicity of the *Bacillus thuringiensis* Cry1Ab toxin on the predator *Chrysoperla carnea* (Neuroptera: Chrysopidae) using diet incorporated bioassays. Environ. Entomol. 27:1255–1263.

Hilbeck, A., W. J. Moar, M. Pusztai-Carey, A. Filippini, and F. Bigler. 1999. Preymediated effects of Cry1Ab toxin and protoxin and Cry2A protoxin on the predator *Chrysoperla carnea*. Entomol. Exp. Appl. 91:305–316.

Hilder, V. A., A. M. R. Gatehouse, S. E. Sheerman, R. F. Barker, and D. Boulter. 1987. A novel mechanism of insect resistance engineered into tobacco. Nature. 330:160–163.

Huang, F., R. A. Higgins, and L. L. Buschman. 1997. Baseline susceptibility and changes in susceptibility to *Bacillus thuringiensis* subsp. *kurstaki* under selection pressure in European corn borer, *Ostrinia nubilalis* Hübner (Lepidoptera: Pyralidae). J. Econ. Entomol. 90:1137–1143.

Huang, F., L. L. Buschman, R. A. Higgins, and W. H. McGaughey. 1999a. Inheritance of resistance to *Bacillus thuringiensis* toxin (Dipel ES) in the European corn borer. Science. 284:965–967.

Huang, F., R. A. Higgins, and L. L. Buschman. 1999. Transgenic Bt-plants: Successes, challenges, and strategies. Pestology: Proceedings, II Asia and Pacific Plant Protection Conference. Mumbai, India. 23: 2–29.

Iannacone, R., P. D. Grieco, and F. Cellini. 1997. Specific sequence modifications of ac endotoxin gene result in high levels of expression and insect resistance. Plant Mol. Biol. 34:485–496.

Ishimoto, M., and K. Kitamura. 1989. Growth inhibitory effects of an α-amylase inhibitor from kidney bean, *Phaseolus vulgaris* (L.) on three species of bruchids (Coleoptera : Bruchidae). Appl. Entomol. Zool. 24:281–286.

Ishimoto, M., T. Sato, M. J. Chrispeels, and K. Kitamura. 1996. Bruchid resistance of transgenic azuki bean expressing seed alpha-amylase inhibitor of common bean. Entomol. Exp. Appl. 79: 309–315.

James, C. 2000. Global status of transgenic crops: Challenges and opportunities. In: A. D. Arencibia (Ed.), Plant Genetic Engineering: Towards the Third Millennium. Elsevier, New York. pp. 1–7.

James, C. 2003. Preview: Global Status of Commercialized Transgenic Crops: 2003. ISAAA Briefs. No. 30. ISAAA: Ithaca, NY.

James, C. 2004. Future Global Potential for Bt Cotton: Opportunities and Challenges. ISAAA: Ithaca, NY. http://www.isaaa.org/

Jansens, S., M. Cornelissne, R. De-Clercq, A. Reynaerts, and M. Peferoen. 1995. *Phthorimaea operculella* (Lepidoptera: Gelechiidae) resistance in potato by expression of the *Bacillus thuringiensis* cryIA(b) insecticidal crystal protein. J. Econ. Entomol. 88:1469–1476.

Jansens, S., A. Van-Vliet, C. Dickurt, L. Buyssel, C. Piens, B. Saey, A. De-Wulf, V. Gossele, A. Paez, E. Goebel, and M. Peferoen. 1997. Transgenic corn expressing a cry9C insecticidal protein from *Bacillus thuringiensis* protected from European corn borer damage. Crop Sci. 37:1616–1624.

Jurat-Fuentes, J. L., and M. J. Adang. 2001. Importance of Cry1 δ-endotoxin domain II Loops for binding specificity in *Heliothis virescens* (L.). Appl. Environ. Microbiol. 67:323–329.

Jurat-Fuentes, J. L., F. L. Gould, and M. J. Adang. 2003. Dual resistance to *Bacillus thuringiensis* Cry1Ac and Cry2Aa toxins in *Heliothis virescens* suggests multiple mechanisms of resistance. Appl. Environ. Microbiol. 69:5898–5906.

Jelenkovic, G., S. Billing, Q. Chen, J. Lashomb, G. Hamilton, and G. Ghidiu. 1998. Transformation of eggplant with synthetic cryIIIA gene produces a high level of resistance to the Colorado potato beetle. J. Am.Soc. Hort. Sci. 123:19–25.

Jenkins, J. N., J. C. Jr. McCarty, R. E. Buehler, J. Kiser, C. Williams, and T. Wofford. 1997. Resistance of cotton with delta-endotoxin genes from *Bacillus thuringiensis* var. *kurstaki* on selected Lepidopteran insects. Agron. J. 89:768–780.

Johnson, D. E., G. L. Brookhart, K. J. Kramer, B. D. Barnett, and W. H. McGaughey. 1990. Resistance to *Bacillus thuringiensis* by the Indian meal moth, *Plodia interpunctella*: comparison of midgut proteinases from susceptible and resistant larvae. J. Invertebr. Pathol. 55:235–244.

Johnson, M. T., and F. Gould. 1992. Interaction of genetically engineered host plant resistance and natural enemies of *Heliothis virescens* (Lepidoptera: Noctuidae) in tobacco. Environ. Entomol. 21:586–597.

Johnson, T. M., A. S. Rishi, P. Nayak, and S. K. Sen. 1996. Cloning of ac endotoxin gene of *Bacillus thuringiensis* var. *tenebrionis* and its transient expressing in indica rice. J. Biosci. 21:673–685.

Jongsma, M. A., and C. Bolter. 1997. The adaptation of insects to plant protease inhibitors. J. Insect Physiol. 43:885–895.

Jouanin, L., M. Bonade-Bottino, C. Girard, G. Morrot, and M. Giband. 1997. Transgenic plants for insect resistance. Plant Sci. 131:1–11.

Kanrar, S., J. Venkateswari, P. B. Kirti, and V. L. Chopra. 2002. Transgenic Indian mustard (*Brassica juncea*) with resistance to the mustard aphid (*Lipaphis erysimi* Kalt.). Plant Cell Rep. 20:976–981.

Kar, S., D. Basu, S. Das, N. A. Ramkrishnan, P. Mukherjee, P. Nayak, and S. K. Sen. 1997. Expression of CryIA(c) gene of *Bacillus thuringiensis* in transgenic chickpea plants inhibits development of pod-borer (*Heliothis armigera*) larvae. Transgen. Res. 6:177–185.

Kirsten, J., and M. Gouse. 2002. Bt cotton in South Africa: Adoption and impact on farm incomes amongst small- and large-scale farmers. ISB News Report October 2002. http://www.isb.vt.edu/articles/oct0204.htm.

Kleiner, K. W., D. D. Ellis, B. H. McCown, and K. F. Raffa. 1995. Field evaluation of transgenic poplar expressing a *Bacillus thuringiensis* cry1A(a) delta-endoxin gene against forest tent caterpillar (Lepidoptera: Lasiocampidae) and gypsy moth (Lepidoptera: Lymantriidae) following winter dormancy. Environ. Entomol. 24:1358–1364.

Knight, J. 2003. Agency ignoring its advisers' over *Bt* maize. Nature. 422:5.

Knols, B. G., and M. Dicke. 2003. *Bt* crop risk assessment in the Netherlands. Nature Biotechnol. 21:973–974.

Koziel, M. G., G. L. Beland, C. Bowman, N. B. Carozzi, R. Crenshaw, L. Crossland, J. Dawson, N. Desai, M. Hill, S. Kadwell, K. Lacinis, K. Lewis, D. Maddox, K. McPherson, M. R. Meghji, E. Merlin, R. Rhodes, G. W. Warren, M. Wright, and S. V. Evola. 1993. Field performance of elite transgenic maize plants expressing an insecticidal protein derived from *Bacillus thuringiensis*. Bio/Technol. 11:194–200.

Kramer, K. J., C. Dziadik-Turner, and D. Koga. 1985. Chitin metabolism in insects. In: G. A. Kerkut and L. I. Gilbert (Eds.), Comprehensive Insect Physiology, Biochemistry and Phramacology Volume 3. Pergamon Press, New York. pp. 75–115.

Lawrence, P. K., and K. R. Koundal. 2002. Plant protease inhibitors in control of phytophagous insects, *Electronic Journal of Biotechnology*, 5, April 15. http://www.ejbiotechnology.info/content/vol5/issue1/full/3/index.html

Lecardonnel, A., L. Chauvin, L. Jouanin, A. Beaujean, G. Prevost, and B. Sangwan-Norreel. 1999. Effects of rice cystatin I expression in transgenic potato on Colorado potato beetle larvae. Plant Sci. 140:71–79.

Lee, S. I., S.-H. Lee, J. C. Koo, H. J. Chun, C. O. Lim, J. H. Mun, Y. H. Song, and M. J. Cho. 1999. Soybean Kunitz trypsin inhibitor (SKTI) confers resistance to the brown planthopper (*Nilaparvata lugens* Stal) in transgenic rice. Mol. Breed. 5:1–9.

Lee, M. K., R. M. Aguda, M. B. Cohen, F. L. Gould, and D. H. Dean. 1997. Determination of binding of *Bacillus thuringiensis* δ -endotoxin receptors to rice stem borer midguts. Appl. Environ. Microbiol. 63:1453–1459.

Lee M. K., F. Rajamohan, F. Gould, and D. H. Dean. 1995. Resistance to *Bacillus thuringiensis* Cry1A δ-endotoxins in a laboratory-selected *Heliothis virescens* strain is related to receptor alteration. Appl. Environ. Microbiol. 61:3836–3842.

Leple, J. C., M. Bonadebottino, S. Augustin, G. Pilate, V. D. Letan, A. Delplanque, D. Cornu, and L. Jouanin. 1995. Toxicity to *Chrysomela tremulae* (Coleoptera: Chrysomelidae) of transgenic poplars expressing a cysteine proteinase inhibitor. Mol. Breed. 1:319–328.

Liang, C., G. Brookhart, G. H. Feng, G. R. Reeck, and K. J. Kramer. 1991. Inhibition of digestive proteinases of stored grain Coleoptera by oryzacystatin a cystein proteinase inhibitor from rice seed. FEBS Letters. 278:139–142.

Liu, Y., B. E. Tabashnik, and M. W. Johnson. 1995. Larval age affects resistance to *Bacillus thuringiensis* in diamondback moth (Lepidoptera: Plutellidae). J. Econ. Entomol. 88:788–792.

Liu, Y., B. E. Tabashnik, and M. Pusztai-Carey. 1996. Field-evolved resistance to *Bacillus thuringiensis* toxin CryIC in diamondback moth (Lepidoptera: Plutellidae). J. Econ. Entomol. 89:798–804.

Liu, Y., B. E. Tabashnik, B. E. Moar, and R. A. Smith. 1998. Synergism between *Bacillus thuringiensis* spores and toxins against resistant and susceptible diamondback moths (*Plutella xylostela*). Appl. Environ. Microbiol. 64:1385–1389.

Liu, Y., B. E. Tabashnik, S. K. Meyer, Y. C. Carrière, and A. C. Bartlett. 2001. Genetics of pink bollworm resistance to *Bacillus thuringiensis* toxin Cry1Ac. J. Econ. Entomol. 94:248–252.

Losey, J. E., L. S. Rayor, and M. E. Carter. 1999. Transgenic pollen harms monarch larvae. Nature. 399:214.

Lynch, R. E., B. R. Wiseman, D. Plaisted, and D. Warnick. 1999. Evaluation of transgenic sweet corn hybrids expressing cryIA(b) toxin for resistance to corn earworm and fall armyworm (Lepidoptera: Noctuidae). J. Econ. Entomol. 92:246–252.

Maqbool, S. B. S. Riazuddin, N. T. Loc., A. M. R. Gatehouse, J. A. Gatehouse, and P. Christou. 2001. Expression of multiple insecticidal genes confers broad resistance against a range of different rice pests. Mol. Breed. 7:85–93.

Machuka, J., E. J. M. Van Damme, W. J. Peumans, and L. E. N. Jackai. 1999. Effect of plant lectins on larval development of the legume pod borer, *Maruca vitrata*. Entomol. Exp. Appl. 93:179–187.

Marfà, V., E. Merlé, R. Gabarra, J. M. Vassal, E. Guiderdoni, and J. Messeguer. 2002. Influence of developmental stage of transgenic rice plants (cv. Senia) expressing the cry1B gene on the level of protection against the striped stem borer *(Chilo suppressalis)*. Plant Cell Rep. 20:1167–1172.

Masoero, F., M. Moschini, F. Rossi, A. Prandini, and A. Pietri. 1999. Nutritive value, mycotoxin contamination and *in vitro* rumen fermentation of normal and genetically modified corn (Cry1A9b) grown in northern Italy. Maydica. 44:205–209.

Mazier, M., J. Chaufaxu, V. Sanchis, D. Lereclus, M. Giband, and J. Tourneur. 1997. The cryptic gene from *Bacillus thuringiensis* provides protection against *Spodoptera littoralis* ih young transgenic plants. Plant Sci. 127:179–190.

McGaughey, W. H. 1985. Insect Resistance to the biological insecticide *Bacillus thuringiensis*. Science 229:193–195.

McGaughey, W. H., and R. W. Beeman. 1988. Resistance to *Bacillus thuringiensis* in colonies of Indianmeal moth and almond moth (Lepidoptera : Pyralidae). J. Econ. Entomol. 81:28–33.

McGaughey, W. H., and D. E. Johnson. 1994. Influence of crystal protein composition of *Bacillus thuringiensis* strains on cross-resistance in Indianmeal moths (Lepidoptera: Pyralidae). J. Econ. Entomol. 87:535–540.

McGaughey, W. H., and D. E. Johnson. 1992. Indianmeal moth (Lepidoptera: Pyralidae) resistance to different strains and mixtures of *Bacillus thuringiensis*. J. Econ. Entomol. 85:1594–1600.

Metz, T.D., R. Dixit, and E.D. Earle. 1995. Agrobacterium tumefaciens-mediated transformation of broccoli (*Brassica oleracea* var. *italica*) and cabbage (*B. oleracea* var. *capitata*). Plant Cell Reports 15:287–292.

Michaud, D. 1997. Avoiding protease-mediated resistance in herbivorous pests. Trends Biotech. 15:3–6.

Moar, W. J., M. Pusztai-Carey, H. van Faassen, D. Bosch, R. Frutos, C. Rang, K. Luo, and M. J. Adang 1995. Development of *Bacillus thuringiensis* CryIC resistance by *Spodoptera exigua* (Hübner) (Lepidoptera: Noctuidae). Appl. Environ. Microbiol. 61:2086–2092.

Mochizuki, A., Y. Nishizawa, H. Onodera, Y. Tabei, S. Toki, Y. Habu, M. Ugaki, and Y. Ohashi. 1999. Transgenic rice plants expressing a trypsin inhibitor are resistant against rice stem borers, *Chilo supressalis*. Entomol. Exp. Appl. 93:173–178.

Morin, S., R. W. Biggs, M. S. Sisterson, L. Shriver, C. Ellers-Kirk, D. Higginson, D. Holley, L. J. Gahan, D. G. Heckel, Y. Carriere, T. J. Dennehy, J. K. Brown, and B. E. Tabashnik. 2003. Three cadherin alleles associated with resistance to *Bacillus thuringiensis* in pink bollworm. Proc. Natl. Acad. Sci. USA. 100:5004–5009.

Müller-Cohn, J., J. Chaufaux, C. Buisson, N. Gilois, V. Sanchis, and D. Lereclus. 1996. *Spodoptera littoralis* (Lepidoptera: Noctuidae) resistance to CryIC and cross-resistance to other *Bacillus thuringiensis* crystal toxins. J. Econ. Entomol. 89:791–797.

Munkvold, G. P., R. L. Hellmich, and W. B. Showers. 1997. Reduced fusarium ear rot and symptomless infection in kernels of maize genetically engineered for European corn borer resistance. Phytopathol. 87:1071–1077.

Murai, N., D. W. Sutton, M. G. Murray, J. L. Slightom, D. J. Merlo, N. A. Reichert, C. Sengupta-Gopalan, C. A. Stock, R. F. Barker, J. D. Kemp, and T. C. Hall. 1983. Phaseolin gene from bean is expressed after transfer to sunflower via tumor-inducing plasmid vectors. Science. 222:476–482.

Murdock, L. L., and R. E. Shade. 2002. Lectins and proteinase inhibitors as plant defenses against insects. J. Agric. Food Chem. 50:6605–6611.

Nayak, P., D. Basu, S. Das, A. Basu, D. Ghosh, N. A. Ramakrishnan, M. Ghosh, and S. K. Sen. 1997. Transgenic elite indica rice plants expressing cryIAc delta-endotoxin of *Bacillus thuringiensis* are resistant against yellow stem borer (*Scirpophaga incertulas*). Proc. Natl. Acad. Sci. USA. 94:2111–2116.

Noteborn, H. P. J. M., M. E. Bienenmann-Ploum, J. H. J. van den Berg, G. A. Alink, L. Zolla, and H. A. Kuiper. 1993. Food safety of transgenic tomatoes expressing the insecticidal crystal protein CryIA(b) from *Bacillus thuringiensis* and the marker enzyme APH(3')11. Mededelingen-van-de-Faculteit-Landbouwwetenschappen,-Universiteit-Gent. 58:(4b)1851–1858.

Nutt, K. A., P. G. Allsopp, T. K. McGhie, K. M. Shepherd, P. A. Joyce, G. O. Taylor, R. B. McQualter, and G. R. Smith. 1999. Transgenic sugarcane with increased resistance to canegrubs. Proc. Aust. Soc. Sugar Cane Technol. 21:171–176.

Omer, A. D., J. Granett, A.M. Dandekar, J. A. Driver, S. L. Uratus, and F.A. Tang 1992. Effects of transgenic petunia expressing *Bacillus thuringiensis* toxin on selected lepidopteran insects. Biocontrol Sci. Technol. 7:437–488.

Oppert, B. S. 2001. Transgenic plants expressing enzyme inhibitors and the prospects for biopesticide development. In: O. Koul and G. S. Dhaliwal. (Eds.), Phytochemical Biopesticides, Harwood Academic, Amsterdam, pp. 83–95.

Oppert, B., K. J. Kramer, R. W. Beeman, D. Johnson, and W. H. McGaughey. 1997. Proteinase-mediated insect resistance to *Bacillus thuringiensis* toxins. J. Biol. Chem. 272:23473–23476.

Ostlie, K. R., W. D. Hutchison, and R. L. Hellmich. 1997. Bt-corn and European corn borer. Long term success through resistance management. North Central Region Extension Publication NCR 602. University of Minnesota, St. Paul, MN USA

Peck, S. L., F. Gould, and S. P. Ellner. 1999. Spread of resistance in spatially extended regions of transgenic cotton: Implications for management of *Heliothis virescens* (Lepidoptera: Noctuidae). J Econ. Entomol. 92:1–16.

Peferoen, M. 1997. Insect control with transgenic plants expressing *Bacillus thuringiensis* crystal proteins. In: N. Craozzi and M. Koziel (Eds.), Advances in Insect Control: The Role of Transgenic Plants, Taylor and Francis, London. pp. 21–48.

Peferoen, M., S. Jansens, A. Reynaerts, and L. Leemans. 1990. Potato plants with engineered resistance against insect attack. In: M. E. Vayda and W. C. Park (Eds.), Molecular and Cellular Biology of the Potato, CAB International, Wallingford, United Kingdom. pp. 193–204.

Perez, C., and A. M. Shelton. 1997. Resistance of *Plutella xylostella* (Lepidoptera: Plutellidae) to *Bacillus thuringiensis* Berliner in Central America. J. Econ. Entomol. 90:87–93.

Perlak, F. J., R. W. Deaton, T. A. Armstrong, R. L. Fuchs, S. R. Rims, J. T. Greenplate, and D. A. Fischhoff. 1990. Insect resistant cotton plants. Bio/Technol. 8:939–943.

Perlak, F. J., T. B. Stone, Y. M. Muskopf, L. J. Peterson, G. B. Parker, S. A. McPherson, J. Wyman, S. Love, G. Reed, D. Biever, and D. A. Fischoff. 1993. Genetically improved potatoes: protection from damage by Colorado potato beetles. Plant Mol. Bio. 22:313–321.

Pew Initiative on Food and Biotechnology . 2002. Three years later: Genetically Engineered Corn and the Monarch Butterfly Controversy. Pew Initiative on Food and Biotechnology, Washington, DC., 19p.

Phipps R., and J. R. Park. 2002. Environmental benefits of genetically modified crops: Global and European perspectives on their ability to reduce pesticide use. J. Animal Feed Sci. 11:1–18.

Pilcher, C. D., M. E. Rice, R. A. Higgins, K. L. Steffey, R. L. Hellmich, J. Witkowski, D. Calvin, K. R. Ostlie, and M. Gray. 2002. Biothechnology and the European corn borer: Measuring historical farmer perceptions and adoption of transgenic Bt corn as a pest management strategy. J. Econ. Entomol. 95:878–892.

Pilcher, C.D., J.J. Obrycki, M.E. Rice, and L.C. Lewis. 1997a. Preimaginal development, survival, field abundance of insect predators on transgenic *Bacillus thuringiensis* corn. Environ. Entomol. 26:446–454.

Pilcher, C.D., M.E. Rice, J.J. Obrycki, and L.C. Lewis. 1997. Field and laboratory evaluations of transgenic *Bacillus thuringiensis* corn on secondary lepidopteran pests (Lepidoptera: Noctuidae). J. Econ. Entomol. 90:669–678.

Pimentel, D. S., and Raven, P. H. 2000. *Bt* corn pollen impacts on nontarget Lepidoptera: Assessment of effects in nature. Proc. Natl. Acad. Sci. USA. 97:8198–8199.

Powell, K. S., A. M. R. Gatehouse, V. A. Hilder, and J. A. Gatehouse. 1993. Antimetabolic effects of plant lectins and plant and fungal enzymes on the nymphal stages of two important rice pests, *Nilaparvata lugens* and *Nephotettix cinciteps*. Entomol. Exp. Appl. 66:119–126.

Powell, K. S., J. Spence, M. Bharathi, J. A. Gatehouse, and A. M. R. Gatehouse. 1998. Immunohistochemical and developmental studies to elucidate the mechanism of action of the snow drop lectin on the rice brown planthopper, *Nilaparvata lugens* (Stal). J. Insect Physiol. 44:529–539.

Qaim, M., and D. Zilberman. 2003. Yield effects of genetically modified crops in developing countries. Science. 299:900–902.

Rahardja, U., and M. E. Whalon. 1995. Inheritance of resistance to *Bacillus thuringiensis* subsp. *tenebrionis* d-endotoxin in the Colorado potato beetle (Coleoptera : Chrysomelidae) J. Econ. Entomol. 88:21–26.

Ramachandran, S., G. D. Buntin, J. N. All, P. L. Raymer, and C. N. Stewart. 1998. Greenhouse and field evaluations of transgenic canola against diamondback moth, *Plutella xylostella*, and corn earworm, *Helicoverpa zea*. Entomol. Exp. Appl. 88:17–24.

Rao, K. V., K. S. Rathore, T. K. Hodges, X. Fu, E. Stoger, D. Sudhakar, S. Williams, P. Christou, M. Bharathi, D. P. Bown, K. S. Powell, J. Spence, A. M. R. Gatehouse, and J. Gatehouse. 1998. Expression of snowdrop lectin (GNA) in transgenic rice plants confers resistance to rice brown planthopper. Plant J. 15: 469–477.

Rhim, S. L., H. J. Cho, B. D. Kim, W. Schnetter, and K. Geider. 1995. Development of insect resistance in tomato plants expressing the delta–endotoxin gene of *Bacillus thuringiensis* subsp. *tenebrionis*. Mol. Breed. 1:229–236.

Rice, M. E., and C. D. Pilcher. 1998. Potential benefits and limitations of transgenic Bt corn for management of the European corn borer (Lepidoptera: Crambidae). Am. Entomol. 44:75–78.

Riddick, E. W., and P. Barbosa. 1998. Impact of Cry3A-intoxicated *Leptinotarsa decemlineata* (Coleoptera: Chrysomelidae) and pollen on consumption, development, and fecundity of *Coleomegilla maculata* (Coleoptera: Coccinellidae). Entomol. Soc. Amer. 91:303–307.

Rissler, J., and M. Mellon. 1996. The Ecological Risks of Engineered Crops, MIT Press, Cambridge, Mass.

Robison, D. J., B. H. McCown, and K. F. Raffa. 1994. Responses of gypsy moth (Lepidoptera: Lymantriidae) and forest tent caterpillar (Lepidoptera: Lasiocampidae) to transgenic poplar, *Populus* spp., containing a *Bacillus thuringiensis* delta-endotoxin gene. Environ. Entomol. 23:1030–1041.

Roush, R. T. 1997. Bt-transgenic crops: just another pretty insecticide or a chance for a new start in resistance management? Pesticide Sci. 51:328–334.

Sacchi, V. F., P. Parenti, G. M. Hanozet, B. Giordana, P. Luthy, and M. G. Wolfersberger. 1986. *Bacillus thuringiensis* toxin inhibits K$^+$-gradient-dependent amino acid transport across the brush border membrane of *Pieris brassicae* midgut cells. FEBS Lett. 204:213–218.

Sachs, E. S., J. H. Benedict, J. F. Taylor, D. M. Stelly, S. K. Davis, and D. W. Altman. 1996. Pyramiding *CryIA(b)* insecticidal protein and terpenoids in cotton to resist tobacco budworm (Lepidoptera: Noctuidae). Environ. Entomol. 25:1257–1266.

Salama, H. S., and M. M. Matter. 1991. Tolerance level to *Bacillus thuringiensis* Berliner in the cotton leafworm *Spodoptera littoralis* (Boisduval) (Lep., Noctuidae). J. Appl. Entomol. 111:225–230.

Sanchis, V., and D. J. Ellar. 1993. Identification and partial purification of a *Bacillus thuringiensis* Cry1C δ-endotoxin binding protein from *Spodoptera littoralis* gut membranes. FEBS Lett. 316:264–268.

Saraswathy, N., and P. A. Kumar. 2004. Protein engineering of d-endotoxins of *Bacillus thuringiensis*. Electronic J. Biotechnol. 7, August 15. http://www.ejbiotechnology.info/content/vol7/issue2/full/3/

Schroeder, H. E., S. Gollasch, A. Moore, L. M. Tabe, S. Craig, D. C. Hardie, D. Spencer, T. J. V. Higgins, and M. J. Chrispeels. 1995. Bean alpha-amylase inhibitor confers resistance to the pea weevil (*Bruchus pisorum*) in transgenic peas (*Pisum sativum* L). Plant Physiol. 107:1233–1239.

Schuler, T. H., G. M. Poppy, B. R. Kerry, and I. Denholm. 1998. Insect-resistant transgenic plants. Trends Biotechnol. 16:175–182.

Sears, M. K., R. L. Hellmich, D. F. Stanley-Horn, K. S. Oberhauser, J. M. Pleasants, H. R. Mattila, B. D. Siegfried, and G. P. Dively. 2001. Impact of *Bt* corn pollen on monarch butterfly populations: A risk assessment. Proc. Natl. Acad. Sci. USA. 98:11937–11942.

Service, R. F. 2001. Arson strikes research labs and tree farm in Pacific Northwest. Science. 292:1622–1223.

Setamou, M., J. S. Bernal, J. C. Legaspi, T. E. Mirkov, and B. C. Legaspi, Jr. 2002. Evaluation of lectin-expressing transgenic sugarcane against stalkborers (Lepidoptera: Pyralidae): Effects on life history parameters. J. Econ. Entomol. 95:469–477.

Shade, R. E., H. E. Schroeder, J. J. Pueyo, L. M. Tabe, L. L. Murdock, T. J. V. Higgins, and M. J. Chrispeels. 1994. Transgenic pea seeds expressing the α-amylase inhibitor of the common bean are resistant to bruchid beetles. Biotechnol. 12:793–796.

Sharma, H. C., K. K. Sharma, N. Seetharama, and R. Ortiz. 2000. Prospects for using transgenic resistance to insects in crop improvement. Electron. J. Biotechnol. 3:August 15. http://www.ejbiotechnology.info/content/vol3/issue2/full/3/index.html.

Shelton, A.M., J.-Z. Zhao, and R. T. Roush. 2002. Economic, ecological, food safety and social consequences of the deployment of Bt transgenic plants. Ann. Rev. Entomol. 47:845–881.

Shelton, A. M., J. L. Robertson, J. D. Tang, C. Perez, S. D. Eigenbrode, H. K. Perisler, W. T. Wilsey, and R. J. Cooley. 1993. Resistance of diamondback moth to *Bacillus thuringiensis* subspecies in the field. J. Econ. Entomol. 86:697–705.

Shin, D. I., G. K. Podila, Y. Huang, and D. F. Karnosky. 1994. Transgenic larch expressing genes for herbicide and insect resistance. Can. J. For. Res. 24:2059–2067.

Sims, S. R., and T. B. Stone. 1991. Genetic basis of tobacco budworm resistance to an engineered *Pseudomonas fluoresces* expressing the delta-endotoxin of *Bacillus thuringiensis kurstaki*. J. Invert. Pathol. 57:206–210.

Sims, S.R., and J.W. Martin. 1997. Effects of the *Bacillus thuringiensis* insecticidal proteins CryIA(b), CryIA(c), CryIIA, and CryIIIA on *Folsomia candida* and *Xenylla grisea* (Insects: Collembola). Pedobiologia. 41:412–416.

Singsit, C., M. L. Adang, R.E. Lynch, W. F. Anderson, A. Wang, G. Cardineau, and P. Ozias-Akins. 1997. Expression of a *Bacillus thuringiensis* cryIA(c) gene in transgenic peanut plants and its efficacy against lesser cornstalk borer. Transgen. Res. 6:169–176.

Schnepf, E., N. Crickmore, J. Van Rie, D. Lereclus, J. Baum, J. Feitelson, D. R. Zeigler, and D. H. Dean. 1998. *Bacillus thuringiensis* and its pesticidal crystal proteins. Microbiol. Mol. Biol. Rev. 62:775–806.

Song, S.S. 1991. Resistance of diamondback moth (*Plutella xylostella* L.: Ypoonomeutidae: Lepidoptera) against *Bacillus thuringiensis* Berliner. Korean J. Appl. Entolomol. 30:291–293.

Stanley-Horn, D. E., G. P. Dively, R. L. Hellmich, H. R. Mattila, M. K. Sears, R. Rose, L. C. H. Jesse, J. E. Losey, J. J. Obrycki, and L. Lewis. 2002. Assessing the impact of Cry1Ab-expressing corn pollen on monarch butterfly larvae in field studies. Proc. Natl. Acad. Sci. USA. 98:11931–11936.

Stewart, C. N. Jr, M. J. Adang, J. N. All, H. R. Boerma, G. Cardineau, D. Tucker, and W. A. Parrott. 1996a. Genetic trasmsformatoin, recovery, and characterization of fertile soybean transgenic for a synthetic *Bacillus thuringiensis* CryIA(c) gene. Plant Physiol. 112:121–129.

Stewart, C. N. Jr, M. J. Adang, J. N. All, P. L. Raymer, S. Ramachandran, and W. A. Parrott. 1996b. Insect control and dosage effects in transgenic canola containing a synthetic *Bacillus thuringiensis* CryIA(c) gene. Plant Physiol. 112:115–120.

Stoger, E., S. Williams, P. Christou, R. E. Down, and J. A. Gatehouse. 1999. Expression of the insecticidal lectin from snowdrop (*Galanthus nivalis* agglutinin; GNA) in transgenic wheat plants: effects on predation by the grain aphids *Sitobion avaenae*. Mol. Breed. 5:65–73.

Stone, T., S. R. Smis, and P. G. Marone. 1989. Selection of tobacco budworm for resistance to a genetically engineered *Pseudomonus fluorescens* containing the delta-endotoxin of *Bacillus thuringiensis* subsp. *kurstaki*. J. Invertebr. Pathol. 53:228–234.

Sutton, D. W., P. K. Havstad, and J. D. Kemp. 1992. Synthetic cryIIIA gene *Bacillus thuringiensis* improved for high expression in plants. Transgen. Res. 1:228–236.

Tabashnik, B. E., N. L. Cushing, N. Finson, and M. W. Johnson. 1990. Field development of resistance to *Bacillus thuringiensis* in diamondback moth (Lepidoptera : Plutellidae). J. Econ. Entomol. 83:1671–1676.

Tabashnik, B. E., N. Finson, C. F. Chilcutt, N. L. Cushing, and M. W. Johnson. 1993. Increasing efficiency of bioassays: evaluation of resistance to *Bacillus thuring*iensis in diamondback moth (Lepidoptera: Plutellidae). J. Econ. Entomol. 86:635–644.

Tabashnik, B. E., N. Finson, and M. W. Johnson. 1991. Managing resistance to *Bacillus thuringiensis*: lessons from the diamondback moth (Lepidoptera: Plutellidae). J. Econ. Entomol. 84:49–55.

Tabashnik, B. E., N. Finson and F. R. Groeters, M. J. Moar, M. W. Johnson, K. Luo, and M. J. Adang. 1994a. Reversal of resistance to *Bacillus thuringiensis* in *Plutella xylostella*. Proc. Natl. Acad. Sci. USA. 91:4120–4124.

Tabashnik, B. E., N. Finson and M. W. Johnson, and D. G. Heckel. 1994b. Cross-resistance to *Bacillus thuringiensis* toxin CryIF in diamondback moth (*Plutella xylostella*). Appl. Environ. Microbiol. 60: 4627–4629.

Tabashnik, B. E., N. Finson, M. W. Johnson, and D. G. Heckel. 1995. Prolonged selection affects stability of resistance to *Bacillus thuringiensis* in diamondback moth (Lepidoptera: Plutellidae). J. Econ. Entomol. 88:219–224.

Tabashnik, B. E., Y. B. Liu, T. Malvar, D. G. Heckel, L. Masson, V. Ballester, F. Granero, J. L. Mensua, and J. Ferré. 1997. Global variation in the genetic and biochemical basis of diamondback moth resistance to *Bacillus thuringiensis*. Proc. Natl. Acad. Sci. USA. 94:12780–12785.

Tabashnik, B. E., J. M. Schwartz, N. Finson, and M. W. Johnson. 1992. Inheritance of resistance to *Bacillus thuringiensis* in diamondback moth (Lepidoptera: Plutellidae). J. Econ. Entomol. 85:1046–1055.

Tang, J. D., S. Gilboa, R. T. Roush, and A. S. Shelton. 1997. Inheritance, stability, and lack-of-fitness costs of field-selected resistance to *Bacillus thuringiensis* in diamondback moth (Lepidoptera: Plutellidae) from Florida. J. Econ. Entomol. 90:732–741.

Tang, T., and Y. Tian. 2003. Transgenic loblolly pine (*Pinus taeda* L.) plants expressing a modified δ - endotoxin gene of *Baciullus thuringiensis* with enhanced resistance to *Dendrolimus punctatus* Walker and *Crypyothelea formosicola* Staud. J. Exp. Bot. 54:835–844.

Tao, R., A. M. Dandekar, S. L. Uratsu, P. V. Vail, and J. S. Tebbets. 1997. Engineering genetic resistance against insects in Japanese persimmon using the CryIA(c) gene of *Bacillus thuringiensis*. J. Am. Soc. Hort. Sci. 122:764–771.

Vadlamudi, R. K., T. H. Ji, and L. A. Bulla, Jr. 1993. A specific binding protein from *Manduca sexta* for the insecticidal toxin of *Bacillus thuringiensis* subsp. *berliner*. J. Biol. Chem. 268:12334–12340.

Van Aarssen, R., R. P. Soetaert, M. Stam, J. Dockx, V. Gossele, J. Seurinck, A. Reynaerts, and M. Cornelissen. 1995. CryIA(b) transcript formation in tobacco is inefficient. Plant Mol. Biol. 28:513–524.

Van Der-Salm, T., D. Bosch, G. Honee, L. Feng, E. Munsterman, P. Bakker, W. J. Stiekema, and B. Visser. 1994. Insect resistance of transgenic plants that express modified *Bacillus thuringiensis* CryIA(b) and CyIC gene: A resistance management strategy. Plant Mol. Biol. 26:51–59.

Van Rie, J., S. Jansens, H. Höfte, D. Degheele, and H. Van Mellaert. 1990a. Receptors on the brush border membrane of the insect midgut as determinants of the specificity of *Bacillus thuringiensis* delta-endotoxins. Appl. Environ. Microbiol. 56:1378–1385.

Van Rie J., W. H. McGaughey, D. E. Johnson, B. D. Barnett, and H. Van Mellaert. 1990b. Mechanism of insect resistance to the microbial insecticide *Bacillus thuringiensis*. Science. 247:72–74.

Van Tol, N.B., and G. L. Lentz. 1998. Influence of Bt cotton on benficial arthropod populations. Proc. U. S. Beltwide Cotton Conf. pp.1052–1054.

Van Damme, E. J. M., W. J. Peumans, A. Barre, and P. Rouge. 1998. Plant lectins: A composite of several distinct families of structurally and evolutionary related proteins with diverse biological roles. Crit. Rev. Plant Sci. 17:575–692.

Walker, D., H. R. Boerma, J. All, and W. Parrott. 2002. Combining *cry1Ac* with QTL alleles from PI 229358 to improve soybean resistance to lepidopteran pests. Mol. Breed. 9:43–51.

Wang, X., X. Ding, B. Gopalakrishnan, T. D. Morgan, L. B. Johnson, F. F. White, and K. J. Kramer. 1996. Characterization of a 46 kDa insect chitinase from transgenic tobacco. Insect Biochem. Mol. Biol. 26:1055–1064.

Warren, G. W., N. B. Carozzi, N. Desai, and M. G. Koziel. 1992. Field Evaluation of transgenic tobacco containing a *Bacillus thuringiensis* insecticidal protein gene. J. Econ. Entomol. 85:1651–1659.

Watt, K., J. Graham, S. C. Gordon, M. Woodhead, and R. J. McNicol. 1999. Current and future transgenic control strategies to vine weevil and other insect resistance in strawberry. J. Hort. Sci. & Biotechnol. 74:409–421.

Whalon, M. E., D. L. Miller, R. M. Hollingworth, E. J. Grafius, and J.R. Miller. 1993. Selection of a Colorado potato beetle (Coleoptera : Chrysomelidae) strain resistant to *Bacillus thuringiensis*. J. Econ. Entomol. 86:226–233.

Williams, W. P., J. B. Sagers, J. A. Hanten, F. M. Davis, and P. M. Burkley. 1997. Transgenic corn evaluated for resistance to fall armyworm and southwestern corn borer. Crop Sci. 37:957–962.

Wilson, F. D., H. M. Flint, W. R. Deaton, D. A. Fischhoff, F. J. Perlak, T. A. Armstrong, R. L. Fuchs, S. A. Berberich, N. J. Parks, and B. R. Stapp. 1992. Resistance of cotton lines containing a *Bacillus thuringiensis* toxin to pink bollworm (Lepidoptera: Gelechiidae) and other insects. J. Econ. Entomol. 85:1516–1521.

Wraight, C. L., A. R. Zangerl, M. J. Carroll, and M. R. Berenbaum. 2000. Absence of toxicity of *Bacillus thuringiensis* pollen to black swallowtails under field conditions. Proc. Nat. Acad. Sci. USA. 97:7700–7703.

Wright, D.J., M. Iqbal, and R.H.J. Verkerk. 1995. Resistance to *Bacillus thuringiensis* and abamectin in the diamondback moth, *plutella xylostella*: A major problem for integrated pest management. Mededelingen. Faculteit Landbouwkundige en toegepaste Biologische Wetenschappen Universiteit Gent 60(3B): 927–933.

Wright, D. J., M. Iqbal, F. Granero, and J. Ferré, J. 1997. A change in a single midgut receptor in *Plutella xylostella* is only in part responsible for field resistance to *Bacillus thuringiensis* subspp. *kurstaki* and *aizawai*. Appl. Environ. Microbiol. 63:1814–1819.

Wu, C., Y. Fan, C. Zhang, N. Oliva, and S. K. Datta. 1997. Transgenic fertile japonica rice plants expressing a modified cryIA(b) gene resistant to yellow stem borer. Plant Cell Rep. 17:129–132.

Wünn, J., A. Kloti, P. K. Burkhardt, G. C. G. Biswas, K. Launis, V. A. Iglesias, and I. Potrykus. 1996. Transgenic indica rice breeding line IR58 expressing a synthetic cryIA(beta) gene from *Bacillus thuringiensis* provides effective insect pest control. Bio-Technol. 14:171–176.

Xu, D., Q. Xue, D. McElroy, Y. Mawal, V. A. Hilder, and R. Wu. 1996. Consititutive expression of a cowpea trypsin inhibitor gene, *CpTi*, in transgenic rice plants confers reistance to two major rice insect pets. Mol. Breed. 2:167–173.

Ye, G. Y., Q.-Y. Shu, H.-W. Yao, H.-R. Cui, X.-Y. Cheng, C. Hu, Y.-W. Xia, M.-W. Gao, and I. Altosaar. 2001. Field evaluation of resistance of transgenic rice containing a synthetic cry1Ab gene from *Bacillus thuringiensis* Berliner to stem borers. J. Econ. Entomol. 94:271–276.

Yeh, K.-W., M.-I. Lin, S.-J. Tuan, Y.-M. Chen, C.-Y. Lin, and S.-S. Kao. 1997. Sweet potato (*Ipomoea batatas*) trypsin inhibitors expressed in transgenic tobacco plants confer resistance against *Spodoptera litura*. Plant Cell Reports. 16:696–699.

Yorobe, J. M., C. B. Quicoy, E. P. Alcantara, and B. R. Sumayao. 2004. Impact assessment of Bt corn in the Philippines. University of the Philippines Los Banos, College, Laguna, Philippines.

Yu, L., R.E. Berry, and B.A. Croft. 1997. Effects of *Bacillus thuringiensis* toxins in transgenic cotton and potato on *Folsomia candida* (Collembola: Isotomidae) and *Oppia nitens* (Acari: Oribatidae). J. Econ. Entomol. 90:113–118.

Zhu, K. Y., J. E. Huesing, R. E. Shade, and L. L. Murdock. 1994. Cowpea trypsin inhibitor and resistance to cowpea weevil (Coleoptera: Bruchidae) in cowpea variety TVu-2027. Environ. Entomol. 23:987–991.

CHAPTER 11

ARTHROPOD BIOTYPES

1. SIGNIFICANCE OF BIOTYPES TO THE DEVELOPMENT OF ARTHROPOD-RESISTANT CULTIVARS

Resistance genes in arthropod-resistant cultivars may be overcome by the development of resistance-breaking arthropod biotypes that possess an inherent genetic capability to overcome the properties of plant resistance genes. Biotypes have been defined as populations within an arthropod species that differ in their ability to utilize a particular trait in a particular plant genotype (Gallun and Khush 1989, Wilhoit 1992, Pedigo 1999). Biotypes are routinely detected by exposing a set of plant cultivars, each possessing a different arthropod resistance gene or gene combination, that react differentially to a given arthropod biotype (Starks and Burton 1972, Saxena and Barrion 1983, Tomar and Prasad 1992, Ratcliffe and Hatchett 1997). With the exception of the brown planthopper, *Nilaparvata lugens* (Stål), existing arthropod biotypes occur in association with individual plant resistance genes inherited as dominant traits, and exhibit virulence genes expressed after selection, recombination, or mutation, to nullify plant resistance genes.

The gene-for-gene relationship between arthropod virulence genes and plant resistance genes is similar to that described by described by Flor (1971) for plant pathogen resistance genes and pathogen virulence genes. In this hypothesis, a race-specific resistance reaction of the host plant is triggered by an interaction between the product of a dominant resistance gene of the plant and the product of a corresponding, dominant avirulence gene in the pathogen. Interactions between plant arthropod resistance genes and arthropod virulence genes occur in a similar manner. The virulence or avirulence of an arthropod biotype to a plant resistance gene depends on how resistance genes in the host plant and virulence genes in the arthropod interact. When gene products of the avirulent arthropod are recognized by the defense system of the resistant plant, the arthropod is unable to infest a resistant plant. Conversely, when a normally resistant plant does not recognize an arthropod's gene products, the virulent arthropod biotype overcomes the plant resistance gene or genes (Figure 11.1).

Puterka and Burton (1990) reasoned that biotypes develop from mutations or a pre-existing variability for virulence, from sexual recombination, or plant resistance gene selection pressure that may change arthropod virulence gene frequency. The intensity and duration of virulence gene expression in turn depend on the category of resistance exhibited by the plant resistance gene, the initial virulence gene frequency, and the interaction between the cultivar, the pest, and the environment.

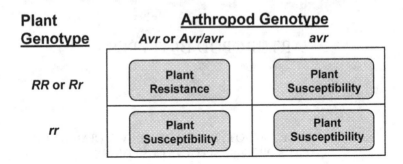

Plant Genotype	Arthropod Genotype	
	Avr or *Avr/avr*	*avr*
RR or *Rr*	Plant Resistance	Plant Susceptibility
rr	Plant Susceptibility	Plant Susceptibility

Figure 11.1. Hypothetical quadratic check of gene-for-gene interactions between arthropod avirulence and virulence genes with plant arthropod resistance and susceptibility genes, based on the gene-for-gene hypothesis of Flor (1971).

The use of the term biotype is viewed as problematic by some researchers because of the lack of knowledge about the genetic structure of arthropod biotypes. A major tenant of Flor's (1971) concept assumes the single gene relationships shown above between the pest and the host. However, arthropod biotypes may be populations of individuals exhibiting a particular set of virulence genes or arthropod populations responding similarly to a set of plant differentials with more than one gene (Wilhoit 1992). For example, *Mi-1.2* in tomato, *Lycopersicon esculentum* Mill., controls resistance to three different organisms, including some biotypes of the potato aphid, *Macrosiphum euphorbiae* (Thomas), one biotype of the whitefly, *Bemisia tabaci* (Gennadius), as well as three *Meloidogyne* spp. nematodes (Rossi et al. 1998, Goggin et al. 2001, Nombela et al. 2003) (see Chapter 9). As described in the following sections of this chapter, genes for virulence in the Hessian fly, *Mayetiola destructor* (Say), the gall midge, *Orseolia oryzae* (Wood Mason), and the greenbug, *Schizaphis graminum* Rondani, are inherited as simply recessive traits, fitting the classical gene-for-gene pattern. However, *N. lugens* resistance in rice exists in several different genes inherited as either dominant or recessive traits, depending on the biotype involved, and virulence in *N. lugens* is polygenic.

Mitchell-Olds and Bergelson (2000) suggest that because of the current developments in plant genomics, the use of a gene-for-gene concept may be oversimplified, and that a "gene-for-genome" concept will allow simultaneous evaluation of several resistance genes involved in potentially overcoming a pest virulence gene. Until an understanding is developed of the genome-wide changes in the response of several plant resistance genes to an arthropod pest however, researchers must rely on existing genetic models and a working definition of biotypes that includes both individuals and populations that exhibit virulence to different genes in arthropod-resistant plant cultivars. As pointed out by Harris et al. (2003) the current grass resistance gene-gall midge virulence gene model, though perhaps simplistic, serves as an excellent working hypothesis from which to study the genetic interactions between grasses and gall midge pests.

2. TAXONOMIC DISTRIBUTION OF BIOTYPES

Eighteen different species of Diptera, Homoptera, Acari and one species of Coleoptera have developed virulent biotypes to plant resistance genes (Table 11.1). Ten of the eighteen are aphid species, in which parthenogenic reproduction contributes greatly to the successful development of resistance-breaking biotypes.

Three of the other five species, *M. destructor, O. oryzae*, and the apple maggot fly, *Rhagoletis pomonella* (Walsh) are sexually dimorphic Diptera with very high reproductive potentials that infest large crop production areas. The green leafhopper, *Nephotettix virescens* (Distant), green rice leafhopper, *Nephotettix cincticeps* Uhler, and *N. lugens* occur continuously on large monocultures of rice, *Oryza sativa* (L.), in much of South and Southeast Asia. For additional information about specific biotypes beyond the following discussions, see Granett et al. (2001), Pani and Sahu (2000), Tanaka (1999) or Webster and Inayatulluh (1985).

3. THE DEVELOPMENT OF BIOTYPES

3.1. Fruit Crops

It is not surprising that one of the most notable instances of arthropod resistance - that of grape, *Vitus* spp., resistant to the grape phylloxera, *Daktulosphaira vitifoliae* (Fitch), (see Chapter 1) has an equally extensive history of phylloxera biotype development. After the re-establishment of the wine grape industry in France in the late 1880s, studies of the *Vitus*-phylloxera interaction became intensely serious. A recent review by Granett et al. (2001) indicated that over 480 articles were published on the phylloxera from 1868 to 1871. Galet (1982) cited over 2,000 citations on phylloxera in the approximate 100 years since it became a major economic pest. Painter (1951) cited over 140 articles on various aspects of *Vitus* resistance to phylloxera and phylloxera virulence alone.

Among these are several early reports on the possible existence of phylloxera biotypes, termed races at the time. C. V. Riley (1872) was the first to hypothesize that phylloxera biotypes would develop, after observing high levels of resistance in the *Vitus vinifera* x *Vitus labrusca* hybrid 'Concord' grape, to phylloxera in the U. S. in Missouri. Although biotypes did not develop at the time, Riley's theory proved prophetic. Soon after the turn of the century, reports of biotypic variation began to be published in Europe. Borner (1914) and Marchal (1931) in France, Grassi and Topi (1917) in Italy, Vasilev (1929) in the Ukraine, and Schilder (1947) in Germany, all reported two presumed phylloxera biotypes, based on biological, geographic, or morphological parameters. However, none of their claims could be substantiated.

Table 11.1. *Arthropod biotypes associated with plant resistance genes*

Crop	Arthropod	Number of Biotypes	Reference(s)
Malus spp.	Eriosoma lanigerum	2	Sen Gupta & Miles 1975
			Young et al. 1982
	Dysaphis devecta	3	Alston & Briggs 1977
	Rhagoletis pomonella	2	Prokopy et al. 1988
Medicago	Acyrthosiphon kondoi	2	Zarrabi et al. 1995
sativa	Acyrthosiphon pisum	4	Auclair 1978, Frazer 1972
	Therioaphis maculata	6	Nielson & Lehman 1980
Oryza	Nephotettix cincticeps	2	Sato & Sogawa 1981
sativa	Nephotettix virescens	3	Heinrichs & Rapusas 1985
			Takita & Hashim 1985
	Nilaparvata lugens	4	Verma et al. 1979
	Orseolia oryzae	4	Heinrichs & Pathak 1981
Rubus	Amphorophora idaei	5	Birch et al. 1996
idaeus			Gordon et al. 1999
			Jones et al. 2000
Rubus nigrum	Dasineura tetentsi	?	Hellqvist 2001
Triticum	Aceria toschiella	5	Harvey et al. 1997a,b,1999,2001
aestivum	Macrosiphum avenae	3	Lowe 1981
	Mayetiola destructor	16	Gallun & Reitz 1977
			Ratcliffe et al. 1994
	Diuraphis noxia	2	Basky et al. 2002, Haley et al. 2004, Smith et al. 2004
	Schizaphis graminum	8	Harvey & Hackerott 1969
			Kindler & Spomer 1986
			Porter et al. 1982, Puterka et al. 1988, Teetes et al. 1975, Wood 1961
Vegetables	Brevicoryne brassicae	2-4	Dunn & Kempton 1972
			Lammerink 1968
	Bemisia tabaci	2	Brown et al. 1995
			Nomella et al. 2003
	Trialeurodes vaporariorum	2	Lei et al. 1998
Vitus spp.	Daktulosphaira vitifoliae	2	Granett et al. 1985
			Williams & Shambaugh 1988
			Song & Granett 1990
			Martinez-Peniche 1999
Zea mays	Rhopalosiphum maidis	5	Singh & Painter 1964
			Wilde & Feese 1973

In the late 1970s, phylloxera infestations intensified on 'Concord' and 'Niagara' grapes in the northeastern U. S., reaching pest status. To determine if biotypes were developing, Williams and Shambaugh (1988) used biological studies to demonstrate that two phylloxera biotypes existed - the 'Clinton' biotype unable to feed on 'Concord' and six other grape species, and the 'Concord' biotype, unable to feed on 'Clinton' and nine other grape species (Table 11.2). Results of isozyme analyses supported the existence of two biotypes, as electrophoretic banding patterns for each was different for three different enzymes. As stated by Garnett et al. (2001), these results "fulfilled Riley's century old prediction of a Concord biotype".

The rootstock AXR#1, a *V. vinifera* x *Vitus rupestris* hybrid, was initially resistant to phylloxera, and was grown in many wine grape production areas of the world since the 1890s. This resistance has been overcome by virulent phylloxera populations in Europe, South Africa, and in the U. S.. AXR#1 was the rootstock in the majority of the Napa and Sonoma county wine producing areas of California in the 1960 and 1970s. Granett et al. (1985) used life-table studies to show that phylloxera from these areas grow much better on AXR#1 than phylloxera from other areas. Phylloxera populations avirulent to AXR#1 were named biotype A and the population virulent to AXR#1 was named biotype B (Table 11.2). The rootstock '5C', a *Vitus berlandieri* x *Vitus riparia* hybrid, possesses both antibiosis and antixenosis to biotypes A and B (Omer et al. 1999).

In France, Song and Granett (1990) identified a phylloxera biotype originating on the *V. berlandieri* x *V. vinifera* hybrid '41B'. The California A and B phylloxera biotypes are unaffected by the resistance in '41B' and grow equally well on it (Granett et al. 2001) (Table 11.2).

Table 11.2. Genes from Vitus *germplasm for resistance to* Daktuloshpaira vitifolliae [a]

Genes	Biotype		
	1	*2*	*3*
A_1	R	S	R
A_2	S	R	R
A_3A_4	S	R	S
A_5	R	S	S
A_6	R	S	S
A_7	R	S	S
$A_1 + A_2$	R	R	R
$A_1 + A_3$	R	R	R
$A_1 + A_4$	R	S	R
A10	R	R	R

[a] *R = resistant, S = susceptible. (from Granett et al. 1985, 2001*

Martinez-Peniche (1999) established populations of the 41B biotype from specific regions of France and identified RAPD (randomly amplified DNA polymorphism) primers (see Chapter 8) that differentially amplified DNA from both 41B virulent and avirulent populations.

At least 12 major genes in raspberry, *Rubus phoenicolasius* Maxim., express resistance to five different biotypes of the large raspberry aphid, *Amphorophora idaei* Borner, in the United Kingdom. Schwartz and Huber (1937) published one of the first reports of resistance *to A. idaei* in raspberry and Briggs (1965) proposed an original gene-for-gene hypothesis involving four *A. idaei* biotypes and specific *A. idaei* resistance genes. The A_1 gene and several uncharacterized minor genes (Jones et al. 2000) from red raspberry are present in approximately one-third of all U. K. cultivars. These genes confer resistance to biotype 1, but not biotype 2 or X, the three major biotypes. The A_{10} gene from North American black raspberry is also present in approximately one-third of all cultivars and is effective against all three biotypes. The prolonged use of A_1-based cultivars led to a shift in *A. idaei* virulence to A_1 from 3% in the 1960s to 75% in the 1990s (Birch et al. 1996). *A. idaei* clones virulent to A_{10} have also been developed (Gordon et al. 1999). Ribosomal DNA probes of biotypes 1, 2, and X have identified large genetic variation both within and between the three biotypes, suggesting that virulence to A_{10} may lead to the selection of further *A. idaei* biotypes (Birch et al. 1994).

Two biotypes of the wooly aphid, *Eriosoma lanigerum* (Hausmann), exist on the foliage of different clones of apple, *Malus domestica* Borkh., in south Australia. The Clare (native) biotype is avirulent to resistant apple cultivars such as 'Northern Spy', but the Blackwood biotype is virulent to several cultivars normally resistant to the Clare biotype (Sen Gupta and Miles 1975). *E. lanigerum* biotypes have also been shown to exist in response to different North American apple cultivars (Young et al. 1982).

Resistance in English apple cultivars to the rosy leaf curling aphid, *Dysaphis devecta* (Walker), follows a gene-for-gene relationship (Alston and Briggs 1977). Four resistance genes govern resistance to three aphid biotypes. Biotypes 1 and 2 are avirulent on 'Cox's Orange Pippin', which is protected by the *Sd1* gene (Roche et al. 1997). Resistance to biotype 1 only exists in 'Northern Spy', via the *Sd2* gene, and a single gene (*Sd3*) in *Malus robusta* (Carr.) Rehd., and *Malus* x *zumi* selections controls resistance to biotype 3. Several cultivars also possess precursor genes necessary for biotype-specific genes for resistance to be effective. Prokopy et al. (1988) identified races of *R. pomonella* that exhibit biotype-like characteristics.

3.2. Legumes

As early as 1942, Harrington (1943) recognized differences in populations of the pea aphid, *Acyrthosiphon pisum* (Harris), and Cartier (1959, 1963) demonstrated the existence of three *A. pisum* biotypes on peas in the field and greenhouse. Frazer (1972) and Auclair (1978) determined that morphological differences exist between two of the three biotypes in North America, and that a fourth biotype exists in a pink color morph from France. Differences among the four biotypes are also evident when they are reared on an artificial diet.

The first biotype of the spotted alfalfa aphid, *Therioaphis maculata* (Buckton), was identified by Pesho et al. (1960) in Arizona. Since then, six additional biotypes (A, B, E, F, H, N) have been elucidated, based on differential responses to clones of 'Moapa' alfalfa, *Medicago sativa* L. (Table 11.3) (Nielson and Lehman 1980). Biotypes C, D, and G are resistant to organophosphorous insecticides (Nielson and Don 1974). Biotype development in *T. maculata* also follows a gene-for-gene progression, although genetic analysis of alfalfa is difficult due to its heterozygosity and tetraploidy.

The blue alfalfa aphid, *Acyrthosiphon kondoi* Shinji, became a pest of alfalfa in the U. S. during the decade of the 1970s. In response, the first alfalfa cultivar with *A. kondoi* resistance, 'CUF 101' was released in 1977 (Lehman et al. 1983). Germplasm previously resistant *A. kondoi* was later observed to be susceptible in 1989 in New Mexico and in Oklahoma in 1991. Zarrabi et al. (1995) compared the response of 'CUF 101' to *A. kondoi* from 1983 to 1992 and found that the frequency of resistance in 'CUF 101' declined from 50% to 19%. The virulent *A. kondoi* biotype was named BAOK90.

Table 11.3. Resistance of 'Moapa' Medicago sativa *clones to biotypes of* Therioaphis maculata [a]

Medicago clone	Aphid biotype [b]					
	A	B	E	F	H	N [c]
C-903	S	R	S	S	S	R
C-904	R	R	S	I	I	R
C-905	R	R	S	S	S	R
C-906	R	R	R	S	S	R
C-907	R	R	R	R	S	R
C-908	S	R	S	S	S	R
C-909	R	R	R	I	I	R
C-910	R	R	R	I	S	R
C-911	S	R	S	S	S	S

[a] *R - resistant; I - Intermediate; S - susceptible;* [b] *biotypes C, D, and G are resistant to organophosphorous insecticides;* [c] *Nebraska population. Adapted from Breeding approaches in alfalfa, pp. 279-311, Nielson, M. W., and W. F. Lehman, In: F. G. Maxwell and P. R. Jennings (Eds.), Breeding Plants Resistant to Insects, Copyright 1980, Reprinted with permission of John Wiley & Sons, Inc.*

3.3. Maize

Biotype development in the corn leaf aphid, *Rhopalosiphum maidis* (Fitch), was first detected by Cartier and Painter (1956) who observed a population denoted KS-2, capable of surviving on cultivars of sorghum, *Sorghum bicolor* (L.) Moench, previously resistant to the parent aphid population KS-1. KS-3, a biotype capable of infesting wheat, *Triticum aestivum* L., and KS-4, a biotype capable of reproduction on barley, *Hordeum vulgare* L., at low (13° C) temperature were later detected by

Painter and Pathak (1962) and Singh and Painter (1964). KS-5, detected in *R. maidis* field populations by Whalon et al. (1973), displays high temperature (29° C) tolerance and the ability to feed on the wheat species *Triticum timopheevi* Zhuk., that is resistant to biotypes 1-4.

3.4. Rice

Nilaparvata lugens is the most serious arthropod pest of rice in the world, due to direct feeding damage, as well as the effects of the grassy stunt and ragged stunt viruses vectored by this pest. Four *N. lugens* biotypes exist due to their abilities to destroy different rice cultivars with specific genes or gene combinations for hopper resistance (Table 11.4). Selection studies by Pathak and Heinrichs (1982) isolated a Philippine *N. lugens* population unable to destroy the cultivar 'Mudgo', (biotype 1); a population damaging to 'Mudgo' but avirulent to the cultivar 'ASD7' (biotype 2); and a laboratory population created in the Philippines that destroys 'ASD7' (biotype 3). Biotypes 1 and 2 were originally termed Philippine biotypes, but are now widely distributed in southeast Asia. *N. lugens* biotype 4 occurs in Bangladesh, Sri Lanka and Southern India, based on virulence to the resistance genes in 'Mudgo' and 'ASD7'. *N. lugens* biotypes vary considerably in both physiological and behavioral aspects of feeding. Biotypes 1, 2, and 3 have been differentiated based on the weight of honeydew collection, amino acids excretion, and morphological characters (Paguia et al. 1980, Sogawa 1978a, Saxena and Rueda 1982, Saxena and Barrion 1983). Biotype 2 also exhibits a distinct electrophoretic variant (Sogawa 1978b).

The dominant gene *Bph-1* in 'Mudgo' confers resistance to Philippine biotypes 1 and 3 (Khush and Brar 1991) (Table 11.4). 'ASD7' contains a single recessive gene *bph-2*, and is resistant to biotypes 1 and 2, but susceptible to biotype 3. *Bph-3* a single dominant gene in the cultivar 'Rathu Heenati' and *bph-4*, a single recessive gene in the cultivar 'Babawee', provide resistance to all known biotypes. 'ARC 10550', a cultivar susceptible to biotypes 1, 2, and 3, is resistant to biotype 4, and contains the single recessive gene *bph-5*, that segregates independently of *Bph-1*, *bph-2, Bph-3* or *bph-4* (Table 9.4). A single dominant gene in 'Swarnalata', also resistant to biotype 4, is independent of *bph-5* and has been termed *Bph-6*. A recessive gene in the cultivar 'T12', also resistant to biotype 4, is non-allelic with *bph-5*, and has been termed *bph-7*. A single recessive gene in 'Chin Saba', controls resistance to biotype 1, 2, and 3 and is non-allelic to *bph-2* or *bph-4*. This gene has been designated *bph-8*. Similarly, a single dominant gene in the cultivar 'Balamawee' that controls resistance to biotypes 1, 2, and 3, is non-allelic to *Bph-1* and *Bph-3*. The Balamawee gene has been designated *Bph-9*.

Shifts in the frequency of different *N. lugens* biotypes have been observed since the beginning of the Green Revolution of cultivar improvement in response to changes in cultivars and cultural practices. N. lugens populations in China and Japan shifted from biotype 1 to biotype 2 between 1987 and 1992 (Sogawa 1992, Takahashi et al. 1994). Ito et al. (1994) determined that Malaysian *N. lugens* populations are a mixture of biotypes 2 and 3. The biotype composition of populations from Japan, Malaysia, and Thailand were further compared by Wada et al. (1994) who determined that populations in tropical areas of Malaysia, Thailand and South Vietnam are predominantly biotype 4 (virulent to *bph-2*) and that populations in Japan and North Vietnam are biotype 2 (avirulent to *bph-2).*

After intensive cultivation of 'IR 26', which contained the *Bph-1* gene, in the Philippines and Indonesia in the 1970s, biotype 2 became predominant. 'IR 36' was then cultivated on about 10 million hectares in the Philippines, Indonesia, the Solomon Islands, and Vietnam, from the late 1970s to the early 1980s. During this period, 'IR36' displayed greater durability to *N. lugens* infestations than many other rice cultivars (Pathak and Heinrichs 1982). However, that resistance has now been overcome and IR36 is now fully susceptible to biotype 2 (Cohen et al. 1997).

Alam and Cohen (1998) noted a similar trend in an evaluation of 'IR64', which was released in 1985 and is presently the predominant rice cultivar in the world (Khush 1995). 'IR64' contains *Bph-1* and several minor genes (Khush 1989) and maintained a moderate level of resistance to biotype 2 in the field for more than 10 years (Cohen et al. 1997) and in greenhouse experiments involving forced selection for virulence (Alam and Cohen 1998). The stability of both 'IR 36' and 'IR64' resistance is thought to be due to the effects of *Bph-1* ('IR36') or *bph-2* ('IR64') with several minor genes.

Table 11.4. Relationships between genes from different Oryza sativa *cultivars to damage by biotypes of* Nilaparvata lugens [a]

Cultivar	Gene	Biotype			
		1	2	3	4
Taichung Native 1	None	S	S	S	S
Mudgo	*Bph1*	R	S	R	S
ASD7	*bph2*	R	R	S	S
Rathu Heenati	*Bph3*	R	R	R	R
Babawee	*bph4*	R	R	R	R
ARC 10550	*bph5*	S	S	S	R
Swarnalata	*Bph6*	S	S	S	R
T12	*bph7*	S	S	S	R
Chin Saba	*bph8*	R	R	R	--
Balamawee	*Bph9*	R	R	R	--

[a] *R - resistant ; MR - moderately resistant, S - susceptible, (--) - reaction not known; Reprinted from Advances in Agronomy, Vol. 45, Khush, G. S., and D. S. Brar. 1991. Genetics of resistance to insects in crop plants. Pages 223 - 274, Copyright 1991, Academic Press, with permission from Elseiver.*

Biotypes 1, 2, and 3 can be effectively converted from themselves to each of the other two biotypes, depending on the rice cultivar on which they are reared (Claridge and Den Hollander 1982). These findings are supported by results that indicate *N. lugens* virulence to different rice cultivars is variable within and among populations and is controlled by a system of polygenes (Claridge and Den Hollander 1983, Cohen et al. 1997, Gallagher et al. 1994, Guo-rui et al. 1983, Roderick 1994, Tanaka 1999). This is not surprising, given the multigenic nature of rice resistance to *N. lugens*, as well as the involvement of both the dominant and recessive modes of inheritance of resistance in rice.

There is little evidence to support the idea that virulence characteristics are used by one *N. lugens* biotype to become another. Results of several studies have shown that there is little correlation between virulence to *Bph1* and *bph2* in different *N. lugens* populations, and that shifts from biotype 1 to biotype 2 and 3 occur independently (den Hollander and Pathak 1981, Sogawa 1981, Claridge and den Hollander 1982, Pathak and Heinrichs 1982, Tanaka 1999). Ketipearachchi et al. (1998) have selected populations virulent to cultivars carrying either *bph-8* or *Bph-9*.

Nephotettix virescens, a vector of rice tungro virus, has shown the ability to overcome genes for hopper resistance in several improved rice cultivars in both laboratory experiments (Takita and Hashim 1985, Heinrichs and Rapusas 1990) and after field infestations for several years (Saito and Sogawa 1981, Rapusas and Heinrichs 1986).

Orseolia oryzae is a major pest in rice production in Asia. Infestations cause grain panicles to form galls, referred to as 'onion shoots' or 'silver shoots', resulting in annual grain yield losses of more than $500 million (Herdt 1991). As the planting of *Orseolia*-resistant cultivars has increased, so has the development of *Orseolia* biotypes. Two *O. oryzae* biotype groups exist; one in China and one in India (Table 11.5).

Table 11.5. Reaction of Oryza sativa *cultivars with various genes for resistance to biotypes of* Orseolia oryzae *in India and China* [a, b]

Resistance gene	Cultivar origin	Orseolia oryzae country of origin									
		India					China				
Gm1	Samridhi	R	S	R	S	R	R	MR	S	R	S
Gm2	Phalguna	R	R	S	S	R	S	R	R	S	S
Gm3	RP20681853	R	-	S	S	R	-	-	-	-	-
Gm4	Abhaya	R	R	S	R	S	-	-	-	-	-
Gm5	ARC5984	R	R	S	S	R	-	R	S	-	R
Gm6(t)	Duckang #1	S	S	S	S	-	-	R	R	R	R
Gm7	ARC10659	R	R	S	R	-	S	S	S	S	-
Gm?	IR36	R	R	R	R	S	S	S	S	S	S

[a] *R = resistant ; MR = moderately resistant, S = susceptible, - reaction not known. Reprinted from Katiyar, S. K., Y. Tan, B. Huang, G. Chandel, Y. Xu, Y. Zhang, Z. Xie, and J. Bennett. 2001.* Molecular mapping of gene Gm-6(t) which confers resistance against four biotypes of Asian rice gall midge in China. *Theor. Appl. Genet. 103:953-961, Copyright 2001 Springer Verlag, and Sardesai, N., K. R. Rajyashri, S. K. Behura, S. Nair, and M. Mohan. 2001. Genetic. physiological and molecular interactions of rice and its major diptera pest, gall midge. Plant Cell Tissue Organ Cult. 64:115-131, Copyright 2001, Kluwer Academic Publishers, with kind permission of Springer Science and Business Media.*

Six biotypes from India have been identified and four biotypes exist in China. Two gall midge biotypes exist in Sri Lanka and others may exist in Cambodia, Indonesia, Laos, Thailand and Vietnam (Katiyar and Bennett 2001). The original biotype determinations were based on midge development and survival on a differential set of cultivars with various midge resistance genes (Bentur et al. 1994, Bentur et al. 1996).

At least seven distinct rice *Orseolia* resistance genes have been identified but none are resistant to all midge biotypes (Katiyar et al. 2001, Sardesai et al. 2001, 2002) (Table 11.5). The *Gm6* gene in 'Duokang #1' is resistant to all Chinese biotypes. An unnamed gene in 'IR36' (*N. lugens* resistant) is resistant to four of six Indian biotypes. All *Gm* genes are inherited as dominant traits with the exception of *gm3*.

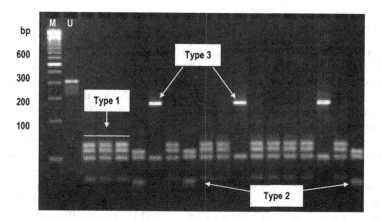

Figure 11.2. Mitotypes of the Indian biotype of Orseolia oryzae. *DraI restriction enzyme digests of the PCR amplification products of adult midge genomic DNA using primers specific to the 12S rRNA gene. M -100 bp ladder, U - undigested PCR product. (Courtesy Modan Mohan, International Centre for Genetic Engineering and Biotechnology)*

Biotypes have begun to be identified using several types of midge DNA analyses (Figure 11.2). Behura et al. (1999) developed a PCR-based assay to differentiate between the Indian *Orseolia* biotypes, based in DNA polymorphisms related to amplification by RAPD primers. Amplification products were sequenced and converted to SCAR (sequence characterized amplified region) primers. These SCAR primers differentially amplify the DNA of the six Indian biotypes, as well as that of the African gall midge, *Orseolia oryzivora* Harris and Gagné. The composition of the Chinese and Indian *O. oryzae* groups has been more closely evaluated using AFLP (amplified fragment length polymorphism) cluster analyses (Katiyar et al. 2000). The China group is more accurately composed of members from southern China, Laos, and northeast India, while the second group contains midges from multiple locations in India, Nepal and Sri Lanka. These data indicate that Indian biotypes 2 and 4 and Chinese biotypes 1 and 4 have likely arisen from recent mutation and selection.

AFLP technology has also been used for even more fine-scale divisions of *O. oryzae* biotypes. Behura et al. (2000) identified an AFLP marker specifically amplified in Indian biotypes 1, 2 and 5 (avirulent to *Gm2*), but absent in biotype 4 (virulent to *Gm2*). The AFLP marker sequence was used to develop a SCAR primer for PCR assay. *O. oryzae* avirulence to *Gm2* is inherited as a sex-linked, recessive trait. When the progeny of crosses between biotypes 1, 2, and 4 are assessed, the SCAR marker is present only in male progeny whose mother is avirulent to *Gm2*. Additional correlations have been identified between *Orseolia* biotypes to some members of the *Stowaway* miniature inverted repeat transposable element superfamily in rice (Behura et al. 2001a) and to intracellular infection by *Wolbachia* bacteria (Behura et al. 2001b).

3.5. Vegetables

Though not as pronounced as biotypes occurring on some other crops, biotypes exist to various resistant cultivars of different vegetables. The best known example is that of a biotype of the cowpea weevil, *Callosobruchus maculatus* (F.), occurring on a landrace of cowpea, *Vigna unguiculata* (L.) Walp., with weevil resistance in Africa. Shade et al. (1996) identified the first evidence of an African *C. maculatus* biotype in 1983, when they noted that approximately 15% of a weevil population from Niger could overcome the resistance in TVu2027, a weevil-resistant cowpea landrace. Several additional studies (Dick and Credland 1986a,b, Ofuyan and Credland 1996, Shade et al. 1999, Appleby and Credland 2003) have since determined that *C. maculatus* populations from Nigeria and Yemen are virulent to the resistance in TVu2027 but that Brazilian populations are avirulent. Shade et al. (1996) conducted extensive selection and intercrossing experiments to demonstrate that TVu2027 can be effectively overcome by the virulent *C. maculatus* Niger biotype in 53 generations in the laboratory. Three biotypes of the cowpea aphid, *Aphis craccivora* (Koch), also occur in Africa. Biotypes A and B occur on different sources of aphid resistance in Niger and biotype K occurs on resistant germplasm in Upper Volta (IITA 1981).

De Kogel et al. (1997) detected evidence of Dutch and Italian biotypes of the western flower thrips, *Frankliniella occidentalis* (Pergande), on accessions of cucumber, *Cucurbita sativus* L., normally resistant to *F. occidentalis*. Biotypes of the cabbage aphid, *Brevicoryne brassicae* (L.), exist among different cultivars of rape, *Brassica napus* L., in New Zealand (Lammerink 1968) and among clones of Brussels sprouts, *Brassica oleracea* var. *gemmifera* Zenker., in England (Dunn and Kempton 1972).

3.6. Wheat

Development of improved wheat cultivars in the U. S. began in New York in the early 1900's, and has been closely paralleled by the development of morphologically indistinguishable biotypes of *M. destructor*. The first biotypes were recognized by Painter (1930) who referred to them as 'biological strains'. Sixteen current *M. destructor* biotypes are differentiated by how various combinations of fly virulence genes interact with wheat genes for *M. destructor* resistance (Table 11.6). These interactions are the most thoroughly studied gene-for-gene relationships between an arthropod and its host plant.

Table 11.6. Cereal genes for resistance to Mayetiola destructor and virulence genes in M. destructor biotypes [a]

Resistance gene(s), Source, Cultivar, and PI or CI Number	Biotype [b]												
	GP	A	B	C	D	E	F	G	J	L	M	O	SP
None, *Triticum aestivum*, Turkey	S [c]	S	S	S	S	S	S	S	S	S	S	S	-
H1H2, *T. aestivum*, Dawson, CI3342	R	R	S	S	S	S	S	S	R	S	S	S	-
H3, *T. aestivum*, Monon, CI12061	R	R	S	R	S	R	R	S	S	S	S	S	R
H4, *T. aestivum*, Java, CI10051	?	?	?	?	?	?	?	?	?	?	?	?	?
H5, *T. aestivum*,Magnum, PI562068	R	R	MR	S	R	S	R	S	S	S	S	S	R
H6, *T. turgidum*,Caldwell, PI94587	R	R	MR	S	S	MR	MR	R	S	S	R	R	-
H7H8,*T. aestivum*, Seneca & CI12529	R	S	S	S	R	R	S	S	S	S	R	R	R
H9, *T. turgidum*, CI17714	R	R	R	R	R	R	R	MR	-	R	MR	MR	R
H10, *T. turgidum*, CI17714	R	R	MR	R	R	S	MR	S	-	R	MR	R	-
H11, *T. turgidum*, PI94587	R	R	R	R	R	R	S	R	S	R	R	S	-
H12, *T. aestivum*, Lusco	R	R	MR	R	R	S	MR	R	-	MR	MR	MR	R
H14, *T. turgidum*, CI1764	R	-	R	R	R	R	R	R	S	S	MR	MR	R
H15, *T. turgidum*, CI1764	R	-	R	R	R	R	R	-	S	R	MR	MR	-
H16, *T. turgidum*, PI94587	R	-	R	R	R	R	-	R	-	R	R	R	-
H17, *T. turgidum*, PI428435	R	R	-	R	R	R	R	R	R	R	R	R	-
H18, *T. turgidum*, CI6887	R	R	MR	R	R	R	R	MR	-	R	MR	R	R
H19, *T. turgidum*, PI422297	R	R	MR	-	R	MR	MR	MR	-	R	S	S	-
H20, *T. turgidum*, Jori	-	-	-	-	-	-	-	-	?	-	-	-	-
H21, *Secale cereale*, Chaupon	-	-	-	-	-	-	-	-	R	R	-	-	R
H22, *T. tauschii*, TA1644	-	-	-	R	-	-	-	-	R	R	-	-	-
H23, *T. tauschii*, TA1642	-	-	-	R	-	-	-	-	R	R	-	-	-
H24, *T. tauschii*, TA2452	-	-	-	R	-	-	-	-	R	R	-	-	-
H25, *S. cereale*, Chaupon	-	-	-	-	-	-	-	-	R	R	-	-	-
H26, *T. tauschii*, TA2473	-	-	-	R	-	-	-	-	R	R	-	-	-
H29, *T. turgidum*, PI422297	-	-	-	-	-	-	-	-	R	R	-	-	-
H30, *Ae. triuncialis*, TR-3531	-	-	-	-	-	-	-	-	-	-	-	-	R

[a] Adapted from Patterson et al. 1992, J. Econ. Entomol. 85:307-311, Ratcliffe et al. 1994, J. Econ. Entomol. 87:1113-1121, Ratcliffe et al. 1996, J. Econ. Entomol. 89:1309-1317, Copyright 1992, 1994, 1996, Entomological Society of America, with permission of the Entomological Society of America; Martin-Sánchez, J. A., M. Gómez-Colmenarejo, J. Del Moral, E. Sin, M. J. Montes, C. González-Belinchón, I. López-Braña, and A. Delibes. 2003. A new Hessian fly resistance gene (H30) transferred from the wild grass Aegilops triuncialis to hexaploid wheat. Theor. Appl. Genet. 106:1248-1255, Copyright 2003, Springer-Verlag, with kind permission of Springer Science and Business Media; and Ratcliffe, R. H., and J. H. Hatchett. 1997. Biology and genetics of the Hessian fly and resistance in wheat, In: K. Bobdari, (Ed.), New Developments in Entomology; Research Signpost, Scientific Information Guild, with permission of Research Signpost. GP - Great Plains of North America, SP - Spain; [c] R = resistant; MR = moderately resistant; S = susceptible

The Great Plains (GP) fly biotype, initially isolated in western Kansas, now occurs in Texas, and carries the homozygous recessive condition for virulence only in association with the wheat cultivar 'Turkey'. The GP biotype is avirulent to cultivars with the *H3, H5,* and *H6* genes, or the *H7H8* gene combination. Biotype A, also found primarily in Texas, is similar to the GP biotype, but is virulent to *H7H8.* Biotype B now occurs in the northeastern U. S. and possesses an additional gene for virulence to *H3.* Biotype C, found occasionally in New York and Texas, has the opposite reaction of biotype B to *H3* and *H6.* Biotype D, found in the Midwestern, mid-Atlantic, and northeastern U. S., is similar to biotype B, but is also virulent to *H6.* Biotype E, detected in Georgia by Hatchett (1969), occurs in Georgia, Florida and Texas, and is avirulent to *H7H8.*

Biotypes F and G occur in the south and southeastern U. S., and are avirulent to *H7H8* (Ratcliffe et al. 1994, 1996). Biotype F is avirulent to *H3, H5,* and *H7H8,* while biotype G is avirulent to *H7H8,* but virulent to *H3* and *H6.* Biotypes J and L developed in field populations in Indiana in response to *H5* resistance in the cultivar 'Arthur 71' (Sosa 1978, 1981). Biotype J is avirulent to *H1H2* and *H7H8.* Biotype L now occurs in the eastern two-thirds of the United States wheat producing areas, and is avirulent to *H9, H10, H12-19,* and *H21-27* (Ratcliffe et al. 1996). Biotypes M and O occur in the southeastern U. S and have similar patterns of virulence to *H3, H5, H6, H11* and *H19* (Ratcliffe et al. 1994, 1996, 2000). Based on results of United States national surveys, biotypes I and N do not presently occur at detectable levels. A Spanish *M. destructor* biotype recently characterized by Martin-Sanchez (2003) is avirulent to *H3, H6, H9, H11, H12, H13, H18, H21,* and *H30,* from the wild wheat *Aegilops triuncialis* L. (See Chapter 8).

Biotype surveys have documented major shifts in the occurrence and degree of *M. destructor* virulence since the mid-1980s (Ratcliffe et al. 1994, 1996). The once resistant genes *H3, H5, H6,* and *H7H8* are ineffective, as virulence exists to each in biotype L, which is now predominant in the eastern United States. Virulence to *H3* and *H6* has also been detected in *M. destructor* populations in Idaho and Washington (Clement 2003, Ratcliffe et al. 2001). Numerous fly populations in the southeastern and Midwestern United States have also begun to exhibit virulence to several un-deployed resistance genes, including *H9, H10, H11, H12, H13, H14, H15, H18* and *H19.* Although potentially ineffective across wide geographic areas, some of these genes may be useful against avirulent biotypes in selected geographic areas (Berzonsky et al. 2003). The *H16* and *H17* genes from durum wheat, *Triticum turgidum* L., remain resistant to all *M. destructor* biotypes currently in the United States. Patterson et al. (1992) proposed a system of biotype nomenclature based on sets of three different wheat cultivars, with each set allowing eight combinations of resistant and susceptible reactions. This designation system appears to add flexibility for the future addition of new plant resistance genes, as well as the deletion of sets of obsolete genes. However, it has not been adapted on a large-scale.

Virulence in a *M. destructor* biotype depends on the existence of the homozygous recessive condition in the virulence gene of the fly at a locus corresponding to a specific dominant plant gene for resistance (Gallun 1977). In each combination of plant resistance and fly avirulence, a particular *M. destructor* biotype lacks the homozygous recessive condition at the virulence gene locus. El-Bouhssini et al. (2001) determined that the dominance and dose of both the plant resistance allele and the fly virulence allele determine the expression of a biotype.

An in-depth understanding of virulence has begun to be gained from results of research with biotypes virulent to *H6*, *H9*, and *H13*. Formusoh et al. (1996) demonstrated that virulence is controlled by a single completely recessive, sex-linked gene (*vH9*) after analyzing progeny from reciprocal matings of a virulent and avirulent biotype. Linkage of *vH9* to the *w* (white eye) trait suggested that *vH9* was on one of the two *M. destructor* sex chromosomes. Zantoko and Shukle (1997) obtained similar results with *vH13* for virulence to *H13*. Stuart et al. (1998) used DNA polymorphisms between *vH6* and *H6* avirulent flies, as well as *in situ* hybridization, to demonstrate that *vH6* is sex-linked and located on *M. destructor* chromosome X1. Schulte et al. (1999) established the relative orientation of *vH6*, *vH9, and vH13* on chromosome X1. In experiments to gain more information about the linkage of *vH13* to molecular markers, Rider et al. (2002) used bulk segregant analyses to identify molecular markers that genetically map to within 13cM of *vH13* on the short arm of *M. destructor* chromosome X2, rather than X1. Attempts are underway to use chromosome walking to locate and clone *vH13* on fly chromosome X2.

The fact that *M. destructor* virulence is controlled by the individual sex-linked genes *vH6*, *vH9* and *vH13* provides the most compelling evidence to date that *M. destructor* resistance genes from cereal plants and *M. destructor* virulence genes exemplify the gene-for-gene relationship described by Flor (1971).

Schizaphis graminum infest barley, oat, *Avena sativa* (L.), rye, *Secale cereale* L., sorghum, wheat, and numerous grasses (Michels 1986, Teetes et al. 1999) and have been regarded as a cereal pest in the U. S. since the late 1800s. Although there were no specific biotype designations, Dahms (1948) observed differences in Mississippi and Oklahoma *S. graminum* populations as early as 1946. Wood (1961) named the first *S. graminum* biotype B, due to its ability to overcome the resistance in 'Dickinson 28A' wheat to biotype A (Table 11.7). Biotype B tolerates higher temperatures than biotype A (Singh and Wood 1963) but cannot survive on 'Piper' sudan grass, *Sorghum vulgare* var. *sudanese* Hitchc. In the late 1960's, *S. graminum* populations began to severely damage sorghum in the southwestern U. S. These populations were named biotype C (Harvey and Hackerott 1969), and forced the development of biotype C-resistant sorghum and wheat cultivars. 'Amigo' wheat, possessing the *Gb2* gene from rye, is resistant to biotypes B and C (Sebesta and Wood 1978). Biotype D is resistant to the insecticide disulfoton (Teetes et al. 1975, Peters et al. 1975).

In the late 1970's, biotype E, capable of feeding on 'Amigo' wheat, became prevalent in the high plains of Texas (Porter et al. 1982). Some biotype C resistant wheat cultivars have retained antixenosis resistance to biotype E, but most barley and sorghum cultivars are susceptible (Starks et al. 1983). Biotype E resistance also exists in sorghum plant introductions with the bloomless character (Starks et al. 1983), in 'Largo' wheat containing *Gb3* and CI17959 wheat containing *Gb4* (Porter et al. 1982, Tyler et al. 1985). Biotype E resistance also exists in CI17882, which contains *Gb5*, presumably donated by the *Ae. speltoides* parent of CI17882 (Tyler et al. 1987).

Biotype E exhibits normal feeding behavior and fecundity on wheat cultivars resistant to biotype B (Niassy et al. 1987) but has increased digestive enzyme activity, enabling it to reproduce at approximately twice the rate of biotype C (Campbell and Dreyer 1985, Dreyer and Campbell 1984, Montllor et al. 1983).

Schizaphis graminum biotypes exhibit significant morphometric differences, the most prominent of which is the length of the first flagellar segment (Inayatullah et al. 1987a). Biotype E individuals also require a longer scotophase to induce the production of males than do biotype C (Eisenbach and Mittler 1987). Few isozyme differences exist between biotypes B, C, or E (Beregovoy and Starks 1986). Inayatullah et al. (1987b) used multivariate analyses of morphometric characteristics to demonstrate close relationships between biotypes C and E and between biotypes C and B. Results of mitochondrial DNA analyses by Powers et al. (1989) provide additional evidence to support these results.

Table 11.7. Wheat genes governing resistance to biotypes of Schizaphis graminum *in North America and Asia*

Source	Origin	Gene	Reaction to biotype [a]									
			B	C	E	F	G	H	I	J	CH[b]	PK[c]
Dickinson	*Triticum turgidum*	*gb1*	R[d]	S	S	R	S	S	-	R	S	S
Amigo	*Secale cereale*	*Gb2*	R	R	S	S	S	S	-	R	R	R
Largo	*Aegilops tauschii*	*Gb3*	S	R	R	S	S	R	R	R	R	S
CI17959	*Aegilops tauschii*	*Gb4*	S	R	R	S	S	S	R	R	S	-
CI17882	*Aegilops speltoides*	*Gb5*	S	R	R	S	S	S	-	R	MR	-
GRS1201	*Secale cereale*	*Gb6*	R	R	R	S	R	S	R	R	MR	-

[a] *Biotype B-J, North America;* [b] *China;* [c] *Pakistan;* [d] *R- resistant, MR - moderately resistant, S- susceptible*

Biotype F, isolated from turfgrass in Ohio, lacks a dorsal stripe and is avirulent to *gb1* (Kindler and Spomer 1986). Biotype F can feed on and kill Canada bluegrass, but biotype E cannot (Kindler and Hays 1999). Biotypes G (from Oklahoma) and H (from Texas), identified by Puterka et al. (1988), have the appearance of biotype F, but differ in their reactions to barley and wheat. Biotype G survives on all known sources of resistant wheat, but is avirulent to *Gb6* in 'GRS 1201', a *T. aestivum/S. cereale* hybrid (Porter et al. 1994). Biotype H has the same effect on wheat cultivars as biotype E, but is avirulent to *Gb3*. Biotypes I and K are virulent to *S. graminum*-resistant sorghums (Harvey et al. 1991, 1997) but avirulent to *Gb3, Gb4,* and *Gb6* (Porter et al. 1982). Biotype J, a unique population from Idaho, is avirulent to all *Gb* genes but feeds effectively on 'Post' barley, which is normally resistant to all *S. graminum* biotypes (Beregovoy and Peters 1994).

Schizaphis graminum biotypes have also been identified in China and Pakistan. The Chinese biotype (CHN-1) is unique in contrast to U. S. biotypes E and I, in that it is virulent to *Gb2* but avirulent to *Gb4* (Liu and Jin 1998) (Table 11.7). The Pakistan biotype (PK-1) differs from U. S. biotypes E and I in that it is avirulent to *Gb3* (Inayatullah et al. 1993).

The question of how *S. graminum* biotypes occur has been addressed at the population, organism and gene levels. At the aphid-plant interface, *S. graminum* biotypes must possess two qualities to overcome *Gb* genes in wheat. A biotype must first produce successful necrotic lesions on wheat leaves, resulting in chlorosis and the break down of leaf cellular components for aphid nutrition. Second, a biotype must assimilate these nutrients efficiently enough to allow a high rate of *S. graminum* reproduction that ultimately results in plant death (Beregovoy and Peters 1993). Resistance in wheat to biotype E depends on the ability of a plant to resist alterations in the plant cell during biotype E feeding, regardless of *S. graminum* feeding site or the type of mechanical damage caused to other plant structures during feeding (Morghan et al. 1994). Knowledge of theses types of interactions in *S. graminum* is limited to biotype E.

At the genetic level, Puterka and Peters (1995) demonstrated that resistance to *S. graminum* biotypes C and E in sorghum germplasm SA 7536-1 and PI264453 is simply inherited as an incompletely dominant trait, and that *S. graminum* virulence to these genes is inherited as a recessive trait. A similar relationship exists between the dominant inheritance of *Gb2* and *Gb3* in wheat and the recessively inherited virulence in biotypes E and F (Puterka and Peters 1989). The interactions between *S. graminum* virulence genes with wheat and sorghum resistance genes are also similar to the gene-for-gene interactions mentioned previously for *M. destructor, O. oryzae* and *N. lugens.*

At the population level, the question shifts from how *S. graminum* biotypes are defined to determining what factors have brought about their occurrence. Biotype development in several insects discussed previously is linked to changes in the composition of the resistance genes in the deployed resistant cultivar. However, there are limited data to support this argument for *S. graminum* biotype development. Porter et al. (1997) reviewed the history of wheat and sorghum resistance to *S. graminum* and found no relationship between genes in *S. graminum* resistant wheat and *S. graminum* biotype development (Table 11.7). The gene-for-gene relationships in Table 11.7 have had no effect on *S. graminum* biotype evolution, because the resistance in *Gb3, Gb4, Gb5* and *Gb6* has never existed in a wheat cultivar produced in the field.

The history of *S. graminum* resistant sorghum cultivars and *S. graminum* biotypes does not preclude development of biotypes being driven by sorghum *S. graminum* resistance genes. However, the types of resistance and the extent of plantings of resistant cultivars do not make a strong case of sorghum resistance genes driving *S. graminum* biotype evolution. Biotype C-resistant sorghum genes were expressed as tolerance, which as described in Chapter 4, provided durable resistance, but did not force biotype development. The biotype E-resistant sorghum cultivars released contain intermediate levels of antixenosis, antibiosis and tolerance (Dixon et al. 1990) and biotype I-resistant cultivars contain antibiosis and antixenosis (Bowling and Wilde 1996). In addition, the level of all three sources of resistance is at best intermediate, as described by Kofoid et al. (1991) and Harvey et al. (1997). Finally, only about one-half of the sorghum producing area of the U. S. southern plains was ever planted with *S. graminum* resistant hybrids prior to the identification of virulent biotypes (Porter et al. (1997). Thus, none of the cultivars were planted on a large scale or contained high levels of resistance based on antibiosis, the situation that has promoted biotype development in other arthropods.

Porter et al. (1997) cited several non-crop cultivar factors as those more likely involved in promoting biotype development. These include extensive *S. graminum* clonal diversity (Shufran et al. 1992, Shufran and Wilde 1994), *S. graminum* non-crop host adaptation (Powers et al. 1989), and *S. graminum* autumn sexual reproduction on cool season grasses, especially blue grass (Puterka et al. 1990). The latter is thought to be a particularly significant factor, because *S. graminum* summer populations on wheat die before sexual forms are produced, essentially eliminating the probability that individuals produced on summer crops plants result in biotypes. The identification of a biotype on western wheat grass (Anstead et al. 2003) with a unique virulence profile adds additional support to the idea that non-cultivated grasses are intimately involved in the evolution of what have become recognized as *S. graminum* biotypes.

Shifts in *S. graminum* biotype composition in the U. S. southern plains have occurred since 1990. The once predominant biotype E has been replaced by biotype I (Peters et al. 1997), presumably because of a higher level of virulence in biotype I than in biotype E (Bowling et al. 1998). In the absence of strong supporting evidence that these changes are plant resistance gene-driven, Porter et al. (1997) contend that changes in sorghum genetic composition in the late 1960s, in combination with preexisting *S. graminum* clonal diversity, resulted in rapid increases of clones with virulence genes on sorghum. However, as stated by Porter et al. (1997) "The question of what shaped the genetic pool of *S. graminum* that resulted in the occurrence of so many biotypes virulent to resistant cultivars remains unanswered."

The Russian wheat aphid, *Diuraphis noxia* (Mordvilko), is a serious worldwide pest of wheat and barley (Quisenberry and Peairs 1998) that has caused yield losses in North America of nearly $1 billion since 1987 (Webster et al. 2000). The injection of *D. noxia* salivary enzymes into plant tissues results in plant leaves rolling longitudinally around the main leaf vein, forming tubular refuge for feeding aphids from predators, parasites, and insecticides (Fouché et al. 1984).

Eleven *Dn* (*Diuraphis noxia*) genes from *Aegilops tauschii* (Coss. Schmal), rye, or wheat control resistance to *D. noxia* (Table 11.8). *Dn1, Dn2, Dn4, Dn5, Dn6* and *Dnx* are from wheat. A recessive gene *dn3* is present in SQ24, a line derived from crosses between *Ae. tauschii* and durum wheat (Nkongolo et al. 1991). *Dn7*, is a rye gene derived from a whole arm rye chromosome translocation to wheat (Marais et al. 1994).

Dn8 and *Dn9* are co-expressed with *Dn5* (Liu et al. 2001). The chromosome location of *Dny*, in the aphid resistant cultivar 'Stanton' (Martin et al. 2001) is not known. The first *D. noxia*-resistant U. S. wheat cultivar, 'Halt', containing *Dn4*, was released in Colorado 1994 (Quick et al. 1996).

Table 11.8. Reaction of Diuraphis noxia *biotypes to wheat genes controlling* D. noxia *resistance* [a]

			D. noxia source [b]						
Source	Origin	Gene	Czr	Chi	Eth	Hun	USA (CO)	USA (KS)	ZA
PI137739	Triticum aestivum	Dn1	-	-	-	S [c]	S	R	R
PI262660	Triticum aestivum	Dn2	R	-	R	S	S	R	R
SQ24	Aegilops tauschii	dn3	-	-	-	-	S	R	-
PI372129	Triticum aestivum	Dn4	S	S	S	S	S	R	R
PI294994	Triticum aestivum	Dn5	R	-	R	S	S	R	R
PI243781	Triticum aestivum	Dn6	R	R	R	-	S	R	-
Turkey 77	Secale cereale	Dn7	-	-	-	-	R	R	R

[a] *Adapted from Basky et al. 2002 and Nkongolo et al. 1989 with permission of the Cereal Research Non-Profit Company; from Haley, S. D., F. B. Peairs, C. B. Walker, J. B. Rudolph, and T. L. Randolph. 2004. Occurrence of a new Russian wheat aphid biotype in Colorado, Crop Sci. 44: 1589-1592, Copyright 2004, Crop Science Society of America, with permission of the Crop Science Society of America; and Smith et al. 1991 with permission of the Entomological Society of America;* [b] *Czr - Czech Republic, Chi - Chile, Eth - Ethiopia, Hun - Hungary, USA(CO) - USA Colorado. USA (KS) - USA Kansas, ZA - South Africa;* [c] *R - resistant, S - susceptible*

Bush et al. (1989) was the first to note differences in *D. noxia* populations on aphid resistant wheat lines in Texas. Puterka et al. (1992) compared populations from different countries using plant differential cultivar responses and identified virulence to *Dn4* in aphids from Syria and Russia (Table 11.8). Virulence to *Dn1, Dn2, Dn4,* and *Dn5* has been extensively documented in Hungary (Basky 2002, 2003, Basky and Jordan 1997), although the aphid is presently not a pest of cereals there. Populations occurring in areas of Colorado, Nebraska and Texas were identified in 2003 as virulent to all *Dn* genes except *Dn7* (Haley et al. 2004). Populations from Chile, the Czech Republic and Ethiopia are also virulent to *Dn4* but avirulent to *Dn6* (Smith et al. 2004) (Table 11.8). The Czech population is also virulent to *Dnx* and the Ethiopian population is virulent to *Dny*.

The Chilean, Ethiopian and Hungarian populations are unaffected by the antibiosis resistance expressed by *Dn4* plants (Basky 2003, Smith et al. 2004). Wheat cultivars and germplasm containing *Dn1, Dn2* and *Dn5* maintain resistance to the *D. noxia* in South Africa (Prinsloo 2000). Research is in progress to develop advanced breeding lines containing genes from new sources of resistance as well as from *Dn7*. *D. noxia* population differences based on molecular marker polymorphism have shown limited variation (Puterka et al. 1993, Robinson et al. 1993, Shufran et al. 1997).

Table 11.9. Reaction of Aceria tosichella biotypes to cereal genes controlling A. tosichella resistance [a]

Gene Source (Germplasm name)	Gene	*A. tosichilla colony origin* [b]					
		KS	MT	NE	SD	TX	CN
Aegilops tauschii (ACPGR 16635)	*Cmc1*	R[c]	R	S	R	R	R
Thinopyrum ponticum (PI525452)	*Cmc2*	R	R	S	R	R	R
Secale. cereale (TAM107)	*Cmc3*	S	R	R	S	S	R
Aegilops tauschii (WGRC40)	*Cmc4*	R	S	R	R	R	R
Triticum aestivum (PI222655)	----	S	S	R	S	S	R
Triticum aestivum (Larned)	----	S	S	S	S	S	S

[a] *Reprinted from Harvey, T. L., D. L. Seifers, T. J. Martin, G. L. Brown-Guedira, and B. S. Gill. 1999. Survival of wheat curl mites on different sources of resistance in wheat. Crop Sci. 39:1887-1889, and Malik, R., C. M. Smith, T. L. Harvey, and G. L. Brown-Guedira. 2003. Genetic mapping of wheat curl mite resistance genes Cmc3 and Cmc4 in common wheat. Crop Sci. 43:644-650, copyright 1999 & 2003 Crop Society of America, with permission of the Crop Science Society of America ;* [b] *KS - Kansas USA, MT - Montana USA, NE - Nebraska USA, SD - South Dakota, USA, TX - Texas USA, CN - Canada;* [c] *Resistance (R) and susceptibility (S) based on numbers of mites per plant*

The wheat curl mite, *Aceria tosichella* Keifer, vectors wheat streak mosaic virus, the most economically important disease of wheat in North America. Genetic plant resistance to the mite has proven effective in controlling both *A. tosichella* and the virus (Conner et al. 1991, Harvey et al. 1994). Resistance was first demonstrated in *Triticum x Thinopyrum* (wheatgrass) hybrids, where the *Thinopyrum* gene, designated *Cmc2*, is inherited as a dominant trait (Whelan and Hart 1988). Mite resistance from the *Ae. tauschii* gene *Cmc1* is also inherited as a dominant trait (Thomas and Conner 1986). Several wheats with translocations from rye also effectively reduce *A. tosichella* populations and the incidence of wheat streak mosaic virus. However, *A. tosichella* adapts quickly to new hosts. Harvey et al. (1995, 1997a,b) observed an *A. tosichella* biotype that became virulent in only 7 years to TAM107, a popular mite-resistant cultivar grown in western Kansas.

Aceria tosichella biotypes from the U. S. states of Kansas, Nebraska, Montana, Alberta Canada, South Dakota, and Texas display differential virulence expression to a differential set of resistance genes in different grasses listed in Table 11.9. Results of Harvey et al. (1999) indicate that *Cmc1* and *Cmc2* are resistant to *A. tosichella* biotypes from Kansas, Montana, South Dakota, Texas, and Alberta, Canada, but *A. tosichella* from Nebraska are virulent to *Cmc1* and *Cmc2*. The rye-based resistance gene in TAM107 and PI475772 (*Cmc3*) is effective against biotypes from Montana, Nebraska, and Canada, but biotypes from South Dakota, Texas, and Kansas, where *A. tosichella* has adapted to TAM107, are virulent to *Cmc3*. The *Cmc4* gene from *Ae. tauschii* in TA2397 (Malik et al. 2003) and a *Triticum timopheevi* gene in wheat PI222655 are resistant to all *A. tosichella* biotypes. Wheat PI222655 contains a gene for resistance to *A. tosichella* from Canada and Nebraska, as well as a gene for resistance to *D. noxia*.

Malik (2001) identified a nuclear ribosomal DNA marker in the rDNA ITS-1 region that differentiates the Nebraska *A. tosichella* biotype from the Kansas, Montana, and Canadian biotypes. ITS gene probes have also been used to demonstrate clonal diversity in *S. graminum* (Shufran et al. 1992, Shufran and Wilde 1994), *A. idaei* (Birch et al. 1994), and the green peach aphid, *Myzus persicae* Sulzer (Fenton et al. 1998).

4. FACTORS CONTRIBUTING TO BIOTYPE AVOIDANCE

Combinations of plant breeding and pest management practices are necessary to avoid the selection of arthropod biotypes. Many studies have been conducted to determine the categories of arthropod resistance in crop plants. However, few have compared the advantages of different resistance categories during long-term exposure to several generations of a pest arthropod.

Discussions presented in Chapter 4 demonstrate how cultivars resistant to arthropods by means of tolerance exert limited selection pressure on pest populations to evolve virulence (Heinrichs et al. 1986). Comparatively, tolerant cultivars are often more stable than those exhibiting antibiosis, where high levels of chemical and physical factors have resulted in selection for virulent individuals. Biotypes of *D. vitifoliae, M. destructor, A. idaei, O. oryzae, D. noxia,* and *A. toschiella* have developed in response to the effects of different single genes inherited as dominant traits whose effects are manifested as high levels of antibiosis. Therefore, one effective management practice is the use of a cultivar with moderate levels of antibiosis or with a combination of antibiosis, antixenosis and tolerance. Results of Basky (2003) provide evidence that virulent *D. noxia* populations overcome the antibiosis component of several different wheat resistance genes, but are unable to overcome tolerance.

There is some indication that horizontal, multigene resistance may be more durable than single gene resistance. As mentioned previously, the prolonged resistance of 'IR36' rice (approximately 5 years) and 'IR64' rice (approximately 10 years) to *N. lugens* is thought to be related to the horizontal resistance effects expressed by several minor genes in these cultivars (Cohen et al. 1997). Developing horizontal resistance however, is a long-term process that can be an extremely challenging objective for plant breeders and entomologists. Outstanding successes in plant resistance to arthropods have been achieved using vertical resistance based on the sequential release of different cultivars with major antibiosis genes. These genes are often short-lived, as in the case of *M. destructor* and *D. noxia*, and may be obsolete before deployment.

Biotype selection is also related to the geographic extent to which resistant cultivars are planted throughout an arthropods' host range. Improper management practices such as elimination of alternate (weed) hosts, lack of crop rotation, or improper insecticide application may also contribute to the selection of arthropod biotypes on previously resistant cultivars. The occurrence of hopper biotypes in Southeast Asia was at one time related to nearly continuous production of large-scale plantings of the same rice cultivars in several countries, as well as prophylactic applications of insecticides for hopper population reduction. However, improved arthropod pest management techniques were developed and adopted, resulting in enhanced *N. lugens* natural enemy fauna and delayed biotype development.

Finally, well-defined sampling programs should be used to monitor arthropod populations from different geographic locations and from various host plants for biotype development. The methods used to determine differences vary considerably. For many arthropods, biotypes continue to be effectively monitored and detected by the response of a differential set of test cultivars to an arthropod population. PCR-based assays such as those developed by Behur et al. (1999, 2000) for detection of *O. oryzae* biotypes hold great promise for use with other arthropods. Ultimately, the most useful method of differentiation is that which gives the most accurate, efficient delineation of biotypes in an arthropod population.

Arthropods biotypes should be anticipated when developing plant resistance to arthropod pests with a high reproductive potential. The parthenogenic reproduction, high reproductive potential and clonal diversity of several aphid pests discussed in this chapter suggest they will continue to generate biotypes. In some cases, genes controlling antibiosis expressed at high levels have promoted the development of virulence. Planting a resistant cultivar over a wide geographic range, as in the case of early *M. destructor* resistant cultivars, may also contribute to the development of virulence. Planting different cultivars with resistance genes to specific biotypes in different geographic areas is a sound pest management approach that will slow the development of arthropod biotypes in crop plants.

REFERENCES CITED

Alston, F. H., and J. B. Briggs. 1977. Resistance genes in apple and biotypes of *Dysaphis devecta*. Ann. Appl. Biol. 87:75–81.

Alam, S. N., and M. B. Cohen. 1998. Durability of brown planthopper, *Nilaparvata lugens*, resistance in rice variety IR64 in greenhouse selection studies. Entomol. Exp. Appl. 89:71–78.

Anstead, J. A., J. D. Burd, and K. A. Shufran. 2003. Over-summering and biotypic diversity of *Schizaphis graminum* (Homoptera: Aphididae) populations on noncultivated grass hosts. Environ. Entomol. 32:662–667.

Appleby, J. H., and P. F. Credland. 2003. Variation in responses to susceptible and resistant cowpeas among west African populations of *Callosobruchus maculatus* (Coleoptera: Bruchidae). J. Econ. Entomol. 96: 489–502.

Auclair, J. L. 1978. Biotypes of the pea aphid *Acyrthrosiphon pisum* in relation to host plants and chemically defined diets. Ent. Exp. Appl. 24:12–16.

Basky, Z. 2002. Biotypic variation in Russian wheat aphid (*Diuraphis noxia* Kurdjumov Homoptera: Aphididae) between Hungary and South Africa. Cereal Res. Commun. 30:133–139.

Basky, Z. 2003. Biotypic variation and pest status differences between Hungarian and South African populations of Russian wheat aphid, *Diuraphis noxia* (Kurdjumov) (Homoptera: Aphididae). Pest Manag. Sci. 59:11152–1158.

Basky, Z., and J. Jordaan. 1997. Comparison of the development and fecundity of Russian wheat aphid (Homoptera: Aphididae) in South Africa and Hungary. J. Econ. Entomol. 90: 623–627.

Behura, S. K., S. Nair, and M. Mohan. 2001a. Polymorphisms flanking the mariner integration sites in the rice gall midge (*Orseolia oryzae* Wood-Mason) genome are biotype-specific. Genome. 44:947–954.

Behura, S. K., S. Nair, S. C. Sahu, and M. Mohan. 2000. An AFLP marker that differentiates biotypes of the Asian rice gall midge (*Orseolia oryzae*, Wood-Mason) is sex-linked and also linked to avirulence. Mol. Genet. Genomics. 263:328–334.

Behura, S. K., S. C. Sahu, M. Mohan, and, S. Nair. 2001b. *Wolbachia* in the Asian rice gall midge, *Orseolia oryzae* (Wood-Mason): correlation between host mitotypes and infection status. Insect Mol. Biol. 10:163–171.

Behura, S. K., S. C. Sahu, S. Rajamani, A. Devi, R. Mago, S. Nair, and M. Mohan. 1999. Differentiation of Asian rice gall midge, *Orseolia oryzae* (Wood-Mason), biotypes by sequence amplified regions (SCARs). Insect Mol. Biol. 8:391–397.

Bentur, J. S., and S. Amudhan. 1996. Reaction of differentials of different populations of Asian rice gall midge (*Orseolia oryzae*) under greenhouse conditions. Ind. J. Agr. Sci. 66:197–199.

Bentur, J. S., M. B. Kalode, and P. S. Prakash-Rao. 1994. Reaction of rice (*Oryza sativa*) varieties to different biotypes of rice gall midge (*Orseolia oryzae*). Ind. J. Agr. Sci. 64:419–420.

Beregovoy, V. H., and K. J. Starks. 1986. Enzyme patterns in biotypes of the greenbug, *Schizaphis graminum* (Rondani) (Homoptera: Aphididae). J. Kansas Entomol. Soc. 59:517–523.

Beregovoy, V. H., and D. C. Peters. 1993. Variation of two kinds of wheat damage characters in seven biotypes of the greenbug, *Schizaphis graminum* (Rondani) (Homoptera: Aphididae). J. Kansas Entomol. Soc. 66:237–244.

Beregovoy, V. H., and D. C. Peters. 1994. Biotype J, a unique greenbug (Homoptera: Aphididae) distinguished by plant damage characteristics. J. Kansas Entomol. Soc. 66:248–252.

Berzonsky, W. A., H. Ding, S. D. Haley, M. O. Harris, R. J. Lamb, R. I. H. McKenzie, H. W. Ohm, F. L. Patterson, F. B. Peairs, D. R. Porter, R. H. Ratcliffe, and T. G. Shanower. 2003. Breeding wheat for resistance to insects. Plant Breed. Rev. 22:221–296.

Birch A. N. E., B. Fenton, G. Malloch, A. T. Jones, M. S. Phillips, B. E. Harower, J. A. T. Woodford, and M. A. Catley. 1994. Ribosomal spacer length variability in the large raspberry aphid, *Amphorophora idaei* (Aphidinae: Macrosiphini). Insect Mol. Biol. 3: 239– 245.

Birch A. N. E., I. E. Geoghegan, M. E. N. Majerus, C. Hackett, and J. Allen. 1996. Interactions between plant resistance genes, pest aphid populations and beneficial aphid predators. Scot. Crop Res Instit. Ann. Rept. pp. 1–4.

Borner, C. 1914. On the susceptibility and immunity of vines to the attack of the vine louse. Biol. Zentbl. 34:1–8.

Bowling, R., and G. Wilde. 1996. Mechanisms of resistance in three sorghum cultivars resistant to greenbug (Homoptera: Aphididae) biotype I. J. Econ. Entomol. 89:558–561.

Bowling, R., G. Wilde, and D. Margolies. 1998. Relative fitness of greenbug (Homoptera: Aphididae) biotypes E and I on sorghum, wheat, rye, and barley. J. Econ. Entomol. 91: 1219–1223.

Briggs, J. B. 1965. The distribution, abundance, and genetic relationships of four strains of the *Rubus* aphid (*Amphorophora rubi*) in relation to raspberry breeding. J. Hort. Sci. 49:109–117.

Brown, J. K., D. R. Frohlich, and R. C. Rosell. 1995. The sweet potato or silverleaf whiteflies: Biotypes of *Bemisia tabaci* or a species complex? Ann. Rev. Entomol. 4:511–534.

Bush, L., J. E. Slosser, and W. D. Worrall. 1989. Variations in damage to wheat caused by Russian wheat aphid (Homoptera: Aphididae). J. Econ. Entomol. 82:466–471.

Campbell, B. C., and D. L. Dreyer. 1985. Host-plant resistance of sorghum: Differential hydrolysis of sorghum pectic substance by polysaccharases of greenbug biotypes (*Schizaphis graminum*, Homoptera: Aphididae). Arch. Biochem. Physiol. 2:203–215.

Cartier, J. J. 1959. Recognition of three biotypes of the pea aphid from southern Quebec. J. Econ. Entomol. 52:293–294.

Cartier, J. J. 1963. Varietal resistance of peas to pea aphid biotypes under field and greenhouse conditions. J. Econ. Entomol. 56:205–213.

Cartier, J. J., and R. H. Painter. 1956. Differential reactions of two biotypes of the corn leaf aphid to resistant and susceptible varieties, hybrids, and selections of sorghums. J. Econ. Entomol. 49:498–508.

Claridge, M. F., and J. Den Hollander. 1982. Virulence to rice cultivars and selection for virulence in populations of the brown planthopper, *Nilaparvata lugens*. Entomol. Exp. Appl. 32:213–221.

Claridge, M. F., and J. Den Hollander. 1983. The biotype concept and its application to insect pests of agriculture. Crop Prot. 2:85–95.

Clement, S. L., L. R. Elberson, F. L. Young, J. R. Alldridge, R. H. Ratcliffe, and C. Hennings. 2003. Variable Hessian fly (Diptera: Cecidomyiidae) populations in cereal production systems in eastern Washington. J. Kansas Entomol. Soc. 76:567–577.

Cohen, M. B., S. N. Alam, E. B. Medina, and C. C. Bernal. 1997. Brown planthopper, *Nilaparvata lugens*, resistance in rice cultivar IR64: mechanism and role in successful *N. lugens* management in Central Luzon, Philippines. Entomol. Exp. Appl. 85:221–229.

Conner, R. L., J. B. Thomas, and E. D. P. Whelan. 1991. Comparison of mite resistance for control of wheat streak mosaic. Crop Sci. 31:315–1318.

Dahms, R. G. 1948. Comparative tolerance of small grains to greenbugs from Oklahoma and Mississippi. J. Econ. Entomol. 41:825.

DenHollander, J., and P. K. Pathak. 1981. The genetics of the 'biotypes' of the rice brown planthopper, *Nilaparvata lugens*. Entomol. Exp. Appl. 29:76–86.

Dick, K. M., and P. F. Credland. 1986a. Variation in the response of *Callosobruchus maculatus* (F.) to a resistant variety of cowpea. J. Stored Prod. Res. 22:43–48.

Dick, K. M., and P. F. Credland. 1986b. Changes in the response of *Callosobruchus maculatus* (Coleoptera: Bruchidae) to a resistant variety of cowpea. J. Stored Prod. Res. 22:227–233.

Dixon, A., G. Olonju, P. J. Bramel-Cox, and T. L. Harvey. 1990. Diallel analysis of resistance in sorghum to greenbug biotype E: antibiosis and tolerance. Crop Sci. 30:1055–1059.

Dreyer, D. L., and B. C. Campbell. 1984. Degree of methylation of intercellular pectin associated with plant resistance to aphids and with induction of aphid biotypes. Experientia 40: 224.

Dunn, J. A., and D. P. H. Kempton. 1972. Resistance to attack by *Brevicornye brassicae* among plants of Brussels sprouts. Ann. Appl. Biol. 72:1–11.

Eisenbach, J., and T. E. Mittler. 1987. Polymorphism of biotypes E and C of the aphid *Schizaphis graminum* (Homoptera: Aphididae) in response to different scotophases. Environ. Entomol. 16:519–523.

El Bouhssini, M. E., J. H. Hatchett, T. S. Cox, and G. E. Wilde. 2001. Genotypic interaction between resistance genes in wheat and virulence genes in the Hessian fly *Mayetiola destructor* (Diptera: Cecidomyiidae). Bull. Entomol. Res. 91:327–331.

Fenton B., G. Malloch, and F. Germa. 1998. A study of variation in rDNA ITS regions show that two haplotypes coexist within a single aphid genome. Genome. 41:337–345.

Flor, H. H. 1971. The current status of the gene-for-gene concept. Ann. Rev. Phytopathol. 9:275–296.

Formusoh, E. S., J. H. Hatchett, W. C. Black IV, and J. J. Stuart. 1996. Sex-linked inheritance of virulence against wheat resistance gene *H9* in the Hessian fly (Diptera: Cecidomyiidae). Ann. Entomol. Soc. Am. 89:428–434.

Fouche, A., R. L. Verhoeven, P. H. Hewitt, M. C. Walters, C. F. Kriel, and J. De Jager. 1984. Russian aphid (*Diuraphis noxia*) feeding damage on wheat, related cereals and a *Bromus* grass species. In Walters, M. C. (ed.), Progress in Russian Wheat Aphid (Diuraphis noxia Mordw.) Research in the Republic of South Africa. Republic of South Africa. Dept. of Agriculture Technical Communication 191. pp. 22–33.

Frazer, B. D. 1972. Population dynamics and recognition of biotypes in the pea aphid (Homoptera: Aphididae). Can. Entomol. 10:1729–33.

Gallagher, K. D., P. E. Kenmore, and K. Sogawa. 1994. Judicial use of insecticides deter planthopper outbreaks and extend the life of resistant varieties in Southeast Asian rice. In: R. F. Denno and J. T. Perfect (Eds.), Planthoppers: Their Ecology and Management. Chapman & Hall, New York, pp. 599–614.

Gallun, R. L. 1977. The genetic basis of Hessian fly epidemics. Ann. N. Y. Acad. Sci. 287:223–229.

Gallun, R. L., and G. S. Khush. 1980. Genetic factors affecting expression and stability of resistance. In F. G. Maxwell and P. R. Jennings (Eds.), Breeding Plants Resistant to Insects. John Wiley. pp. 64–85.

Galet, P. 1982. Phylloxera. In. Les Maladies et le Parasites de la Vigne, Tome II Les Pararites Animaux. ontpelier: Paysan du Midi. pp. 1059–1313.

Goggin, F. L., V. M. Williamson, and D. E. Ullman. 2001. Variability in the response of *Macrosiphum euphorbiae* and *Myzus persicae* (Hemiptera: Aphididae) to the tomato resistance gene *Mi*. Environ. Entomol. 30:101–106.

Gordon, S. C., J. A. T. Woodford, B. Williamson, A. N. E. Birch, and A. T. Jones. 1999. Progress towards integrated crop management (ICM) for European raspberrry production. Scot. Crop Res. Instit. Ann. Rept. pp. 153–156.

Granett, J., P. Timper, and L. A. Lider. 1985. Grape phylloxera (*Daktulosphaira vitifoliae*) (Homoptera: Phylloxeridae) biotypes in California. J. Econ. Entomol. 78:81463–1467.

Granett, J., M. A. Walker, L. Kocsis, and A. D. Omer. 2001. Biology and management of grape phylloxera. Ann. Rev. Entomol. 46:387–412.

Grassi, B., and M. Topi. 1917. The number of races of vine *Phylloxera*. Bull. Agr. Intell. Plant Dis. Rome 9:1322–1327. (Rev. Appl. Entomol. 6:56).

Guo-rui, W. C. Fu-yan, T. Lin-yong, H. Ci-Wei, and F. Bin-can. 1983. Studies on the biotypes of the brown planthopper, *Nilaparvata lugens* (Stal). Acta Entomol. Sinica 26:154–160.

Haley, S. D., F. B. Peairs, C. B. Walker, J. B. Rudolph, and T. L. Randolph. 2004. Occurrence of a new Russian wheat aphid biotype in Colorado. Crop Sci. 44:1589–1592.

Harrington, C. D. 1943. The occurrence of physiological races of the pea aphid. J. Econ. Entomol. 36:118–119.

Harris, M.O., J. J. Stuart, M. Mohan, S. Nair, R. J. Lamb, and O. Rohfritsch. 2003. Grasses and gall midges: Plant defense and insect adaptation. An. Rev. Entomol. 48:549–577.

Harvey, T. L., and H. L. Hackerott. 1969. Recognition of a greenbug biotype injurious to sorghum. J. Econ. Entomol. 62:776–779.

Harvey, T. L., K. D. Kofoid, T. J. Martin, and P. E. Sloderbeck. 1991. A new greenbug biotype virulent to E-biotype resistant sorghum. Crop Sci. 31:1689–1691.

Harvey, T. L., T. J. Martin, and D. L. Seifers. 1994. Importance of resistance to insects and mite vectors in controlling virus diseases of plants: resistance to WCM. J. Agric. Entomol. 11:271–277.

Harvey, T. L., T. J. Martin, and D. L. Seifers. 1995. Survival of five wheat curl mite, *Aceria tosichella* Keifer (Acari: Eriophyidae), strains on mite resistant wheat. Exp. Appl. Acarol. 19: 459–463.

Harvey, T. L., T. J. Martin, D. L. Seifers, and P. E. Sloderbeck. 1997a. Change in virulence of wheat curl mite detected on TAM 107 wheat. Crop Sci. 37: 624 – 625.

Harvey, T. L., T. J. Martin, D. L. Seifers, and P. E. Sloderbeck. 1997b. Adaptations of wheat curl mite (Acaria: Eriophytidae) to resistant wheat in Kansas. J. Agric. Entomol. 12:119–125.

Harvey, T. L., D. L. Seifers, and T. J. Martin. 2001. Host range differences between two strains of wheat curl mites (Acaria: Eriophytidae). J. Agric. Urban Entomol. 18:35–41.

Harvey, T. L., D. L. Seifers, T. J. Martin, G. L. Brown-Guedira, and B. S. Gill. 1999. Survival of wheat curl mites on different sources of resistance in wheat. Crop Sci. 39:1887–1889.

Harvey, T. L., G. E. Wilde, and K. D. Kofoid. 1997. Designation of a new greenbug biotype, biotype K, injurious to resistant sorghum. Crop Sci. 37:989–991.

Hatchett, J. H. 1969. Race E, sixth race of the Hessian fly, *Mayetiola destructor*, discovered in Georgia wheat fields. Ann. Entomol. Soc. Am. 62:677–678.

Heinrichs, E. A. 1986. Perspectives and directions for the continued development of insect-resistant rice varieties. Agric. Ecosyst. Environ. 18:9–36.

Heinrichs, E. A., and H. R. Rapusas. 1985. Cross-virulence of *Nephotettix virescens* (Homoptera: Cicadellidae) biotypes among some rice cultivars with the same major-resistance gene. Environ. Entomol. 14:696–700.

Heinrichs, E. A., and H. R. Rapusas. 1990. Response to selection for virulence of *Nephotettix virescens* (Homoptera: Cicadellidae) on resistant rice cultivars. Environ. Entomol. 19:167–175.

Heinrichs, E. A., and P. K. Pathak. 1981. Resistance to the rice gall midge, *Orseolia oryzae* in rice. Insect Sci. Applic. 1:123–132.

Hellqvist, S. 2001. Biotypes of *Dasineura tetensi*, differing in ability to gall and develop on black currant genotypes. Entomol. Exp. Appl. 98: 85–94.

Herdt, R. W. 1991. Research priorities for rice biotechnology. In: G. S. Khush and G. H. Toenniessen (Eds.) Rice Biotechnology. CAB International, Oxon, UK. pp. 19–54.

Inayatullah, C., J. A. Webster, and W. S. Fargo. 1987a. Morphometric variation in the alates of greenbug (Homoptera: Aphididae) biotypes. Ann. Entomol. Soc. Am. 80:306–311.

Inayatullah, C., W. S. Fargo, and J. A. Webster. 1987b. Use of multivariate models in differentiating greenbug (Homoptera: Aphididae) biotypes and morphs. Environ. Entomol. 16:839–846.

Inayatullah, C., M. N. Ehsan-Ul-Haq, and M. F. Chaudhry. 1993. Incidence of greenbug, *Schizaphis graminum* (Rondani) (Homoptera: Aphididae) in Pakistan and resistance in wheat against it. Insect Sci. Applic. 14:247–254.

IITA. International Institute of Tropical Agriculture. 1981. Research Highlights. IITA. Ibadan, Nigeria.

Ito, K., T. Wada, A. Takashi, N. M. Noor Nik Salleh, and H. Hassim. 1994. Brown planthopper *Nilaparvata lugens* Stal (Homoptera: Delphacidae) biotypes capable of attacking resistant rice varieties in Malaysia. Appl. Entomol. Zool. 29:523–532.

Jones, A. T., W. J. McGavin, and A. N. E. Birch. 2000. Effectiveness of resistance genes to the large raspberry aphid, *Amphorophora idaei* Borner, in different raspberry (*Rubus idaeus* L.) genotypes and under different environmental conditions. Ann. Appl. Biol. 136:107–113.

Katiyar, S. K., G. Chandel, Y. Tan, Y. Zhang, B. Huang, L. Nugaliyadde, K. Fernando, J. S. Bentur, S. Inthavong, S. Constantino, and J. Bennett. 2000. Biodiversity of Asian rice gall midge (*Orseolia oryzae* Wood Mason) from five countries examined by AFLP analysis. Genome. 43: 322–332.

Katiyar, S. K., and J. Bennett. 2001. Biotechnology for gall midge resistance: from molecular tagging to gene pyramiding. In: S. Peng and B. Hardy (Eds.), Rice Research for Food Security and Poverty Alleviation. Proc. Intl. Rice Research Conf., 31 March-3 April 2000, Los Banos, Philippines. Los Banos (Philippines): International Rice Research Institute. pp. 369–378.

Katiyar, S. K., Y. Tan, B. Huang, G. Chandel, Y. Xu, Y. Zhang, Z. Xie, and J. Bennett. 2001. Molecular mapping of gene *Gm–6(t)* which confers resistance against four biotypes of Asian rice gall midge in China. Theor. Appl. Genet. 103:953–961.

Ketipearachchi, Y., C. Kaneda, and C. Nakamura. 1998. Adaptation of the brown planthopper (BPH), *Nilaparvata lugens* (Stal) (Homoptera: Delphacidae), to BPH resistant rice cultivars carrying *bph8* or *Bph9*. Appl. Entomol. Zool. 33:497–505.

Khush, G. S. 1995. Modern cultivars - their real contribution to food supply and equity. Geojourn. 35:275–284.

Khush, G. S., and D. S. Brar. 1991. Genetics of resistance to insects in crop plants. Adv. Agron. 45:223–274.

Kimura, M. 1980. A simple method for estimating evolutionary rate of base substitutions through comparative studies of nucleotide sequences. J. Mol. Evol. 16: 111–120.

Kindler, S. D., and S. M. Spomer. 1986. Biotypic status of six greenbug (Homoptera: Aphididae) isolates. Environ. Entomol. 15:567–572.

Kindler, S. D., and D. B. Hays. 1999. Susceptibility of cool-season grasses to greenbug biotypes. J. Agric. Urban Entomol. 16:235–243.

de Kogel, W. J., M. van der Hoek, and C. Mollema. 1997. Variation in performance of western flower thrips populations on susceptible and partially resistant cucumber. Entomol. Exp. Appl. 83:73–80.

Lammerink, J. 1968. A new biotype of cabbage aphid, *Brevicornye brassicae* L. on aphid resistant rape (*Brassica napus* L.). N. Z. J. Agric. Res. 11:341–344.

Lehman, W. F., M. W. Nielson, V. L. Marble, and E. H. Stanford. 1983. Registration of CUF 101 alfalfa. Crop Sci. 23:398.

Lei, H., W. F. Tjallingii, and J. C. van Lenteren. 19998. Probing and feeding characteristics of the greenhouse whitefly in association with host-plant acceptance and whitefly strains. Entomol. Exp. Appl. 88:73–80.

Liu, X. M., and D. S. Jin. 1998. A new biotype (CHN-1) of greenbug, *Schizaphis graminum*, found in Beijing. Acta Entomol. Sinica. 41: 141–144.

Liu, X., C. M. Smith, B. S. Gill, and V. Tolmay. 2001. Microsatellite markers linked to six Russian wheat aphid resistance genes in wheat. Theor. Appl. Genet. 102:504–510.

Lowe, H. J. B. 1981. Resistance and susceptibility to colour forms of the aphid *Sitobion avenae* in spring and winter wheats (*Triticum aestivum*). Ann. Appl. Biol. 99:87–98.

Malik, R. 2001. Molecular genetic characterization of wheat curl mite, *Aceria tosichella* Keifer (Acari:Eriophyidae), and wheat genes conferring wheat curl mite resistance. Ph.D. Dissertation, Kansas State University, 102 pages.

Malik, R., C. M. Smith, T. L. Harvey, and G. L. Brown-Guedira. 2003. Genetic mapping of wheat curl mite resistance genes *Cmc3* and *Cmc4* in common wheat. Crop Sci. 43:644–650.

Marchal, P. 1931. La question des races du *Phylloxera* de la vigne. Ann. Epiphyt. 16:232–234. (Rev. Appl. Entomol. 33:124).

Marais, G. F., M. Horn, and F. du Toit. 1994. Intergeneric transfer (rye to wheat) of a gene(s) for Russian wheat aphid resistance. Plant Breed. 113:265–271.

Martin, T. J., A. Fritz, and J. P. Shroyer. 2001. Stanton Hard Red Winter Wheat. Kansas State University Agricultural Experiment Station and Cooperative Extension Service Publication L–921.

Martinez-Peniche, R. 1999. Effect of different phylloxera (*Daktulosphaira vitifoliae* Fitch) populations from South France, upon resistance expression of rootstocks 41 B and Aramon x Rupestris Ganzin No. 9. Vitis. 38:167–178.

Martín-Sánchez, J. A., M. Gómez-Colmenarejo, J. Del Moral, E. Sin, M. J. Montes, C. González-Belinchón, I. López-Braña, and A. Delibes. 2003. A new Hessian fly resistance gene (*H30*) transferred from the wild grass *Aegilops triuncialis* to hexaploid wheat. Theor. Appl. Genet. 106:1248–1255.

Michels, G. J., Jr. 1986. Graminaceous North American host plants of the greenbug with notes on biotypes. Southwest. Entomol. 11: 55–65.

Mitchell-Olds, T., and J. Bergelson. 2000. Biotic interactions Genomics and coevolution. Curr. Opin. Plant Biol. 3:273–277.

Montllor, C. B., B. C. Campbell, and T. E. Miller. 1983. Natural and induced differences in probing behavior of two biotypes of the greenbug, *Schizaphis graminum*, in relation to resistance in sorghum. Entomol. Exp. Appl. 34:99–106.

Morgham, A. T., P. E. Richardson, R. K. Campbell, J. R. Burd, R. D. Eikenbary, and L. C. Sumner. 1994. Ultrastructural responses of resistant and susceptible wheat to infestation by greenbug biotype E (Homoptera: Aphididae). Ann. Entomol. Soc. Am. 87:908–917.

Niassy, A., J. D. Ryan, and D. C. Peters. 1987. Variations in feeding behavior, fecundity, and damage of biotypes B and E of *Schizaphis graminum* (Homoptera: Aphididae) on three wheat genotypes. Environ. Entomol. 16:1163–1168.

Nielson, M. W., and H. Don. 1974. A new virulent biotype of the spotted alfalfa aphid in Arizona. J. Econ. Entomol. 67:64–66.

Nielson, M. W., and R. O. Kuehl. 1982. Screening efficacy of spotted alfalfa aphid biotypes and genic systems for resistance in alfalfa. Environ. Entomol. 11:989–996.

Nielson, M. W., and W. F. Lehman. 1980. Breeding approaches in alfalfa. In: F. G. Maxwell and P. R. Jennings (Eds.), Breeding Plants Resistant to Insects. John Wiley, New York. pp. 279–311.

Nkongolo, K. K., J. S. Quick, W. L. Meyers, and F. B. Peairs. 1989. Russian wheat aphid resistance of wheat, rye, and triticale in greenhouse tests. Cer. Res. Comm. 17:227–232.

Nkongolo, K. K., J. S. Quick, A. E. Limin, and D. B. Fowler. 1991. Sources and inheritance of resistance to Russian wheat aphid in *Triticum* species amphiploids and *Triticum tauschii*. Can. J. Plant Sci. 71:703–708.

Nombela, G., V.M. Williamson, and M. Muñiz. 2003. The root-knot nematode resistance gene *Mi-1.2* of tomato is responsible for resistance against the whitefly *Bemisia tabaci*. Mol. Plant-Microbe Interact. 16:645–649.

Ofuya, T. I., and P. F. Credland. 1996. Responses of three populations of the seed beetle, *Callosobruchus maculatus* (F.) (Coleoptera:Bruchidae), to seed resistance in selected varieties of cowpea, *Vigna unguiculata* (L.) Walp. J. Stored Prod. Res. 31:17–27.

Omer, A. D., J. Granett, L. Kocsis, and D. A. Downie. 1999. Preference and performance responses of California grape phylloxera to different *Vitis* rootstocks. J. Appl. Ent. 123:341–346.

Paguia, P., M. D. Pathak, and E. A. Heinrichs. 1980. Honeydew excretion measurement techniques for determining differential feeding activity of biotypes of *Nilaparvata lugens* on rice varieties. J. Econ. Entomol. 73:35–40.

Painter, R. H. 1930. Biological strains of Hessian fly. J. Econ. Entomol. 23:322–326.

Painter, R. H. 1951. Insect Resistance in Crop Plants. University of Kansas Press, Lawrence, KS. 520 pp.

Painter, R. H., and M. D. Pathak. 1962. The distinguishing features and significance of the four biotypes of the corn leaf aphid *Rhopalosiphum maidis* (Fitch). Proc. XI Int. Cong. Entomol. 2:110–115.

Pani, J., and S.C. Sahu. 2000. Inheritance of resistance against biotype 2 of the Asian rice gall midge, *Orseolia oryzae*. Entomol. Exp. Appl. 95: 15–19.

Pathak, P. K., and E. A. Heinrichs. 1982. Selection of biotype populations 2 and 3 of *Nilaparvata lugens* by exposure to resistant rice varieties. Environ. Entomol. 11:85–90.

Patterson, F. L., J. E. Foster, H. W. Ohm, J. E. Hatchett, and P. L. Taylor. 1992. Proposed system of nomenclature for biotypes of Hessian fly (Diptera: Cecidomyiidae) in North America. J. Econ. Entomol. 85:307–311.

Pedigo, L. 1999. Entomology and Pest Management. Prentice Hall, Upper Saddle River, NJ. 742 pp.

Pesho, G. R., F. V. Lieberman, and W. F. Lehman. 1960. A biotype of the spotted alfalfa aphid on alfalfa. J. Econ. Entomol. 53:146–150.

Peters, D. C., F. Ullah, M. A. Karner, W. B. Massey, P. G. Mulder, and V. H. Beregovoy. 1997. Greenbug (Homoptera:Aphididae) biotype surveys in Oklahoma, 1991–1996. J. Kansas Entomol. Soc. 70:120–128.

Peters, D. C., and A. Wood, Jr., and K. J. Starks. 1975. Insecticide resistance in selections of the greenbug. J. Econ. Entomol. 75:339–340.

Porter, D. R., J.D. Burd, K.A. Shufran, J.A. Webster, and G.L. Teetes. 1997. Greenbug (Homoptera: Aphididae) biotypes: selected by resistant cultivars or preadapted opportunists?. J. Econ. Ent. 90:1055–1065.

Porter, D. R., J. A. Webster, and B. Friebe. 1994. Inheritance of greenbug biotype G resistance in wheat. Crop Sci. 34:625–628.

Porter, K. B., G. L. Peterson, and O. Vise. 1982. A new greenbug biotype. Crop Sci. 22:847–850.

Powers, T. O., S. G. Jensen, S. D. Kindler, C. J. Stryker, and L. J. Sandall. 1989. Mitochondrial DNA divergence among greenbug (Homoptera: Aphididae) biotypes. Ann. Entomol. Soc. Am. 82:298–302.

Prinsloo, G. J., 2000. Host and host instar preference of *Aphelinus* sp. nr. *varipes* (Hymenoptera: Aphelinidae), a parasitoid of cereal aphids (Homoptera: Aphididae) in South Africa. African Entomol. 8:57–61.

Prokopy, R. J., S. R. Diehl, and S. S. Cooley. 1988. Behavioral evidence for host races in *Rhagoletis pomonella* flies. Oecologia. 76:138–147.

Puterka, G. J., J. D. Burd, and R. L. Burton. 1992. Biotypic variation in a worldwide collection of Russian wheat aphid (Homoptera: Aphididae). J. Econ. Entomol. 85:1497–1506.

Puterka, G. J., and R. L. Burton. 1990. Aphid genetics in relation to host plant resistance. In: D. C. Peters, J. A. Webster and C. S. Chlouber (Eds.), Proc. Aphid-Plant Interactions: Populations to Molecules. Okla. Agric. Exp. Stn. MP-132: 59–69.

Puterka, G. J., and D. C. Peters. 1989. Inheritance of greenbug virulence to *Gb2* and *Gb3* resistance genes in wheat. Genome. 32:109–114.

Puterka, G. J., and D. C. Peters. 1995. Genetics of greenbug (Homoptera: Aphididae) virulence to resistance in sorghum. J. Econ. Entomol. 88:421–429.

Puterka, G. J., D. C. Peters, D. L. Kerns, J. E. Slosser, L. Bush, D. W. Worrall, and R. W. McNew. 1988. Designation of two new greenbug (Homoptera: Aphididae) biotypes G and H. J. Econ. Entomol. 81:1754–1759.

Puterka, G. J., W. C. Black IV, W. M. Steiner, and R. L. Burton. 1993. Genetic variation and phylogenetic relationships among worldwide collections of Russian wheat aphid, *Diuraphis noxia* (Mordvilko), inferred from allozyme and RAPD-PCR markers. Heredity. 70:604–618.

Quick, J. S., G. E. Ellis, R. M. Normann, J. A. Stromberger, J. F. Shanahan, F. B. Peairs, J. B. Rudolph, K. Lorenz. 1996. Registration of 'Halt' wheat. Crop Sci. 36:210.

Quisenberry, S. S., and F. B. Peairs. 1998. A response model for an introduced pest- the Russian wheat aphid. Thomas Say Publications in Entomology, Entomological Society of America, Lanham, MD. 442 pp.

Raspusas, H. R., and E. A. Heinrichs. 1986. Virulence of green leafhopper (GLH) colonies from Luzon, Philippines, on IR36 and IR42. Int. Rice Res. Newsl. 11:15.

Ratcliffe, R. H., and J. H. Hatchett. 1997. Biology and genetics of the Hessian fly and resistance in wheat, In: K. Bobdari, (Ed.), New Developments in Entomology, Research Signpost, Scientific Information Guild, Trivandrum.

Ratcliffe, R. H., S. E. Cambron, K. L. Flanders, N. A. Bosque-Perez, S. L. Clement, and H. W. Ohm. 2001. Biotype composition of Hessian fly (Diptera: Cecidomyiidae) populations from the southeastern, midwestern, and northwestern United States and virulence to resistance genes in wheat. J. Econ. Entomol. 93:1319–1328.

Ratcliffe, R. H., H. W. Ohm, F. L. Patterson, S. E. Cambron, and G. G. Safranski. 1996. Response of resistance genes H9 - H19 in wheat to Hessian fly (Diptera: Cecidomiidae) laboratory biotypes and field populations from the Eastern United States. J. Econ. Entomol. 89:1309–1317.

Ratcliffe, R. H., G. C. Safranski, F. L. Patterson, H. W. Ohm, and P. L. Taylor. 1994. Biotype status of Hessian fly (Diptera: Cecidomyidae) populations from the eastern United States and their responses to 14 Hessian fly resistance genes. J. Econ. Entomol. 87:1113–1121.

Rider Jr., S. D., W. Sun, R. H. Ratcliffe, and J. J. Stuart. 2002. Chromosome landing near avirulence gene vH13 in the Hessian fly. Genome. 45:812–822.

Riley, C. V. 1872. Grape disease. In. Fourth Annual Report of the Noxious, Beneficial, and Other Insects of the State of Missouri. Regan & Edwards, Jefferson City, MO. pp. 55–71.

Robinson, J., M. Fischer, and D. Hoisington. 1993. Molecular characterization of Diuraphis spp. using random amplified polymorphic DNA. Southwest. Entomol. 18:121–127.

Roche, P. A., F. H. Alston, C. A. Maliepaard, K. M. Evans, R. Vrielink, F. Dunemann, T. Markussen, S. Tartarini, L. M. Brown, C. Ryder, and G. J. King. 1997. RFLP and RAPD markers linked to the rosy leaf curling aphid resistance gene (Sd1) in apple. Theor. Appl. Genet. 94:528–533.

Roderick, G. K. 1994. Genetics of host plant adaptation in delphacid planthoppers. In: R. F. Denno and J. T. Perfect (Eds.), Planthoppers: Their Ecology and Management. Chapman & Hall, New York, pp. 551–570.

Rossi, M., F. L. Goggin, S. B. Milligan, I. Klaoshian, D. E. Ullman, and V. M. Williamson. 1998. The nematode resistance gene Mi of tomato confers resistance against the potato aphid. Proc. Natl. Acad. Sci. USA. 95:9750–9754.

Sardesai, N., A. Kumar, K. R. Rajyashri, S. Nair, and M. Mohan. 2002. Identification and mapping of an AFLP marker linked to Gm7, a gall midge resistance gene and its conversion to a SCAR marker for its utility in marker aided selection in rice. Theor. Appl. Genet. 105:691–698.

Sardesai, N., K. R. Rajyashri, S. K. Behura, S. Nair, and M. Mohan. 2001. Genetic. physiological and molecular interactions of rice and its major diptera pest, gall midge. Plant Cell Tissue Organ Cult. 64:115–131.

Sato, A., and K. Sogawa. 1981. Biotypic variations in the green rice leafhopper, Nephotettix virescens Uhler (Homoptera : Deltocephalidae). Appl. Ent. Zool. 16:55–57.

Saxena, R. C., and A. A. Barrion. 1983. Biotypes of the brown planthopper, Nilaparvata lugens (Stal). Korean J. Plant Prot. 22:52–66.

Saxena, R. C., and L. M. Rueda. 1982. Morphological variations among three biotypes of the brown planthopper, Nilaparvata lugens in the Philippines. Insect Sci. Applic. 3:193–210.

Scharwtz, C. D., and G. A. Hubner. 1937. Aphis resistance in breeding mosaic-escaping red raspberries. Science. 86:15–159.

Schilder, F. A. 1947. The biology of phylloxera races. Züchter. 17/18:413–415.

Schulte, S. J., S. D. Rider, J. H. Hatchett, and J. J. Stuart. 1999. Molecular genetic mapping of three X-linked avirulence genes, *vH6*, *vH9* and *vH13*, in the Hessian fly. Genome. 42:821–828.

Sebesta, E. E., and E. A. Wood, Jr. 1978. Transfer of greenbug resistance from rye to wheat with x-rays. Agron. Absts. 61–62.

Sen Gupta, G. C., and P. W. Miles. 1975. Studies on the susceptibility of varieties of apple to the feeding of two strains of wooly aphids (Homoptera) in relation to the chemical content of the tissues of the host. Aust. J. Agric. Res. 26:157–68.

Shade, R. E., L. W. Kitch, P. Mentzer, and L. L. Murdock. 1996. Selection of a cowpea weevil (Coleoptera: Bruchidae) biotype virulent to cowpea weevil resistant landrace TVu 2027. J. Econ. Entomol. 89:1325–1331.

Shade, R. E., L. L. Murdock, and L. W. Kitch. 1999. Interactions between cowpea weevil (Coleoptera: Bruchidae) populations and *Vigna* (Leguminosae) species. J. Econ. Entomol. 92:740–745.

Shufran K. A., J. D. Burd, and J. A. Webster. 1997. Biotypic status of Russian wheat aphid (Homoptera: Aphididae) populations in the United States. J. Econ. Entomol. 90:1684–1689.

Shufran K. A., and G. Wilde. 1994. Clonal diversity in overwintering populations of *Schizaphis graminum* (Homoptera: Aphididae). Bull. Entomol. Res. 84:105–114.

Shufran K. A., D. C. Margolies, and W. C. Black IV. 1992. Variation between biotype E clone of *Schizaphis graminum* (Homoptera: Aphididae). Bull. Entomol. Res. 82:407–416.

Singh, S. R., and E. A. Wood, Jr. 1963. Effect of temperature on fecundity of two strains of the greenbug. J. Econ. Entomol. 56:109–110.

Singh, S. R., and R. H. Painter. 1964. Effect of temperature and host plants on progeny production of four biotypes of corn leaf aphid. J. Econ. Entomol. 75:348–350.

Smith, C. M., T. Belay, C. Stauffer, P. Stary, I. Kubeckova, and S. Starkey. 2004. Identification of Russian wheat aphid (Homoptera: Aphididae) biotypes virulent to the *Dn4* resistance gene. J. Econ. Entomol. 97:112–117.

Sogawa, K. 1978a. Variations in gustatory response to amino acid-sucrose solutions among biotypes of the brown planthopper. International Rice Res. Newslttr. 3(5):9.

Sogawa, K. 1978b. Electrophoretic variations in esterase among biotypes of the brown planthopper. International Rice Res. Newslttr. 3(5):8–9.

Sogawa, K. 1981. Biotypic variations in the brown planthopper, *Nilaparvata lugens* (Homoptera: Delpahacidae) at the IRRN, the Philippines. Appl. Entomol. Zool. 16: 129–137.

Sogawa, K. 1992. A change in biotype property of brown plant-hopper populations immigrating into Japan and their probable source areas. Proc. Assoc. Plant Prot. Kyushu Univ. 38:63–68.

Song, G.-C., and J. Granett. 1990. Grape phylloxera (Homoptera: Phylloxeridae) biotypes in France. J. Econ. Entomol. 83:489–493.

Sosa, O., Jr. 1978. Biotype L, ninth biotype of the Hessian fly. J. Econ. Entomol. 71:458–460.

Sosa, O., Jr. 1981. Biotypes J and L of the Hessian fly discovered in an Indiana wheat field. J. Econ. Entomol. 74:180–182.

Starks, K. J., and R. L. Burton. 1972. Greenbugs: Determining biotypes, culturing, and screening for plant resistance. USDA-ARS Tech. Bull. No. 1556. 12 pp.

Starks, K. J., R. L. Burton, and O. G. Merkle. 1983. Greenbug (Homoptera: Aphididae) plant resistance in small grains and sorghum to biotype E. J. Econ. Entomol. 76:877–880.

Stuart, J. J., S. J. Schulte, P. S. Hall, and K. M. Mayer. 1998. Genetic mapping of Hessian fly avirlence gene vH6 using bulked segregant analysis. Genome. 41:702–708.

Takita, T., and H. Hashim. 1985. Relationship between laboratory-developed biotypes of green leafhopper and resistant varieties of rice in Malaysia. Japan. Agric. Res. Quart. 19: 219–223.

Takahashi, A., K. Ito, J. Tang, G. Hu, and T. Wada. 1994. Biotypical property in the populations of the brown planthopper, Nilaparvata lugens Stal (Homoptera: Delphacidae), collected in China and Japan. Appl. Entomol. Zool. 29:461–463.

Tanaka, K. 1999. Quantitative genetic analysis of biotypes of the brown planthopper Nilaparvata lugens: heritability of virulence to resistant rice varieties. Entomol. Exp. Appl. 90: 279–287.

Teetes, G. L., C. A. Schaefer, J. R. Gipson, R. C. McIntyre, and E. E. Latham. 1975. Greenbug resistance to organophosphorous insecticides on the Texas high plains. J. Econ. Entomol. 68:214–216.

Teetes, G. L., G. C. Peterson, K. F. Nwanze, and B. B. Pendelton. 1999. Genetic diversity of sorghum: A source of insect-resistant germplasm. In: S. L. Clement and S. S. Quisenberry (Eds.), Global Plant Genetic Resources for Insect Resistant Crops. CRC Press, Boca Raton, FL. pp. 63–85.

Thomas, J. B., and R. L. Conner. 1986. Resistance to colonization by the wheat curl mite in Aegilops squarrosa and its inheritance after transfer to common wheat. Crop Sci. 26:527–530.

Tomar, J. B., and S. C. Prasad. 1992. Genetic analysis of resistance to gall midge (Orseolia oryzae Wood Mason) in rice. Plant Breed. 109:159–167.

Tyler, J. M., J. A. Webster, and E. L. Smith. 1985. Biotype E greenbug resistance in wheat streak mosaic virus-resistant wheat germplasm lines. Crop Sci. 25:686–688.

Tyler, J. M., J. A. Webster, and O. G. Merkle. 1987. Designations for genes in wheat germplasm conferring greenbug resistance. Crop Sci. 27:526–527.

Vasilev, I. V. 1929. On the race of the Ukranian phylloxera (Russian). Vestn. Vinodel Ukrainui. 30:13–14. (Rev. Appl. Entomol. 17:306).

Verma, S. K., P. K. Pathak, B. N. Singh, and M. N. Lal. 1979. Indian biotypes of the brown planthopper. International Rice Res. Newslttr. 4(6):7.

Wada, T., K. Ito, and A. Takahashi. 1994. Biotype comparisons of the brown planthopper, Nilaparvata lugens (Homoptera: Delphacidae) collected in Japan and the Indochina peninsula. Appl. Entomol. Zool. 29:477–484.

Webster, J. A., and C. Inayatullah. 1985. Aphid biotypes in relation to host plant resistance: a selected bibliography. Southwest. Entomol. 10:116–125.

Webster, J. A., P. Treat, L. Morgan, and N. Elliott. 2000. Economic impact of the Russian wheat aphid and greenbug in the western United States. 1993–94, 1994–95, and 1995–96. USDA ARS Service Report PSWCRL Rep. 00–001.

Whalon, M. E., G. Wilde, and H. Feese. 1973. A new corn leaf aphid biotype and its effect on some cereal and small grains. J. Econ. Entomol. 66:570–571.

Whelan, E. D., and G. E. Hart. 1988. A spontaneous translocation that transfers wheat curl mite resistance from decaploid *Agropyron elongatum* to common wheat. Genome. 30:289–292.

Wilhoit, L. R. 1992. Evolution of virulence to plant resistance: Influence of variety mixtures. In: R. S. Fritz and E. L. Simms (Eds.), Plant Resistance to Herbivores and Pathogens: Ecology, Evolution, and Genetics, University of Chicago Press, Chicago, pp. 91–119.

Williams, R. N., and G. F. Shambaugh. 1988. Grape phylloxera (Homoptera: Phylloxeridae) biotypes confirmed by electrophoresis and host susceptibility. Ann. Entomol. Soc. Am. 81:1–5.

Wood, E. A., Jr. 1961. Biological studies of a new greenbug biotype. J. Econ. Entomol. 54:1171–1173.

Young, E., G. C. Rock, and D. C. Zeiger. 1982. Infestation of some *Malus* cultivars by the North Carolina woolly apple aphid biotype. HortSci. 17:787–788.

Zantoko, L., and R. H. Shukle. 1997. Genetics of virulence in the Hessian fly to resistance gene H13 in wheat. J. Hered. 88:120–123.

Zarrabi, A. A., R. C. Berberet, and J. L. Caddell. 1995. New biotype of *Acyrthosiphon kondoi* (Homoptera: Aphididae) on alfalfa in Oklahoma. J. Econ. Entomol. 88:1461–1465.

BIBLIOGRAPHY

CHAPTER 12

PLANT RESISTANCE IN ARTHROPOD PEST MANAGEMENT SYSTEMS

1. SIGNIFICANCE OF PLANT RESISTANCE IN ARTHROPOD PEST MANAGEMENT

1. 1 Resistance Categories

The type of insect resistance deployed in an integrated arthropod pest management (IPM) system has a direct influence on the stability and ultimate success of an arthropod-resistant cultivar. Different categories of resistance have different degrees of effectiveness in an IPM system, depending on the movement and host preferences of the pest arthropod(s). Simulation modeling studies by Kennedy et al. (1987) indicate that low levels of antixenosis, antibiosis, and tolerance can be effective in controlling resident arthropod pests that invade plantings early in the development of a crop and increase gradually during the growing season. Since the movement of resident arthropods is inherently limited, antixenosis may be adequate to reduce populations of this type of arthropod in an IPM system. The results of Alvarado-Rodriguez et al. (1986) indicate that feeding antixenosis resistance in cultivars of common bean, *Phaseolus vulgaris* L., to the lygus bug, *Lygus hesperus* Knight, are effective against both migratory and endemic populations. Leszczynski (1987) drew similar conclusions concerning antixenosis in European wheat to the grain aphid, *Sitobion avenae* F. Antixenosis due to the presence of awns on grain heads may be more beneficial than antibiosis resistance to migrating populations of *S. avenae*.

Cropping systems attacked by highly mobile pests however, require different categories and levels of resistance for their effective management. In further research, Kennedy et al. (1987) used simulation modeling to compare the effects of resistance in tomato, *Lycopersicon esculentum* Mill., and soybean, *Glycine max* (L.) Merr., on the corn earworm, *Heliocoverpa zea* Boddie. These results indicate that high mortality in early instar larvae from tomato antibiosis or oviposition antixenosis will reduce *H. zea* populations, but insecticidal control is still necessary. However, since soybean plants can withstand greater defoliation than tomato plants, *H. zea* populations reach economically damaging levels less frequently on soybean than on tomato. For this reason, low or moderate levels of antibiosis or antixenosis are useful in reducing *H. zea* infestations on soybean.

The use of tolerant cultivars in arthropod pest management systems also offers several advantages. Pest population levels are not diminished from exposure to tolerant plants, as they are on plants exhibiting antibiosis and antixenosis, but their virulence gene frequencies remain diluted, because the selection pressure placed on them by high levels of antibiosis is reduced or absent. Thus, the potential for development of resistance-breaking biotypes (see Chapter 9) is greatly diminished through the use of tolerant cultivars (Teetes 1980). Some cultivars of maize, *Zea mays* L., tolerant of foliar damage by *H. zea* and stem boring by the European corn borer, *Ostrinia nubilalis* Hübner, actually harbor larger larval populations than susceptible cultivars, due presumably to increased plant biomass (Wiseman et al. 1972, Hudon et al. 1979). Nevertheless, they provide significantly greater yields than susceptible cultivars.

Tolerance also enhances the effects of biological control agents in crop protection systems. Tolerant cultivars do not expose beneficial arthropods to the adverse effects of plant morphological or allelochemical factors in cultivars that exhibit antibiosis or antixenosis (see Chapters 2 and 3). From the perspective of the total effect of the resistant plant on the arthropod population, cultivars with tolerance require less antixenosis or antibiosis than cultivars without tolerance. Brewer et al. (1986) suggest that production of alfalfa, *Medicago sativa* L., cultivars exhibiting tolerance, antibiosis, and antixenosis to the potato leafhopper, *Empoasca fabae* (Harris), may provide stable *E. fabae* control by raising alfalfa economic injury levels.

1. 2 Relationship of Plant Resistance to Crop Economic Injury Level

The crop economic injury level (EIL), the pest density at which the cost of control equals the value of crop injury, has been defined by Pedigo (1986) as: EIL = C/VDIK, where C = the cost of pest control per unit of crop production, D = crop damage per pest per production unit, I = the economic injury per unit of pest damage, and K = pest arthropod mortality. A companion term, the economic threshold (ET), describes the pest population density at which some control action should be taken to prevent population density from reaching the EIL.

As discussed by Eigenbrode and Trumble (1994), different categories of resistance affect EILs in different ways. Both antibiosis and antixenosis affect EILs and ETs. Pest populations on cultivars with antibiosis develop more slowly and are at least moderately reduced. On cultivars with antixenosis, pest population densities are lowered and less likely to reach the critical density of either the ET or the EIL. In the example shown in Figure 12.1, antibiosis and antixenosis resistance in *Sorghum bicolor* (L.) Moench, resistant to the spotted stalk borer, *Chilo partellus* (Swinhoe), shift the ET later into the growing season. Economic damage occurs on the susceptible cultivar early in the growing season, but does not occur on the moderately resistance cultivar until approximately one month later. On the highly resistant cultivar, pest populations remain below the ET throughout the season. Tolerance resistance also affects the EIL. On tolerant plants, the EIL is raised, because D (crop damage per pest per production unit) is increased, as is I (crop economic injury per unit of pest damage).

Section 5 of Chapter 1 described instances of all three categories operating simultaneously in some cereal crops, in sugarcane, and in yellow mustard. The occurrence of at least two categories of resistance in a resistant genotype is typical (Singh 1987, Birch 1988, Sharma 1993). Cultivar-specific ETs have been developed for sorghum hybrids resistant to the sorghum midge, *Stenodiplosis sorghicola* (Coquillett) (Hallman 1984, Sharma et al. 1993) and several other sorghum pests (Sharma 1993). Cultivar specific EILs have been developed for cultivars of cotton, *Gossypium hirsutum* L., resistant to the bollworm, *Heliothis virescens* (F.) (Ring et al. 1993). Van den Berg et al. (1997) evaluated the effect of resistance to *C. partellus* on the ET of sorghum hybrids in South Africa and found that the ET on resistant hybrids was increased by 20- to 30- fold over the ET on the susceptible hybrid.

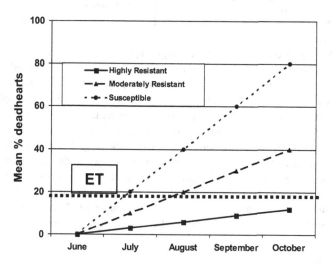

Figure 12.1. Effect of resistance (reduced stem deadheart) on the economic threshold (ET) of sorghum hybrids highly resistant, moderately resistant or susceptible to the spotted stalk borer, Chilo partellus. *Reprinted from Crop Protection, Vol. 12, Sharma , H. C. 1993. Host-plant resistance to insects in sorghum and its role in integrated pest management. Pages 11 - 34, Copyright 1993, Butterworth Heinemann, Inc., with permission from Elseiver.*

2. INTEGRATION WITH CULTURAL CONTROL TECHNIQUES

Variations in cultural practices in production agriculture is a management tactic that was first used by Chinese rice farmers over 2,000 years ago, when adjusted planting dates and crop stubble burning were used to reduce the buildup of arthropod pest populations (Flint and VandenBosch 1981). Undoubtedly, many other early agricultural systems implemented similar types of cultural practices in their development. There are few documented examples of the integration of cultural practices with plant resistance, but many cultural practices are effective when used in combination with other pest arthropod pest management tactics in integrated pest management systems. As mentioned in Chapter 1, the use of resistant cultivars to maximize cultural control tactics such as early-planted cultivars, trap crops, and early maturing cultivars is well-documented in several crops (Maxwell 1985).

2. 1 Trap Crops

Trap crops that attract pest arthropod populations, so that they may be destroyed, are synergistic when used in combination with arthropod resistant cultivars of cotton, soybean and rice. The combination of an early maturing, okra-leaved (see below) cotton breeding line and a non-preferred cotton breeding line resistant to the boll weevil, *Anthomonous grandis grandis* Boheman, is effective in suppressing *A. grandis* populations (Burris et al. 1983). Treatment of *A. grandis* in the concentrated area of the trap crop causes a 20% reduction in overall insecticide application and increases yields by 14-33%.

Trap cropping also has potential for integration with the growth of resistant rice and soybean cultivars. Rice fields planted adjacent to a trap crop planted 20 days ahead of the main crop attract more brown planthoppers, *Nilaparvata lugens* (Stål), yield significantly more, and preserve more natural enemies than control fields without trap crops (Saxena 1982). Trap crops of cowpea, *Vigna unguiculata* (L.) Walp., planted prior and adjacent to soybean plantings are also effective in concentrating populations of the southern green stink bug, *Nezara viridula* (L.), allowing their control with insecticides and reduced insecticidal treatment of the main crop (Newsom et al. 1975). Infestations of the greenbug, *Schizaphis graminum* Rondani, were shown to be effectively reduced by combining the planting of a *S. graminum*-resistant sorghum cultivar at a later than normal planting date, in combination with no tillage cultivation of the preceding crop stubble (Burton et al. 1990).

2.2 Early Maturity

The planting of early-maturing, arthropod-resistant cultivars has been shown to reduce populations of several key pests in rice. Planting the rice crop in a manner that avoids the peak pest insect population generally reduces insect pest damage. Heinrichs et al. (1986a) demonstrated that *N. lugens* populations and *N. lugens*: predator ratios on very early- and early-maturing rice cultivars are significantly lower than those on mid-season maturing cultivars. These results indicate that the incorporation of *N. lugens* resistance into early maturing cultivars can increase rice crop protection in the event of *N. lugens* biotype development. Similarly, early maturing cotton cultivars that contain the nectariless trait (lack of extrafloral nectaries) have yields in untreated plots that are only slightly less than those of insecticide treated plots (Bailey et al. 1980). The improved cotton cultivar 'Gumbo 500,' matures earlier than other adapted cultivars, eliminating the need for one to two insecticide applications (Jones et al. 1981).

3. INTEGRATION WITH CHEMICAL CONTROL

3. 1 Cereal Crops

Wheat cultivars resistant to the Hessian fly, *Mayetiola destructor* (Say), may benefit from the application of insecticides, depending on the level of fly resistance in the cultivar involved. Zelarayan et al. (1991) noted that applications of disulfoton insecticide to the partially resistant cultivar 'Morrison' enhanced the yields of this cultivar in areas of heavy fly infestation. However, Buntin et al. (1992) found that cultivars with higher levels of resistance did not require disulfoton treatment to yield economic returns similar to or better than those obtained from susceptible cultivars treated with insecticide.

The production of cultivars of sorghum with moderate resistance to *C. sorghicola* raises the economic thresholds for insecticidal control of the midge. Untreated *C. sorghicola*-resistant hybrids yield more than treated susceptible hybrids, and at moderate and high *C. sorghicola* densities, resistant hybrids respond more efficiently to insecticide application in terms of net crop yield and value (Teetes et al. 1986). van den Berg et al. (1994) found striking increases in insecticide use efficiency in sorghum inbred lines with resistance to *C. partellus* and the maize stalk borer, *Busseola fusca* (Fuller) in South Africa. Insecticide efficacy for control of both borer species is increased to the extent that yield losses in the untreated resistant inbred line are similar to those in susceptible inbred line treated with insecticides (Figure 12.2).

Maize hybrids resistant to *H. zea* require less insecticide than susceptible hybrids to achieve equivalent levels of *H. zea* population reduction. One application of insecticide at one half of the normal rate on the resistant maize hybrid '471-U6X81-1' controls *H. zea* larvae at a level equal to seven applications of the same insecticide at the normal rate on a susceptible hybrid (Wiseman et al. 1975). However, application of insecticides to maize hybrids with intermediate and high levels of resistance to *O. nubilalis* is of little benefit in reducing borer damage in the field (Robinson et al. 1978).

The susceptibility of *N. lugens* and the whitebacked planthopper, *Sogatella furcifera* (Horvath), to insecticides increases when these pests are reared on only moderately resistant rice cultivars (Heinrichs et al. 1984). Application of insecticide to *N. lugens* resistant-rice cultivars increases hopper mortality for prolonged periods after transplanting and decreases insecticide usage. Gross profit, net gain on investment and yields are also greater when arthropod-resistant rice cultivars are grown than when susceptible rice cultivars are produced (Heinrichs et al. 1979, 1986b).

In some instances, rice arthropod resistance is great enough that insecticides are of no practical value. Early on, Kalode (1980) indicated that cultivars with resistance to the gall midge, *Orseolia oryzae* (Wood-Mason), yielded similarly; with or without insecticides. Though the same trend was true for insecticide treatments to the susceptible cultivar, yields were nearly 50% less than those of the resistant cultivar. Similar studies by Reissig et al. (1981) in the Philippines indicated that applications of insecticide to IR36, a rice cultivar resistant to *N. lugens* and the green leafhopper, *Nephotettix virescens* (Distant), did not increase yields and actually decreased farmer net profits. Heinrichs et al. (1984) demonstrated that *N. lugens* mortality from insecticide application was no different between hoppers reared on resistant and moderately resistant cultivars. In areas of low insecticide use for *N. lugens* control, high levels of resistance are unnecessary to maintain *N. lugens* populations below the economic threshold (Cohen et al. 1997). Thus, the use of moderate resistance is of real economic value and helps reduce the selection pressure for *N.lugens* biotype development.

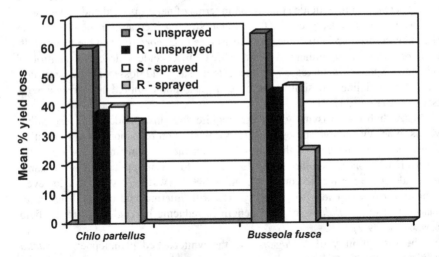

Figure 12.2. Yield losses of resistant and susceptible sorghum inbred lines under infestation by the spotted stalk borer, Chilo partellus, *and the maize stalk borer,* Busseola fusca, *in South Africa. Reprinted from Crop Protection, Vol. 13, van den Berg, J., G. D. J. van Rensburg, and M. C. van der Westhuizen. 1994, Host-plant resistance and chemical control of* Chilo partellus *(Swinhoe) and* Busseola fusca *(Fuller) in an integrated pest management system on grain sorghum, Pages 308 - 310, Copyright 1994, Butterworth Heinemann, Inc., with permission from Elseiver.*

3. 2 Cotton

Two morphological characteristics of improved cotton cultivars increase light penetration into the plant canopy and increase the efficiency of insecticides applied to control cotton insect pests. The frego or open bract condition of cotton buds (squares) (Figure 2.4) allows increased penetration of insecticides into the areas around each boll and increases mortality in treated infestations of *A. grandis* (Parrott et al. 1973). The okra leaf (open leaf) character (Figure 12.3) improves insecticide penetration by increasing coverage on all plant parts (James and Jones 1985, Jones et al. 1981) and imparts resistance to the pink bollworm, *Pectinophora gossypiella* (Saunders), (Wilson 1991, Naranjo and Martin 1993) and the two-spotted spider mite, *Tetranychus urticae* Koch (Wilson 1994). Reductions in insecticide applications of as great as 40% have been reported on some *P. gossypiella*-resistant cotton lines (Wilson et al. 1991). In Australia, okra leaf cotton comprises approximately 50% of the total cotton hectarage (Thompson 1995).

In some cases, the effects of early maturity and *P. gossypiella* resistance together exceed the benefit of insecticide application. George and Wilson (1983) found no significant reduction in yield between insecticide-treated and untreated plantings of the resistant line 'AET-5,' even though the resistant line had a greater *P. gossypiella* infestation than a susceptible cultivar. In the same experiment, the susceptible cultivar sustained a 16% yield reduction when not treated with insecticide.

Figure 12.3. Normal leaves of an arthropod-susceptible cotton cultivar (left) and open, non-lobed okra leaves of a cultivar resistant to foliar feeding by Pectinophora gossypiella *(right).*

3. 3 Legumes

Chalfant (1965) noted a positive, synergistic interaction between cultivars of common bean, *Phaseolus vulgaris* L., resistant to *E. fabae* and toxicity of the insectide carbaryl, when both were used in combination for field suppression of *E. fabae* populations. Campbell and Wynne (1985) determined that insecticide applications for control of the southern corn rootworm, *Diabrotica undecimpunctata howardi* LeConte, on a resistant cultivar of peanut, *Arachis hypogea* L., were reduced by 80% of the normally applied amount of insecticide. Similarly, insecticide applications for control of thrips and *E. fabae* on resistant peanut cultivars were reduced by approximately 60%. These reductions in insecticide volumes significantly reduce peanut production costs.

As in rice, legume resistance to arthropods may replace insecticidal control completely in a management system. Research by Cuthbert and Fery (1979) in the southern U. S. indicates that resistance in southern pea, *Pisum sativum* var. *macrocarpon* L., to the cowpea curculio, *Chalcodermus aeneus* Boheman, is actually more effective in reducing *C. aeneus* damage than applications of insecticide.

Resistance in soybean is equally synergistic with insecticides. In laboratory experiments, Rose et al. (1988) found that the susceptibility of larvae of the soybean looper, *Pseudoplusia includens* (Walker), and the velvetbean caterpillar, *Anticarsia gemmatalis* (Hübner), to poisoning by the insecticides fenvalerate and acephate is enhanced by prior consumption of foliage of the insect resistant soybean plant introduction 'PI227687'. Larvae of *P. includens* feeding on foliage of the insect-resistant soybean breeding line 'ED73-371' are more susceptible to poisoning by methyl parathion insecticide than larvae feeding on the susceptible cultivar 'Bragg', in both laboratory and field assays (Kea et al. 1978). Gazzoni (1995) noted a similar synergism between insecticide applications and soybean cultivars resistant to feeding *N. viridula*. The most apparent positive effects were noted at low rates of insecticide application, where the increased seed quality and seed yield of the resistant cultivar are significantly greater than those of susceptible cultivars.

3. 4 Vegetables

Resistance in carrot, *Daucus carota* L., to the carrot fly, *Psila rosae* F., has been extensively documented (Ellis 1999). Soil insecticide applications in plantings of the resistant cultivar 'Sytan' have additive effects on *P. rosae* populations, and these plants require only one-third the amount of insecticide normally used on a susceptible carrot cultivar. Similar effects are apparent in the resistance of swedes, *Brassica napus* L. spp. *rapifera* [Metz.] Sinsk., to the turnip root fly, *Delia floralis* (Fallen), and insecticide applications. Taksdal (1992) demonstrated applications of the insecticide chlorfenvinphos could be reduced by 50% on resistant swede cultivars compared to susceptible cultivars.

Even greater possible insecticide application rates are possible. Verkerk et al. (1998) found that pirimicarb doses to a cultivar of cabbage, *Brassica oleracea* var. *capitata* DC., with moderate resistance to the green peach aphid, *Myzus persicae* Sulzer, could be reduced by 85% of the recommended dose, compared to susceptible cultivars that required a three-fold greater dose. Resistance in sweet potato, *Ipomoea batatas* [(L.) Lam.], is also more effective or as effective as insecticide, depending on the use of a highly or moderately resistant cultivar, in reducing injury by an insect complex consisting of *D. howardi*, the southern potato wireworm, *Conoderus falli* Lane, the banded cucumber beetle, *Diabrotica balteata* LeConte, the elongate flea beetle, *Systena elongata* (F.), the white grub, *Plectris aliena* Chapin and the sweet potato beetle, *Cylas formicarius elegantulus* (Summers) (Cuthbert and Jones 1978).

3. 5 Potential Negative Insecticide Interactions

In spite of many positive interactions between plant resistance and insecticides, negative interactions between the two tactics do occur. Arthropods possess a variety of detoxification mechanisms that allow them to survive on plants containing defensive allelochemicals. Several allelochemicals induce increases in the activity of digestive enzymes such as mixed function oxidases, glutathione s-transferases and hydrolases (Brattsen 1979, Yu 1982, Yu and Hsu 1985). Arthropods with such induced enzyme systems may be more tolerant of the toxic effects of insecticides normally detoxified by these same enzymes.

Several studies have investigated the effects of selected allelochemicals from different host plants on arthropod metabolism. These include the allelochemical sinigrin (Yu 1983) and xanthotoxin (Yu 1984). However, information about the interactions that exist between allelochemicals controlling the resistance of crop plants and the metabolic capabilities of associated pests is an under-developed area of plant resistance investigation.

As indicated in Chapter 3, 2-tridecanone from *Lycopersicon hirsutum* f. *glabratum* C. H. Mull., increases the ability of the tobacco hornworm, *Manduca sexta* (L.), to detoxify the insecticide carbaryl (Kennedy 1984). In a similar manner, gossypol (Chapter 3) increases the activity of N-demethylase enzymes in the cotton leafworm, *Spodoptera littoralis* (Boisd.) (El-Sebae et al. 1981). In both cases, the potential exists for a negative interaction between allelochemically-based resistance to pest Lepiodoptera and conventional insecticidal control of these pests. These effects should be viewed on a case-by-case basis however, as indicated by results of Rector et al. (2003) on the effects of the flavone glycoside maysin (Chapter 3) on the resistance of the armyworm, *Helicoverpa armigera* (Hübner), to insecticides. *H. armigera* strains resistant to deltamethrin insecticide or to the Cry1Ac transgene are unable to detoxify maysin, suggesting that the physiological mechanisms of *H. armigera* resistance to insecticides and to resistant plant allelochemicals are very different.

4. INTEGRATION WITH BIOLOGICAL CONTROL

Many examples demonstrate the effects of arthropod resistant cultivars in improving the efficiency of predators, parasites and arthropod pathogens by decreasing the vigor and physiological state of the pest arthropod. In most resistant crop cultivars evaluated in the past, the effects are beneficial and in some cases, the effects have additive or synergistic effects on the actions of pest arthropod predators and parasites. (Eigenbrode and Trumble 1994, Quisenberry and Schotzko 1994, Smith 1999). However, instances of incompatible effects have been documented (van Emden 1986, Cortesero et al. 2000, Grot and Dicke 2002). The following discussions provide insights into both the beneficial and detrimental aspects of predator, parasite and arthropod pathogen interactions with arthropod resistant crop cultivars.

4. 1 Entomopathogens

The interactions of viruses and fungi with arthropod resistant cultivars have been relatively unexplored. Results of Hamm and Wiseman (1986) however, confirm the existence of a synergistic interaction between maize cultivars with resistance to leaf feeding by the fall armyworm, *Spodoptera frugiperda* (J. E. Smith), a nuclear polyhedrosis virus (NPV). Larvae feeding on artificial diet containing freeze-dried silks from *S. frugiperda* resistant inbred lines or on the intact silks of field grown plants are more susceptible to infection and mortality from NPV than larvae fed similarly with silks from susceptible inbred lines. As indicated in Figure 1.3, more recent results of Wiseman and Hamm (1993) indicate that a commercial formulation of NPV also significantly increases the mortality suffered by *H. zea* larvae feeding on artificial diets containing silks of the resistant maize cultivar 'Zapalote Chico', compared to larvae feeding on diets containing silks of a susceptible maize cultivar.

However, allelochemicals mediating plant resistance may adversely affect the synergism of resistance with the effects of NPV. The phenolics rutin and chlorogenic acid that mediate tomato resistance to some arthropods, can inhibit the rate of infection of NPV in *H. zea* larvae and extend the survival time of infected larvae (Felton et al. 1987). Most recently, Johnson et al. (1997) determined that in some instances, exposure of populations of *H. virescens* to the fungus, *Noumuraea rileyi* (Farlow), increases the rate of *H. virescens* adaptation to the Cry1Ab toxin in transgenic tobacco plants over several generations of selection for adaptation.

4. 2 Predators and Parasites

In general, resistant plants support the actions of arthropod parasites and predators. The positive synergism of predators, parasites and arthropod resistance has been documented in cultivars of many different crop plant species (Bergman and Tingey 1979, Boethel and Eikenbary 1986). As pointed out by Bottrell et al. (1998) and Verkerk et al. (1998a), most researchers view the combined actions of arthropod resistant cultivars and biological control organisms as synergistic. Numerous published studies, as well as published mathematical models involving resistant and susceptible cultivars, support this perspective (Gutierrez 1986, van Emden 1986).

However, several studies have documented specific adverse effects of the physical and chemical changes in arthropod-resistant plants that impart high levels of resistance toward beneficial arthropods (Hare 1992, Groot and Dicke 2002). The following narrative provides commentary on both the positive and negative interactions of different resistant crop plants with biological control organisms and provides insights for optimizing their interaction in integrated past management systems. Results of these effects are summarized in Tables 12.2 and Tables 12.3.

4. 2. 1 Cereal crops

One of the first examples of the positive interaction of plant resistance and biological control was by Starks et al. (1972), who determined that the effects of barley and sorghum cultivars resistant to *S. graminum* are complemented by the activity of the parasite, *Lysiphlebus testaceipes* (Cresson). The effectiveness of *L. testaceipes* remains unaffected, even after the development of *S. graminum* biotypes (Salto et al. 1983). Since then several other instances of the interactions between arthropod resistance cereal crops cultivars and biological control organisms have been documented, and most, but not all of these interactions are synergistic and positive. Synergistic effects of parasitic Hymenoptera have been shown to occur with cultivars of wheat with resistance to the wheat stem sawfly, *Cephus cinctus* Norton, (Morrill et al. 1994) the aphid *Metopolophium dirhodum* (Walker), (Gowling and van Emden 1994) and the Russian wheat aphid, *Diuraphis noxia* (Mordvilko), (Reed et al. 1991, Farid et al. 1998). Wheat cultivars with moderate resistance to *S. avenae* are compatible with the cereal aphid parasitoid *Aphidius rhopalosiphi* DeStephani-Perez. The resistance prolongs the developmental time of the aphid and parasite (Fuentes-Contreras and Niemeyer 1998).

However, results from studies conducted on highly antibiotic wheat cultivars indicate that high levels of antibiosis have adverse effects on the reproduction and population development of both the pest and the parasite arthropod. Chen et al. (1991) found that *M. destructor* mortality due to parasitism by *Platygaster hiemalis* Forbes was greatly reduced on the antibiotic wheat cultivar Knox 62, compared to the resistant cultivars Caldwell and Monon. Reed et al. (1991) noted that a triticale plant introduction with high levels of *D. noxia* resistance reduced growth and development of the parasite *Diaeretilla rapae* (McIntosh) significantly more than a *D. noxia* resistant wheat plant introduction. Reed et al. (1992) also evaluated the effects of plant introductions of slender wheatgrass, *Agropyron trachycaulum* (Link) Maltex H. F. Lewis, with high levels of antibiosis to *D. noxia* on the biology of *D. rapae* and found that parasitoid weights and numbers were significantly lower on resistant *A. trachycaulum* introductions than on the susceptible wheat cultivar 'TAMW-101'. The effect of *C. sorghicola*-resistant sorghum cultivars on the *C. sorghicola* hymenopterous parasites *Eupelmus popa* Gir. and *Tetrastichus* spp. is also detrimental in some instances (Sharma 1994). Highly resistant cultivars lower the rate of *C. sorghicola* parasitization, but on moderately resistant cultivars the rate is much less affected.

In some of the first studies of the interactions of maize arthropod-resistant cultivars with biological control organisms, Lynch and Lewis (1976) found that the protozoan parasite, *Nosema pyrausta* and maize cultivars resistant to leaf and sheath-collar feeding by *O. nubilalis*, interact to significantly reduce by *O. nubilalis* populations. At about the same time, Wiseman et al. (1976) noted that maize hybrids that are tolerant to *H. zea* damage maintain high populations of the predator *Orious insidiosus* Say, throughout the course of the silking stage, and these predator populations contribute to a greater suppression of *H. zea* larval populations on the resistant hybrids than on susceptible hybrids.

Positive interactions have also been noted for parasitic hymenoptera in maize. In field studies, Isenhour and Wiseman (1987) found a synergistic interaction between genotypes of maize resistant to *S. frugiperda* and the hymenopterous parasite, *Campoletis sonorensis* (Cameron), which resulted in further reductions in *S. frugiperda* larval weights over those caused by consumption of resistant foliage alone. No adverse affects on parasite development were noted. In a two-year field study, Riggin et al. (1992) also noted a similar absence of adverse effects of *S. frugiperda*-resistant maize hybrids on parasitism by *Cotesia marginiventris* (Cresson) or *Archytas marmoratus* (Townsend). In some cases parasitism was significantly higher on the resistant hybrid than on the susceptible hybrid. Laboratory studies involving *S. frugiperda* fed meridic diets containing *S. frugiperda*-resistant or susceptible maize genotypes provided similar results (Riggin et al. 1994). Further laboratory studies of the effects of maize resistance to *H. zea* on *A. marmoratus,* as well as *Ichneumon promissorius* (Erich), involved feeding *H. zea* larvae meridic diets containing silks of resistant and susceptible hybrids (Mannion et al. 1994). Parasites feeding on larvae consuming diet containing the resistant maize silks exhibit lower weights, reduced emergence (*A. marmoratus*) and reduced longevity (*I. promissorius*), yet none of these effects lower the fecundity or increase the development time of either parasite.

Table 12.1. Mortality of Nilaparvata lugens *due to the combined effects of predators with moderately resistant (MR) and susceptible (S) rice cultivars.*

	Rice Cultivar			
	Triveni (MR)		*Taichung Native 1 (S)*	
Predator	*No predators*	*Predators*	*No predators*	*Predators*
Lycosa pseudoannulata	30 bc [a]	47 a	15 d	36 ab
Cyrtohinus lividipennis	8 b	34 a	0 c	10 b

[a] *- means in each row differ significantly, P = 0.05, from Kartohadjono and Heinrichs 1984.*
(Adapted with permission from Environ. Entomol., Vol. 13:359-365. Copyright 1984,
Entomological Society of America)

The synergistic effect of rice cultivar resistance to Homoptera pests with biological control organisms is well established. Predation of *N. lugens* and *N. virescens* by the spider, *Lycosa pseudoannulata* Boes. et Str., or the mirid bug *Cyrtorhinus lividipennis* Reuter, is higher on rice cultivars resistant to both *N. lugens* and *N. virescens* (Kartohardjono and Heinrichs 1984, Myint et al. 1986). This combination of moderate plant resistance and predation has been shown to keep pest homopteran population levels below the economic threshold on resistant and moderately resistant rice cultivars (Table 12.1).

Results of additional experiments by Cuong et al. (1997) indicate that both moderate and high levels of rice plant resistance are compatible with biological control of *N. lugens, S. furcifera,* and *Nephotettix* spp. by *C. lividipennis,* the water bugs *Mesovelia orientalis* Kirkaldy, and *Microvelia* spp.; and several species of spiders.

4. 2. 2 Cotton

Improved cotton cultivars with the frego bract or okra leaf traits also complement biological control organisms. *A. grandis* larvae infesting cotton frego bract cultivars sustain approximately 40% greater parasitism by the braconid parasite, *Bracon mellitor* Say, than larvae infesting bolls on cultivars with the normal closed bract condition (McGovern and Cross 1976).

Combinations of the okra leaf trait with the nectariless character (lack of extrafloral nectaries) exhibit a mixture of results on beneficial arthropods in the cotton agroecosystem. Nectariless genotypes suppress oviposition and ensuing infestations of cotton by lepidopterous larvae, but the removal of the extrafloral nectaries, a high-energy insect food source, from the cotton plant also reduces the effectiveness of several predators and parasites (Schuster et al. 1976, Treacy et al. 1987).

However, results of more recent field experiments by Flint et al. (1992) indicate that populations of the predators *Orius tristicolor* (White), *Geocoris* spp., *Nabis* spp., and *Lygus* spp. are no different on okra leaf-nectariless cotton genotypes that are resistant to the pink bollworm, *Pectinpophora gossypiella* (Saunders), and genotypes that are susceptible to *P. gossypiella.* Populations of *Hippodamia convergens* (Guérin-Méneville), were significantly reduced on the okra leaf-nectariless genotypes, but Flint et al. (1992) point out that reductions in populations of beneficial arthropods due to the nectariless character are small when compared to the major reductions that occur in all beneficial arthropod populations after application of insecticides, which are unnecessary with *P. gossypiella* resistant cultivars.

4. 2. 3 Soybean

High levels of resistance in the foliage of soybean plants to several insects are also detrimental to parasites (Table12.3). Parasitization of *P. includens* larvae by the hymenopterous parasite, *Microplitis demolitor* (Wilkinson), and the resistance of soybean plant introduction (PI) 227687 reduce *P. includens* larval consumption of soybean foliage more than resistance alone. However, the allelochemically-based antibiosis resistance in PI227687, which causes high mortality of *P. includens* larvae (Reynolds et al. 1985), also decreases *M. demolitor* survival (Yanes and Boethel 1983).

Development of the parasite, *Microplitis croceipes* (Cresson), in *H. virescens* larvae fed foliage of PI227687 and PI229358 is also greatly delayed (Powell and Lambert 1984). More successful *M. croceipes* development occurs among those utilizing *H. virescens* larvae fed the moderately resistant plant introduction PI171451 and the susceptible cultivar 'Davis'. Consumption of the foliage of these genotypes by *H. virescens* larvae is greatly reduced due to *M. croceipes* parasitism. Similar synergistic tritrophic level interactions occur between *P. includens* larvae fed moderately resistant soybean cultivar foliage and *C. marginventralis* (McCutcheon et al. 1991), or the hymenopterous parasite *Copidosoma truncatellum* (Dalman) (Orr and Boethel 1986, Beach and Todd 1986).

The effect of parasitism on the Mexican bean beetle, *Epilachna varivestis* Mulsant, by *Pediobius foveolatus* Crawford, a larval endoparasite, is greater on *E. varivestis* adults fed the resistant soybean cultivar 'Cutler 71' than in those fed the susceptible cultivar 'Bonus' (Kauffman and Flanders 1986). The higher the level of soybean resistance, the greater the negative impact on parasite survival and development. Orr et al. (1985) demonstrated that emergence, fecundity, and progeny production of the parasite *Telenomus chloropus* Thompson are reduced when parasites develop on eggs of *N. viridula* that are fed pods of the resistant soybean cultivar PI171444 (Kester et al. 1984) compared to stink bugs fed pods of the susceptible cultivar 'Davis'.

Orr and Boethel (1985) also evaluated a four trophic level interaction between PI227687 foliage, *P. includens* larvae, the pentatomid predator *Podisus maculiventris* (Say) and the pentatomid parasite, *Teleonomus podisi* (Ashmead). As in tritrophic interactions, predator development was adversely affected after consumption of *P. includens* larvae fed resistant soybean foliage. The development of *T. podisi* after development on *P. maculiventris* reared on *P. includens* larvae fed PI227687 foliage was unaffected, but the reproductive potential of the parasite did decrease. Thus, the high level of antibiosis resistance to pest Lepidoptera expressed by soybean PI227687 is expressed through four trophic levels.

4. 2. 4 Vegetable crops

Beyond the previously discussed studies in cereal crops and soybean, research by several investigators has demonstrated the beneficial effect of plant arthropod resistance characters on predator behavior, biology and survival. The waxless, glossy-leaved characters in broccoli, *Brassica oleracea* L., and cabbage, *Brassica oleracea* var. *capitata* L., resistant to larvae of the diamondback moth, *Plutella xylostella* (L.), described in Chapters 2 and 3 have been shown to have a positive effect on several different *P. xylostella* predators. Glossy-leaved cabbage plants increase the effectiveness of *P. xylostella* larval predation by *Chrysoperla carnea* (Stephens), *Orius insidiosus* (Say), and *Hippodamia convergens* Guerin-Meneville (Eigenbrode et al. 1995) (Table 12.2). Wax crystals on plants susceptible to *P. xylostella* impede the mobility of all three predators. Conversely, their movement and effectiveness are greatly improved on glossy-leaved plants (Eigenbrode et al. 1996). Similar beneficial effects of the glossy-leaf character in broccoli plants have been observed in the predation of *P. xylostella* larvae by *H. convergens* and *Chrysoperla plorabunda* (Fitch) (Eigenbrode et al. 1999a,b).

Resistance to the pea aphid, *Acyrthosiphon pisum* (Harris), in the foliage of pea plants, *Pisum sativum* L., occurs on plants with greatly reduced leaf area and increased area of leaf stipules, leafy outgrowths at the base of the petiole, which provide a substrate for locomotion by *H. convergens* and *Coccinella septempunctata* L. This change in plant architecture greatly increases the predation of *A. pisum* (Kareiva and Sahakian 1990). Verkerk et al. (1998b) evaluated the tritrophic interactions between cabbage cultivars with moderate resistance to the cabbage aphid, *Brevicoryne brassicae* (L.), *B. brassicae*, and the predatory gall midge, *Aphidoletes aphidimyza* Rondani. In laboratory and field experiments, the *B. brassicae*-resistant cultivars had no significant effect on predator growth or consumption. Similar results were observed by Bottenberg et al. (1998) in assessments of the compatibility of predation by an *Orius* spp. complex with moderate levels of resistance in the wild cowpea, *Vigna vexillata* A. Richard, to a pest complex consisting of the pod borer, *Maruca vitrata* F., the bean fly, *Ophiomyia phaseoli* (Tryon), and the pod-sucking bug, *Clavigralla tomentosicollis* Stål. There were no indications that the high levels of resistance in *V. vexillata* have any adverse effects on the *Orius* spp. predator complex in two separate, season-long field experiments. Resistance in other cowpea germplasm exhibiting resistance to the cowpea aphid, *Aphis craccivora* Koch, also complements aphid predation by the coccinellid *Cheilomenes lunata* (F.) (Ofuya 1995) (Table 12 .2).

Progress in the integration of biological control agents and plant arthropod resistance continues to expand over a wide range of plant taxa. Resistance in alfalfa is synergistic with the actions of several beneficial arthropods. Cultivars with resistance to *A. pisum* have no negative effect on *A. pisum* parasites (Pimentel and Wheeler 1973), or the predatory convergent lady beetle, *Hippodamia convergens* Guérin-Méneville (Karner and Manglitz 1985). Giles et al. (2002) reported that an alfalfa cultivar resistant to the blue alfalfa aphid, *Acyrthosiphon kondoi* Shinji, had no detrimental effects on *H. convergens* or *Coccinella septempunctata* L. Finally, arthropod resistant cultivars of gerbera daisy, *Gerbera jamesonii* Adlam, and sugarcane have also been shown to have additive or neutral effects on the ability of predators to effectively utilize pest arthropods as prey (Bessin and Reagan 1993, Krips et al. 1999) (Table 12 .2).

4. 2. 5 Effects of plant pubescence on predators and parasites

The review of Bergman and Tingey (1979) presented initial indications that high levels of plant pubescence in arthropod resistant cultivars may be detrimental to predators and parasites. More recent reviews of Hare (1992), Cortesero et al. (2000) and Groot and Dicke (2002) provide examples of how high levels of resistance, especially those mediated by glandular trichomes in wild species of potato, *Solanum* spp. and tomato *Lycopersicon* spp., adversely affect beneficial arthropods. The density of trichomes on the stems of tomato plants also affects the ability of the predatory mite, *Phytoseiulus persimilis* Athias-Henriot, to control the phytophagous two-spotted spider mite, *Tetranychus urticae* Koch (Van Haren et al. 1987) (Table 12.2).

Table 12.2. *Effects of physical characters and uncharacterized factors from arthropod resistant plants on natural enemies*

Plant	Pest Arthropod	Natural Enemy	Effect [a]
Frego bract			
Gossypium hirsutum	Anthonomous grandis	Bracon mellitor	+
Leaf stipule			
Pisum sativum	Acyrthosiphon pisum	Coccinella septempunctata Hippodamia convergens	+
Nectariless leaf			
G. hirsutum	Helicoverpa zea Heliothis virescens	Trichogramma minutum	-
	Pectinpophora gossypiella	H. convergens Orius tristicolor	- +
Solid stem			
Triticum aestivum	Cephus cinctus	Bracon spp.	neutral
Trichomes			
Cucurbita sativus	Trialeuodes vaporariorum	Encarsia formosa	+ (moderate resistance)
Glycine max	H. zea	Geocoris punctipes	neutral
G. hirsutum	H. virescens	Chrysopa rufilabris Trichogramma pretiosum	-
L. esculentum	Bemisia argentifolii	Delphastus pusillus	+
	Tetranychus urticae	Phytoseiulus persimilis	-
Solanum berthaultii	Aphididae spp.	Aphidius spp. Praon spp.	+ (moderate resistance)
Uncharacterized resistance factor(s)			
Agropyron trachycaulum	Diuraphis noxia	Diaeretilla rapae	-
Brassica oleracea var. capitata	Brevicoryne brassicae	Aphidoletes aphidimyza	+
Gerbera jamesonii	Tetranychus urticae	P. persimilis	+
Glycine max	Nezara viridula	Telenomus chloropus	-

Table 12.2 continued

Medicago sativa	*A. pisum*	*H. convergens*	+
	Acyrthosiphon kondoi	*C. septempunctata* *H. convergens*	+
Oryza sativa	*Nilaparvata lugens* *Nephotettix virescens*	*Cyrtorhinus lividipennis* *Lycosa pseudoannulata*	+
Saccharum spp.	*Diatraea saccharalis*	*Solenopsis invicta*	+
Sorghum bicolor	*Schizaphis graminum*	*Lysiphlebus testaceipes*	+
	Stenodiplosis sorghicola	*Eupelmus popa*	- (high), + (moderate) resistance
Triticale	*D. noxia*	*D. rapae*	- (high resistance)
T. aestivum	*D. noxia*	*D. rapae*	+
	Mayetiola destructor	*Platygaster hiemalis*	-
	Metopolophium dirhodum	*Aphidius rhopalosiphi*	+
	Sitobion avenae	*A. rhopalosiphi*	+
Vigna spp.	*Aphis craccivora*	*Cheilomenes lunata*	+
Vigna vexillata	*Maruca vitrata*	*Orius* spp.	+
Waxless leaf			
B. oleracea var. *botrytis* & *capitata*	*P. xylostella*	*Chrysoperla carnea* *Chrysoperla plorabunda* *H. convergens* *Orius insidiosus*	+

^{a} - *see text for references*

T. urticae is also affected by the trichomes, but to a lesser extent, due to the ability to reach plant tissues by descent on silk threads and to spin silken mats over areas of glandular trichomes. Tomato cultivars without trichomes allow leaf-to-leaf dispersal of *P. persimilis* enabling them to effectively control the pest mite.

Obrycki and Tauber (1984) and Obrycki et al. (1983) determined that moderate levels of glandular pubescence and associated adhesive trichome exudates in potato hybrids derived from the wild potato, *Solanum berthaultii* Hawkes, integrate effectively with predators and parasites in the management of populations of the green peach aphid, *Myzus persicae* (Sulzer), potato aphid, *Macrosiphon euphorbiae* (Thomas), and pea aphid, *Acyrthosiphon pisum* (Harris), infesting potato.

Treacy et al. (1985) studied the effects of the parasite *Trichogramma pretiosum* (Riley) and the predator *Chrysopa rufilabris* (Burmeister) on larvae of the bollworm, *Heliothis virescens* (F.), feeding on glabrous-leaved, pubescent, and densely pubescent cotton cultivars. Although glabrous-leaved cultivars reduce *H. virescens* populations, they are low-yielding, necessitating continued production of pubescent leaved cotton cultivars. The effects of both beneficial arthropods are reduced with increasing degrees of cotton leaf pubescence. Schuster and Calderon (1986) observed that the number of predators actually increases on pubescent cottons, but that predator searching efficiency decreases.

A similar relationship exists in pubescent genotypes of cucumber, *Cucurbita sativus* L., and their interaction with the pest greenhouse whitefly, *Trialeuodes vaporariorum* (Westwood), and the parasite *Encarsia formosa* Gahan. Densely pubescent genotypes impede the movement of *E. formosa* to locate *T. vaporariorum* and glabrous genotypes allow parasites to move too quickly, resulting in "missed" searches for *T. vaporariorum* larvae. Cucumber hybrids with one-half the normal density of pubescence provide an environment that maximizes parasite search efficiency (Van Lenteren 1991).

In contrast, Heinz and Zalom (1996) found that glabrous *L. esculentum* cultivars sustain greatly reduced damage by the silverleaf whitefly, *Bemisia argentifolii* Bellows & Perring, compared to cultivars with pubescent foliage. The ability of the predator *Delphastus pusillus* LeConte to suppress *B. argentifolii* populations is unaffected by the lack of tomato pubescence. In this instance, the effects of resistance and biological control are synergistic, since both contribute to reduction of the pest population without adverse effects on each other. In one of the few studies involving trichome-based resistance in a cereal crop, Lampert et al. (1983) determined that resistance in small grain cultivars to the cereal leaf beetle, *Oulema melanopus* (L.), based on increased leaf pubescence has very little negative effect on cereal leaf beetle parasites. Powell and Lambert (1993) evaluated the effect of *H. zea*-resistant soybean cultivars with reduced foliar pubescence on predation by the bigeyed bug, *Geocoris punctipes* (Say), in the laboratory and found no adverse effects on the ability of *G. punctipes* to prey on *H. zea* eggs (Table 12 .2).

4. 2. 6 Effects of plant allelochemicals on predators and parasites

Suppression of the effects of beneficial arthropod populations may also result from the effects of allelochemicals in resistant plant cultivars being transferred to the predator or parasite. The first example of this interaction was observed in the toxicity of α-tomatine (see Chapter 3), an alkaloid from resistant tomato cultivars, to *Hyposoter exiguae* (Viereck), an endoparasite of *H. zea* (Campbell and Duffey 1979).

The most comprehensive evaluations of the effects of allelochemicals in an arthropod resistant plant on beneficial arthropods have been conducted with the methyl ketone-based resistance in glandular trichomes from *L. hirsutum* f. *glabratum* accession PI134417 (Table 12.3). As described in Chapters 2 and 3, the methyl ketones 2-tridecanone and 2-undecanone (Figure 3.1) are produced in vacuoles on the tip of foliar glandular trichomes (Figure 2.8) of *L. hirsutum* f. *glabratum* foliage and mediate resistance to *H. zea*, *M. sexta* and the Colorado potato beetle, *Leptinotarsa decemlineata* (Say).

Initial results of Farrar and Kennedy (1991) and Kashyap et al. (1991a,b) indicated that parasitism of *H. zea* and *M. sexta* eggs by *Telenomus sphingis* (Ashmead) and *Trichogramma pretiosum* Riley is reduced when eggs are laid on PI134417 foliage. These reductions in parasitism are due to the effects of exposure to 2-tridecanone and 2-undecanone volatiles, as well as reduced parasite mobility after entrapment in trichome adhesive exudates. Laboratory studies conducted by Farrar et al. (1992) and Farrar and Kennedy (1993) with the *H. zea* dipterous parasitoids *Archytas marmoratus* (Townsend) and *Eucelatoria bryani* (Sabrosky) indicated that the rate of parasitization of each was reduced by foliage with glandular trichomes in the laboratory, but in field cage experiments, *E. bryani* was unaffected by the methyl ketones in glandular trichomes.

More recently, Farrar et al. (1994) assayed the incidence of naturally occurring *H. zea* and *H. virescens* parasitism in season-long field studies on plants of PI134417, susceptible plants lacking 2-tridecanone and an F_1 hybrid with moderate glandular trichome density and moderate to high levels of 2-tridecanone and 2-undecanone, from the cross PI134417 x 'Walter' (no 2-tridecanone).   Egg parasitism by *T. sphingis* was reduced on plants of PI134417 and the hybrid, as was parasitism of larvae by *C. sonorensis* and *Cotesia congregata* (Say).   However, there was no consistent effect of plant type on larval parasitism by *C. marginiventris,* and parasitism rates of *Cardiochiles nigriceps* Viereck were unaffected by glandular trichomes.   In this series of studies, parasite size was a determining factor in susceptibility to glandular trichomes, and the least vulnerable parasite species evaluated by Farrar et al. (1994), *C. nigriceps,* was the largest. An additional field study that monitored the differences in populations of the *H. zea* and *M. sexta* predators *G. punctipes* and *Coleomegilla maculata* (DeGeer) found no incompatability of glandular trichome-based resistance with the effects of these beneficial arthropods. However, the increase of the beneficial arthropod population occurred to late in the growing season to be of practical benefit (Barbour et al. 1997).

As described in Section 4.2.3, high levels of isoflavone-based resistance in soybean are detrimental to a number of natural enemies. The higher the level of resistance, the greater the negative impact on parasite biology (Table 12.3). In addition to differences between resistant and susceptible cultivars in toxin content and morphological factors, differences in the volatile chemistry of resistant and susceptible cultivars (Chapter 2) may also affect natural enemies (Price 1986).

Resistant cultivars containing volatiles that attract beneficial insects draw greater numbers of parasites to the pest host plant (Agelopoulos et al. 1994, Udayagiri and Jones 1992, Ngi-Song et al. 1996).   Conversely, reduced concentrations of attractants, the lack of attractants or the presence of compounds repellent to beneficial insects will result in them being driven away from the plant and their prey (van Emden 1986).

Table 12.3. Effects of allelochemicals from arthropod resistant plants on natural enemies

Plant	Pest Arthropod	Natural Enemy	Effect [a]
Lycopersicon esculentum (α-tomatine)	*Helicoverpa zea*	*Hyposoter exiguae*	-
L. hirsutum f. glabratum methyl ketones	*H. zea* *Heliothis virescens*	*Archytas marmoratus* *Campoletis sonorensis*	- - (laboratory)
		Cardiochiles nigriceps	neutral
		Coleomegilla maculata	neutral (field)
		Cotesia congregata	-
		Cotesia marginiventris	neutral
		Eucelatoria bryani	neutral (field)
		Geocris punctipes	neutral (field)
		Telenomus sphingis	-
		Trichogramma pretiosum	-
Zea mays flavone glycosides	*H. zea*	*Orious insidiosus*	+
		Ichneumon promissorius	+/- [b]
		A. marmoratus	+/- [b]
	Spodoptera frugiperda	*A. marmoratus*	+
		C. sonorensis	+
		C. marginiventralis	+
Glycine max isoflavones	*Epilachna varivestis*	*Pediobius foveolatus*	- (high), + (moderate) resistance
	H. virescens	*Copidosoma truncatellum*	- (high), + (moderate) resistance
		C. marginventralis *Microplitis croceipes*	
	Pseudoplusia includens	*Microplitis demolitor*	-

[a] - *see text for references;* [b] *(+/-) - some aspects of biology affected*

An example of this relationship has been observed in parasitism of *B. brassicae* by *D. rapae* on cultivars of *B. brassicae*-resistant cruciferous crops. Reduced concentrations of allyl isothiocynate impart resistance to *B. brassicae* but because allyl isothiocynate is also a *D. rapae* volatile host habitat stimulus, the rate of parasitism is reduced (Reed et al. 1970, Andow and Rosset 1990). As discussed in Chapter 1, associational resistance related to the volatile content of companion crops grown with resistant cultivars may also have positive (attractant) or negative (repellent) effects on beneficial insects seeking prey on the target crop plant (see reviews of Price 1986, and Cortesero et al. 2000).

4.3 Optimizing Biological Control and Plant Resistance

General theories describing the optimum level of plant resistance and pest arthropod predation or parasitism may not only be very difficult to devise, but unnecessary as well. The extant literature discussed represents a wide variety of chemical and physical arthropod resistance factors occurring across a diversity of crop plants (Graminae, Leguminosae, Malvaceae, Solanaceae) to a broad range of pest arthropods (Acarina, Diptera, Hemiptera, Homoptera, Lepidoptera). In addition, there is great variation in the biology and life history of the many species of beneficial arthropods involved in plant-arthropod pest interactions.

However, in the instances where plant resistance has been shown to have negative effects, beneficial arthropods were exposed to high levels of resistance mediated by allelochemicals, glandular trichomes or surface waxes. Feeny (1976) observed that cultivars with moderate arthropod resistance are beneficial in pest management by reducing the intrinsic rate of pest population increase and providing a longer duration of host availability for beneficial arthropods.

Where a moderate degree of resistance is employed, positive and beneficial effects on natural enemies have resulted in numerous instances of reduced populations of the pest arthropod. Cortesero et al. (2000) recommend the intentional development of crop plants that weaken the pest population, in order to insure the survival of the natural enemy population, instead of plants with high levels of resistance that kill the pest population and, as a result, destroy the natural enemy population as well.

In addition to plant factors affecting the success or failure of plant resistance and biological control, natural enemy life history characters may also determine the outcome of the interactions of plants, pests and natural enemies. These factors may include the type and degree of density dependence of the natural enemy to the pest arthropod, natural enemy alternate food preferences (host switching tendencies), and other causes of pest mortality, such as entomopathogens.

Cortesero et al. (2000) reviewed the effects of chemical and physical characters bred into (and out of) crops for arthropod resistance on natural enemies, and noted that plant domatia (refugia) in leaves, stems and roots; and foliar pubescence and volatile arthropod attractants act in a positive manner to enhance populations of beneficial beneficial arthropods. To improve the efficiency of predators and parasites, the authors proposed that future crop improvement efforts include attempts to breed plants with characteristics that enhance natural enemy populations. Unfortunately, this concept remains largely unaccepted as an objective of plant breeding companies using either conventional or transgenic methodologies.

As pointed out by Cortesero et al. (2000) an ideal plant characteristic to enhance biological control does not exist. Rather, the relevance of each such character depends on the plant, the pest arthropod, and the beneficial arthropod involved, as well as the characteristics of the relationship among them. Real future progress towards integrating plant resistance and biological control will first involve identifying which natural enemy species play a major role in regulating a pest's population before plant breeders select characters that mediate arthropod resistance, in order to derive the maximum synergism from plant resistance and biological control. From the perspective of integrated pest management, the ideal interaction of plant resistance and biological control is one that lowers pest arthropod population density sufficiently to allow beneficial arthropods to reduce pest population densities to slightly below the crop economic damage threshold.

5. CONCLUSIONS

Plant resistance to arthropods has been successfully integrated with allied pest management tactics in many of the world's major food and fiber crops. However, in order to be of greatest utility, arthropod-resistant cultivars must fit efficiently into existing crop pest management systems that involve biological, chemical or cultural pest management tactics. The fit of an arthropod -resistant cultivar into the management system is also enhanced if it exhibits resistance to several arthropod species, as opposed to a single species.

Multiple species resistance is especially beneficial, because it may result in a greater reduction in the total amount of insecticides applied to the system than reductions resulting from the production of a cultivar with resistance to only one pest. Multi-species resistance is also helpful in avoiding the emergence of a secondary arthropod pest species as a primary pest. Secondary pests often take on primary pest status as a result of reductions in the amounts of pesticides applied for control of the original primary pest.

In spite of these obvious advantages, multiple arthropod resistance is often difficult to develop. Neverthelss, multiple arthropod species resistance has been developed in cultivars of alfalfa (Sorenson et al. 1983), cotton (Bird 1982), maize (Wiseman and Widstrom 1986, Mihm 1985), and rice (Khush 1984).

The variations in the effects of allelochemical and morphological resistance on both pest and beneficial arthropods discussed previously demonstrate the need for the development of crop cultivars with moderate levels of arthropod resistance for use in crop pest management systems. From a cropping systems perspective, the development of individual crop cultivars with resistance to individual pest arthropods is of limited utility. The development of moderate levels of resistance to the arthropod pest complex in each of the different crops of the system is ultimately of much greater economic and ecological value to the system, its producers and its users, than resistance to a single pest arthropod in a single crop.

REFERENCES CITED

Agelopoulos, N.A., and M. A. Keller. 1994. Plant-natural enemy association in the tritrophic system *Cotesia rebecula-Pieris rapae*-Brassicae (Cruciferae). I. Sources of infochemicals. J. Chem. Ecol. 20:1725–1734.

Alvarado-Rodriquez, B., T. F. Leigh, K. W. Foster, and S. S. Duffey. 1986. Resistance in common bean (*Phaseolus vulgaris*) to *Lygus hesperus* (Heteroptera: Miridae). J. Econ. Entomol. 79:484–489.

Andow, D. A., and P. M. Rosset. 1990. Integrated pest management. In: C. R. Carroll, J. H. Vandermeer, and P. M. Rosset (Eds.) Agroecology. McGraw-Hill, New York. 641.

Bailey, J. C., B. W. Hanny, and W. R. Meredith, Jr. 1980. Combinations of resistant traits and insecticides: effect on cotton yield and insect populations. J. Econ. Entomol. 73:58–60.

Barbour, J. D., R. R. Farrar, and G. G. Kennedy. 1997. Populations of predaceous natural enemies developing on insect-resistant and susceptible tomato in North Carolina. Biol. Control. 9:173–184.

Beach, R. M., and J. W. Todd. 1986. Foliage consumption and larval development of parasitized and unparasitized soybean looper, *Pseudoplusia includens* [Lep.: Noctuidae], reared on a resistant soybean genotype and effects on an associated parasitoid, *Copidosoma truncatellum* [Hym. : Encyrtidae]. Entomophaga 31:237–242.

Bergman, J. M. and W. M. Tingey. 1979. Aspects of interaction between plant genotypes and biological control. Bull. Entomol. Soc. Am. 25:275–279.

Bessin, R. T., and T. E. Reagan. 1993. Cultivar resistance and arthropod predation of sugarcane borer (Lepidoptera: Pyralidae) affects incidence of deadhearts on Louisiana sugarcane. J. Econ. Entomol. 86:929–932.

Birch, A.N. 1988. Field and glasshouse studies on components of resistance to root fly attack in swedes. Ann. Appl. Biol. 113:89–100.

Bird, L. S. 1982. Multi-adversity (diseases, insects and stresses) resistance (MAR) in cotton. Pl. Dis. 66:173–176.

Boethel, D. J., and R. D. Eikenbary (Eds.) 1986. Interactions of Plant Resistance and Parasitoids and Predators of Insects. Ellis Horwood Ltd., Chichester 224 pp.

Bottenberg, H., M. Tamo, and B. B. Singh. 1998. Occurrence of phytophagous insects on wild *Vigna* sp. and cultivated cowpea: comparing the relative importance of host-plant resistance and millet intercropping. Agric., Ecosyst. and Environ. 70:217–229.

Brattsen, L. B. 1979. Biochemical defense mechanisms in herbivores against plant allelochemics. In G. A. Rosenthal and D. H. Janzen (Eds.). Herbivores: Their Interaction with Secondary Plant Metabolites. Academic Press Inc. New York. pp. 199–270.

Brewer, G. J., E. Horber, and E. L. Sorensen. 1986. Potato leafhopper (Homoptera: Cicadellidae) antixenosis and antibiosis in *Medicago* species. J. Econ. Entomol. 79:421–425.

Buntin, G. D., S. L. Ott, and J. W. Johnson. 1992. Integration of plant resistance, Insecticides, and planting date for management of the Hessian fly (Diptera: Cecidomyiidae) in winter wheat. J. Econ. Entomol. 85:530–538.

Burris, E., D. F. Clower, J. E. Jones, and S. L. Anthony. 1983. Controlling boll weevil with trap cropping and resistant cotton. La Agric. 26:22–24.

Burton, R. L., J. D. Burd, O. R. Jones, and G. A. Wicks. 1990. Crop production strategies for managing greenbug (Homoptera: Aphididae) in sorghum. J. Econ. Entomol. 82:2476–2479.

Campbell, B. C., and S. S. Duffey. 1979. Tomatine and parasitic wasps: Potential incompatability of plant antibiosis with biological control. Sci. 205:700–702.

Campbell, W. V., and J. C. Wynne. 1985. Influence of the insect–resistant peanut cultivar NC6 on performance of soil insecticides. J. Econ. Entomol. 78:113–116.

Chalfant, R. B. 1965. Resistance of bunch bean varieties to the potato leafhopper and relationship between resistance and chemical control. J. Econ. Entomol. 58:681–682.

Chen, B. H., J. E. Foster, J. E. Araya, and P. L. Taylor. 1991. Parasitism of *Mayetiola destructor* (Diptera: Cecidomyiidae) by *Platygaster hiemalis* (Hymenoptera: Platygasteridae) on Hessian fly-resistant wheats. J. Entomol. Sci. 26:237–243.

Cohen, M. B., S. N. Alam, E. B. Medina, and C.C. Bernal. 1997. Brown planthopper, *Nilaparvata lugens*, resistance in rice cultivar IR64: mechanism and role in successful *N. lugens* management in Central Luzon, Philippines. Entomol. Exp. Appl. 85:221–229.

Cortesero, A. M., J. O. Stapel, and W. J. Lewis. 2000. Understanding and manipulating plant attributes to enhance biological control. Biological Control. 17:35–49.

Cuong, N. L., P. T. Ben, L. T. Phuong, L. M. Chau, and M. B. Cohen. 1997. Effect of host plant resistance and insecticide on brown planthopper *Nilaparvata lugens* (Stal) and predator population development in the Mekong Delta, Vietnam. Crop Protection 16:707–715.

Cuthbert, F. P., and R. L. Fery. 1979. Value of plant resistance for reducing cowpea curculio damage to the southernpea (*Vigna unguiculata* (L.) Walp.). J. Amer. Soc. Hort. Sci. 104:199–201.

Cuthbert, F. P., Jr. and A. Jones. 1978. Insect resistance as an adjunct or alternative to insecticides for control of sweet potato soil insects. J. Am. Soc. Hort. Sci. 103:443–445.

Eigenbrode, S. D., T. Castagnola,M.-B. Roux, and L. Steljes. 1996. Mobility of three generalist predators is greater on cabbage with glossy leaf wax than on cabbage with a wax bloom. Entomol. Exp. Appl. 81:335–343.

Eigenbrode, S. D., S. Moodie, and T. Castagnola. 1995. Predators mediate host plant resistance to a phytophagous pest in cabbage with glossy leaf wax. Entomol. Exp. Appl. 77:335–342.

Eigenbrode, S. D., and N. N. Kabalo. 1999. Effects of *Brassica oleracea* waxblooms on predation and attachment by *Hippodamia convergens* Entomol. Exp. Appl. 91:125–130.

Eigenbrode, S. D., N. N. Kabalo, and K. A. Stoner. 1999. Predation, behavior, and attachment by *Chrysoperla plorabunda* larvae on *Brassica oleracea* with different surface waxblooms. Entomol. Exp. Appl. 90:225–235.

Eigenbrode, S. D., and J. T. Trumble. 1994. Host plant resistance to insects in integrated pest management in vegetable crops. J. Agric. Entomol. 11:201–224.

Ellis, P. R. 1999. The identification and exploitation of resistance in carrots and wild Umbelliferae to the carrot fly *Psila rosae* (F.). Integrated Pest Management Reviews. 4:259–268.

El-Sabae, A. H., S. I. Sherby, and N. A. Mansour. 1981. Gossypol as an inducer or inhibitor in *Spodoptera littoralis*. J. Environ. Sci. Health. B16:167–178.

Farid, A., S. S. Quisenberry, J. B. Johnson, and B. Shafi. 1998. Impact of wheat resistance on Russian wheat aphid and a parasitoid. J. Econ. Entomol. 91:334–339.

Farrar, R. R., Jr., J. D. Barbour and G. G. Kennedy. 1994. Field evaluation of insect resistance in a wild tomato and its effects on insect parasitoids. Entomol. Exp. Appl. 71: 211–226.

Farrar, R. R. Jr., and G. G. Kennedy. 1991. Inhibition of *Telenomus sphingis* an egg parasitoid of *Manduca* spp. by trichome/2-tridecanone-based host plant resistance in tomato. Entomol. Exp. Appl. 60:157–166.

Farrar, R. R. Jr., and G. G. Kennedy. 1993. Field cage performance of two tachinid parasitoids of the tomato fruitworm on insect resistant and susceptible tomato lines. Entomol. Exp. Appl. 67:73–78.

Farrar, R. R. Jr., G. G. Kennedy, and R. K. Kashyap. 1992. Influence of life history differences of two tachinid parasitoids of *Helicoverpa zea* (Boddie) (Lepidoptera: Noctuidae) on their interactions with glandular trichome/methyl ketone-based resistance in tomato. J. Chem. Ecol. 18:499–515.

Feeny, P. 1976. Plant apparency and chemical defense. In: J. W. Wallace and R. L. Mansell (Eds.) Biochemical Interaction Between Plants and Insects. Plenum Press, New York. pp. 1–40.

Felton, G. W., S. S. Duffey, P. V. Vail, H. K. Kaya, and J. Manning. 1987. Interaction of nuclear polyhedrosis virus with catechols: Potential incompatibility for host-plant resistance against Noctuid larvae. J. Chem. Ecol. 13:947–957.

Flint, M. L., and R. van den Bosch. 1981. Introduction to Integrated Pest Management. Plenum Press, New York, 240 pp.

Flint, H. M., F. D. Wilson, N. J. Parks, R. Y. Reynoso, B. R. Stapp, and J. L. Szaro. 1992. Suppression of pink bollworm and effect on beneficial insects of a nectariless okra-leaf cotton germplasm line. Bull. Entomol. Res. 82:379–384.

Fuentes-Contreras, E. and H. M. Niemeyer. 1998. DIMBOA-glucoside, a wheat chemical defence, affects *Sitobion avenae* acceptance and suitability to the cereal aphid parasitoid *Aphidius rhopalosiphi*. J. Chem Ecol. 24:371–381.

Gazzoni, D. L. 1995. Associated effect of host plant resistance and insecticides to control stink bugs (Hemiptera: Pentatomidae) on soybean. An. Soc. Entomol. Brasil. 24:105–114.

George, B. W., and F. D. Wilson. 1983. Pink bollworm (Lepidoptera: Gelechiidae) effects of natural infestation on upland and pima cottons untreated and treated with insecticides. J. Econ. Entomol. 76:1152–1155.

Giles, K. L., R. C. Berberet, A. A. Zarrabi, and J. W. Dilwith. 2002. Influence of alfalfa cultivar on suitability of *Acyrthosiphon kondoi* (Homoptera: Aphididae) for survival and development of

Hippodamia convergens and *Coccinella septempunctata* (Coleoptera: Coccinellidae). J. Econ. Entomol. 95:552–557.

Gowling, G. R., and H. F. van Emden. 1994. Falling aphids enhance impact of biological control by parasitoids on partially aphid-resistant plant varieties. Ann. Appl. Biol. 125:233–242.

Groot, A. T., and M. Dicke. 2002. Insect transgenic plants in a multi-trophic context. The Plant Journal. 31:387–406.

Gutierrez, A.P. 1986. Analysis of the interactions of host plant resistance, phytophagous and entomophagous species. In: D. J. Boethel and R. D. Eikenbary (Eds.) Interactions of Plant Resistance and Parasitoids and Predators of Insects. Ellis Horwood Ltd., Chichester. pp. 198–215.

Hallman, G. J., G. L. Teetes, and J. W. Johnson. 1984. Relationship of sorghum midge (Diptera: Cecidomyiidae) density to damage to resistant and susceptible sorghum hybrids. J. Econ. Entomol. 77:83–87.

Hamm, J. J., and B. R. Wiseman. 1986. Plant resistance and nuclear polyhedrosis virus for suppression of the fall armyworm (Lepidoptera: Noctuidae). Fla. Entomol. 69:549–559.

Hare, D. J. 192. Effects of plant variation on herbivore-enemy interactions. In: R. S. Fritz and E. L. Simms (Eds.) Plant Resistance to Herbivores and Pathogens. University of Chicago Press. pp. 278–298.

van Haren, R. J. F., M. M. Steenhuis, M. W. Sabelis, and O. M. B. dePonti. 1987. Tomato stem trichomes and dispersal success of *Phytoseiulus persimilis* relative to its prey, *Tetranychus urticae*. Exp. Appl. Acarol. 3:115–121.

Heinrichs, E. A., R. C. Saxena, and S. Chelliah. 1979. Development and implementation of insect pest management systems for rice in tropical Asia. Food and Fertilizer Technology Center Extension Bull. No. 127. 38 pp.

Heinrichs, E. A., L. T. Fabellar, R. P. Basilio, Tu-Cheng Wen, and F. Medrano. 1984. Susceptibility of rice planthoppers *Nilaparvata lugens* and *Sogatella furcifera* (Homoptera: Delphacidae) to insecticide as influenced by level of resistance in the host plant. Environ. Entomol. 13:455–458.

Heinrichs, E. A., G. B. Aquino, S. L. Valencia, S. DeSagun, and M. B. Arceo. 1986a. Management of the brown planthopper, *Nilaparvata lugens* (Homoptera: Delphacidae), with early maturing rice cultivars. Environ. Entomol. 15:93–95.

Heinrichs, E. A., H. R. Rapusas, G. B. Aquino, and F. Palis.1986b. Integration of host plant resistance and insecticides in the control of *Nephotettix virescens* (Homoptera: Cicadellidae), a vector of rice tungro virus. J. Econ. Entomol. 79:437–443.

Heinz, K. M., and F. G. Zalom. 1996. Performance of the predator *Delphastus pusillus* on *Bemisia* resistant and susceptible tomato lines. Entomol. Exp. Appl. 81:345–352.

Hudon, M., M. S. Chiang, and D. Chez. 1979. Resistance and tolerance of maize inbred lines to the European corn borer *Ostrinia nubilalis* (Hubner) and their maturity in Quebec. Phytoprotection 60:1–22.

Isenhour, D. J., and B. R. Wiseman. 1987. Foliage consumption and development of the fall armyworm (Lepidoptera: Noctuidae) as affected by the interactions of a parasitoid, *Campoletis sonorensis* (Hymenoptera: Ichneumonidae), and resistant corn genotypes. Environ. Entomol. 16:1181–1184.

James, D., and J. E. Jones. 1985. Effects of leaf and bract isolines on spray penetration and insecticidal efficacy. In: T. C. Nelson and J. M. Brown (Eds.) Proc. Beltwide Cotton Prod. Res. Conf., New Orleans, LA. 6–11 Jan. 1985. Natl. Cotton Counc. Am, Memphis , TN. pp. 395–396.

Johnson, M. T., F. Gould, and G. G. Kennedy. 1997. Effect of an entomopathogen on adapttion of *Heliothis virescens* populations to transgenic host plants. Entomol. Exp. Appl. 83:121–135.

Jones, J. E., W. D. Caldwell, D. T. Bowman, J. W. Brand, A. Coro, J. G. Marshall, D. J. Boquet, R. Hutchinson, W. Aguillard, and D. F. Clower. 1981. Gumbo: An improved open-canopy cotton. La. Agric. Exp. Stn. Cir. 114.

Kalode, M. B. 1980. The rice gall midge-varietal resistance and chemical control. In: Rice Improvement in China and Other Asian Countries. International Rice Research Institute and Chinese Academy of Agricultural Sciences. pp. 173–193.

Kareiva, P., and R. Sahakian. 1990. Tritrophic effects of a simple architectural mutation in pea plants. Nature (Lond.) 345:433–434.

Karner, M. A., and G. R. Manglitz. 1985. Effects of temperature and alfalfa cultivar on pea aphid (Homoptera: Aphididae) fecundity and feeding activity of convergent lady beetle (Coleoptera: Coccinellidae). J. Kansas Entomol. Soc. 58:131–136.

Kartohardjono, A., and E. A. Heinrichs. 1984. Populations of the brown planthopper, *Nilaparvata lugens* (Stal) (Homoptera: Delphaciidae), and its predators on rice varieties with differing levels of resistance. Environ. Entomol. 13: 359–365.

Kauffman, W. C., and R. V. Flanders. 1986. Effects of variably resistant soybean and lima bean cultivars on *Pediobius foveolatus* (Hymenoptera: Eulophidae), A parasitoid of the Mexican bean beetle, *Epilachna varivestis* (Coleoptera: Coccinellidae). Environ. Entomol. 14:678–682.

Kashyap, R. K., G. G. Kennedy, and R. R. Farrar Jr. 1991a. Behavioral response of *Trichogramma pretiosum* Riley and *Telenomus sphingis* (Ashmead) to trichome/methyl ketone mediated resistance in tomato. J. Chem. Ecol. 17:543–556.

Kashyap, R. K., G. G. Kennedy, and R. R. Farrar Jr. 1991b. Mortality and inhibition of *Helicoverpa zea* egg parasitism rates by *Trichogramma* in relation to trichome/methyl ketone-mediated insect resistance of *Lycopersicon hirsutum* f. *glabratum* accession PI 134417. J. Chem. Ecol. 17:2381–2395.

Kea, W. C., S. C. Turnipseed, and G. R. Carner. 1978. Influence of resistant soybeans on the susceptibility of lepidopterous pests to insecticides. J. Econ. Entomol. 71:58–60.

Kennedy, G. G. 1984. 2-tridecanone, tomatoes, and *Heliothis zea*: Potential incompatibility of plant antibiosis with insecticidal control. Entomol. Exp. Appl. 35:305–311.

Kennedy, G. G., F. Gould, O. M. B. DePonti, and R. E. Stinner. 1987. Ecological, agricultural, genetic and commercial considerations in the deployment of insect resistant germplasm. Environ. Entomol. 16:327–338.

Kester, K. M., C. M. Smith, and D. F. Gilman. 1984. Mechanisms of resistance in soybean (*Glycine max* (L.) Merrill) genotype PI171444 to the southern green stink bug, *Nezara viridula* (L.). Environ. Entomol. 13:1208– 1215.

Khush, G. S. 1984. Breeding rice for resistance to insects. Protect. Ecol. 7:147–165

Krips, O. E., P. E. L. Willems, and M. Dicke. 1999. Compatibility of host plant resistance and biological control of the two-spotted spider mite *Tetranychus urticae* in the ornamental crop gerbera. Biological Control. 16:155–163.

Lampert, E. P., D. L. Haynes, A. J. Sawyer, D. P. Jokinen, S. G. Wellso, R. L. Gallun, and J. J. Roberts. 1983. Effects of regional releases of resistant wheats on the population dynamics of the cereal leaf beetle (Coleoptera: Chrysomelidae). Ann. Entomol. Soc. Am. 76:972–980.

Leszczynski, B. 1987. Winter wheat resistance to the grain aphid *Sitobion avenae* (Fabr.) (Homoptera, Aphididae). Insect Sci.Applic. 8:251–254.

Lynch, R. E., and L. C. Lewis. 1976. Influence on the European corn borer of *Nosema pyrausta* and resistance in maize to sheath-collar feeding. Environ. Entomol. 5:143–146.

Mannion, C. M., J. E. Carpenter, B. R. Wiseman, and H. R. Gross. 1994. Host corn earworm (Lepidoptera: Noctuidae) reared on meridic diet containing silks from a resistant corn genotype on *Archytas marmoratus* (Diptera: Tachinidae) and *Ichneumon promissorius* (Hymenoptera: Ichneumonidae). Environ. Entomol. 23:837–845.

Maxwell, F. G. 1985. Utilization of host plant resistance in pest management. Insect Sci. Applic. 6:437–442.

McCutcheon, G. S., M.J. Sullivan, and S. G. Turnipseed. 1991. Preimaginal development of *Cotesia marginiventris* (Hymenoptera: Braconidae) in soybean looper (Lepidoptera: Noctuidae) in insect-resistant soybean genotypes. J. Agric. Entomol. 26:381–389.

McGovern, W. L., and W. H. Cross. 1976. Affects of two cotton varieties on levels of boll weevil parasitism (Coleoptera: Curculionidae). Entomophaga. 21:123–125.

Mihm, J. A. 1985. Breeding for host plant resistance to maize stem-borers. Insect Sci. Applic. 6:369–377.

Morrill, W. L., G. D. Kushnak, P. L. Bruckner, and J. W. Gabor. 1994. Wheat stem sawfly (Hymenoptera: Cephidae) damage, rates of parasitism, and overwinter survival in resistant wheat lines. J. Econ. Entomol. 87:1373–1376.

Myint, M. M., H. R. Rapusas, and E. A. Heinrichs. 1986. Integration of varietal resistance and predation for the management of *Nephottetix virescens* (Homoptera: Cicadellidae) populations on rice. Crop Protection. 5:259–265.

Naranjo, S. E., and J. M. Martin. 1993. Comparative development, reproduction, and oviposition of pink bollworm (Lepidoptera: Gelechiidae) on a resistant okra-leaf cotton and commercial upland and Pima cottons. J. Econ. Entomol. 86:1094–1103.

Newsom, L. D., R. L. Jensen, D. C. Herzog, and J. W. Thomas. 1975. A pest management system for soybeans in Louisiana. La. Agric. 18:10–11.

Ngi-Song, A. J., W. A. Overholt, P. G. N. Njagi, M. Dicke, J. N. Ayertey, and W. Lwande. 1996. Volatile infochemicals used in host and host habitat selection by *Cotesia flavipes* (Cameron) and *Cotesia sesamiae* (Cameron) (Hymenoptera: Braconidae) larval parasitoids of stemborers on Graminae. J. Chem. Ecol. 22:307–323.

Obrycki, J. J., and M. J. Tauber. 1984. Natural enemy activity on glandular pubescent potato plants in the greenhouse: An unreliable predictor of effects in the field. Environ. Entomol. 13:679–683.

Obrycki, J. J., M. J. Tauber, and W. M. Tingey. 1983. Predator and parasitoid interaction with aphid-resistant potatoes to reduce aphid densities: a two year field study. J. Econ. Entomol. 76:456–462.

Ofuya, T. I. 1995. Studies on the capability of *Cheilomenes lunata* (Fabricius) (Coleoptera: Coccinellidae) to prey on the cowpea aphid, *Aphis craccivora* Koch (Homptera: Aphididae) in Nigeria. Agric., Ecosyst. Environ. 52:35–38.

Orr, D. B., and D. J. Boethel. 1985. Comparative development of *Copidosoma truncatellum* (Hymenoptera:Eucyrtidae) and its host, *Pseudoplusia includens* (Lepidoptera:Noctuidae), on resistant and susceptible soybean genotypes. Environ. Entomol. 14:612–616.

Orr, D. B., and D. J. Boethel. 1986. Influence of plant antibiosis through four trophic levels. Oecologia 70:242–249.

Orr, D. B., D. J. Boethel, and W. A. Jones. 1985. Biology of *Telenomus chloropus* (Hymenoptera: Scelionidae) from eggs of *Nezara viridula* (Hemiptera:Pentatomidae) reared on resistant and susceptible soybean genotypes. Can. Entomol. 117:1137–1142.

Parrott, W. L., J. N. Jenkins, and D. B. Smith. 1973. Frego bract cotton and normal bract cotton: How morphology affects control of boll weevils by insecticides. J. Econ. Entomol. 66:222–225.

Pedigo, L. P., S. H. Hutchins, and L. G. Higley. 1986. Economic injury levels in theory and practice. Ann. Rev. Entomol. 31:341–368.

Pimentel, D. S., and A. G. Wheeler, Jr. 1978. Influence of alfalfa resistance on a pea aphid population and its associated parasites, predators, and competitors. Environ. Entomol. 2:1–11.

Powell, J. E., and L. Lambert. 1984. Effects of three resistant soybean genotypes on development of *Microplitis croceipes* and leaf consumption by its *Heliothis* spp. hosts. J. Agric. Entomol. 1:169–176.

Powell, J. E., and L. Lambert. 1993. Soybean genotype effects on bigeyed bug feeding on feeding on corn earworm in the laboratory. Crop Sci. 33:556–559.

Price, P. W. 1986. Ecological aspects of host plant resistance and biological control: Interactions among three trophic levels. In: D. J. Boethel and R. D. Eikenbary (Eds.) Interactions of Plant Resistance and Parasitoids and Predators of Insects. Ellis Horwood Ltd., Chichester. pp. 11–30.

Rector, B. G., G. Liang, and Y. Guo. 2003. Effect of maysin on wild-type, deltamethrin-resistant, and Bt-resistant *Helicoverpa armigera* (Lepidoptera: Noctuidae). J. Econ. Entomol. 96:909–913.

Reed, D. P., P. P. Feeny, and R. B. Root. 1970. Habitat selection by the aphid parasite *Diaeretiella rapae* (Hymenoptera: Braconidae) and hyperparasite *Charips brassicae* (Hymenoptera: Cynipidae). Can. Entomol. 102:1567–1568.

Reed, D. K., S. D. Kindler, and T. L. Springer. 1992. Interactions of Russian wheat aphid, a hymenopterous parasitoid and resistant and susceptible slender wheatgrasses. Ent. Exp. Appl. 64:239–246.

Reed, D. K., J. A. Webster, B. C. Jones, and J. D. Burd. 1991. Tritrophic relationships of Russian wheat aphid (Homoptera: Aphididae), a hymenopterous parasitoid (*Diaeretiella rapae* McIntosh), and resistant and susceptible small grains. Biol. Con. 1:35–41.

Reissig, W. H., E. A. Heinrichs, L. Antonio, M. M. Salac, A. C. Santiago, and A. M. Tenorio. 1981. Management of pest insects of rice in farmers fields in the Philippines. Protect. Ecol. 3:203–218.

Reynolds, G. W., C. M. Smith, and K. M. Kester. 1985. Reductions in consumption, utilization and growth rate of soybean looper (Lepidoptera: Noctuidae) larvae fed foliage of soybean genotype PI227687. J. Econ. Entomol. 77:1371–1375.

Riggin, T. M., B. R. Wiseman, D. J. Isenhour, and K. E. Espelie. 1992. Incidence of fall armyworm (Lepidoptera: Noctuidae) parasitoids on resistant or susceptible corn genotypes. Environ. Entomol. 21:888–895.

Riggin, T. M., B. R. Wiseman, D. J. Isenhour, and K. E. Espelie. 1994. Functional response of *Cotesia marginiventris* (Cresson) (Hym., Braconidae) to *Spodoptera frugiperda* (J. E. Smith) (Lep., Noctuidae) on meridic diet containing resistant or susceptible corn genotypes. J. Appl. Entomol. 117:144–150.

Ring, D. R., J. H. Benedict, J. A. Landivar, and B. R. Eddleman. 1993. Economic injury levels and development and application of response surfaces relating insect injury, normalized yield, and plant physiological age. Environ. Entomol. 22:273–282.

Robinson, J. F., E. C. Berry, L. C. Lewis, and R. E. Lynch. 1978. European corn borer: Host-plant resistance and use of insecticides. J. Econ. Entomol. 71:109–110.

Salto, C. E., R. D. Eikenbary, and K. J. Starks. 1983. Compatibility of *Lysiphlebus testaceipes* (Hymenoptera: Braconidae) with greenbug (Homoptera: Aphididae) biotypes "C" and "E" reared on susceptible and resistant oat varieties. Environ. Entomol. 12:603–604.

Saxena, R. C. 1982. Colonization of rice fields by *Nilaparvata lugens* (Stal) and its control using a trap crop. Crop Protection. 1:191–198.

Schuster, M. F., and M. Calderon. 1986. Interactions of host plant resistant genotypes and beneficial insects in cotton ecosystems. In: D. J. Boethel and R. D. Eikenbary (Eds.) Interactions of Plant Resistance and Parasitoids and Predators of Insects. Ellis Horwood Ltd., Chichester. pp. 84–97.

Schuster, M. F., M. J. Lukefahr, and F. G. Maxwell. 1976. Impact of nectariless cotton on plant bugs and natural enemies. J. Econ. Entomol. 69:400–402.

Sharma, H. C. 1993. Host-plant resistance to insects in sorghum and its role in integrated pest management. Crop Prot. 12:11–34.

Sharma, H. C., P. Vidyasagar and K. F. Nwanze. 1993. Effect of host-plant resistance on economic injury levels for the sorghum midge, *Contarinia sorghicola*. Intl. J. Pest Management. 39:435–444.

Sharma, H. C. 1994. Effect of insecticide application and host plant resistance on parasitization of sorghum midge, *Contarinia sorghicola* Coq. Biocontrol Sci. Technol. 4:53–60.

Singh, S. R. 1987. Host plant resistance for cowpea insect management. Insect Sci. Appl. 8:765–769.

Sorenson, E. L., E. K. Horber, and D. L. Stuteville. 1983. Registration of KS 80 alfalfa germplasm resistant to the blue alfalfa aphid, pea aphid, spotted alfalfa aphid, anthracnose, and downy mildew. (Reg. No. GP 126). Crop Sci. 23:599.

Starks, K. J., J. R. Muniappan, and R. D. Eikenbary. 1972. Interaction between plant resistance and parasitism against greenbug on barley and sorghum. Ann. Entomol. Soc. Am. 65:650–655.

Taksdal, G. 1992. The complementary effects of plant resistance and reduced pesticide dosage in field experiments to control the turnip root fly, *Delia floralis*, in swedes. Ann. Appl. Biol. 120:117–125.

Teetes, G. L. 1980. Breeding sorghums resistant to insects. In: F. G. Maxwell and P. R. Jennings (Eds.) Breeding Plants Resistant to Insects. John Wiley and Sons, New York. pp. 459–485.

Teetes, G. L., M. I. Becerra, and G. C. Peterson. 1986. Sorghum midge (Diptera: Cecidomyiidae) management with resistant sorghum and insecticide. J. Econ. Entomol. 79:1091–1095.

Thompson, N. J. 1995. Commercial utilisation of the okra leaf mutant cotton. The Australian experience. In G. A. Constable and N. W. Forrester (Eds.) Challenging the Future. Proc. World Cotton Res. Conf. I. Brisbane, Queensland. 24–28 Fe. 1994. CSIRP, Australia. pp. 393–410.

Treacy, M. F., G. R. Zummo, and J. H. Benedict. 1985. Interactions of host-plant resistance in cotton with predators and parasites. Agric. Ecosystems and Environ. 13:151–158.

Treacy, M. F., J. H. Benedict, M. W. Walmsley, J. D. Lopez, and R.K. Morrison. 1987. Parasitism of bollworm (Lepidoptera: Noctuidae) eggs on nectaried and nectariless cotton. Environ. Entomol. 16:420–423.

Udayagiri, S., and R. L. Jones. 1992. Role of plant odor in parasitism of European corn borer by braconid specialist parasitoid *Macrocentrus grandii* Goidanich: Isolation and characterization of plant synomones eliciting parasitoid flight response. J. Chem. Ecol. 18:1841–1855.

van den Berg, J., G. D. J. van Rensburg, and M. C. van der Westhuizen. 1997. Economic threshold levels for *Chilo partellus* (Lepidoptera:Pyralidae) control on resistant and susceptible sorghum plants. Bull. Entomol. Res. 87:89–93.

van den Berg, J., G. D. J. van Rensburg, and M. C. van der Westhuizen. 1994. Host-plant resistance and chemical control of *Chilo partellus* (Swinhoe) and *Busseola fusca* (Fuller) in an integrated pest management system on grain sorghum. Crop Prot. 13:308–310.

van Emden, H. F. 1986. The interaction of plant resistance and natural enemies: Effects on populations of sucking insects. In: D. J. Boethel and R. D. Eikenbary (Eds.) Interactions of Plant Resistance and Parasitoids and Predators of Insects. Ellis Horwood Ltd., Chichester. pp. 138–150.

Van Lenteren, J. C. 1991. Biological control in a tritrophic system approach. In: D. C. Peters, J. A. Webster, and C. S. Chlouber (Eds.) Proceedings, Aphid-plant interactions: populations to molecules, 12–17 Aug. 1990, Oklahoma Agricultural Experiment Station, Stillwater, pp. 2–28.

Verkerk, R. H. J., S. R. Leather, and D. J. Wright. 1998a. The potential for manipulating crop-pest-natural enemy interactions for improved insect pest management. Bull. Ent. Res. 88:493–501.

Verkerk, R. H. J., K. R. Neugebauer, P. R. Ellis, and D. J. Wright. 1998b. Aphis on cabbage: tritrophic and selective insecticide interactions. Bull. Entomol. Res. 88:343–349.

Yu, S. J. 1982. Induction of microsomal oxidases by host plants in the fall armyworm, *Spodoptera frugiperda* (J. E. Smith). Pestic. Biochem. Physiol. 17:59–67.

Yu, S. J. 1983. Induction of detoxifying enzymes by allelochemicals and host plants in the fall armyworm. Pestic. Biochem. Physiol. 19:330–336.

Yu, S. J. 1984. Interactions of allelochemicals with detoxication enzymes of insecticide-susceptible and resistant fall armyworms. Pestic. Biochem. Physiol. 22:60–68.

Yu, S. J., and E. L. Hsu. 1985. Induction of hydrolases by allelochemicals and host plants in fall armyworm (Lepidoptera: Noctuidae) larvae. Environ. Entomol. 512–515.

Wilson, F. D. 1991. Relative resistance of cotton lines to pink bollworm. Crop Sci. 30:500–504.

Wilson, L. J. 1994. Resistance of okra-leaf cotton genotypes to twospotted spider mites (Acari: Tetranychidae). J. Econ. Entomol. 87:1726–1735.

Wilson, F. D., H. M. Flint, L. A. Bariola, and C. C. Chu. 1991. Reduction in insecticide use associated with cotton resistance to pink bollworm. Crop Sci. 31:363–366.

Wiseman, B. R., and J. J. Hamm. 1993. Nuclear polyhedrosis virus and resistant corn silks enhance mortality of corn earworm (Lepidoptera: Noctuidae) larvae. Biol. Control 3:337–342.

Wiseman, B. R., E. A. Harrell, and W. W. McMillian. 1975. Continuation of tests of resistant sweet corn hybrid plus insecticides to reduce losses from corn earworm. Environ. Entomol. 2:919–920.

Wiseman, B. R., W. W. McMillian, and N. W. Widstrom. 1972. Tolerance as a mechanism of resistance in corn to the corn earworm. J. Econ. Entomol. 65:835–837.

Wiseman, B. R., W. W. McMillian, and N. W. Widstrom. 1976. Feeding of corn earworm in the laboratory on excised silks of selected corn entries with notes on *Orius insidiosus*. Fla. Entomol. 59:305–308.

Wiseman, B. R., and N. W. Widstrom. 1986. Mechanisms of resistance in 'Zapalote Chico' corn silks to fall armyworm (Lepidoptera: Noctuidae) larvae. J. Econ. Entomol. 79:1390–1393.

Yanes, J., Jr. and D. J. Boethel. 1983. Effect of a resistant soybean genotype on the development of the soybean looper (Lepidoptera: Noctuidae) and an introduced parasitoid, *Microplitis demolitor* Wilkinson (Hymenoptera: Braconidae) Environ. Entomol. 12:1270–1274.

Zelarayan, E. L., G. D. Buntin, J. W. Johnson, P. L. Bruckner, and P. L. Raymer. 1991. Integrated management for Hessian fly in Triticale. J. Prod. Agric. 4:629–633.

INDEX